Lecture Notes in Computer Science   14189

Founding Editors

Gerhard Goos
Juris Hartmanis

Editorial Board Members

The series Lecture Notes in Computer Science (LNCS), including its subseries Lecture Notes in Artificial Intelligence (LNAI) and Lecture Notes in Bioinformatics (LNBI), has established itself as a medium for the publication of new developments in computer science and information technology research, teaching, and education.

LNCS enjoys close cooperation with the computer science R & D community, the series counts many renowned academics among its volume editors and paper authors, and collaborates with prestigious societies. Its mission is to serve this international community by providing an invaluable service, mainly focused on the publication of conference and workshop proceedings and postproceedings. LNCS commenced publication in 1973.

Gernot A. Fink · Rajiv Jain · Koichi Kise ·
Richard Zanibbi
Editors

Document Analysis and Recognition – ICDAR 2023

17th International Conference
San José, CA, USA, August 21–26, 2023
Proceedings, Part III

Springer

Editors
Gernot A. Fink
TU Dortmund University
Dortmund, Germany

Rajiv Jain
Adobe
College Park, MN, USA

Koichi Kise
Osaka Metropolitan University
Osaka, Japan

Richard Zanibbi
Rochester Institute of Technology
Rochester, NY, USA

ISSN 0302-9743   ISSN 1611-3349 (electronic)
Lecture Notes in Computer Science
ISBN 978-3-031-41681-1   ISBN 978-3-031-41682-8 (eBook)
https://doi.org/10.1007/978-3-031-41682-8

This Springer imprint is published by the registered company Springer Nature Switzerland AG
The registered company address is: Gewerbestrasse 11, 6330 Cham, Switzerland

Foreword

We are delighted to welcome you to the proceedings of ICDAR 2023, the 17th IAPR International Conference on Document Analysis and Recognition, which was held in San Jose, in the heart of Silicon Valley in the United States. With the worst of the pandemic behind us, we hoped that ICDAR 2023 would be a fully in-person event. However, challenges such as difficulties in obtaining visas also necessitated the partial use of hybrid technologies for ICDAR 2023. The oral papers being presented remotely were synchronous to ensure that conference attendees interacted live with the presenters and the limited hybridization still resulted in an enjoyable conference with fruitful interactions.

ICDAR 2023 was the 17th edition of a longstanding conference series sponsored by the International Association of Pattern Recognition (IAPR). It is the premier international event for scientists and practitioners in document analysis and recognition. This field continues to play an important role in transitioning to digital documents. The IAPR-TC 10/11 technical committees endorse the conference. The very first ICDAR was held in St Malo, France in 1991, followed by Tsukuba, Japan (1993), Montreal, Canada (1995), Ulm, Germany (1997), Bangalore, India (1999), Seattle, USA (2001), Edinburgh, UK (2003), Seoul, South Korea (2005), Curitiba, Brazil (2007), Barcelona, Spain (2009), Beijing, China (2011), Washington, DC, USA (2013), Nancy, France (2015), Kyoto, Japan (2017), Sydney, Australia (2019) and Lausanne, Switzerland (2021).

Keeping with its tradition from past years, ICDAR 2023 featured a three-day main conference, including several competitions to challenge the field and a post-conference slate of workshops, tutorials, and a doctoral consortium. The conference was held at the San Jose Marriott on August 21–23, 2023, and the post-conference tracks at the Adobe World Headquarters in San Jose on August 24–26, 2023.

We thank our executive co-chairs, Venu Govindaraju and Tong Sun, for their support and valuable advice in organizing the conference. We are particularly grateful to Tong for her efforts in facilitating the organization of the post-conference in Adobe Headquarters and for Adobe's generous sponsorship.

The highlights of the conference include keynote talks by the recipient of the IAPR/ICDAR Outstanding Achievements Award, and distinguished speakers Marti Hearst, UC Berkeley School of Information; Vlad Morariu, Adobe Research; and Seiichi Uchida, Kyushu University, Japan.

A total of 316 papers were submitted to the main conference (plus 33 papers to the ICDAR-IJDAR journal track), with 53 papers accepted for oral presentation (plus 13 IJDAR track papers) and 101 for poster presentation. We would like to express our deepest gratitude to our Program Committee Chairs, featuring three distinguished researchers from academia, Gernot A. Fink, Koichi Kise, and Richard Zanibbi, and one from industry, Rajiv Jain, who did a phenomenal job in overseeing a comprehensive reviewing process and who worked tirelessly to put together a very thoughtful and interesting technical program for the main conference. We are also very grateful to the members of the Program Committee for their high-quality peer reviews. Thank you to our competition chairs, Kenny Davila, Chris Tensmeyer, and Dimosthenis Karatzas, for overseeing the competitions.

The post-conference featured 8 excellent workshops, four value-filled tutorials, and the doctoral consortium. We would like to thank Mickael Coustaty and Alicia Fornes, the workshop chairs, Elisa Barney-Smith and Laurence Likforman-Sulem, the tutorial chairs, and Jean-Christophe Burie and Andreas Fischer, the doctoral consortium chairs, for their efforts in putting together a wonderful post-conference program.

We would like to thank and acknowledge the hard work put in by our Publication Chairs, Anurag Bhardwaj and Utkarsh Porwal, who worked diligently to compile the camera-ready versions of all the papers and organize the conference proceedings with Springer. Many thanks are also due to our sponsorship, awards, industry, and publicity chairs for their support of the conference.

The organization of this conference was only possible with the tireless behind-the-scenes contributions of our webmaster and tech wizard, Edward Sobczak, and our secretariat, ably managed by Carol Doermann. We convey our heartfelt appreciation for their efforts.

Finally, we would like to thank for their support our many financial sponsors and the conference attendees and authors, for helping make this conference a success. We sincerely hope those who attended had an enjoyable conference, a wonderful stay in San Jose, and fruitful academic exchanges with colleagues.

August 2023   David Doermann
                                                   Srirangaraj (Ranga) Setlur

Preface

Welcome to the proceedings of the 17th International Conference on Document Analysis and Recognition (ICDAR) 2023. ICDAR is the premier international event for scientists and practitioners involved in document analysis and recognition.

This year, we received 316 conference paper submissions with authors from 42 different countries. In order to create a high-quality scientific program for the conference, we recruited 211 regular and 38 senior program committee (PC) members. Regular PC members provided a total of 913 reviews for the submitted papers (an average of 2.89 per paper). Senior PC members who oversaw the review phase for typically 8 submissions took care of consolidating reviews and suggested paper decisions in their meta-reviews. Based on the information provided in both the reviews and the prepared meta-reviews we PC Chairs then selected 154 submissions (48.7%) for inclusion into the scientific program of ICDAR 2023. From the accepted papers, 53 were selected for oral presentation, and 101 for poster presentation.

In addition to the papers submitted directly to ICDAR 2023, we continued the tradition of teaming up with the International Journal of Document Analysis and Recognition (IJDAR) and organized a special journal track. The journal track submissions underwent the same rigorous review process as regular IJDAR submissions. The ICDAR PC Chairs served as Guest Editors and oversaw the review process. From the 33 manuscripts submitted to the journal track, 13 were accepted and were published in a Special Issue of IJDAR entitled "Advanced Topics of Document Analysis and Recognition." In addition, all papers accepted in the journal track were included as oral presentations in the conference program.

A very prominent topic represented in both the submissions from the journal track as well as in the direct submissions to ICDAR 2023 was handwriting recognition. Therefore, we organized a Special Track on Frontiers in Handwriting Recognition. This also served to keep alive the tradition of the International Conference on Frontiers in Handwriting Recognition (ICFHR) that the TC-11 community decided to no longer organize as an independent conference during ICFHR 2022 held in Hyderabad, India. The handwriting track included oral sessions covering handwriting recognition for historical documents, synthesis of handwritten documents, as well as a subsection of one of the poster sessions. Additional presentation tracks at ICDAR 2023 featured Graphics Recognition, Natural Language Processing for Documents (D-NLP), Applications (including for medical, legal, and business documents), additional Document Analysis and Recognition topics (DAR), and a session highlighting featured competitions that were run for ICDAR 2023 (Competitions). Two poster presentation sessions were held at ICDAR 2023.

As ICDAR 2023 was held with in-person attendance, all papers were presented by their authors during the conference. Exceptions were only made for authors who could not attend the conference for unavoidable reasons. Such oral presentations were then provided by synchronous video presentations. Posters of authors that could not attend were presented by recorded teaser videos, in addition to the physical posters.

    Three keynote talks were given by Marti Hearst (UC Berkeley), Vlad Morariu (Adobe Research), and Seichi Uchida (Kyushu University). We thank them for the valuable insights and inspiration that their talks provided for participants.

    Finally, we would like to thank everyone who contributed to the preparation of the scientific program of ICDAR 2023, namely the authors of the scientific papers submitted to the journal track and directly to the conference, reviewers for journal-track papers, and both our regular and senior PC members. We also thank Ed Sobczak for helping with the conference web pages, and the ICDAR 2023 Publications Chairs Anurag Bharadwaj and Utkarsh Porwal, who oversaw the creation of this proceedings.

August 2023

Gernot A. Fink
Rajiv Jain
Koichi Kise
Richard Zanibbi

Organization

General Chairs

David Doermann   University at Buffalo, The State University of New York, USA

Srirangaraj Setlur   University at Buffalo, The State University of New York, USA

Executive Co-chairs

Venu Govindaraju   University at Buffalo, The State University of New York, USA

Tong Sun   Adobe Research, USA

PC Chairs

Gernot A. Fink   Technische Universität Dortmund, Germany (Europe)

Rajiv Jain   Adobe Research, USA (Industry)

Koichi Kise   Osaka Metropolitan University, Japan (Asia)

Richard Zanibbi   Rochester Institute of Technology, USA (Americas)

Workshop Chairs

Mickael Coustaty   La Rochelle University, France

Alicia Fornes   Universitat Autònoma de Barcelona, Spain

Tutorial Chairs

Elisa Barney-Smith   Luleå University of Technology, Sweden

Laurence Likforman-Sulem   Télécom ParisTech, France

Competitions Chairs

Kenny Davila	Universidad Tecnológica Centroamericana, UNITEC, Honduras
Dimosthenis Karatzas	Universitat Autònoma de Barcelona, Spain
Chris Tensmeyer	Adobe Research, USA

Doctoral Consortium Chairs

Andreas Fischer	University of Applied Sciences and Arts Western Switzerland
Veronica Romero	University of Valencia, Spain

Publications Chairs

Anurag Bharadwaj	Northeastern University, USA
Utkarsh Porwal	Walmart, USA

Posters/Demo Chair

Palaiahnakote Shivakumara	University of Malaya, Malaysia

Awards Chair

Santanu Chaudhury	IIT Jodhpur, India

Sponsorship Chairs

Wael Abd-Almageed	Information Sciences Institute USC, USA
Cheng-Lin Liu	Chinese Academy of Sciences, China
Masaki Nakagawa	Tokyo University of Agriculture and Technology, Japan

Industry Chairs

Andreas Dengel	DFKI, Germany
Véronique Eglin	Institut National des Sciences Appliquées (INSA) de Lyon, France
Nandakishore Kambhatla	Adobe Research, India

Publicity Chairs

Sukalpa Chanda	Østfold University College, Norway
Simone Marinai	University of Florence, Italy
Safwan Wshah	University of Vermont, USA

Technical Chair

Edward Sobczak	University at Buffalo, The State University of New York, USA

Conference Secretariat

University at Buffalo, The State University of New York, USA

Program Committee

Senior Program Committee Members

Srirangaraj Setlur
Richard Zanibbi
Koichi Kise
Gernot Fink
David Doermann
Rajiv Jain
Rolf Ingold
Andreas Fischer
Marcus Liwicki
Seiichi Uchida
Daniel Lopresti
Josep Llados
Elisa Barney Smith
Umapada Pal
Alicia Fornes
Jean-Marc Ogier
C. V. Jawahar
Xiang Bai
Liangrui Peng
Jean-Christophe Burie
Andreas Dengel
Robert Sablatnig
Basilis Gatos

Apostolos Antonacopoulos
Lianwen Jin
Nicholas Howe
Marc-Peter Schambach
Marcal Rossinyol
Wataru Ohyama
Nicole Vincent
Faisal Shafait
Simone Marinai
Bertrand Couasnon
Masaki Nakagawa
Anurag Bhardwaj
Dimosthenis Karatzas
Masakazu Iwamura
Tong Sun
Laurence Likforman-Sulem
Michael Blumenstein
Cheng-Lin Liu
Luiz Oliveira
Robert Sabourin
R. Manmatha
Angelo Marcelli
Utkarsh Porwal

Program Committee Members

Harold Mouchere	Jean-Yves Ramel
Foteini Simistira Liwicki	Haikal El Abed
Vernonique Eglin	Alireza Alaei
Aurelie Lemaitre	Xiaoqing Lu
Qiu-Feng Wang	Sheng He
Jorge Calvo-Zaragoza	Abdel Belaid
Yuchen Zheng	Joan Puigcerver
Guangwei Zhang	Zhouhui Lian
Xu-Cheng Yin	Francesco Fontanella
Kengo Terasawa	Daniel Stoekl Ben Ezra
Yasuhisa Fujii	Byron Bezerra
Yu Zhou	Szilard Vajda
Irina Rabaev	Irfan Ahmad
Anna Zhu	Imran Siddiqi
Soo-Hyung Kim	Nina S. T. Hirata
Liangcai Gao	Momina Moetesum
Anders Hast	Vassilis Katsouros
Minghui Liao	Fadoua Drira
Guoqiang Zhong	Ekta Vats
Carlos Mello	Ruben Tolosana
Thierry Paquet	Steven Simske
Mingkun Yang	Christophe Rigaud
Laurent Heutte	Claudio De Stefano
Antoine Doucet	Henry A. Rowley
Jean Hennebert	Pramod Kompalli
Cristina Carmona-Duarte	Siyang Qin
Fei Yin	Alejandro Toselli
Yue Lu	Slim Kanoun
Maroua Mehri	Rafael Lins
Ryohei Tanaka	Shinichiro Omachi
Adel M. M. Alimi	Kenny Davila
Heng Zhang	Qiang Huo
Gurpreet Lehal	Da-Han Wang
Ergina Kavallieratou	Hung Tuan Nguyen
Petra Gomez-Kramer	Ujjwal Bhattacharya
Anh Le Duc	Jin Chen
Frederic Rayar	Cuong Tuan Nguyen
Muhammad Imran Malik	Ruben Vera-Rodriguez
Vincent Christlein	Yousri Kessentini
Khurram Khurshid	Salvatore Tabbone
Bart Lamiroy	Suresh Sundaram
Ernest Valveny	Tonghua Su
Antonio Parziale	Sukalpa Chanda

Mickael Coustaty
Donato Impedovo
Alceu Britto
Bidyut B. Chaudhuri
Swapan Kr. Parui
Eduardo Vellasques
Sounak Dey
Sheraz Ahmed
Julian Fierrez
Ioannis Pratikakis
Mehdi Hamdani
Florence Cloppet
Amina Serir
Mauricio Villegas
Joan Andreu Sanchez
Eric Anquetil
Majid Ziaratban
Baihua Xiao
Christopher Kermorvant
K. C. Santosh
Tomo Miyazaki
Florian Kleber
Carlos David Martinez Hinarejos
Muhammad Muzzamil Luqman
Badarinath T.
Christopher Tensmeyer
Musab Al-Ghadi
Ehtesham Hassan
Journet Nicholas
Romain Giot
Jonathan Fabrizio
Sriganesh Madhvanath
Volkmar Frinken
Akio Fujiyoshi
Srikar Appalaraju
Oriol Ramos-Terrades
Christian Viard-Gaudin
Chawki Djeddi
Nibal Nayef
Nam Ik Cho
Nicolas Sidere
Mohamed Cheriet
Mark Clement
Shivakumara Palaiahnakote
Shangxuan Tian

Ravi Kiran Sarvadevabhatla
Gaurav Harit
Iuliia Tkachenko
Christian Clausner
Vernonica Romero
Mathias Seuret
Vincent Poulain D'Andecy
Joseph Chazalon
Kaspar Riesen
Lambert Schomaker
Mounim El Yacoubi
Berrin Yanikoglu
Lluis Gomez
Brian Kenji Iwana
Ehsanollah Kabir
Najoua Essoukri Ben Amara
Volker Sorge
Clemens Neudecker
Praveen Krishnan
Abhisek Dey
Xiao Tu
Mohammad Tanvir Parvez
Sukhdeep Singh
Munish Kumar
Qi Zeng
Puneet Mathur
Clement Chatelain
Jihad El-Sana
Ayush Kumar Shah
Peter Staar
Stephen Rawls
David Etter
Ying Sheng
Jiuxiang Gu
Thomas Breuel
Antonio Jimeno
Karim Kalti
Enrique Vidal
Kazem Taghva
Evangelos Milios
Kaizhu Huang
Pierre Heroux
Guoxin Wang
Sandeep Tata
Youssouf Chherawala

Reeve Ingle
Aashi Jain
Carlos M. Travieso-Gonzales
Lesly Miculicich
Curtis Wigington
Andrea Gemelli
Martin Schall
Yanming Zhang
Dezhi Peng
Chongyu Liu
Huy Quang Ung
Marco Peer
Nam Tuan Ly
Jobin K. V.
Rina Buoy
Xiao-Hui Li
Maham Jahangir
Muhammad Naseer Bajwa

Oliver Tueselmann
Yang Xue
Kai Brandenbusch
Ajoy Mondal
Daichi Haraguchi
Junaid Younas
Ruddy Theodose
Rohit Saluja
Beat Wolf
Jean-Luc Bloechle
Anna Scius-Bertrand
Claudiu Musat
Linda Studer
Andrii Maksai
Oussama Zayene
Lars Voegtlin
Michael Jungo

Program Committee Subreviewers

Li Mingfeng
Houcemeddine Filali
Kai Hu
Yejing Xie
Tushar Karayil
Xu Chen
Benjamin Deguerre
Andrey Guzhov
Estanislau Lima
Hossein Naftchi
Giorgos Sfikas
Chandranath Adak
Yakn Li
Solenn Tual
Kai Labusch
Ahmed Cheikh Rouhou
Lingxiao Fei
Yunxue Shao
Yi Sun
Stephane Bres
Mohamed Mhiri
Zhengmi Tang
Fuxiang Yang
Saifullah Saifullah

Paolo Giglio
Wang Jiawei
Maksym Taranukhin
Menghan Wang
Nancy Girdhar
Xudong Xie
Ray Ding
Mélodie Boillet
Nabeel Khalid
Yan Shu
Moises Diaz
Biyi Fang
Adolfo Santoro
Glen Pouliquen
Ahmed Hamdi
Florian Kordon
Yan Zhang
Gerasimos Matidis
Khadiravana Belagavi
Xingbiao Zhao
Xiaotong Ji
Yan Zheng
M. Balakrishnan
Florian Kowarsch

Mohamed Ali Souibgui
Xuewen Wang
Djedjiga Belhadj
Omar Krichen
Agostino Accardo
Erika Griechisch
Vincenzo Gattulli
Thibault Lelore
Zacarias Curi
Xiaomeng Yang
Mariano Maisonnave
Xiaobo Jin
Corina Masanti
Panagiotis Kaddas
Karl Löwenmark
Jiahao Lv
Narayanan C. Krishnan
Simon Corbillé
Benjamin Fankhauser
Tiziana D'Alessandro
Francisco J. Castellanos
Souhail Bakkali
Caio Dias
Giuseppe De Gregorio
Hugo Romat
Alessandra Scotto di Freca
Christophe Gisler
Nicole Dalia Cilia
Aurélie Joseph
Gangyan Zeng
Elmokhtar Mohamed Moussa
Zhong Zhuoyao
Oluwatosin Adewumi
Sima Rezaei
Anuj Rai
Aristides Milios
Shreeganesh Ramanan
Wenbo Hu

Arthur Flor de Sousa Neto
Rayson Laroca
Sourour Ammar
Gianfranco Semeraro
Andre Hochuli
Saddok Kebairi
Shoma Iwai
Cleber Zanchettin
Ansgar Bernardi
Vivek Venugopal
Abderrhamne Rahiche
Wenwen Yu
Abhishek Baghel
Mathias Fuchs
Yael Iseli
Xiaowei Zhou
Yuan Panli
Minghui Xia
Zening Lin
Konstantinos Palaiologos
Loann Giovannangeli
Yuanyuan Ren
Shubhang Desai
Yann Soullard
Ling Fu
Juan Antonio Ramirez-Orta
Chixiang Ma
Truong Thanh-Nghia
Nathalie Girard
Kalyan Ram Ayyalasomayajula
Talles Viana
Francesco Castro
Anthony Gillioz
Huawen Shen
Sanket Biswas
Haisong Ding
Solène Tarride

Contents – Part III

Posters: Data and Synthesis

Posters: Document NLP

Evaluation of Different Tagging Schemes for Named Entity Recognition in Handwritten Documents

David Villanova-Aparisi[1]([⊠]) [iD], Carlos-D. Martínez-Hinarejos[1] [iD],
Verónica Romero[2] [iD], and Moisés Pastor-Gadea[1]

[1] PRHLT Research Center, Universitat Politècnica de València, Camí de Vera, s/n,
València 46021, Spain
davilap@inf.upv.es
[2] Departament d'Informàtica, Universitat de València, València 46010, Spain

Abstract. Performing Named Entity Recognition on Handwritten Documents results in categorizing particular fragments of the automatic transcription which may be employed in information extraction processes. Different corpora employ different tagging notations to identify Named Entities, which may affect the performance of the trained model. In this work, we analyze three different tagging notations on three databases of handwritten line-level images. During the experimentation, we train the same Convolutional Recurrent Neural Network (CRNN) and n-gram character Language Model on the resulting data and observe how choosing the best tagging notation depending on the characteristics of each task leads to noticeable performance increments.

Keywords: Named Entity Recognition · Historical documents · Tagging notation · Coupled approach

1 Introduction

Historical Handwritten Text Recognition (HTR) [19] aims to build robust systems capable of obtaining accurate transcriptions from scanned pages of historical documents. However, there are many tasks in which the goal is to perform some kind of information retrieval from the images, targeting specific fields within the records [17]. A way to solve this problem is to rely on Named Entity Recognition (NER) [13], which is a process that allows the identification of parts of text based on their semantic meaning, such as proper names or dates.

When trying to perform HTR and NER over scanned documents, the first approach that may come to mind would be first to obtain the transcription

This work was supported by Grant PID2020-116813RB-I00 funded by MCIN/AEI /10.13039/501100011033, by Grant ACIF/2021/436 funded by Generalitat Valenciana and by Grant PID2021-124719OB-I00 funded by MCIN/AEI/10.13039/ 501100011033 and by ERDF, EU A way of making Europe.

G. A. Fink et al. (Eds.): ICDAR 2023, LNCS 14189, pp. 3–16, 2023.
https://doi.org/10.1007/978-3-031-41682-8_1

and then apply NER techniques over such output to obtain the tagged digital text [13]. The alternative to this decoupled method is developing a method that obtains the tagged output from the scanned documents in one step (coupled approximation) [4].

Recent results indicate that applying coupled models can improve the performance on such task [3,21] by avoiding error propagation, which arises as a consequence of the decoupled approach. Error propagation occurs because the NER model expects clean input, similar to that with which it has been trained. However, the output of the HTR process is not exempt from mistakes, hindering the performance of the NER model. Nevertheless, many works in the recent literature still resort to the decoupled approximation [1,14,23] and do so while obtaining good results. These works usually encode the HTR output via pre-trained word embeddings (e.g., BERT [5]) and then utilize the sequence of vectors to determine the tag associated with each word. This last step is usually performed by a Conditional Random Field (CRF) [9]. For a more thorough overview of NER, see [28,31].

In the coupled approximation, it is usual to apply an HTR architecture directly to the task, only modifying the alphabet by including the tagging symbols. While some recent works still perform HTR at word level [8], it is more common to find tasks where only the line segmentation is available. Therefore, we usually see Convolutional Recurrent Neural Networks (CRNN) being employed, although it could be possible to utilize Transformer architectures [29]. In our work, we obtain line-level transcriptions and perform NER in the same step. The output of this coupled approach is the transcription of the text along with the tags that help identify and categorize the Named Entities, including the so-called nested Named Entities, i.e., Named Entities that appear inside another Named Entity (for example, "Mary queen of Scotland" is a person name that includes the place name "Scotland"). The correct addition of such tags may allow further information extraction processes.

To the best of our knowledge, the impact of the tagging notation on the performance of a coupled model has only been studied in [4]. However, this study was conducted on a single structured dataset [17] and the considered notations were specific to the task. Therefore, one of our main contributions is to evaluate three different tagging notations to determine which one obtains better results according to the relevant features of a dataset (number of different entities, occurrence of nested entities, size and complexity, etc.). This study has been performed over three corpora: HOME, George Washington and IAM-DB. HOME is a multilingual historical handwritten document database [3] which was recently used to benchmark different NLP libraries in [14]. In HOME, data is available at line level and words pertaining to a Named Entity are grouped using parenthesized notation. The George Washington corpus [22] and the IAM database [11] were manually tagged at word level while including sentence segmentation (see more in Sect. 3), as described in [24], although IAM-DB was previously used for NER in [18]. These corpora are described with greater detail in Sect. 3.

Since our system works with lines and only HOME has line-level tagged transcriptions available, we processed the George Washington and IAM-DB datasets to generate line-level tagged transcriptions, which we made available at [25]. To validate our experiments, we will compare the obtained results with these two last datasets to those presented in [24]. However, since our model does not employ the word segmentation, a lower performance is expected.

The rest of the paper is structured as follows. Section 2 overviews the employed architecture, presents the different tagging notations that have been studied, and briefly introduces the evaluation metrics that have been considered. Section 3 describes the experimental methodology followed and discusses the obtained results. Lastly, Sect. 4 concludes the paper by remarking the key takeaways.

2 Framework

2.1 HTR and NER via a Coupled Model

Before introducing the chosen approach to deal with the problem, it may be helpful to formalize its definition. Considering that the input to our system is an image corresponding to a text line, the HTR problem can be seen as the search for the most likely word sequence, $\hat{w} = \hat{w}_1\hat{w}_2...\hat{w}_l$, given the representation of the input line image, a feature vector sequence $x = x_1x_2...x_m$. This leads to a search in the probability distribution $p(w \mid x)$:

$$\hat{w} = \underset{w}{\mathrm{argmax}}\, p(w \mid x) \tag{1}$$

If we consider the sequences of NE tags $t = t_1t_2...t_l$ related with the word sequence w as a hidden variable in Eq. 1, we obtain the following equation:

$$\hat{w} = \underset{w}{\mathrm{argmax}} \sum_t p(w, t \mid x) \tag{2}$$

If we follow the derivation presented in [3] and explicitly search for the most likely tagging sequence $\hat{t}$ during the decoding process, the obtained equation is:

$$(\hat{w}, \hat{t}) \approx \underset{w,t}{\mathrm{argmax}}\, p(x \mid w, t) \cdot p(w, t) \tag{3}$$

If the tagging sequence t is combined with the transcription w we would obtain the tagged transcription h, resulting in the following equation:

$$\hat{h} \approx \underset{h}{\mathrm{argmax}}\, p(x \mid h) \cdot p(h) \tag{4}$$

Equation 4 is similar to that of the original HTR problem. The main difference is that the hypothesis h to be generated contains the most likely transcription and tagging sequence. Therefore, we estimate both the optical probability, $p(x \mid h)$, and the syntactical probability, $p(h)$, to perform the search for the best hypothesis. We decided to implement a decoding architecture based on a Convolutional Recurrent Neural Network (CRNN) [30] to estimate the optical probability and on a character n-gram to estimate the syntactical one. Both models are combined following the approach introduced in [2].

2.2 Tagging Notation

It is usual in the NER literature to come upon different ways to annotate data, i.e., to have different vocabulary options that would be used to obtain a tag sequence t. It is also usual to have this discrepancy between tasks where NER in scanned images must be performed. While the tagging notation may have little impact on NLP models, it may be something to consider when we try to perform both HTR and NER in one step.

In our experiments we have analyzed the impact of using three different tagging notations in a coupled model: parenthesized notation (PN), continuous notation with a reject class (CN+RC) and continuous notation without a reject class (CN).

Parenthesized notation:	\<persName\> Miss Locke, duchess of \<placeName\> Sussex \</placeName\> \</persName\> was found running.
Continuous notation with reject class:	Miss@persName Locke@persName,@persName duchess@persName of@persName Sussex@placeName@persName was@O found@O running@O.@O
Continuous notation without reject class:	Miss@persName Locke@persName,@persName duchess@persName of@persName Sussex@placeName@persName was found running.

Fig. 1. Different types of tagging notations for the sentence "Miss Locke, duchess of Sussex was found running". Note that the tag @placeName@persName constitutes a single symbol and that the tag @O represents the reject class.

Parenthesized Notation. Parenthesized notation refers to the usage of opening and closing tags for the group of words that composes each Named Entity. An example of this notation can be found in Fig. 1.

As we can see, in the case of nested Named Entities, the tags must be closed in a particular order. This notation imposes, as well, the usage of two symbols for each kind of Named Entity. It is also worth mentioning that, when training n-gram Language Models on transcriptions with parenthesized tagging, the prior probability of opening tags will be based on the words written before the Named Entity, whereas the prior probability of closing tags will account for the contents of the Named Entity.

Continuous Notation with a Reject Class. We refer to continuous notation as using a tagging symbol for each word that composes a Named Entity. In the case of using a reject class, every word in the transcription has a symbol indicating whether it pertains to a Named Entity or not. In the case that it pertains to a Named Entity, the symbol marks its type. An example of this notation can be found in Fig. 1.

When dealing with nested Named Entities, we may choose to generate symbols that indicate that the word is associated with multiple kinds of Named Entities or to add one tag for each Named Entity to which the word pertains. In the example in Fig. 1, we used the first option. We will do so throughout our experiments for consistency reasons, as we can only find nested Named Entities in one of the considered corpora [3].

The main advantage brought by this notation is that only one symbol is required for each kind of Named Entity, reducing the number of characters in the vocabulary. Moreover, having a tag for each word inside a Named Entity increases the number of samples associated with each symbol, which may result in a more robust Language Model. Also, in contrast to parenthesized notation, this notation does not introduce syntactical constraints. However, it is worth mentioning that, in the case of finding two consecutive Named Entities of the same type without any kind of separation between them, it would be impossible to determine how to split the sequence of words. In that particular case, it would be necessary to assume that the whole sequence constitutes a single Named Entity.

Continuous Notation Without a Reject Class. In the case of not using a reject class, only words that compose a Named Entity have a symbol that indicates their category. In the case of nested Named Entities, we follow the same policy as in the previous notation, where we add a single symbol that represents the whole stack of nested Named Entities. Following the previous example, the tagged transcription can be found in Fig. 1.

This tagging notation has the same virtues and problems as the previous syntax. However, the fact that only some words carry a tag may influence the patterns that the optical model has to learn. Before, the model had to choose which tag to add after a word, whereas now, it has to choose whether or not to append a symbol and its type.

2.3   Evaluation Metrics

In order to assess the performance of our system, we will be using different metrics: Character Error Rate (CER); Word Error Rate (WER); Micro Precision, Recall and F1 scores; Entity Character Error Rate (ECER); and Entity Word Error Rate (EWER). Micro Precision, Recall, and F1 scores are calculated considering the number of True Positives, False Positives, and False Negatives across all the Named Entity categories. ECER and EWER scores are computed, for each line, as the edit distance between the sequence of recognized Named Entities and the sequence of Named Entities in the ground truth, weighted by the edit distance between the reference and obtained transcriptions of the words inside the Named Entity (see [26] for more details).

When evaluating at Named Entity level, sequences of words pertaining to the same Named Entity are evaluated as a single item. Therefore, tagging mistakes or character errors affect the whole Named Entity for some metrics. The alternative is to evaluate at word level, considering the tag and the transcription for each word belonging to a Named Entity separately. In our experiments, the evaluation of the NER process will be done at word level and we will also project the outputs to continuous tagging notation without reject class to have comparable results across different notations. Hence, the evaluation will be more optimistic than considering the whole word sequence of the NE.

In order to validate our experimental method, we will also compute Macro Precision, Recall, and F1 scores on the sequences of tagging symbols, similarly to how it is done in [24]. However, as we work at line level, the number of tags in the hypothesis and the ground truth may differ. Therefore, we have implemented an edit distance that works with sequences of tags and calculates the number of True Positives, False Positives, and False Negatives for each Named Entity category. Having this information, we can compute the mean Precision, Recall, and F1 scores, which we refer to as the Macro Precision, Recall, and F1 scores.

In order to give some insight on how this edit distance is computed, we present an example of its application in Fig. 2. The only valid edit operations are an insertion of a ground truth tag (False Negative), a deletion of a hypothesized tag (False Positive) and a substitution with no cost when both tags are equal (True Positive). In order to force these three operations, the cost of a substitution of two different tags is raised to a number bigger than two and the cost of both the insertion and deletions is set to one.

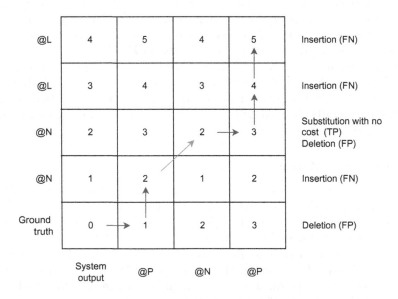

Fig. 2. Computation of the custom edit distance between two sequences of word-level Named Entity tags.

3 Experimental Method

3.1 Datasets

HOME. The HOME corpus [3] consists of 499 page images written by different authors in three languages: Czech, Latin, and German. Originally, parenthesized notation was used to tag the Named Entities, which imposes additional challenges. One of those challenges is the appearance of nested Named Entities. As an example of this event, we could have a birth date inside a proper noun. To correctly tag such kind of structures, the model must be able to incorporate some sort of syntactical knowledge in the decoding process. Another unique challenge comes from Named Entities that span over several lines. To properly tag those Named Entities, the model should be able to have contextual information from lines before the one being transcribed. Since we work at line level, we will simplify the task by splitting such Named Entities upon line ending.

We followed an experimental scheme similar to that presented in [3]. Therefore, the available data is split into three parts: a training set containing 80% of the charters, a validation set with 10% of the letters, and a testing set with the remaining 10% of the samples.

Throughout the corpus we can find five types of Named Entities: the name of a person, the name of a place, a date, the name of an organization, and an extra miscellaneous category for entities that do not match any of those types. However, the number of different syntactical structures that can be generated is significant due to the appearance of nested Named Entities.

George Washington. The George Washington corpus [22] is widely used for benchmarking word-spotting models [6]. It is composed of 20 pages written by a single person in historical English. The authors of [24] manually tagged the word images by considering five different categories of Named Entities: Cardinal, Date, Location, Organization, and Person. The tags for the word images were grouped by sentences, following the original sentence segmentation. They also published a custom split for training NER models where 12 pages are used for training, 2 for validation and the remaining 6 for testing the models.

We have extended their work using the available annotated data to generate line-level tagged transcriptions, including capitalization and punctuation symbols as shown in Fig. 3. The resulting corpus can be found at [25]. Our experimentation has been carried out on the line-level corpus.

IAM DB. The IAM Database [11] consists of 1539 scanned text pages written in modern English by 657 different authors. The documents contain sentences from the Lancaster - Oslo - Bergen corpus [7]. As in the George Washington corpus, the authors of [24] annotated the corpus at word level. This time, two sets of tags of different complexity were considered. The first consists of 18 categories of Named Entities: Cardinal, Date, Event, FAC, GPE, Language, Law, Location, Money, NORP, Ordinal, Organization, Person, Percent, Product, Quantity, Time, and

Original line-level transcription

273-03 for Bread; the Commissary having no orders to
273-04 make provisions for you. If any of your men

Sentence word tagging (all words are tagged with reject except Bread)

for bread (PER) the commissary having no orders to make provisions for you

Parenthesized tagged line-level transcription

273-03 for <PER> bread; </PER> the commissary having no orders to
273-04 make provisions for you. if any of your men

Fig. 3. Generation of line-level parenthesized tagged transcriptions (bottom) from line-level transcriptions [22] and tagged words with sentence segmentation [24]. Note that in the sentence tagging the punctuation symbols are not present.

Work of art. However, due to the complexity of the task, they also created a less demanding set of tags with six categories of Named Entities: Location, Time, Cardinal, NORP, Person, and Organization. The tags for the word images were grouped by sentences, following the original sentence segmentation.

We have extended their work using their annotated data to generate line-level transcriptions for both tagging sets, including capitalization and punctuation symbols similarly to what is shown in Fig. 3. The resulting corpus can be found at [25]. We have considered both the 18-categories and 6-categories tagging sets in our experimentation with the line-level corpus.

3.2   Implementation Details

The employed architecture is based on the coupled approach described in Sect. 2.1. We scale the line images to a height of 64 pixels and apply contrast enhancement and noise removal as described in [27]. No additional preprocessing is applied to the input.

The optical model contains a CRNN with four convolutional layers, where the n-th layer has $16n$ 3×3 filters, and a Bidirectional Long Short-Term Memory (BLSTM) unit of three layers of size 256 plus the final layer with a Softmax activation function. The rest of the hyperparameters are the same ones used in [16]. This model is implemented and trained with the PyLaia toolkit [12], which was regarded as one of the best HTR toolkits in [10].

The chosen Language Model is a character 8-gram with Kneser-Ney back-off smoothing. Its probabilities are trained with the SRILM toolkit [20] by considering the tagged transcriptions in the training set.

The CRNN and the character 8-gram are combined into a Stochastic Finite State Automata (SFSA) with the Kaldi toolkit [15]. Additional parameters, such as the Optical Scale Factor (OSF) and Word Insertion Penalty (WIP) are then estimated over the validation partition with the Simplex algorithm provided by the SciPy library.[1] Finally, this SFSA is used to obtain the hypothesized tagged transcription for each line in the test set. Our code is available at: https://github.com/DVillanova/HTR-NER.

3.3   Obtained Results

Table 1 compares the different tagging notations given the performance of the coupled model in the chosen corpora. Note that for the HOME corpus we present the performance obtained by applying an additional decoding technique for parenthesized notation.

Regarding parenthesized notation, it seems that its usage results in worse results. In the HOME corpus, however, we obtain the best Micro F1 score by using said notation and applying the n-best decoding technique presented in [26] to restrict the output. Even then, the ECER and EWER scores are substantially worse than those obtained with continuous notation. It is worth mentioning that we have tried the n-best decoding technique on the rest of the chosen corpora with unpromising results. Therefore, it seems that the parenthesized notation may only be helpful when dealing with nested Named Entities and with a mechanism to constrain the output syntactically.

The results obtained in the George Washington corpus do not favor any particular notation in every metric, as the difference in performance falls within the confidence intervals due to the lack of samples. However, it is worth mentioning that using continuous notation with a reject class yields noticeable improvements in the quality of the transcription for this task.

Our evaluation in the IAM corpus indicates that tagging the data using continuous notation without a reject class improves the performance of the coupled model in almost every metric, with statistically significant improvements in some metrics. This seems to also be the case for the other corpora, where F1 scores improve when using continuous notation instead of parenthesized with the baseline model. This may be due to the decrease in the number of tagging symbols in the vocabulary, which could be leading to a more robust estimation of their probabilities. There is also a noticeable improvement in ECER and EWER rates when employing this notation, especially in the case of 18 Named Entity categories. This improvement is directly related to the increase in F1 score and the decrease in CER and WER rates.

Table 2 shows the Macro Precision, Recall and F1 scores obtained with the best syntax according to Micro-F1 (i.e., CN) for the George Washington and IAM databases. We compare those results with the work done by other authors, even though the tasks are not identical in terms of complexity, to validate our experimental method.

The models proposed by other authors for the George Washington and IAM DB corpora [18,24] employed word and sentence segmentation data, which is costly to annotate and usually unavailable as, for example, in the HOME corpus.

---

[1] The documentation for the employed Simplex implementation is available at: https://docs.scipy.org/doc/scipy/reference/optimize.linprog-simplex.html.

Table 1. Results obtained in each corpus with the considered tagging notations: parenthesized notation (PN), continuous notation with a reject class (CN+RN) and continuous notation without a reject class (CN). For the HOME corpus, the 95% confidence intervals are between 1.82 (CER) and 3.10 (Recall). For the GW corpus, the confidence intervals are between 3.20 (CER) and 6.93 (Precision). For the IAM-DB corpus, the confidence intervals are between 0.82 (CER) and 1.71 (Precision). Note that we show Micro Precision, Micro Recall and Micro F1 scores.

Metric	Notation in HOME			
	PN	PN + n-best [26]	CN+RC	CN
CER (%)	9.23	9.24	9.40	9.55
WER (%)	28.20	28.14	33.05	30.65
Precision (%)	55.77	71.55	71.99	70.84
Recall (%)	60.20	61.04	56.36	56.03
F1 (%)	57.90	65.88	63.22	62.57
ECER (%)	41.31	31.60	26.38	26.82
EWER (%)	52.26	43.99	39.00	39.70

Metric	Notation in GW		
	PN	CN+RC	CN
CER (%)	5.64	3.92	4.58
WER (%)	24.24	13.80	17.14
Precision (%)	68.02	62.35	79.09
Recall (%)	50.00	45.29	47.62
F1 (%)	57.64	52.48	59.45
ECER (%)	38.61	38.56	42.60
EWER (%)	46.41	47.87	48.94

Metric	Notation in IAM-6			Notation in IAM-18		
	PN	CN+RC	CN	PN	CN+RC	CN
CER (%)	7.16	6.72	6.08	7.32	6.92	6.06
WER (%)	25.57	19.93	18.63	26.13	20.42	19.00
Precision (%)	37.44	54.96	53.18	36.24	53.34	48.97
Recall (%)	30.12	31.09	34.61	26.62	26.72	31.17
F1 (%)	33.88	39.72	41.93	30.69	35.60	38.09
ECER (%)	59.37	52.29	49.27	63.61	57.26	53.28
EWER (%)	67.05	60.69	58.44	69.80	64.48	62.17

Table 2. Comparison between the performance of the proposed coupled model with the best tagging notation and the work of previous authors in the chosen corpora.

	Metrics	Line-level CN	HTR - NER Tuselmann et al. [24]	End-to-end Rowtula et al. [18]
GW	Macro Precision (%)	94.84	86.90	76.40
	Macro Recall (%)	54.01	78.30	59.80
	Macro F1 (%)	68.82	81.30	66.60
IAM-6	Macro Precision (%)	73.06	83.30	65.50
	Macro Recall (%)	47.02	71.00	47.60
	Macro F1 (%)	57.22	76.40	54.60
IAM-18	Macro Precision (%)	62.65	64.80	36.90
	Macro Recall (%)	36.99	47.50	28.00
	Macro F1 (%)	46.52	53.60	30.30

Therefore, those models can produce hypotheses with the exact number of words or tags required while still having contextual information to guide the model. Our model, which works at line level, has to deal with an additional hidden variable: the segmentation of the line image. This lack of information results in a slight increase in WER (with CN: in GW from 14.52 to 17.14, in IAM from 18.66 to 19.00), although the difference is not statistically significant considering 95% confidence intervals. The increase in complexity also results in a decrease in F1 scores compared to [24] across all the tasks. However, our model improves the results of [18] in the IAM-18 corpus and works similarly well in the rest of the corpora. Overall, given the additional complexity of the task, we consider that the coupled approach obtains comparatively good results with the suitable tagging notation.

4   Conclusions and Future Work

We have evaluated the impact of different tagging notations on the performance of a coupled line-level HTR-NER model. Even with the increase in complexity due to working at line level instead of at word level with sentence segmentation, the chosen approach obtains comparatively good evaluation scores. This, in turn, validates the proposed experimental method. The obtained results show that using continuous notation without a reject class generally leads to improvements in performance, while including the reject class seems to have a higher deleterious effect as less data is available. However, parenthesized notation should be considered when dealing with nested Named Entities, as it can reduce the number of tagging symbols in the vocabulary. As future work, it could be interesting to replicate the experimentation with a decoupled approach to see how the different tagging notations impact the performance of an NLP model. Another option would be to refine the method proposed by [26] to apply syntactical constraints to the output, possibly employing some form of categorical language model to prune the lattice. We could also extend our experiments to structured documents, such as those present in the ESPOSALLES database [17], to check the relevance of the tagging notation in this kind of data.

References

1. Abadie, N., Carlinet, E., Chazalon, J., Duménieu, B.: A benchmark of named entity recognition approaches in historical documents application to 19th century French directories. In: Uchida, S., Barney, E., Eglin, V. (eds.) Document Analysis Systems, pp. 445–460. Springer International Publishing, Cham (2022). https://doi.org/10.1007/978-3-031-06555-2_30
2. Bluche, T.: Deep Neural Networks for Large Vocabulary Handwritten Text Recognition. Ph.D. thesis, Université Paris Sud-Paris XI (2015)
3. Boroş, E., et al.: A comparison of sequential and combined approaches for named entity recognition in a corpus of handwritten medieval charters. In: 2020 17th International Conference on Frontiers in Handwriting Recognition (ICFHR), pp. 79–84. IEEE (2020)
4. Carbonell, M., Villegas, M., Fornés, A., Lladós, J.: Joint recognition of handwritten text and named entities with a neural end-to-end model. In: 2018 13th IAPR International Workshop on Document Analysis Systems (DAS), pp. 399–404. IEEE (2018)
5. Catelli, R., Casola, V., De Pietro, G., Fujita, H., Esposito, M.: Combining contextualized word representation and sub-document level analysis through bi-LSTM+CRF architecture for clinical de-identification. Knowl.-Based Syst. **213**, 106649 (2021)
6. Fischer, A., Keller, A., Frinken, V., Bunke, H.: Lexicon-free handwritten word spotting using character HMMS. Pattern Recognition Letters **33**(7), 934–942 (2012). https://doi.org/10.1016/j.patrec.2011.09.009, special Issue on Awards from ICPR 2010
7. Johansson, S., Leech, G., Goodluck, H.: Manual of information to accompany the lancaster-oslo-bergen corpus of British English, for use with digital computers (1978). http://korpus.uib.no/icame/manuals/LOB/INDEX.HTM
8. Kang, L., Toledo, J.I., Riba, P., Villegas, M., Fornés, A., Rusiñol, M.: Convolve, attend and spell: an attention-based sequence-to-sequence model for handwritten word recognition. In: Brox, T., Bruhn, A., Fritz, M. (eds.) GCPR 2018. LNCS, vol. 11269, pp. 459–472. Springer, Cham (2019). https://doi.org/10.1007/978-3-030-12939-2_32
9. Lafferty, J.D., McCallum, A., Pereira, F.C.N.: Conditional random fields: Probabilistic models for segmenting and labeling sequence data. In: Proceedings of the Eighteenth International Conference on Machine Learning, ICML 2001, pp. 282–289. Morgan Kaufmann Publishers Inc., San Francisco, CA, USA (2001)
10. Maarand, M., Beyer, Y., Kåsen, A., Fosseide, K.T., Kermorvant, C.: A comprehensive comparison of open-source libraries for handwritten text recognition in norwegian. In: Uchida, S., Barney, E., Eglin, V. (eds.) Document Analysis Systems, pp. 399–413. Springer International Publishing, Cham (2022). https://doi.org/10.1007/978-3-031-06555-2_27
11. Marti, U.V., Bunke, H.: The i am-database: an English sentence database for offline handwriting recognition. Int. J. Doc. Anal. Recogn. **5**(1), 39–46 (2002)
12. Mocholí Calvo, C.: Development and experimentation of a deep learning system for convolutional and recurrent neural networks. Degree's thesis, Universitat Politècnica de València (2018)

13. Mohit, B.: Named entity recognition. In: Zitouni, I. (ed.) Natural Language Processing of Semitic Languages. TANLP, pp. 221–245. Springer, Heidelberg (2014). https://doi.org/10.1007/978-3-642-45358-8_7
14. Monroc, C.B., Miret, B., Bonhomme, M.L., Kermorvant, C.: A comprehensive study of open-source libraries for named entity recognition on handwritten historical documents. In: Uchida, S., Barney, E., Eglin, V. (eds.) Document Analysis Systems, pp. 429–444. Springer International Publishing, Cham (2022). https://doi.org/10.1007/978-3-031-06555-2_29
15. Povey, D., et al.: The kaldi speech recognition toolkit. In: IEEE 2011 Workshop on Automatic Speech Recognition and Understanding. No. CFP11SRW-USB, IEEE Signal Processing Society (2011)
16. Puigcerver, J.: Are multidimensional recurrent layers really necessary for handwritten text recognition? In: 2017 14th IAPR International Conference on Document Analysis and Recognition (ICDAR), vol. 1, pp. 67–72. IEEE (2017)
17. Romero, V., et al.: The Esposalles database: an ancient marriage license corpus for off-line handwriting recognition. Pattern Recognit. **46**(6), 1658–1669 (2013). https://doi.org/10.1016/j.patcog.2012.11.024
18. Rowtula, V., Krishnan, P., Jawahar, C.: Pos tagging and named entity recognition on handwritten documents. In: Proceedings of the 15th International Conference on Natural Language Processing, p. 87–91 (2018)
19. Sánchez, J.A., Bosch, V., Romero, V., Depuydt, K., De Does, J.: Handwritten text recognition for historical documents in the transcriptorium project. In: Proceedings of the First International Conference on Digital Access to Textual Cultural Heritage, pp. 111–117 (2014)
20. Stolcke, A.: Srilm - an extensible language modeling toolkit. In: Proceedings of 7th International Conference on Spoken Language Processing (ICSLP 2002), pp. 901–904 (2002)
21. Tarride, S., Lemaitre, A., Coéasnon, B., Tardivel, S.: A comparative study of information extraction strategies using an attention-based neural network. In: Uchida, S., Barney, E., Eglin, V. (eds.) Document Analysis Systems. DAS 2022. LNCS, vol. 13237, pp. 644–658. Springer, Cham (2022). https://doi.org/10.1007/978-3-031-06555-2_43
22. Tjong Kim Sang, E.F., Buchholz, S.: Introduction to the CoNLL-2000 shared task chunking. In: Fourth Conference on Computational Natural Language Learning and the Second Learning Language in Logic Workshop, pp. 127–132 (2000). https://aclanthology.org/W00-0726
23. Tüselmann, O., Fink, G.A.: Named entity linking on handwritten document images. In: Uchida, S., Barney, E., Eglin, V. (eds.) Document Analysis Systems, pp. 199–213. Springer International Publishing, Cham (2022). https://doi.org/10.1007/978-3-031-06555-2_14
24. Tüselmann, O., Wolf, F., Fink, G.A.: Are end-to-end systems really necessary for NER on handwritten document images? In: Lladós, J., Lopresti, D., Uchida, S. (eds.) ICDAR 2021. LNCS, vol. 12822, pp. 808–822. Springer, Cham (2021). https://doi.org/10.1007/978-3-030-86331-9_52
25. Villanova-Aparisi, D.: Line-level named entity recognition annotation for the George Washington and IAM datasets (2023). https://doi.org/10.5281/zenodo.7805128
26. Villanova-Aparisi, D., Martínez-Hinarejos, C.D., Romero, V., Pastor-Gadea, M.: Evaluation of named entity recognition in handwritten documents. In: Uchida, S., Barney, E., Eglin, V. (eds.) Document Analysis Systems. DAS 2022. LNCS, vol. 13237, pp. 568–582. Springer, Cham (2022). https://doi.org/10.1007/978-3-031-06555-2_38

27. Villegas, M., Romero, V., Sánchez, J.A.: On the modification of binarization algorithms to retain grayscale information for handwritten text recognition. In: Paredes, R., Cardoso, J.S., Pardo, X.M. (eds.) IbPRIA 2015. LNCS, vol. 9117, pp. 208–215. Springer, Cham (2015). https://doi.org/10.1007/978-3-319-19390-8_24

28. Wen, Y., Fan, C., Chen, G., Chen, X., Chen, M.: A survey on named entity recognition. In: Liang, Q., Wang, W., Liu, X., Na, Z., Jia, M., Zhang, B. (eds.) CSPS 2019. LNEE, vol. 571, pp. 1803–1810. Springer, Singapore (2020). https://doi.org/10.1007/978-981-13-9409-6_218

29. Wick, C., Zöllner, J., Grüning, T.: Transformer for handwritten text recognition using bidirectional post-decoding. In: Lladós, J., Lopresti, D., Uchida, S. (eds.) ICDAR 2021. LNCS, vol. 12823, pp. 112–126. Springer, Cham (2021). https://doi.org/10.1007/978-3-030-86334-0_8

30. Xingjian, S., Chen, Z., Wang, H., Yeung, D.Y., Wong, W.K., Woo, W.C.: Convolutional lstm network: a machine learning approach for precipitation nowcasting. In: Advances in Neural Information Processing Systems, pp. 802–810 (2015)

31. Yadav, V., Bethard, S.: A survey on recent advances in named entity recognition from deep learning models. In: Proceedings of the 27th International Conference on Computational Linguistics. pp. 2145–2158. Association for Computational Linguistics, Santa Fe, New Mexico, USA, August 2018. https://aclanthology.org/C18-1182

Analyzing the Impact of Tokenization on Multilingual Epidemic Surveillance in Low-Resource Languages

Stephen Mutuvi[1,2]([⊠])[iD], Emanuela Boros[1][iD], Antoine Doucet[1][iD],
Gaël Lejeune[3][iD], Adam Jatowt[4][iD], and Moses Odeo[2][iD]

[1] University of La Rochelle, L3i, 17000 La Rochelle, France
{stephen.mutuvi,emanuela.boros,antoine.doucet}@univ-lr.fr
[2] Multimedia University of Kenya, Nairobi, Kenya
{smutuvi,modeo}@mmu.ac.ke
[3] Sorbonne Université, STIH Lab, 75006 Paris, France
gael.lejeune@paris-sorbonne.fr
[4] University of Innsbruck, 6020 Innsbruck, Austria
adam.jatowt@uibk.ac.at

Abstract. Pre-trained language models have been widely successful, particularly in settings with sufficient training data. However, achieving similar results in low-resource multilingual settings and specialized domains, such as epidemic surveillance, remains challenging. In this paper, we propose hypotheses regarding the factors that could impact the performance of an epidemic event extraction system in a multilingual low-resource scenario: the type of pre-trained language model, the quality of the pre-trained tokenizer, and the characteristics of the entities to be extracted. We perform an exhaustive analysis of these factors and observe a strong correlation between them and the observed model performance on a low-resource multilingual epidemic surveillance task. Consequently, we believe that providing language-specific adaptation and extension of multilingual tokenizers with domain-specific entities is beneficial to multilingual epidemic event extraction in low-resource settings.

Keywords: Multilingual epidemic surveillance · Low-resource languages · Tokenization

1 Introduction

Disease outbreak news reports are increasingly becoming a rich source of data for early warning systems that aim at detecting such unforeseen events. Detecting emerging and evolving epidemic events in news published over time could support timely analysis of the situation and inform public health responses. In the context of epidemic surveillance, an event can be defined as an occurrence with potential public health importance and impact, such as a disease outbreak [2].

© The Author(s), under exclusive license to Springer Nature Switzerland AG 2023
G. A. Fink et al. (Eds.): ICDAR 2023, LNCS 14189, pp. 17–32, 2023.
https://doi.org/10.1007/978-3-031-41682-8_2

Thus, epidemic event extraction seeks to detect critical information about epidemic events by monitoring and analyzing online information sources, including news articles published in a variety of local languages.

While conventional surveillance methods are becoming increasingly more reliable, they are still limited in terms of geographical coverage as they depend on well-defined data from healthcare facilities [4,9,24,42,43]. To address these shortcomings, data-driven surveillance approaches that complement traditional surveillance methods have been proposed. However, extracting epidemic events from online news texts presents numerous challenges. First, obtaining sufficient high-quality annotations required to train and evaluate the automatic extraction systems is a time-consuming, labor-intensive, and costly process due to the need for the involvement of domain experts [15,29]. The second challenge relates to the multilingual nature of online data, since news is reported in a wide range of languages. In such settings, some languages (high-resource languages) such as English are likely to have readily available data, while others (low-resource languages) may have limited or no existing training data [22]. Furthermore, in a specialized domain such as the epidemiological domain, which is the focus of this work, many domain-specific words could be treated as out-of-vocabulary (OOV) words by the models. While pre-trained language models apply subword tokenization to deal with unseen words, subword tokenization algorithms are less optimal for low-resource morphologically rich languages, with limited training resources [21]. Rather than learning models from scratch, extending the vocabulary of the pre-trained language models could form the basis for improving the performance of epidemic event extraction tasks [31]. The adaptation involves utilizing the original vocabulary of pre-trained language models along with new domain-specific vocabulary, which has been shown to achieve improved performance on downstream tasks in new domains [13].

In this paper, we investigate the adaptation of pre-trained language models to the domain of low-resource epidemic surveillance. Therefore, we seek to answer two key questions: (1) *What impact does the quality of tokenization have on the performance of models on the task of extracting epidemic events from low-resource language texts?* (2) *Does the addition of unseen epidemic-related entities improve the performance of the models?* Answering these questions would be a critical step in adapting Transformer-based pre-trained models to the epidemiological domain, thereby improving the extraction of health-related events from text. Toward that end, we perform a detailed analysis of the quality of pre-trained tokenizer output and assess the impact of adding decompounded epidemic-related entity names to the vocabulary of the pre-trained models. Experimental results show that adapting the pre-trained language models by leveraging domain-specific entities is helpful for the epidemic event extraction task in low-resource multilingual settings.

The rest of the paper is organized as follows. Section 2 discusses the related research. The epidemiological event dataset utilized in the study is presented in Sect. 3. Section 4 describes the models and tasks selected for analysis, while the results are discussed in Sect. 5. Finally, Sect. 6 provides the conclusions and suggestions for future research.

2 Related Work

Recent multilingual pre-trained language models (e.g., multilingual BERT [7], XLM-RoBERTa [6]) demonstrated great performance on various direct zero-shot cross-lingual transfer tasks [10, 26, 36, 37, 44], where a model is first fine-tuned on the source language, and then evaluated directly on multiple target languages that are unseen in the fine-tuning stage [22, 28, 38, 39]. However, despite the remarkable success achieved by the models, a major challenge remains how to effectively adapt the models to new domains.

Typically, domain adaptation of the pre-trained language models is accomplished through unsupervised pre-training on target-domain text. For example, BioBERT [23] was initialized from general-domain BERT and then pre-trained on biomedical corpora comprising scientific publications. ClinicalBERT [17] was trained on clinical text from the MIMIC-III database [18] while PubMedBERT was pre-trained from scratch using abstracts from PubMed [14]. Another domain-specific model is SciBERT, a pre-trained language model based on BERT, which was pre-trained on a large corpus of scientific text [3]. In contrast to the other above-mentioned models, SciBERT uses an in-domain vocabulary (SciVOCAB), whereas the other models use the original BERT vocabulary. Generally, the pre-training approach to domain adaptation requires training the model from scratch, which is prohibitively expensive in terms of time and computational cost.

An alternative approach involves extending the vocabulary of the pre-trained models by including domain-specific words in the vocabulary. This alleviates the problem of directly using the original pre-trained vocabulary on new domains, where unseen domain-specific words are split into several subwords, making the training more challenging [16, 35]. Therefore, extending the vocabulary of pre-trained language models, such as BERT, could be beneficial, particularly in low-resource settings with limited training data. Some studies investigate the extension and adaptation of pre-trained models from general domains to specific domains with new additive vocabulary. The BERT model was adapted to the biomedical domain by adding new vocabulary using an extension module to create the extended BERT (exBERT) [35]. While an extension module is required, the weights of the original BERT model remain fixed, which substantially reduces the number of training resources required.

A similar study [30] focused on the named entity recognition (NER) task where the authors trained static embeddings (i.e., word2vec [27]) on target domain text and aligned the resulting word vectors with the wordpiece vectors of a general domain pre-trained language model while gaining domain-specific knowledge in the form of additional word vectors. Similarly, **A**dapt the **Voca**bulary to downstream **Do**main (AVocaDo) [16] model was proposed. The approach required only the downstream dataset to generate domain-specific vocabulary, which was then merged with the original pre-trained vocabulary for in-domain adaptation. By selecting a subset of domain-specific vocabulary while considering the relative importance of words, the method is applicable to a wide range of natural language processing (NLP) tasks and diverse domains (e.g., biomedical, computer science, news, and reviews).

Despite these recent efforts, studies on the adaptation of pre-trained language models to low-resource multilingual datasets are largely lacking. Moreover, an in-depth analysis of model performance that would take into account salient attributes of the extracted information and the pre-trained language model is needed. While previous studies have investigated the importance of high-quality subword-based tokenizers on performance, the focus was on models whose tokenizers are based on WordPiece [33], such as the multilingual BERT (mBERT). In this study, in addition, we explore the XLM-RoBERTa model that uses Byte-Pair Encoding (BPE) [12,34] subword tokenization algorithm. Unlike BPE, which considers the frequency of the symbol pair to determine whether it is to be added to the vocabulary, WordPiece selects the pair that maximizes the language-model likelihood of the training data. In both algorithms, commonly utilized words remain unsegmented, but rare words are decomposed into known subwords.

3   Epidemiological Event Dataset

Before detailing the experimental setup, we first present the epidemic event-based dataset, DAnIEL[1] that is utilized in this study. The dataset contains articles in typologically diverse languages, namely, English, Greek, Russian, and Polish, it is a subset of the dataset presented in [25]. This dataset is imbalanced, with only about 10% of the documents describing epidemic events. The number of documents in each language, however, is relatively balanced, as shown in Table 1. The task of extracting epidemic events from this dataset entails detecting all occurrences of the disease name and the locations of the reported event. For example, the following excerpt from an English article in the DAnIEL dataset published on January 15th, 2012 reports a potential *norovirus* outbreak in *Victoria*, Canada:

Table 1. Statistical description of the DAnIEL data. DIS = disease, LOC = location.

Language	Doc	Sent	Token	Entity	DIS	LOC
English	474	9,090	255,199	563	349	214
Greek	390	6,402	191,890	300	166	134
Polish	352	9,240	160,887	638	444	194
Russian	426	6,415	138,396	327	224	103

[1] The corpus is freely and publicly available at https://daniel.greyc.fr/public/index.php?a=corpus.

> *[...] Dozens of people ill in a suspected outbreak of norovirus at a student journalism conference in Victoria are under voluntary quarantine in their hotel rooms. [...]*

An epidemiological event extraction system should be able to detect the *norovirus* disease name (DIS) along with the aforementioned location of *Victoria* (LOC).

Table 2. The language family, vocabulary size, and the shared vocabulary between the language-specific and multilingual models.

Language	Family	Vocab. Size	Shared mBERT(%)	Shared XLMR(%)
English	Germanic	30,522	17.814	3.683
Greek	Hellenic	35,000	4.734	2.076
Polish	Balto-Slavic	60,000	11.490	5.077
Russian	Balto-Slavic	119,547	15.981	11.487

We note that all the languages in the DAnIEL dataset belong to the Indo-European family, with our sample including languages from the following genera: Germanic, Balto-Slavic, Romance, and Greek. The Indo-European family is (arguably) the most well-studied language family, comprising the majority of the highest resourced languages in the world, and thus, large pre-trained language models are generally likely to be biased towards such high-resource languages. The Greek language, whose writing system is the Greek alphabet, is an independent branch of the Indo-European family of languages, with a morphology that depicts an extensive set of productive derivational affixes. On the other hand, Russian (East-Slavic) and Polish (West-Slavic) are both Balto-Slavic languages with highly inflectional morphology. This can cause serious problems for nouns or any type of entity that needs to be detected (disease names and locations, in our case) in these languages since there are cases where it is impossible to determine the base form of an entity without appealing to data such as verb-frame sub-categorization, which is usually out of scope for the kind of tasks as ours. While English is typically considered a high-resource language, we include it in our experiments due to the fact that fewer than 500 English documents are present in our dataset.

4   Model and Task Selection

Based on the analysis of related literature, we expect that an epidemic event extraction system in a multilingual low-resource scenario could be impacted by the type of pre-trained language model used and the quality of the applied pre-trained tokenizer.

4.1   Language Models

Thus, first, we comprehensively compare two multilingual models, multilingual BERT (mBERT) [8] and XLM-RoBERTa (XLMR) [6], against their language-specific (monolingual) counterparts, for all the low-resource languages in our dataset. For the language-specific models, we use `bert-base-uncased`, `bert-base-greek-uncased-v1`, `rubert-base-cased`, `bert-base-polish-uncased-v1`[2]. The information about the vocabulary size of the language-specific models is presented in Table 2. The multilingual BERT, which is pre-trained on Wikipedia, has a vocabulary size of 119,547 tokens (same as the monolingual model for Russian) shared across 104 languages. On the other hand, XLM-RoBERTa is pre-trained on 2.5TB of CommonCrawl data covering 100 languages. As a result, when compared to Wikipedia data, CommonCrawl data is significantly larger, effectively increasing the available monolingual training resources for the respective languages, particularly low-resource languages [6,40].

Despite most of the languages having already been covered by the multilingual models, the justification for the language-specific models has been the lack of capacity by the multilingual models to represent all languages in an equitable manner [31]. The models have been pre-trained on either smaller or lower-quality corpora, particularly for most of the low-resource languages. Nonetheless, previous research indicates that both monolingual and multilingual models can produce competitive results, with the advantage of training a single language model, if any, being marginal [11].

4.2   Quality of the Pre-trained Tokenizer

While subword tokenizers [33,34] provide an effective solution to the OOV problem, pre-trained language models have been shown to still struggle to understand rare words [32]. The pre-trained tokenizers, in most instances, fail to represent such words by single tokens but instead do it via a sequence of subword tokens, thus limiting the learning of high-quality representations for rare words. Ensuring effective tokenization for low-resource and morphologically complex languages and thus improving the quality of their corresponding models is even more challenging due to the limited resources available to build tokenizers with sufficient vocabulary. In order to analyze the quality of the tokenizer, we first tokenized the text using the respective tokenizers for the language-specific, mBERT, and XLMR models. We then computed the two metrics proposed by [31]: *fertility* and *continued words*. Both metrics assess the suitability of a tokenizer for texts from different languages.

Fertility measures the average number of subwords generated per tokenized word, with a lower fertility value indicating that a tokenizer splits the tokens aggressively. For example, ["nor", "##ovi", "##rus"] has a fertility of 3. A minimum fertility score of 1 means that the vocabulary of the tokenizer contains

---

[2] All models can be found at Hugging Face website: https://huggingface.co.

just about every single word in the dataset. The fertility values of the pre-trained tokenizers per language are presented in Fig. 1. We can observe that the language-specific models have the lowest fertility scores, followed by XLMR for the low-resource languages (Greek, Polish, and Russian). This implies that the quality of tokenization is superior for language-specific tokenizer than for the other tokenizers, which could contribute to the improved model performance on downstream tasks such as epidemic event extraction.

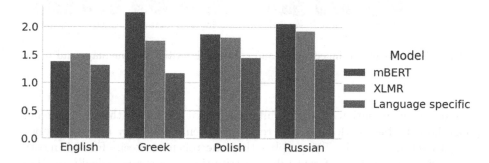

Fig. 1. Tokenizer fertility.

Continued Words and Entities describe words not found in the vocabulary of the pre-trained models and which are tokenized into multiple subword tokens. For example, since the word "norovirus" is not found in the vocabulary of BERT, it is tokenized into "nor", "##ovi" and "##rus" word pieces that are present in the BERT tokenizer vocabulary. The continuation symbol ## means the prefixed token should be attached to the previous one. In the case of XLM-RoBERTa, the word "norovirus" is tokenized into "_no", "ro", "virus", where the first token is prefixed with an underscore (_) and the rest of the tokens (without the underscore symbol) can be appended to form the complete word.

Therefore, the proportion of continued words indicates how frequently a tokenizer splits words on average. As with fertility, a low rate of continued words is desired, which indicates that the majority of the tokens are present in the vocabulary of the tokenizer. The ratio of the continued words is presented in Fig. 2, where we can make the same observation with regard to the language-specific models. Overall, the language-specific tokenizers produce superior results compared to their multilingual counterparts, which can be attributed to the fact that language-specific models have a higher parameter budget compared to the parameters allocated to the various languages in the vocabulary of multilingual models. Additionally, the models are typically prepared by native-speaker experts who are aware of relevant linguistic phenomena exhibited by the respective languages.

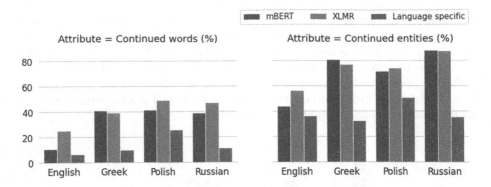

Fig. 2. The ratios of continued words and continued entities per language and model type.

Further examination of the results per language shows that English language had the best performance in terms of tokenizer fertility and the proportion of continued words. This implies that the tokenizer keeps English mostly intact while generating different token distributions in morphologically rich languages. Besides English being morphologically poor, this performance could also be attributed to the models having seen the most data in this language during pre-training, since English is a high-resource language. Notably, a high rate of continued words was observed for the Russian language on both multilingual models (mBERT and XLMR). Moreover, we include the ratios of *continued entities*, which are the locations and disease names in the training set that have been tokenized into multiple subwords. We observe, for all the languages, a high percentage of continued entities. Similar to continued words, and for the same reason (high parameter budget), the language-specific tokenizer yields better results than the multilingual tokenizers (mBERT and XLMR).

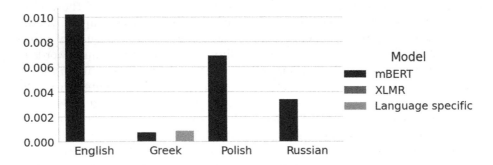

Fig. 3. The ratios of unknown words per language and model type.

OOV Words. We also provide the ratios of out-of-vocabulary (OOV) words in Fig. 3. OOV are words that cannot be tokenized, thus the entire word is mapped to a special token <UNK>. Generally, the proportion should be extremely low, i.e., the tokenizers should typically split the words into known subwords. We notice that mBERT seems to generate a significant number of OOVs, particularly for English and Polish languages. A closer look at the OOV words shows that the majority of the words, especially for English language text, comprises backticks and curly double quotation symbols which could easily be handled through text pre-processing. On the other hand, the XLMR tokenizer is able to recognize and split all tokens in our dataset, without the need for the <UNK> symbol, since byte-level Byte Pair Encoding (BPE) tokenization [41] covers any UTF-8 sequence with just 256 characters, which ensures that every base character is included in the vocabulary.

5 Results and Analysis

In order to analyze the impact of tokenizer quality, we fine-tune both the multilingual and language-specific models on the epidemic event extraction dataset. We train[3] and evaluate the models by averaging the precision (P), recall (R), and F1 score over five runs using different seed values and reporting the standard deviation ($\pm$).

Table 3. Model performance per language and model type.

Lang	Model	P	R	F1
English	mBERT	76.18±1.21	64.52±3.95	69.88±2.36
	XLMR	69.50±9.68	55.48±3.53	61.59±5.16
	Language-specific	**76.83**±3.69	**68.39**±5.30	**72.30**±3.67
Greek	mBERT	**83.94**±7.00	73.85±5.01	78.39±4.07
	XLMR	81.45±7.73	70.00±6.32	74.85±2.52
	Language-specific	72.78±3.57	**91.54**±5.02	**80.96**±1.95
Polish	mBERT	88.02±5.70	89.41±2.63	88.60±3.05
	XLMR	86.94±4.33	87.45±2.24	87.14±2.49
	Language-specific	**89.62**±6.45	89.41±3.56	**89.32**±2.54
Russian	mBERT	**67.18**±3.84	**67.28**±3.32	**67.08**±0.76
	XLMR	57.35±11.23	53.34±11.86	54.47±3.55
	Language-specific	51.83±4.30	65.94±4.16	58.14±1.75

---

[3] In all experiments, we use AdamW [19] with a learning rate of $1e-5$ and for 20 epochs. We also considered a maximum sentence length of 164 [1].

5.1   Results and Error Analysis

As presented in Table 3, the language-specific models attain the best overall performance, recording the highest scores in all the languages considered except for Russian. The high performance of the language-specific model is congruent with the lower fertility and continued word values when compared to the multilingual models, as shown in Figs. 1 and 2, respectively. In terms of per-language performance, the Russian language had the lowest performance, which could be attributed to high values of subword fertility, continued words, and entities.

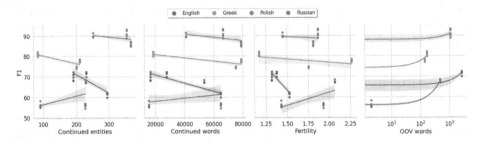

Fig. 4. Relationship between tokenization quality and F1 performance. The languages considered were Polish, Greek, Russian, and English. The assessment of tokenizer quality was based on the real values of continued entities, continued words, OOVs, and fertility.

To determine whether there is a relationship between the number of continued words, entities, and the fertility of the tokenizer and the F1 scores for all models, we draw several scatter plots per language for each of these variables, then fit a regression model and plot the resulting regression line and a 95% confidence interval. As shown in Fig. 4, we observe that, for English, the results are generally stable, but the performance is negatively impacted by the fertility and continued words and entities. Furthermore, the performance score (F1) decreases with an increase in the proportion of unseen entities (continued entities). As shown in Table 2, English has a high amount of shared vocabulary (e.g., 17% for mBERT), thus we would expect that further addition of new entities would have minimal impact on performance.

Similarly, while rather unstable in regard to the tokenizer quality, the results for the Greek and Polish were also impacted by fertility, continued words, and entities. Interestingly, Russian does not seem to be impacted by any of the tokenizer's characteristics. We could attribute this fact to the vocabulary size of its corresponding language-specific tokenizer. Besides, the proportion of shared words or entities is relatively higher for Russian, as shown in Table 2, implying that the language is fairly well represented in the vocabulary of the pre-trained multilingual models, and thus, a few OOVs or continued words would have little or no impact on the performance.

Based on these findings, we can expect English, Greek, and Polish to perform better for both multilingual models, mBERT and XLMR, if the continued entities are correctly tokenized. We would also anticipate a change in performance for all models for Russian, though it is unclear whether this would be a lower or higher performance.

5.2   Repairing Continued Entities

On the basis of the preceding analyses, we evaluate the impact of increasing the model capacity by adding words from the epidemiological domain into an existing tokenizer vocabulary[4]. Concretely, continued (unseen) entities, which are entities from the training corpus of the specialized domain that were not already present in the existing vocabulary and which were split into multiple subwords by the pre-trained tokenizer, were used as the extension vocabulary. We note that it is critical to add to an existing subword tokenizer only whole words rather than their respective subwords. This is because the inclusion of subwords instead of whole entity words could introduce errors into the vocabulary, considering that tokens in the tokenizer vocabulary are ordered by frequency. We refer to the model with the vocabulary extension as the Extended Model or, in short, the E-model.

Table 4 presents the number of unseen entities per language that were included in the vocabulary of the considered pre-trained models. We notice that Polish has the highest number of unseen tokens among all the languages, while Greek has the lowest. The "DIS" entity accounted for the greatest number of unseen entities in comparison to the "LOC" entity. For instance, all the 14 unseen entities identified when a language-specific tokenizer was applied to the Greek text were "DIS" entities.

In Table 5, a notable performance improvement was observed for the Polish language on all the models, while for the Greek and Russian languages, only mBERT was unable to record a performance gain. A significant performance drop is observed for the English language which can be attributed to either the negative influence of rare words or due to the vocabulary of pre-trained and fine-tuned models becoming too distant and the model losing previously learned knowledge, thus hindering the generalization performance of the models, a phenomenon referred to as catastrophic forgetting [5, 20]. While the addition of entities to the vocabulary marginally improves the performance of the different model types evaluated, we presume that the performance could substantially improve when the vocabulary is extended with a sizable number of entities. In

---

[4] The HuggingFace https://huggingface.co/docs/transformers/ library provides a function for adding continued entities to the existing vocabulary of a tokenizer. In addition, the function includes a mechanism for discarding tokens in the extension vocabulary that appear in the original pre-trained vocabulary, ensuring that the extension vocabulary is an absolute complement to the original vocabulary. The size of the extension vocabulary varies depending on the language and pre-trained model.

Table 4. The number of entities added to the vocabulary per language and model. DIS (%) and LOC (%) denote the percentage of unseen disease and location entities, respectively, which were not found in the tokenizer vocabulary.

Language	Model	Unseen	DIS (%)	LOC (%)
	mBERT	58	60.34	41.38
English	XLMR	75	52.00	49.33
	Language-specific	46	63.04	39.13
	mBERT	36	58.33	41.67
Greek	XLMR	33	63.64	36.36
	Language-specific	14	100.0	0.00
	mBERT	146	66.44	34.93
Polish	XLMR	157	61.78	39.49
	Language-specific	124	68.55	33.06
	mBERT	76	56.58	43.42
Russian	XLMR	74	54.05	45.95
	Language-specific	39	74.36	25.64

Table 5. Comparison between the default pre-trained model and E-Model.

Language	Model	Default	E-Model
	mBERT	**69.88 ± 2.36**	50.11 ± 2.30
English	XLMR	**61.59 ± 5.16**	55.30 ± 1.73
	Language-specific	**72.30 ± 3.67**	56.79 ± 2.20
	mBERT	**78.39 ± 4.07**	76.47 ± 2.61
Greek	XLMR	74.85 ± 2.52	**79.88 ± 2.73**
	Language-specific	80.96 ± 1.95	**83.95 ± 4.08**
	mBERT	88.60 ± 3.05	**91.43 ± 2.90**
Polish	XLMR	87.14 ± 2.49	**87.67 ± 2.92**
	Language-specific	89.42 ± 2.54	**91.76 ± 2.39**
	mBERT	**67.08 ± 0.76**	60.40 ± 0.09
Russian	XLMR	54.47 ± 3.55	**58.77 ± 9.90**
	Language-specific	58.14 ± 1.75	**61.81 ± 3.22**

our case, only a few continued entities were available for each language in the dataset, as presented in Table 4.

6   Conclusions and Future Work

In this paper, we performed a systematic analysis of the impact of tokenizer output quality on multilingual epidemic surveillance, by measuring the proportion of unknown words, continued words and entities, and tokenizer fertility. Based on the analysis, unseen entities related to the epidemiology domain were identified and further leveraged to extend the pre-trained vocabulary. Our results suggest that in-domain vocabulary plays an important role in adapting pre-trained models to the epidemiological surveillance domain, and is particularly effective for low-resource settings. More specifically, we make the following key observations: the quality of the tokenizer, as measured by subword fertility, the proportion of continued words, and the OOV words, influence model performance, especially for low-resource languages. This is reflected in the slightly improved performance of language-specific models in the epidemic event extraction task when compared to their multilingual counterparts. Therefore, when performing information extraction in multilingual and low-resource settings, the choice of tokenizer is a crucial factor to consider.

Furthermore, domain adaptation by expanding the vocabulary of pre-trained tokenizers was found to improve model performance. In particular, we discovered that adding domain-specific entities to the vocabulary of the tokenizers, which were not initially present in the generic datasets used to train the tokenizers, is critical in domain adaptation. While domain adaptation using domain-specific entities is a promising approach, the possibility of concept drift and catastrophic forgetting occurring should be considered, since they could potentially result in performance degradation, as has been observed for English. Future work includes further investigating the effectiveness of leveraging both domain adaptation and self-training, an important technique for mitigating the challenges associated with training data scarcity, by leveraging unlabelled data along with the few available labelled training examples.

Acknowledgements. This work has been supported by the ANNA (2019-1R40226), TERMITRAD (2020-2019-8510010) and PYPA (2021-2021-12263410) projects funded by the Nouvelle-Aquitaine Region, France. It has also been supported by the French Embassy in Kenya and the French Foreign Ministry.

References

1. Adelani, D.I., et al.: MasakhaNER: named entity recognition for African languages. Trans. Assoc. Comput. Linguist. **9**, 1116–1131 (2021). https://doi.org/10.1162/tacl_a_00416, https://aclanthology.org/2021.tacl-1.66
2. Balajee, S.A., Salyer, S.J., Greene-Cramer, B., Sadek, M., Mounts, A.W.: The practice of event-based surveillance: concept and methods. Global Secur. Health Sci. Policy **6**(1), 1–9 (2021)
3. Beltagy, I., Lo, K., Cohan, A.: SciBERT: a pretrained language model for scientific text. In: Proceedings of the 2019 Conference on Empirical Methods in Natural Language Processing and the 9th International Joint Conference on Natural Language Processing (EMNLP-IJCNLP), pp. 3615–3620. Association for Computational Linguistics, Hong Kong, China, November 2019. https://doi.org/10.18653/v1/D19-1371, https://aclanthology.org/D19-1371
4. Brownstein, J.S., Freifeld, C.C., Reis, B.Y., Mandl, K.D.: Surveillance sans frontieres: internet-based emerging infectious disease intelligence and the healthmap project. PLoS Med. **5**(7), e151 (2008)

5. Chen, X., Wang, S., Fu, B., Long, M., Wang, J.: Catastrophic forgetting meets negative transfer: batch spectral shrinkage for safe transfer learning. Adv. Neural Inf. Process. Syst. **32** (2019)
6. Conneau, A., et al.: Unsupervised cross-lingual representation learning at scale. In: Proceedings of the 58th Annual Meeting of the Association for Computational Linguistics, ACL 2020, 5-10 July 2020, pp. 8440–8451. Association for Computational Linguistics (2020). https://www.aclweb.org/anthology/2020.acl-main.747/
7. Devlin, J., Chang, M.W., Lee, K., Toutanova, K.: BERT: pre-training of deep bidirectional transformers for language understanding. In: Proceedings of the 2019 Conference of the North American Chapter of the Association for Computational Linguistics: Human Language Technologies, Volume 1 (Long and Short Papers), pp. 4171–4186. Association for Computational Linguistics, Minneapolis, Minnesota, June 2019. https://doi.org/10.18653/v1/N19-1423, https://aclanthology.org/N19-1423
8. Devlin, J., Chang, M.W., Lee, K., Toutanova, K.: BERT: pre-training of deep bidirectional transformers for language understanding. In: Proceedings of the 2019 Conference of the North American Chapter of the Association for Computational Linguistics: Human Language Technologies, Volume 1 (Long and Short Papers), pp. 4171–4186. Association for Computational Linguistics, Minneapolis, Minnesota, June 2019. https://doi.org/10.18653/v1/N19-1423
9. Dórea, F.C., Revie, C.W.: Data-driven surveillance: effective collection, integration and interpretation of data to support decision-making. Front. in Vet. Sci. **8**, 225 (2021)
10. Faruqui, M., Kumar, S.: Multilingual open relation extraction using cross-lingual projection. arXiv preprint arXiv:1503.06450 (2015)
11. Feijo, D.D.V., Moreira, V.P.: Mono vs multilingual transformer-based models: a comparison across several language tasks. arXiv preprint arXiv:2007.09757 (2020)
12. Gage, P.: A new algorithm for data compression. C Users J. **12**(2), 23–38 (1994)
13. Garneau, N., Leboeuf, J.S., Lamontagne, L.: Predicting and interpreting embeddings for out of vocabulary words in downstream tasks. In: Proceedings of the 2018 EMNLP Workshop BlackboxNLP: Analyzing and Interpreting Neural Networks for NLP, pp. 331–333. Association for Computational Linguistics, Brussels, Belgium, November 2018. https://doi.org/10.18653/v1/W18-5439, https://aclanthology.org/W18-5439
14. Gu, Y., et al.: Domain-specific language model pretraining for biomedical natural language processing. ACM Trans. Comput. Healthcare (HEALTH) **3**(1), 1–23 (2021)
15. Hedderich, M.A., Lange, L., Adel, H., Strötgen, J., Klakow, D.: A survey on recent approaches for natural language processing in low-resource scenarios. In: Proceedings of the 2021 Conference of the North American Chapter of the Association for Computational Linguistics: Human Language Technologies, pp. 2545–2568. Association for Computational Linguistics, June 2021. https://doi.org/10.18653/v1/2021.naacl-main.201, https://aclanthology.org/2021.naacl-main.201
16. Hong, J., Kim, T., Lim, H., Choo, J.: AVocaDo: strategy for adapting vocabulary to downstream domain. In: Proceedings of the 2021 Conference on Empirical Methods in Natural Language Processing, pp. 4692–4700. Association for Computational Linguistics, Online and Punta Cana, Dominican Republic, November 2021. https://doi.org/10.18653/v1/2021.emnlp-main.385, https://aclanthology.org/2021.emnlp-main.385
17. Huang, K., Altosaar, J., Ranganath, R.: Clinicalbert: modeling clinical notes and predicting hospital readmission. arXiv preprint arXiv:1904.05342 (2019)

18. Johnson, A.E., et al.: Mimic-iii, a freely accessible critical care database. Sci. Data **3**(1), 1–9 (2016)
19. Kingma, D.P., Ba, J.: Adam: a method for stochastic optimization. arXiv preprint arXiv:1412.6980 (2014)
20. Kirkpatrick, J., et al.: Overcoming catastrophic forgetting in neural networks. Proc. Nat. Acad. Sci. **114**(13), 3521–3526 (2017)
21. Klein, S., Tsarfaty, R.: Getting the## life out of living: How adequate are word-pieces for modelling complex morphology? In: Proceedings of the 17th SIGMOR-PHON Workshop on Computational Research in Phonetics, Phonology, and Morphology, pp. 204–209 (2020)
22. Lauscher, A., Ravishankar, V., Vulić, I., Glavaš, G.: From zero to hero: on the limitations of zero-shot cross-lingual transfer with multilingual transformers. arXiv preprint arXiv:2005.00633 (2020)
23. Lee, J., et al.: Biobert: a pre-trained biomedical language representation model for biomedical text mining. Bioinformatics **36**(4), 1234–1240 (2020)
24. Lejeune, G., Brixtel, R., Doucet, A., Lucas, N.: Multilingual event extraction for epidemic detection. Artif. Intell. Med. **65**(2), 131–143 (2015)
25. Lejeune, G., Brixtel, R., Lecluze, C., Doucet, A., Lucas, N.: Added-value of automatic multilingual text analysis for epidemic surveillance. In: Peek, N., Marín Morales, R., Peleg, M. (eds.) AIME 2013. LNCS (LNAI), vol. 7885, pp. 284–294. Springer, Heidelberg (2013). https://doi.org/10.1007/978-3-642-38326-7_40
26. Lin, Y., Liu, Z., Sun, M.: Neural relation extraction with multi-lingual attention. In: Proceedings of the 55th Annual Meeting of the Association for Computational Linguistics (Volume 1: Long Papers), pp. 34–43 (2017)
27. Mikolov, T., Chen, K., Corrado, G., Dean, J.: Efficient estimation of word representations in vector space. In: Bengio, Y., LeCun, Y. (eds.) 1st International Conference on Learning Representations, ICLR 2013, Scottsdale, Arizona, USA, 2-4 May 2013, Workshop Track Proceedings (2013). http://arxiv.org/abs/1301.3781
28. Mutuvi, S., Boros, E., Doucet, A., Lejeune, G., Jatowt, A., Odeo, M.: Multilingual epidemiological text classification: a comparative study. In: COLING, International Conference on Computational Linguistics (2020)
29. Neves, M., Leser, U.: A survey on annotation tools for the biomedical literature. Briefings Bioinf. **15**(2), 327–340 (2014)
30. Poerner, N., Waltinger, U., Schütze, H.: Inexpensive domain adaptation of pre-trained language models: Case studies on biomedical NER and covid-19 QA. In: Findings of the Association for Computational Linguistics: EMNLP 2020, November 2020
31. Rust, P., Pfeiffer, J., Vulić, I., Ruder, S., Gurevych, I.: How good is your tokenizer? on the monolingual performance of multilingual language models. In: Proceedings of the 59th Annual Meeting of the Association for Computational Linguistics and the 11th International Joint Conference on Natural Language Processing (Volume 1: Long Papers), pp. 3118–3135. Association for Computational Linguistics, August 2021. https://doi.org/10.18653/v1/2021.acl-long.243, https://aclanthology.org/2021.acl-long.243
32. Schick, T., Schütze, H.: Rare words: a major problem for contextualized embeddings and how to fix it by attentive mimicking. In: Proceedings of the AAAI Conference on Artificial Intelligence, vol. 34, pp. 8766–8774 (2020)
33. Schuster, M., Nakajima, K.: Japanese and Korean voice search. In: 2012 IEEE International Conference on Acoustics, Speech and Signal Processing (ICASSP), pp. 5149–5152. IEEE (2012)

34. Sennrich, R., Haddow, B., Birch, A.: Neural machine translation of rare words with subword units. In: Proceedings of the 54th Annual Meeting of the Association for Computational Linguistics (Volume 1: Long Papers), pp. 1715–1725. Association for Computational Linguistics, Berlin, Germany, August 2016. https://doi.org/10.18653/v1/P16-1162, https://aclanthology.org/P16-1162

35. Tai, W., Kung, H., Dong, X.L., Comiter, M., Kuo, C.F.: Exbert: extending pre-trained models with domain-specific vocabulary under constrained training resources. In: Proceedings of the 2020 Conference on Empirical Methods in Natural Language Processing: Findings, pp. 1433–1439 (2020)

36. Tian, L., Zhang, X., Lau, J.H.: Rumour detection via zero-shot cross-lingual transfer learning. In: Oliver, N., Pérez-Cruz, F., Kramer, S., Read, J., Lozano, J.A. (eds.) ECML PKDD 2021. LNCS (LNAI), vol. 12975, pp. 603–618. Springer, Cham (2021). https://doi.org/10.1007/978-3-030-86486-6_37

37. Wang, X., Han, X., Lin, Y., Liu, Z., Sun, M.: Adversarial multi-lingual neural relation extraction. In: Proceedings of the 27th International Conference on Computational Linguistics, pp. 1156–1166 (2018)

38. Wang, Z., Mayhew, S., Roth, D., et al.: Cross-lingual ability of multilingual bert: an empirical study. arXiv preprint arXiv:1912.07840 (2019)

39. Wu, S., Dredze, M.: Beto, bentz, becas: the surprising cross-lingual effectiveness of BERT. In: Proceedings of the 2019 Conference on Empirical Methods in Natural Language Processing and the 9th International Joint Conference on Natural Language Processing (EMNLP-IJCNLP), pp. 833–844. Association for Computational Linguistics, Hong Kong, China, November 2019. https://doi.org/10.18653/v1/D19-1077, https://aclanthology.org/D19-1077

40. Wu, S., Dredze, M.: Are all languages created equal in multilingual BERT? In: Proceedings of the 5th Workshop on Representation Learning for NLP. Association for Computational Linguistics, July 2020. https://doi.org/10.18653/v1/2020.repl4nlp-1.16, https://aclanthology.org/2020.repl4nlp-1.16

41. Wu, Y., et al.: Google's neural machine translation system: bridging the gap between human and machine translation. arXiv preprint arXiv:1609.08144 (2016)

42. Yangarber, R., Best, C., Von Etter, P., Fuart, F., Horby, D., Steinberger, R.: Combining information about epidemic threats from multiple sources. In: Proceedings of the MMIES Workshop, International Conference on Recent Advances in Natural Language Processing (RANLP 2007), Citeseer (2007)

43. Yangarber, R., Jokipii, L., Rauramo, A., Huttunen, S.: Extracting information about outbreaks of infectious epidemics. In: Proceedings of HLT/EMNLP 2005 Interactive Demonstrations, pp. 22–23 (2005)

44. Zou, B., Xu, Z., Hong, Y., Zhou, G.: Adversarial feature adaptation for cross-lingual relation classification. In: Proceedings of the 27th International Conference on Computational Linguistics, pp. 437–448 (2018)

DAMGCN: Entity Linking in Visually Rich Documents with Dependency-Aware Multimodal Graph Convolutional Network

Yi-Ming Chen[1]([✉]), Xiang-Ting Hou[1], Dong-Fang Lou[1], Zhi-Lin Liao[1], and Cheng-Lin Liu[2,3]

[1] Hundsun Technologies Inc., Jiangnan Ave., Binjiang District, Hangzhou 3588, China
`tony_chenheu@hotmail.com`, `HOUX0009@e.ntu.edu.sg`,
`zhilinliao@yeah.net`
[2] State Key Laboratory of Multimodal Artificial Intelligence Systems, Institute of Automation of Chinese Academy of Sciences, Beijing 100190, People's Republic of China
`liucl@nlpr.ia.ac.cn`
[3] School of Artificial Intelligence, University of Chinese Academy of Sciences, Beijing 100049, People's Republic of China
`https://en.hundsun.com/`

Abstract. Linking semantically related entities in visually rich documents (VRDs) is important for understanding the document structure and contents. Existing approaches usually predict the links using layout and textual semantic information simultaneously. They may generate false links or omit real links because of the insufficient utilization of the dependency among adjacent value entities. In this paper, we propose a dependence-aware multimodal graph convolutional network (DAMGCN) for improving the accuracy of entity linking. In DAMGCN, a graph construction approach based on layout-based k-NN is proposed to reduce invalid entity links. Then, the links are determined from a proposed local transformer block (LTB), which utilizes multiple layout information (angle, distance, intersection over union) to identify the adjacent entities with potential dependency with a given key entity. Experiments on the public FUNSD and XFUND datasets show that our DAMGCN achieves competitive results, i.e., F1 scores of 0.8063 and 0.7303 in entity linking tasks, respectively, while the model size is much smaller than the state-of-the-art models.

Keywords: Visually Rich Documents · Entity linking · Dependency · Multimodal Graph Convolutional Network · Relation Extraction

1 Introduction

Entity linking from visually rich documents (VRDs) is an important directions of the recent multimodal document understanding since it extracts relational semantic information from unstructured documents [1–3]. Entity linking for VRDs generally predicts links between any two semantic entities considering both visual and textual information. The visual information usually comes from the document layout (geometric relation

G. A. Fink et al. (Eds.): ICDAR 2023, LNCS 14189, pp. 33–47, 2023.
https://doi.org/10.1007/978-3-031-41682-8_3

between text fields such as table cells), and the textual information is mainly from the recognized text fields using an optical character recognition (OCR) technology [1]. The results of entity linking are usually presented as key-value pairs, as shown in Fig. 1.

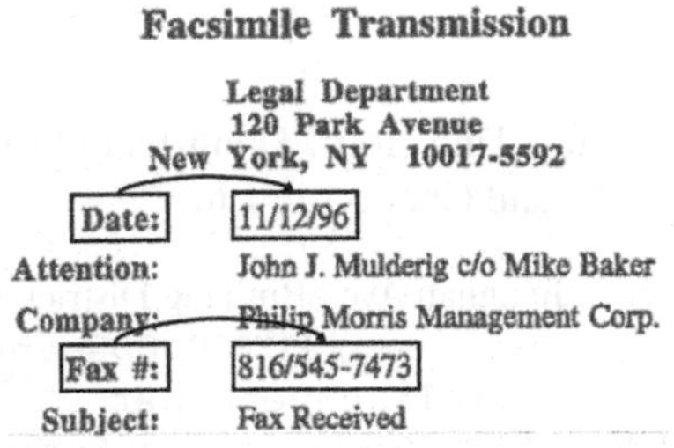

Facsimile Transmission

Legal Department
120 Park Avenue
New York, NY 10017-5592

Date: 11/12/96

Attention: John J. Mulderig c/o Mike Baker

Company: Philip Morris Management Corp.

Fax #: 816/545-7473

Subject: Fax Received

Fig. 1. Illustration of entity linking. The association of entities "Date" and "11/12/96" is presented as a directed link. Likewise, entity "Fax #:" links to entity "816/545-7473".

With the development of Transformer models that fuse visual and textual information with attention mechanism, layout-based models [2, 3] have been proposed to encode the multi-modal layout features in terms of the position of text blocks, text contents and layout by using the BERT [4] as the backbone model. Besides, GNN-based models are also used to learn a document layout representation [1]. Based on the representation, a multi-layer perceptron (MLP) [5] is usually adopted to infer the link given any selected pairs of entities [2, 3, 6–8]. However, there are two challenges of the existing models, as illustrated in Fig. 2. Firstly, invalid links of irrelevant entities can be introduced because each key entity is connected to all the value entities in a document (Challenge 1). Secondly, the discrimination of the links ending with the entities with similar features is difficult because the dependency among the adjacent value entities with a given key entity is not considered (Challenge 2).

In this paper, we propose a dependency-aware multimodal graph convolutional network, named DAMGCN, to model the entity linking task as a graph edge classification problem. Specifically, a graph construction approach using layout-based k-nearest neighbor (k-NN) and heuristic filtering is proposed to reduce the number of invalid links. Moreover, we introduce a local transformer block (LTB) that leverages the dependencies among adjacent value entities to predict the links in the graph. Our proposed approach offers a novel and effective way to model the complex relationships between entities in VRDs.

The contributions of this paper can be summarized as follows:

1) A dependency-aware multimodal graph convolutional network (DAMGCN) is proposed, with the graph constructed based on layout information to greatly reduce the invalid links.
2) A local transformer block (LTB) is proposed to predict the links by considering the dependency among adjacent value entities to improve the discrimination between similar value entities linked to a key entity.
3) We report competitive results on the public datasets FUNSD and XFUND, with F1 scores of 0.8063 and 0.7303, respectively, while the model size of DAMGCN is only 480M compared to more than 1G of existing models.

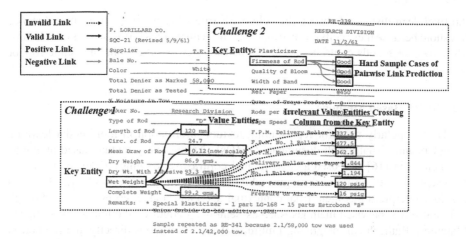

Fig. 2. Challenges of entity linking. Challenge 1: Invalid links exist when key and value entities are clearly irrelevant, such as entities crossing different rows (columns) or entities being spatially far from each other. Challenge 2: Three adjacent entities "Good" have exactly the same semantic features which make them linked independently to the same key "Firmness of Rod".

2 Related Works

The recent methods for entity linking can be divided into two categories: graph neural network (GNN) based [1, 9–13] and BERT-based models [2, 3, 6–8, 16, 17].

2.1 GNN-Based Methods

The methods in [1, 9–15] introduce GNN-based model to fuse the textual and visual feature from documents. The graph edges are considered as potential links among the entities. In [11], the node features are determined from both the spatial information of a given entity and the semantic features extracted from XLM-RoBERTa [13] and BERT-base-multilingual [4] models. In order to enhance the learning ability of GNN models towards the layout information, [1, 12] introduce edge features determined from the x-axis and y-axis distances or the aspect ratio of bounding boxes between two entities. For graph construction in [1, 9, 12], all nodes are fully connected with each other, which introduces a massive graph with invalid edges. To tackle this problem, [10] uses a k-NN-based method to create the graph edges from the distance calculated from the top-left corner of the word bounding boxes. [14] uses beta-skeleton method to construct graph effectively by allowing information propagation with respect to the structural layout of documents. According to [15], line-of-sight heuristic strategy is used to build relationships in the information extraction task. Also, the method in [11] proposed a line-of-sight strategy in the graph construction phase, considering that two nodes are linked if they could "see" each other without any entities in sight regarding their spatial information. However, these strategies are not suitable for tabular documents, where the Euclidean distance between the cells with true links is divergent from a distance threshold because these cells are distributed along with various forms of rows or columns. Moreover, such cells could be invisible when they cross other cells in a same row or column.

2.2 BERT-Based Methods

BERT is popularly used for representing textual information in multimodal learning to understand visually-rich documents (VRDs). Among early textual-based attempts, the method in [16] proposed a cross-lingual contrastive learning model to encourage representations to be more distinguishable. More recently, some works [2, 3, 6, 17] consider interpreting content from contextual texts and image representations through different self-supervised methods, such as Sentence Length Prediction and Text-Image Matching. Rather than integrating multimodal representations directly, the method in [7] encodes the text and layout features separately to ensure that the representations of layout branch remain unchanged with the substitution of languages. In addition, the method in [8] focuses on the proper reading orders of the documents, and proposes XY-cut algorithm to capture and leverage rich layout information and achieve competitive results.

Inspired by the above methods in the literature, we propose a multimodal model fusing layout and textual information for entity linking in VRDs. We use a layout-based K-nearest neighbor (k-NN) and heuristic filtering strategy to reduce invalid links in graph construction, and propose a local transformer block to predict the links by considering the dependency among adjacent value entities of a key entity.

3   Methodology

Figure 3 gives the overview of our proposed DAMGCN, where entity linking is modeled as a graph edge classification task. Like many previous works of information extraction in VRDs, we assume that OCR has been performed, which means the raw textual contents and bounding boxes, as well as the corresponding semantic entity classes (key or value entities) in the document, are already available. Our DAMGCN is composed of two components: *graph construction* and *link prediction*.

3.1 Graph Construction

In our model, the graph structure is determined from entities in terms of spatial and layout information, where the semantic entities are represented as graph nodes, and the potential directional relations between pairs of nodes are represented as graph edges. The feature of each node is determined by concatenating the following two features: (1) Normalized coordinates and bounding box features (width and height), as well as the class of each semantic entity. (2) The CLS vector f_{BERT} of BERT with the input of the textual information of each semantic entity. Therefore, each node feature can be expressed as $f_i = \{x_{c,i}, y_{c,i}, w_i, h_i, x_{0,i}, y_{0,i}, x_{1,i}, y_{1,i}, c_i, f_{BERT,i}\}$, where $x_{c,i}, y_{c,i}$ are the central coordinates of the corresponding bounding box with width of w_i and the height of

h_i, $x_{0,i}$, $y_{0,i}$, $x_{1,i}$, $y_{1,i}$ are the top-left and bottom-right coordinates of the bounding box, c_i is the class semantic entity. To simplify the construction process, we do not assign any edge features.

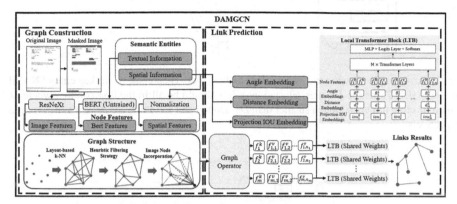

Fig. 3. Proposed DAMGCN architecture.

For constructing the graph structure, we propose a layout-based k-NN strategy to build the graph edges considering that the semantic entities with true links are spatially close to each other. Our proposed layout-based k-NN strategy is a symmetry to the traditional k-NN strategies, that we build edges from a key entity node to its top-k nearest value entity nodes through the layout distance dis_K in Eq. 1. In contrast to the Euclidean distance utilized in traditional k-NN strategies for tabular layout, where a true link lies in the same row/column, as shown in Fig. 4, our proposed layout distance is capable of accommodating such situations by ensuring that the distance between two cells within the same row or column is smaller than that of a pair of cells spanning rows or columns. The layout distance dis_K between the central coordinates $\left(x_{c,i}^k, y_{c,i}^k\right)$ and $\left(x_{c,j}^v, y_{c,j}^v\right)$ of the key entity node N_i^k and value entity node N_j^v is calculated as follows:

$$dis_K\left(N_i^k, N_j^v\right) = \begin{cases} 0, & abs\left(x_{c,i}^k, -x_{c,j}^v\right) < T_t \text{ or } abs\left(y_{c,i}^k - y_{c,j}^v\right) < T_t, \\ dis_e\left(N_i^k, N_j^v\right), & otherwise, \end{cases}$$

(1)

where T_t is the same-row-or-column threshold for entities. In Eq. 1, the dis_K between two entities equals 0 if two entities are in the same row or column, otherwise dis_K is calculated using the Euclidean distance. Unlike the fully-connected edge building method [9], our proposed layout-based k-NN approach is designed to result in a reduced number of edges in the graph (Challenge 1). A table illustrating the constructed links is provided in Sect. 4.

In order to further minimize the occurrence of invalid edges, which may take place when the edge linking the key and value entity pair is crossed by one bounding box of a third key entity (shown in Fig. 5), a heuristic filtering strategy is proposed by leveraging the positions of bounding boxes and entity classes.

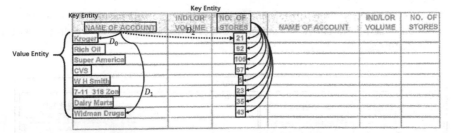

Fig. 4. An example showing the links built in table layout. Theoretically, for key entity "NAME OF ACCOUNT", its top-k nodes selected by k-NN should be all in the first column, where D_1 should be close to D_0 and much smaller than D_2. Dotted and solid line are invalid and true links, respectively.

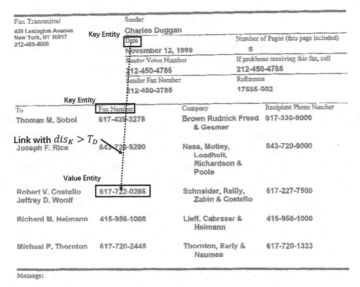

Fig. 5. Demonstration of invalid link generated by proposed layout-based k-NN. The link (dotted line) with $dis_K > T_D$ crosses the key entity "Fax Number" is invalid.

In our heuristic filtering strategy, we use e_i to represent the third semantic entity, and the edge with length dis_K links the two semantic entity nodes N_i^k and N_i^y. The validity of the edge can be determined by following the pseudo code outlined in Algorithm 1. To simplify the heuristic filtering strategy procedure, we propose a projection method to determine the invalid links from line 3–9.

Since the semantic entities are not able to reflect the global layout information of the whole document, we consider the image to provide global layout information according to [18–21]. In this procedure, one document image is masked by the bounding boxes of entities with class-corresponding colors and incorporated into the graph as one graph node. Then the graph node with the features extracted from the masked image through ResNeXt network [22], is connected to all other nodes in the graph (see Fig. 6).

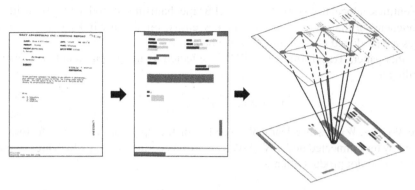

Fig. 6. Demonstration of image incorporation. We mask the semantic entities with their corresponding defined colors, such as red for key entity "header" and green for value entity "answer" etc., in the image.

Algorithm 1: Heuristic Filtering Strategy

Input: Given $\left\{N_i^k | i \in [1, T_l], N_i^k \ni \left\{x_{c,i}^k, y_{c,i}^k, c_i^k\right\}\right\}$ and $\left\{N_i^v | i \in [1, T_l], N_i^v \ni \left\{x_{i,c}^v, y_{i,c}^v, c_i^v\right\}\right\}$,

graph with links $L = \left\{l_i | i \in [1, T_l], starts\ from\ N_i^k\ to\ N_i^v\right\}$ by layout-based k-NN,

and value entity set $S^v = \left\{e_j^v | j \in [1, T_v], \{x_{0,j}, y_{0,j}, x_{1,j}, y_{1,j}, c_j\} \in e_j^v\right\}$,

dis_K threshold T_D, T_A;

Output: Links set L_{HFS};

1.   **for** $i = 1$ to T_l **do**
2.   **for** $j = 1$ to T_v **do**
3. $p_h \leftarrow \min\left(\max\left(x_{c,i}^k, y_{i,c}^v\right), x_{1,j}\right) - \max\left(\min\left(x_{c,i}^k, y_{i,c}^v\right), x_{0,j}\right)$;
4. $p_v \leftarrow \min\left(\max\left(y_{c,i}^k, y_{i,c}^v\right), y_{1,j}\right) - \max\left(\min\left(y_{c,i}^k, y_{i,c}^v\right), y_{0,j}\right)$;
5.   **if** $T_A == 1$:
6.   flag $\leftarrow \left(dis_K\left(N_i^k, N_i^v\right) > T_D\ or\ dis_K\left(N_i^k, N_i^v\right) == 0\right)$;
7.   **else**:
8.   flag $\leftarrow dis_K\left(N_i^k, N_i^v\right) > T_D$;
9.   **if** $\left((p_h == (x_{1,j} - x_{0,j})\ and\ p_v \geq 0)\ or\ (p_v = (y_{1,j} - y_{0,j})\ and\ p_h \geq 0)\right)\ and\ c_j \neq c_j^v$ and flag **then** remove l_i from L;
10.   **end for**
11.   **end for**
12. $L_{HFS} = L$;
13.   **return** L_{HFS};

3.2 Link Prediction

This component is responsible for determining the links of the document. Given the graph from the previous component and the corresponding node features, we first use a graph operator called *TransformerConv* [23] to aggregate the features of nearby nodes for each graph node. Then, to facilitate the matching of key entities with their corresponding value entities, a local transformer block (LTB) mechanism is introduced, to consider the dependencies of the value entities on the key entity. Finally, the links among semantic entities are predicted through the LTB mechanism.

In this paper, TransformerConv [23] is selected as the graph calculation operator according to

$$x_i' = W_1 x_i + \Sigma_{j \in N(i)} \alpha_{i,j} W_2 x_j, \tag{2}$$

where W_1, W_2, W_3, W_4 are learnable weights in the operator. The node feature x_i is updated by the connected nodes $N(i)$ with attention coefficients $\alpha_{i,j}$ which are computed by multi-head dot product attention:

$$\alpha_{i,j} = \text{softmax} \left(\frac{(W_3 x_i)^T (W_4 x_j)}{\sqrt{d}} \right). \tag{3}$$

In DAMGCN, one layer of TransformerConv with a Relu activation function is used to calculate graph embedding vector. To address the challenge of predicting the true links between a key entity and its multiple candidate value entities, which are typically determined pairwise in traditional methods without considering the dependencies among adjacent value entities associated with a given key entity (Challenge 2), we propose the Local Transformer Block to predict the links by matching the key entity and each candidate value entity, as well as the graph node features, where the dependency among the value entities of the key entity can be modeled.

The proposed LTB structure consists of M Transformer layers, and the final layer gives the links among entities by using MLP and logit layers to perform edge classification. Specifically, given a graph node N_i^k (key entity) with its feature f_i^k and its candidate n_i adjacent nodes $N_1^v, N_2^v, \cdots, N_{n_i}^v$ (value entities) with features $f_{i,1}^v, f_{i,2}^v, \cdots, f_{i,n_i}^v$, the pairs of key entity and value entities are factored as follows:

$$h_i = \left\{ \left(f_i^k, f_i^k \right), \left(f_i^k, f_{i,1}^v \right) \cdots \left(f_i^k, f_{i,n_i}^v \right) \right\}, \tag{4}$$

where the pair of $\left(f_i^k, f_i^k \right)$ helps LTB to find the key and its candidate value entities.

To better incorporate the layout information for links prediction, we further introduce the spatial features of the key-value entity pairs into h_i by considering angles, distances, and the projection of intersection over union (IOU) between two normalized entity bounding boxes:

$$\widehat{h_i} = \left\{ \left(f_i^k, f_i^k, a_{i,i}, d_{i,i}, iou_{i,i} \right), \left(f_i^k, f_{i,1}^v, a_{i,1}, d_{i,1}, iou_{i,1} \right) \cdots \left(f_i^k, f_{i,n_i}^v, a_{i,n_i}, d_{i,n_i}, iou_{i,n_i} \right) \right\}. \tag{5}$$

The angles, distances, and the projection of IOU between two normalized entity bounding boxes are given as follows.

Angle: The angle $a_{i,j}$, of any two semantic entity boxes is calculated by

$$A = round \left(\tan^{-1} \frac{y_{c,i} - y_{c,j}}{x_{c,i} - x_{c,j}} \times 180/\pi \right), \tag{6}$$

$$a_{i,j} = \begin{cases} A + 360, A < 0, \\ A, A \geq 0, \end{cases} \tag{7}$$

where $round(\cdot)$ indicates rounding operation, and $a_{i,j}$ presents the integer from 0 to 359 with a scale of 1.

Distance: Given that the maximum distance between the center of any two normalized semantic entity boxes is approximately $\sqrt{2} \approx 1.41$, we partition the distance space into 142 parts to represent the distance range from 0 to 1.41 with a scale of 0.01. The distance embedding index $d_{i,j}$ is therefore defined as

$$d_{i,j} = round\left(\sqrt{(x_{c,i} - x_{c,j})^2 + (y_{c,i} - y_{c,j})^2} \times 100\right). \tag{8}$$

Projection IOU: Projection IOU is defined as the existence of a horizontal or vertical intersection between any two normalized entity bounding boxes, specifically calculated by

$$iou_x = \min(x_{1,i} - x_{1,j}) - \min(x_{0,i} - x_{0,j}), \tag{9}$$

$$iou_y = \min(y_{1,i} - y_{1,j}) - \min(y_{0,i} - y_{0,j}), \tag{10}$$

$$iou_{i,j} = \begin{cases} 0, & i = j, \\ 1, & iou_x > 0 \, or \, iou_y > 0, \\ 2, & otherwise. \end{cases} \tag{11}$$

To reduce the number of parameters, we use LTBs with shared weights instead of multiple independent LTBs for each key node N_i^k and its candidate value nodes.

In order to overcome the imbalance of samples, the following Balance Cross Entropy Loss is used as the total loss in our study,

$$Balance\ CE\ Loss = -\left(\frac{1}{N_{pos}} \sum_{i=1}^{N_{pos}} \log \hat{y}^i + \frac{1}{N_{neg}} \sum_{j=1}^{N_{neg}} \log\left(1 - \hat{y}^j\right)\right), \tag{12}$$

where the mean value of positive samples and negative samples is calculated separately to offset the gradient deviation. ADAM [24] is used to optimize the weight parameters of DAMGCN.

4 Experiments

In our paper, we use two public datasets of FUNSD and XFUND for evaluation. FUNSD is a collection of noisy scanned documents. It contains 199 fully annotated documents (149 in training set and 50 in testing set) that vary widely regarding their structure and appearance. These documents are annotated in a bottom-up approach, enabling them to be utilized for various document analysis tasks, such as text detection, text recognition, spatial layout understanding, as well as the extraction of question-answer pairs [25].

XFUND, which extends the dataset of FUNSD, is a multilingual document understanding benchmark dataset. It contains 1,393 human-labeled documents with key-value pairs in 7 languages (Chinese, Japanese, Spanish, French, Italian, German, Portuguese) [26], and each language has 199 forms, with 149 forms for training and 50 for testing. These two datasets provide the official OCR annotations (bounding boxes and tokens) and the classes of entities.

In this paper, the feature f_{Bert} of each node is obtained from BERT-based models. Depending on language, BERT-base-uncased model is used for English, BERT-base-Chinese for Chinese, and BERT-base-multilingual for other languages. The length of each feature is set to be 768. In ResNeXt, the input image is resized to 512×512. During the training procedure, ADAM optimizer is used with $(\beta_1, \beta_2) = (0.5, 0.999)$. The initial learning rate is $5e^{-5}$, the number of epochs is 600, and the batch size of the document data is 1. The same-row-or-column threshold T_t in Eq. 1 is $5e^{-4}$.

First, to evaluate the graph construction of our proposed layout-based k-NN and heuristic filtering strategy, we conduct experiments using English and Chinese documents from the FUNSD and XFUND training sets, respectively. The experimental results are presented in Fig. 7, Fig. 8 and Table 1. Subsequently, we compare the performance of our DAMGCN model with state-of-the-art models on both the FUNSD and XFUND datasets.

Figure 7 illustrates the total number of constructed links and positive links, utilizing the proposed layout-based k-NN and heuristic filtering strategy with varying values of K, T_D and T_A. In the traditional k-NN approach, the number of constructed links is substantial due to the large value of K required to ensure a sufficient number of positive links. In contrast, the proposed layout-based k-NN combined with the heuristic filtering strategy achieves a significant reduction in the number of constructed links, with a comparable or increased number of positive links. Specifically, when K = 80, our layout-based k-NN with heuristic filtering strategy (with $T_D = 0.2, T_A = 1$) generates only approximate 6×10^4 constructed links with a small percentage of approximately 1% of the total positive links missing. Furthermore, our proposed heuristic filtering strategy and layout-based k-NN approach can significantly decrease the total number of constructed links as the value of K increases. The layout-based k-NN without the heuristic filtering strategy can correctly identify all positive links, but it introduces an excess of 8×10^4 additional links. Similarly, traditional k-NN results in a significantly higher number of total constructed links, approximately 2.2×10^5, and have a higher percentage of missing positive links, approximately 1.5%. Figure 8 shows that our heuristic filtering strategy and layout-based k-NN can also significantly reduce the number of total links in Chinese documents.

Table 1 shows the effects of graph construction method (layout-based k-NN, heuristic filtering, varying Transformer layers) on the number of links and entity linking performance. When K = 80, our layout-based k-NN with heuristic filtering strategy ($T_D = 0.4, T_A = 0$) and 3 Transformer layers in LTB method achieves the best trade-off F1 score since it has only 0.023% (1/4230) missing positive link. The utilization of a layout-based k-NN algorithm (K = 80) without the implementation of heuristic filtering resulted in an F1 score of 0.7703, with a total of 15.83×10^4 links. This demonstrates that the presence of a large number of invalid links can negatively impact the performance of the model (Fig. 8).

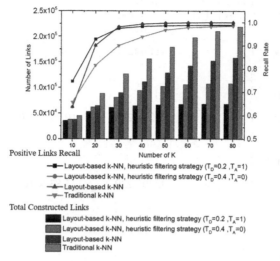

Fig. 7. Results of proposed graph construction methods on English Documents from FUNSD.

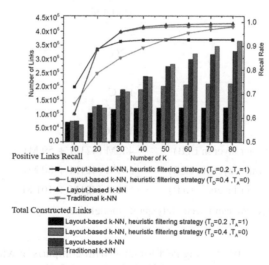

Fig. 8. Results of proposed graph construction methods on Chinese Documents from XFUND.

Finally, we compared the accuracy of DAMGCN with state-of-the-art models on FUNSD and XFUND testing set, as shown in Table 2 and Table 3. The comparison methods as well as the proposed method give results based on the same assumption that OCR has been performed to specify text location and semantic entity classes (key or value entities).

Table 1. Results of layout-based k-NN and heuristic filtering strategy on FUNSD training set.

Graph Construction Method	Number of Transformer Layers in LTB	Constructed Positive Links/Total Positive Links	Total Constructed Links	F1 Score
Layout-based k-NN (K = 80), heuristic filtering strategy ($T_D = 0.2, T_A = 1$)	1	4117/4230	4.82×10^4	0.7716
	2			**0.8063**
	3			0.7873
Layout-based k-NN (K = 80), heuristic filtering strategy ($T_D = 0.4, T_A = 0$)	1	4229/4230	10.78×10^4	0.7431
	2			0.7573
	3			0.7997
Layout-based k-NN (K = 80)	1	4230/4230	15.83×10^4	0.7508
	2			0.7576
	3			0.7703

Table 2. Performance of entity linking task on the FUNSD dataset.

Model	Model Size	F1 Score
GNN + MLP [10]	–	0.39
StrucTexT [17]	1.2G	0.441
SERA [1]	–	0.6596
LayoutLMv2 [3]	1.1 G	0.7057
LayoutXLM [2]	1.1 G	0.7683
BROS [6]	1.34 G	0.7701
DAMGCN (1 Transformer layer in LTB)	480M (421M + 59 M)	0.7716
DAMGCN (2 Transformer layers in LTB)	501M (421M + 80 M)	**0.8063**
DAMGCN (3 Transformer layers in LTB)	522M (421M + 101 M)	0.7873

- stands for close source. We can't accurately get its model size.

Table 2 shows the entity linking performance with comparison with state-of-the-art methods. It is shown that our proposed DAMGCN with 2 Transformer layers in LTB achieves 0.8063 F1 score in the FUNSD benchmark, outperforming other state-of-the-art models. In particular, the GNN + MLP model, which employs a similar graph-based architecture to ours, achieved an F1 score of 0.39. This indicates that the integration of multi-modal information from all adjacent value entities with a given key, enabled by the LTB approach, significantly improves the accuracy of link prediction. Besides the linking accuracy, we compare the size of each open-source model for deployment. Without the consideration of BERT-base-uncased model, which is not used for training, the smallest model (1 Transformer layer in LTB) we proposed for training procedure is only 59M. In Table 3, we conduct language-specific experiments on multilingual dataset of XFUND, which means training and testing is implemented on the same language. We conduct $K = 120$, $T_D = 0.4$, $T_A = 0$ in the graph construction methods for languages with the consideration of the different layout distribution in XFUND. We observe that proposed DAMGCN (with 3 transformer layers in LTB) also performs well and achieves 0.7303 F1 score compared with the state-of-the-art models.

Table 3. Language-specific performance of entity linking task on the XFUND.

Model	F1 AVG	ZH	JA	ES	FR	IT	DE	PT
XLM-RoBERT [13]	0.4769	0.5105	0.5800	0.5295	0.4965	0.5305	0.5041	0.3982
InfoXLM [16]	0.4910	0.5214	0.6000	0.5516	0.4913	0.5281	0.5262	0.4170
LayoutXLM$_{BASE}$ [2]	0.6432	0.7073	0.6963	0.6896	0.6353	0.6415	0.6551	0.5718
XYLayoutLM [8]	0.6779	0.7445	0.7059	0.7259	0.6521	0.6572	0.6703	0.5898
LiLT [7]	0.6781	0.7297	0.7037	0.7195	0.6965	0.7043	0.6558	0.6781
LayoutXLM$_{LARGE}$ [2]	0.7206	**0.7888**	**0.7255**	**0.7666**	0.7102	**0.7691**	0.6843	0.6796
DAMGCN	**0.7303**	0.7563	0.7150	0.7256	**0.7599**	0.7665	**0.7064**	**0.6825**

5 Conclusion

In this paper, we present DAMGCN, a dependency-aware multimodal graph convolutional network for entity linking. Unlike existing graph-based methods, our DAMGCN approach employs a holistic strategy for link prediction that encompasses both graph feature learning and graph construction. Specifically, we introduce a layout-based k-NN algorithm and a heuristic filtering strategy to establish a robust graph structure by eliminating invalid links. Additionally, we leverage the proposed local transformer block (LTB) to capture the dependency among adjacent value entities, thereby improving the distinction between similar value entities linked to a key entity. Experimental results on the FUNSD and XFUND datasets have demonstrated the superiority of DAMGCN with its simple architecture, while the model size of DAMGCN is much smaller than the existing models.

For future work, we will continue to explore models with small size to facilitate more real-world scenarios. Besides, we will explore unsupervised pre-training strategy to improve generalization performance of the model. The joint modeling of semantic entity extraction and link prediction is a further task for document understanding.

References

1. Zhang, Y., Zhang, B., Wang, R., Cao, J., Li, C., Bao, Z.: Entity relation extraction as dependency parsing in visually rich documents. arXiv preprint: arXiv:2110.09915 (2021)

2. Xu, Y., et al.: LayoutXLM: multimodal pre-training for multilingual visually-rich document understanding. arXiv preprint: arXiv:2104.08836 (2021)
3. Xu, Y., et al.: LayoutLMv2: Multi-modal pre-training for visually-rich document understanding. arXiv preprint: arXiv:2012.14740 (2020)
4. Devlin, J., Chang, W., Lee, K., Toutanova, K.: BERT: pre-training of deep bidirectional transformers for language understanding. arXiv preprint: arXiv:1810.04805 (2018)
5. Pinkus, A.: Approximation theory of the MLP model in neural networks. Acta Numer **8**, 143–195 (1999)
6. Hong, T., Kim, D., Ji, M., Hwang, W., Nam, D., Park, S.: BROS: a pre-trained language model focusing on text and layout for better key information extraction from documents. In: Proceedings of the AAAI Conference on Artificial Intelligence, vol. 36, pp. 10767-10775 (2022)
7. Wang, J., Jin, L., Ding, K.: LiLT: a simple yet effective language-independent layout transformer for structured document understanding. arXiv preprint:arXiv:2202.13669 (2022)
8. Gu, Z., et al.: XYLayoutLM: towards layout-aware multimodal networks for visually-rich document understanding. In: Proceedings of the IEEE/CVF Conference on Computer Vision and Pattern Recognition, pp. 4583–4592 (2022)
9. Gemelli, A., Biswas, S., Civitelli, E., Lladós, J., Marinai, S.: Doc2Graph: a task agnostic document understanding framework based on graph neural networks. arXiv preprint: arXiv: 2208.11168 (2022)
10. Carbonell, M., Riba, P., Villegas, M., Fornés, A., Lladós, J.: Named entity recognition and relation extraction with graph neural networks in semi structured documents. In: 25th International Conference on Pattern Recognition, pp. 9622–9627 (2021)
11. Déjean, H., Clinchant, S., Meunier, L.: LayoutXLM vs. GNN: an empirical evaluation of relation extraction for documents. arXiv preprint:arXiv:2206.10304 (2022)
12. Liu, X., Gao, F., Zhang, Q., Zhao, H.: Graph convolution for multimodal information extraction from visually rich documents. arXiv preprint: arXiv:1903.11279 (2019)
13. Conneau, A., et al.: Unsupervised cross-lingual representation learning at scale. arXiv preprint: arXiv:1911.02116 (2019)
14. Lee, C., et al.: Formnet: structural encoding beyond sequential modeling in form document information extraction. arXiv preprint: arXiv:2203.08411 (2022)
15. Davis, B., Morse, B., Cohen, S., Price, B., Tensmeyer, C.: Deep visual template-free form parsing. In: 2019 International Conference on Document Analysis and Recognition (ICDAR), pp. 134–141 (2019)
16. Chi, Z., et al.: InfoXLM: an information-theoretic framework for cross-lingual language model pre-training. arXiv preprint:arXiv:2007.07834 (2020)
17. Li, Y., et al.: StrucText: structured text understanding with multi-modal transformers. In: Proceedings of the 29th ACM International Conference on Multimedia, pp. 1912–1920 (2021)
18. Qiao, L., et al.: LGPMA: Complicated table structure recognition with local and global pyramid mask alignment. In: Lladós, J., Lopresti, D., Uchida, S. (eds.) ICDAR 2021. LNCS, vol. 12821, pp. 99–114. Springer, Cham (2021). https://doi.org/10.1007/978-3-030-86549-8_7
19. Dang, N., Nguyen, D.: End-to-end information extraction by character-level embedding and multi-stage attentional u-net. arXiv preprint: arXiv:2106.00952 (2021)
20. Zhao, W., Gao, L., Yan, Z., Peng, S., Du, L., Zhang, Z.: Handwritten mathematical expression recognition with bidirectionally trained transformer. In: Lladós, J., Lopresti, D., Uchida, S. (eds.) ICDAR 2021. LNCS, vol. 12822, pp. 570–584. Springer, Cham (2021). https://doi.org/10.1007/978-3-030-86331-9_37
21. Lin, W., et al.: ViBERTgrid: a jointly trained multi-modal 2D document representation for key information extraction from documents. In: Lladós, J., Lopresti, D., Uchida, S. (eds.) ICDAR 2021. LNCS, vol. 12821, pp. 548–563. Springer, Cham (2021). https://doi.org/10.1007/978-3-030-86549-8_35

22. Xie, S., Girshick, R., Dollár, P., Tu, Z., He, K.: Aggregated residual transformations for deep neural networks. In: Proceedings of the IEEE Conference on Computer Vision and Pattern Recognition, pp. 1492–1500 (2017)
23. Shi, Y., Huang, Z., Feng, S., Zhong, H., Wang, W., Sun, Y.: Masked label prediction: Unified message passing model for semi-supervised classification. arXiv preprint: arXiv:2009.03509 (2020)
24. Kingma, P., Ba, J.: ADAM: a method for stochastic optimization. arXiv preprint: arXiv:1412.6980 (2014)
25. Jaume, G., Ekenel, K., Thiran, P.: FUNSD: a dataset for form understanding in noisy scanned documents. In: 15th International Conference on Document Analysis and Recognition Workshops, vol. 2, pp. 1–6 (2019)
26. Xu, Y., et al.: XFUND: a benchmark dataset for multilingual visually rich form understanding. In: Findings of the Association for Computational Linguistics, pp. 3214–3224 (2022)

Analyzing Textual Information from Financial Statements for Default Prediction

Chinesh Doshi, Himani Shrotiya, Rohit Bhiogade$^{(\boxtimes)}$, Himanshu S. Bhatt, and Abhishek Jha

AI Labs, American Express, New York, USA
{Chinesh.Doshi,Himani.Shrotiya,Rohit.Bhiogade,Himanshu.S.Bhatt,
Abhishek.Jha1}@Aexp.Com

Abstract. Financial statements provide a view of company's financial status at a specific point in time including the quantitative as well as qualitative view. Besides the quantitative information, the paper asserts that the qualitative information present in the form of textual disclosures have high discriminating power to predict the financial default. Towards this, the paper presents a technique to capture comprehensive 360-° features from qualitative textual data at multiple granularities. The paper proposes a new sentence embedding (SE) from large language models specifically built for financial domain to encode the textual data and presents three deep learning models built on SE for financial default prediction. To accommodate unstructured and non-standard financial statements from small and unlisted companies, the paper also presents a document processing pipeline to be inclusive of such companies in the financial text modelling. Finally, the paper presents comprehensive experimental results on two datasets demonstrating the discriminating power of textual features to predict financial defaults.

Keywords: Financial Statement Analysis · Document Features Extraction · Document classification

1 Introduction

With recent global economic situations and reports around inevitable recession in 2023 [25], creditors and financial institutions have stepped up scrutinizing their credit books. Historically, during such uncertain times, the financial defaults are more likely which leads to prolonged influence on the economy. Financial default reflects a problem with debt due to missing payments or paying late. The problem of analysing corporate financial data and effectively predicting early signals of financial default has become even more pressing during the current global economic unrest.

Traditionally, quantitative information such as accounting-based ratios and stock market data [14] has been extensively studied and used to detect signals on whether a firm is financially healthy or may end up defaulting or going

G. A. Fink et al. (Eds.): ICDAR 2023, LNCS 14189, pp. 48–65, 2023.
https://doi.org/10.1007/978-3-031-41682-8_4

bankruptcy. This quantitative information is used to create standard and structured numeric features which are fed to an ML model (including boosting, discriminant analysis, SVMs, and deep neural networks) to predict financial defaults and has proved to be beneficial. With advancements in AI, creditors and financial institution are looking beyond traditional methods for more effective prediction of financial defaults.

On the other hand, the qualitative information present in the form of textual disclosures in financial statements contains valuable evidence and insights about the credit risk, overall financial health and any ongoing concerns such as revenue concentrations, operations, legal concerns and so on. Thus, it is imperative to include this qualitative information in the financial modelling tasks. Since this qualitative information is unstructured, a review manager go over these disclosures and identify statements or claims that can affect the credit rating of the business. This manual process is very tedious and also suffers from the subjectivity of the human reviewer and is not pragmatic.

Existing approaches [22] suggest that combining qualitative textual features with quantitative accounting and market-based features further boost the performance of the default prediction. While the quantitative data has been extensively researched over the last few decades, the potential of unstructured qualitative textual data has been explored superficially. This gap motivates us to explore new opportunities with the qualitative textual data for financial modelling. Existing approaches that attempts to capture qualitative information either explored one of the traditional features (i.e. dictionary, sentiment or theme based) or the average word embeddings. However, research on the quantitative side suggests that not a single view sufficiently captures the information and it has evolved over time to capture comprehensive 360-° view of accounting variables including absolute P&L variables, aggregate features financial ratios, cumulative or absolute rate of growth etc. Following the evolution in quantitative space, this paper presents a new framework to extract a comprehensive 360-° view from the textual features at multiple granularity.

Moreover, the existing solutions assume that the company financial are available in standard structured format from the market regulatory body; for e.g. in XBRL [26] format from SEC [24] in US. This restricts the application of existing solutions to the real world because a creditor receives financial statements from all sorts of companies. Many of these companies may not be liable to furnish financial in a predefined template/layout and may have completely different format. To address the above limitation, this paper also presents a financial statement processing pipeline to accommodate arbitrary financial statements from any company to be included in the financial modelling task.

Contributions. Given the above mentioned considerations, the main contributions of this paper are summarised below:

- A comprehensive 360-° view from qualitative textual data using curated, contextual embedding, and hybrid features for financial modelling.
- A financial statement processing algorithm to process unstructured and non-standard PDF documents with variations including native vs scanned, large

vs small business, public vs private companies. This is critical as it allows the smaller as unlisted companies to be included in the financial text modelling.

- A novel method of encoding textual data using large language models specifically built for finance domain and three deep learning architectures for learning contextual embeddings.
- A comprehensive empirical evaluation of textual features on two different data sets along with multiple ablation studies to provide insights on discriminating power of textual features for predicting financial default.

2   Relevant Work

Bankruptcy prediction or corporate default prediction is a widely studied problem since 1960s s in the accounting and finance domain. Some of the earlier work performed in this direction focussed on analysing key financial metrics and ratios [5,14,23]. Subsequent work focussed on leveraging newer machine learning algorithms for default prediction. SVM-based analysis [15] demonstrates building non-linear models and ensemble-based analysis [16,17] combines several weak classifiers to learn a strong model. These methods provided good ways to predict default using accounting data and financial metrics, but did not utilize additional information communicated through textual disclosures in the financial statements.

Earlier work leveraging neural network architectures which were aimed to capture feature interactions and obtain feature representations relied on quantitative data only [18,19]. Deeper neural network architectures have been explored recently for numerical data which learn higher-level feature representations [1,2]. Hosaka et al. [4] transforms financial ratios into a grayscale image and trains a Convolutional neural network to predict default for Japanese market. With the majority of work involving numerical data, [20] analyses natural language text to predict trends in the stock market.

A hybrid model utilizing numerical data and text data [22] presents an Average word embedding based architecture for the representation of the textual data contained in Management Discussion and Analysis of SEC 10-K [24] filings. This architecture uses skip-gram Word2Vec embeddings to obtain word representation, but fails to capture contextual information between words. With the introduction of Transformer-based pretrained Large Language models (LLM) which takes into account semantic meaning or context of the word, text is often represented with models like BERT [35], GPT [36] or its variants. While the BERT-based representations are explored in the context of financial sentiment analysis [6,7] and financial numeric entity recognition [8], these haven't been explored much for predicting financial default. Empirical studies also show that finance domain specific LLMs [32] perform slightly better than LLMs pretrained on general vocabulary.

SEC 10-K has been the only source of Management's Discussion and Analysis (MD&A) text for all the prior work. While the SEC provides financial performance data for all public companies, attempts to extract text disclosures from raw financial statements haven't been studied to a great extent. Systems heavily dependent on OCR engines to extract text from financial statements lose structure and layout information present in the document, and this further affects the information retrieval process. In recent years, methods for document layout analysis have been used successfully to detect components of the document [10, 11] into 5 well known components namely, text, title, table, figure, and list. These methods, however, have not been trained to detect components like header and footer (for English language) which are present in abundance in financial statements and are required to be discarded for downstream analysis. For header and footer identification, [21] presents an unsupervised heuristic-based page association method for long form documents that works reasonably well and aims to capture similar body of text across multiple pages.

3 Financial Text Modelling

It is our assertion that financial text data has discriminating cues that can be efficiently leveraged to predict default. This section starts with a pipeline to process any kind of financial data, be it a standard 10K document from SEC or a custom financial statement (either Native or Scanned PDF) to extract the relevant sections and its textual content for the financial modelling task. Next, the paper presents our framework to extract a 360-° view from the textual data by deploying different types of feature extraction methods to capture curated, contextual embeddings, and hybrid features. These features are then used to predict the financial default either combined with a classification method or with an end-to-end deep learning based architecture.

3.1 Financial Statement Processing Pipeline

Company financials for large or listed companies are generally available from a regulatory agency in different countries, such as SEC in US or Companies House in UK. However, small or private companies that are not required to file their financials with the regulatory agencies may have huge variations in the template or layout of their financial statements, which is highly unstructured with the sections names, content flow and hierarchy varying across companies. In the US, the companies which have their financial data available via SEC [24] follow a standard structured format and is available in extensible Business Reporting Language (XBRL) [26] format. The XBRL format clearly marks different sections of the financial documents, and thus, the textual data can be extracted in a standard format.

Item 7: Management's Discussions and Analysis (MD&A) in SEC 10K filings intends to serve as a qualitative disclosure for investors to make more accurate projections of future financial and operating results. It contains important trends, forward-looking statements including disclaimers & projections, and information about various types of risks that the company is exposed to. Extracting MD&A section is of utmost importance as it helps in constructing the accurate representation of the risk profile of a company. This section, however, is not

52   C. Doshi et al.

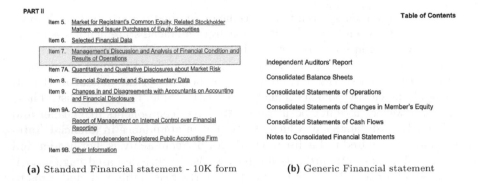

(a) Standard Financial statement - 10K form   (b) Generic Financial statement

Fig. 1. An example of how the Management Discussion & Analysis (MD&A) section is easily available in a 10K-form versus the financial statement document of a small-scale company

well-structured in the PDFs as compared to the XBRL format. Figure 1 shows a comparison between the table of contents of a $10K$ financial statement and a PDF financial statement. It demonstrates that a financial statement may have arbitrary different layout & template, varied naming conventions which make it arduous to process and extract relevant MD&A section.

One of the major gaps in the literature is that most of the existing approaches for financial text mining only caters to company financials available in structured XBRL [26] format from SEC [24]. The creditors and financial institutions need to process a significant volume of financial information from companies which provide details in unstructured PDFs. This paper presents a financial statement processing pipeline to identify the relevant textual data which succinctly describes the performance of a company. The framework to understand, interpret, and segment the PDF (both native & scanned) financial statement into its constituent sections and select the relevant MD&A section for further processing is as follows:

- **Step-1**: We apply pretrained DiT-based Document Layout Analysis (DLA) models [10,11] on a PDF financial statement to detect the document components like titles, text, lists, tables, and figures. The pre-trained model is fine-tuned on financial statements to enhance its performance for detecting tables, as financial tables are peculiar.
- **Step-2**: We then leverage page association algorithm [21] to detect the header and footer from the financial statements. Since the content in a financial statement would span across multiple pages, the header and footer information is used to discard irrelevant content from the header and footer part of a page.
- **Step-3**: The detected components from the financial documents are mapped and used to segment the document into a hierarchical, semi-structured format.
- **Step-4**: Average sentence embedding as described in Sect. 3.2 is computed for all the *title* components.

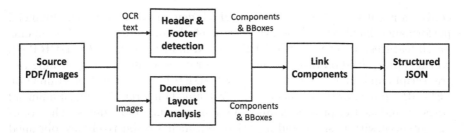

Fig. 2. Financial statement processing pipeline for robust detection of components and its representation into hierarchical structured format. Titles, text, tables, figures and lists are detected with Document Layout analysis models while header and footer components are identified with page association algorithm. Linked components are used to segment the document into a hierarchical semi-structured format.

- **Step-5**: To map MD&A section for a new document, cosine similarity between the section title embeddings and MD&A title embeddings is used. The MD&A title embedding is learned as the average embedding of all sections title that were manually tagged as MD&A on historic financial documents.

While extracting the MD&A section from the XBRL documents is straight forward, the above pipeline is used to extract the MD&A section for the financial statements that are available in PDF (native or scanned) format (Fig. 2).

3.2 Feature Extraction

Once we obtain textual information from the MD&A section of a financial statement either in structured XBRL or unstructured PDF (Scanned or Native) format, the next step is to extract relevant textual features to get a 360-° information view. This paper proposes to extract a broad spectrum of features from the textual data, starting from bespoken features to learning newer feature representations. We broadly group the extracted features into 3 categories; namely, Curated features, Contextual embeddings and Hybrid features.

Curated Features: Expert guided features which are inspired from literature on default prediction. These features include Dictionary-based, Sentiment-based, Theme-based, Time-based and Average Sentence Embedding features.

- **Dictionary based features:** Loughran-McDonald Master dictionary [3] provides a mapping between word tokens and 7 sentiment categories, namely 'negative', 'positive', 'uncertainty', 'litigious', 'strong modal', 'weak modal', and 'constraining'. In total, 7 features are created corresponding to each sentiment category by aggregating relevant token frequencies.
- **Sentiment based features:** Sentiment analysis allows assessing a scale of opinions, attitudes and emotions towards financial risks and performance. In financial domain, text with potential indication of risk would generally

convey a negative sentiment and text which demonstrates a healthy financial performance manifests a positive sentiment. In order to obtain sentiment-based features, we leverage two BERT-based models; namely, **FinBERT** [7], which is pretrained on three financial corpora: corporate reports, earning call transcripts and analyst reports, and **SEC-BERT** [8], which is pretrained on SEC $10K$ filings. Both the models are pretrained on large amount of financial corpora and hence, provide finance domain specific embeddings. The count of sentences with positive and negative sentiments respectively are obtained as 2 of the total number of features.

- **Theme based classification features:** We categorize sentences into 4 themes, viz., bankruptcy, going concern, litigious and crime. Loughran-McDonald Master dictionary [3] and a curated dictionary of keywords is used to label over 10,000 sentences from EDGAR-CORPUS [30] with appropriate theme in a weakly supervised manner. We built 4 different binary BERT-based classifiers [29] which predict whether the sentence is associated with a respective theme or not. These models independently achieve F1 score of over 80% on validation set.

- **Average Sentence Embedding (ASE) features:** For relevant section encoding, AWE [22] averages the embeddings of all words in a sentence while ignoring the fact that these words are contextually linked. Instead, we propose the use of **Sentence Embedding (SE)** for input sentence encoding to preserve the meaning of the entire sentence and **Average Sentence Embedding (ASE)** to encode the entire MD&A section. Figure 3 demonstrates the encoding process where each sentence from MD&A section is fed through the pretrained LLM and the output of the *[CLS]* token from the last hidden state gives the embedding of the sentence (abbreviated as **SE**). Finally, we take the average of all the sentence embeddings and represent it using a 768-dimensional feature vector (abbreviated as **ASE**).
We experiment with **FinBERT** [7] and **SEC-BERT** [8], and obtain two sets of sentence embeddings.

The curated features extracted from the textual information in MD&A section are then combined with a classification model, such as XGBoost [28], to predict the probability of default.

Contextual Embeddings: We use **SE** from LLMs to encode input sentences and learn contextual embeddings from following end-to-end deep learning based architectures:

- **ASE - Feed forward model (ASE-FF):** The average sentence embedding feature vector is computed for MD&A section either from FinBERT or SEC-BERT and is fed as input to a feed-forward deep learning model. The model consists of two fully connected layers. The first layer consists of 256 output nodes followed by RELU activation, batch normalization and dropout. The second layer follows the same architecture except that it contains 128 output nodes. The output layer consists of a single node with sigmoid activation. The model is trained for 50 epochs with binary cross entropy loss.

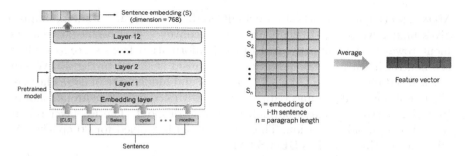

(a) Step 1: Obtain Sentence Embedding   (b) Step 2: Average of Sentence Embeddings

Fig. 3. Illustrates the extraction of sentence embeddings from pretrained LLMs. (a) In BERT-based models, *[CLS]* token is inserted at the beginning of each sentence. We use the output of *[CLS]* token from the last layer as the sentence embedding. (b) We average the sentence embeddings

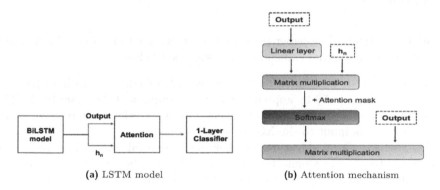

(a) LSTM model   (b) Attention mechanism

Fig. 4. Architecture diagram for LSTM model with attention

- **Fine-tuned BERT-based models:** We fine-tune BERT-based FinBERT and SEC-BERT models. These models are pretrained using the masked language modelling objective. For fine-tuning, the model is first initialized with the pretrained weights, and then we train it for our downstream classification task of default prediction. The input to the model is the MD&A section. We use Hugging Face API [29] for fine-tuning.
- **LSTM with Attention:** The architecture of the LSTM model is shown in Fig. 4a. The embedding of each sentence(**SE**) obtained either using FinBERT or SEC-BERT are fed as input to a bidirectional LSTM model. The output and final hidden state(h_n) from the LSTM, goes as input to the attention layer. Here, output contains the concatenated forward and reverse hidden states from the last layer for each element in the sequence, whereas h_n contains the concatenated final forward and reverse hidden states from the last layer. Finally, the weighted output from the attention layer is fed to a 1-layer classifier, which predicts the probability of default. Not all the sentences in the

MD&A section are important for default prediction. The attention mechanism gives higher weightage to sentences which convey a positive/negative sentiment towards the company, and discard information from sentences which are neutral in nature. The attention mechanism is shown in Fig. 4b. We use stacked bidirectional LSTM with two layers and hidden size of 200. The classifier contains a fully connected (FC) layer with 400 output neurons followed by ReLU activation, batch normalization and dropout. The FC layer is connected to sigmoid activation. The entire model is trained for 30 epochs. We will refer to this model as **SE-LSTM**.

From these deep learning architectures, the *contextual embedding* for MD&A section is obtained from the feature representation of the last hidden layer. With FinBERT as input sentence encoder, we obtain three contextual embeddings corresponding to FinBERT-ASE-FF, FinBERT-Finetuned and FinBERT-LSTM models. Similarly, we get three contextual embeddings using SEC-BERT as input sentence encoder.

Hybrid Features: The paper presents two strategies to combine curated features with the contextual embeddings as described below:

1. *PCA+XGBoost:* The dimensions of contextual embeddings are 768 which are further reduced to 58 by applying Principal Component Analysis(PCA) [27]. The reduced embeddings are then concatenated with the curated features and then passed as input to the XGBoost [28] model.
2. *FC+Sigmoid:* Curated features along with Contextual embeddings are passed as input to a Fully Connected (FC) layer head. The FC layer contains 64 neurons, followed by ReLU activation, batch Normalization and dropout. The FC layer subsequently connects to a final layer with sigmoid activation to predict the probability of default.

The paper presents first-of-its-kind comprehensive 360-° textual feature extraction from MD&A section of financial statements, thus filling a wide gap in the existing literature. These features allow to capture subtle cues and signals that are highly predictive of financial default as validated in Sect. 4.

4   Experimental Results

4.1   Datasets

This study used two independent data sources, SEC $10K$ fillings and a private Financial Statement data set. For both the datasets, we collected the financial documents one year prior to the year of default. In some cases, we collected financial documents up to five years prior to the year of default.

SEC Dataset: We obtain $10K$ documents from SEC EDGAR-CORPUS [30] for the years from 1993 to 2021 and identify default companies using UCLA-LoPucki Bankruptcy Research Database (BRD) [31]. In 10K documents, Item 7 and Item 7A make up the relevant sections, which highlight major risks and discuss future projections. Item 7 conveys Management's Discussion and Analysis of Financial Condition and Results of Operations, and Item 7A talks about Quantitative and Qualitative Disclosures about Market Risks. The number of bankrupt companies is significantly lower than the number of healthy companies in each year, as recorded in SEC data. To simulate the real world scenario, we further constructed two datasets from SEC, namely, *SEC-balanced* and *SEC-imbalanced*. The *SEC-imbalanced* consists of 800 samples from default companies and, 8000 samples from non-default companies. The *SEC-balanced* consists of equal number of companies from default and non-default categories, i.e. around 800 each.

FS Dataset: The paper also presents a real-world data set comprising Annual financial statements received by a global creditor from different private and small businesses across different markets in PDF format (native and scanned). This dataset has a lot of variability in terms of unstructured financial documents from small, mid & large-scale private companies and thus is more challenging as compared to the SEC. To extract the MD&A section from these financial, we deploy the financial statement processing pipeline as described Sect. 3.1. This dataset consists of 63 companies out of which 14 are default and 49 are non-default, and was only used for testing the models trained on SEC dataset.

4.2   Experiment Protocols

We frame the challenge of predicting default as a binary classification problem. The independent variables are obtained for each company as described in Sect. 3.2. The entire data from SEC is split into training, validation and test datasets in the ratio of 80 : 10 : 10. Models are trained using the training dataset and out-of-sample test dataset is used for prediction and determining model performance.

Since company default is an infrequent event and the dataset is imbalanced, metrics for model performance like Accuracy rate and Error rate are not a good choice as they are heavily dependent on the chosen threshold to obtain classification results. Therefore, we report area under the receiver operating characteristic curve (AUC) [34] and the cumulative decile rankings [33].

4.3   Analysis of Results

The paper presents comprehensive experimental evaluations with multiple settings on two datasets to provide a thorough analysis of what factors influence the discriminating power of textual features to predict financial defaults. The paper reports different ablation studies analysing individual contributions of different granular features and their combinations, impact of concatenating textual features across multiple years, and finally, impact of combining the textual features with the traditional accounting variables in financial domain.

Results on SEC Dataset

Analysing Curated Features: To assess the curated features and how they correlate with the default, we capture the event rate for multiple curated features. Event Rate is a measure of how often a particular statistical event occurs within the experimental group of an experiment. In our settings, event rate is given as:

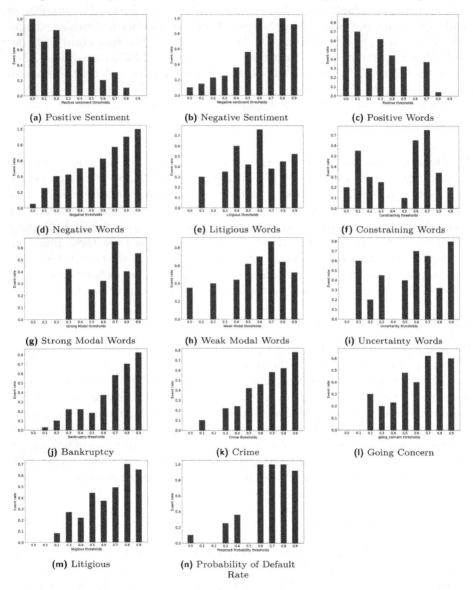

(a) Positive Sentiment   **(b)** Negative Sentiment   **(c)** Positive Words

(d) Negative Words   **(e)** Litigious Words   **(f)** Constraining Words

(g) Strong Modal Words   **(h)** Weak Modal Words   **(i)** Uncertainty Words

(j) Bankruptcy   **(k)** Crime   **(l)** Going Concern

(m) Litigious   **(n)** Probability of Default Rate

Fig. 5. (a–m) are Event Rate plots for dictionary, sentiment and theme based features correlation with defaults. (n) Event rate plot for prediction using FinBERT-Finetuned model with PCA + XGBoost

Table 1. Comparing performance measures for XGBoost model on various combination of curated features evaluated on out-of-sample *SEC-imbalanced* test split.

Curated features						AUC (%)	Cumulative Decile Values					
Dictionary	Sentiment	Theme	Time based	ASE			1	2	3	4	5	6-10
✓	✓					63	0.46	0.62	0.75	0.85	0.90	1
✓	✓	✓				63	0.53	0.76	0.88	0.91	0.93	1
✓	✓	✓	✓			68	0.55	0.78	0.86	0.93	0.96	1
✓	✓	✓		✓		74	**0.73**	**0.91**	**0.96**	**0.99**	**0.99**	1
✓	✓	✓	✓	✓		**77**	0.71	0.90	**0.96**	**0.99**	**0.99**	1

$$EventRate = \frac{\#Defaults}{\#Defaults + \#NonDefaults} \tag{1}$$

Figure 5 shows the event rate across different bins of the curated feature values. An overall increasing or decreasing (may not be monotonic) trend in the event rate suggests that the individual features are correlated with the default event. Results suggest that many of the individual features as well as their interplay learned by the FinBERT-Finetuned model with PCA+XGBoost shows a strong relation with the event rate.

Table 1 compares the performance of models built using multiple combinations of curated features. The results suggest combination of all curated features yields the best performance as compared to individual or any other combination of the curated features. We report the AUC and cumulative decile-rankings for XGBoost model as it performed better in comparison to other models including Logistic Regression, Random Forest and SVM.

Analysing Deep Learning Based Features: We compare the performance of our proposed models with Average Word Embedding(AWE) based model [22]. We report the AUC and cumulative decile-ranking scores obtained using deep learning models on the *SEC-imbalanced* dataset in Table 2. On *SEC-imbalanced* dataset, FinBERT representation of sentences along with SE-LSTM provides a huge increment of at least 5% in AUC in comparison with the AWE model.

Attention mechanism used in LSTM-based models gives higher weightage to useful information from sentence sequences, thus, are slightly better than ASE fine-tuned variants for both FinBERT and SEC-BERT. Both SE-LSTM and ASE-FF capture over 90% of defaults in 20% of financial, providing a significant boost of 10% over AWE-based architecture. The fine-tuned variants of both FinBERT and SEC-BERT has a lower AUC score compared to ASE-FF and SE-LSTM due to maximum input length constraint of 512 tokens in BERT. Overall, FinBERT based embeddings are better than SEC-BERT embeddings.

Results in Table 3 summarizes the performance on the *SEC-balanced* dataset. On this dataset, the AUC of ASE-FF and SE-LSTM models are at least 9% higher than AWE-based model. The performance of SEC-BERT based ASE-FF architecture exceeds all the models.

Table 2. Comparing performance of deep learning models on *SEC-imbalanced* dataset.

Model Strategy	AUC (%)	Cumulative Decile Values					
		1	2	3	4	5	6-10
AWE [22]	91.33	0.54	0.79	0.9	0.94	0.97	1
FinBERT-ASE-FF	96.21	**0.75**	0.9	0.93	**0.99**	0.99	1
FinBERT-Finetuned	88.22	0.63	0.74	0.84	0.88	0.9	1
FinBERT-SE-LSTM	**96.45**	0.72	**0.91**	**0.96**	**0.99**	**1.00**	1
SEC-BERT-ASE-FF	95.3	0.72	0.9	0.94	0.97	0.99	1
SEC-BERT-Finetuned	93.65	0.68	0.87	0.91	0.93	0.96	1
SEC-BERT-SE-LSTM	95.56	0.74	**0.91**	0.93	0.96	0.99	1

Table 3. Comparing performance of deep learning models on *SEC-balanced* dataset.

Model Strategy	AUC (%)	Cumulative Decile Values					
		1	2	3	4	5	6-10
AWE [22]	82.81	0.21	0.37	0.55	0.71	0.76	1
FinBERT-ASE-FF	91.38	**0.24**	**0.45**	0.63	0.73	0.87	1
FinBERT-Finetuned	81.86	0.18	0.39	0.53	0.63	0.79	1
FinBERT-SE-LSTM	90.48	**0.24**	0.42	**0.66**	**0.82**	**0.89**	1
SEC-BERT-ASE-FF	**94.40**	**0.24**	**0.45**	**0.66**	**0.82**	**0.89**	1
SEC-BERT-Finetuned	80.07	0.21	0.34	0.5	0.68	0.74	1
SEC-BERT-SE-LSTM	93.23	**0.24**	0.42	0.61	0.79	**0.89**	1

Analysing Hybrid Features: Table 4 shows AUC & Cumulative decile-rankings for two hybrid strategies evaluated on the *SEC-imbalanced* dataset. FinBERT-ASE-FF model achieves significant boost over AWE based models with PCA+XGBoost as well as FC+Sigmoid combining strategy, yielding a 3% increment in AUC and capturing over 90% defaults in 20% financials. Overall, FC+Sigmoid is a better strategy as comparison to PCA+XGBoost as PCA does not retain all information from the embeddings.

To understand the impact of financial text data over the years, we also performed an ablation study to analyse if additional data from consecutive years improves the performance for predicting the delinquency. In this study, we used the hybrid setting (PCA+XGBoost) and concatenated features from up to consecutive past five years. Table 5 illustrates the performance when the data from multiple years is combined. The results suggest that the textual data from the immediate previous year financial statement has the overall maximum discriminative power to predict the likelihood of default. It further suggests that including textual information from further previous years do not add any significant value as compared to the effort needed to extract the information.

Combining Qualitative & Quantitaive Features: It is our assertion that the qualitative textual information in the financial statements adds complementary value to the quantitative accounting data. To validate this assertion, we analysed the effect of augmenting textual embeddings to the Accounting (ACCG) variables for predicting default. For this study, we use 18 accounting variables from SEC bankruptcy prediction dataset [12, 13]. Table 6 illustrates the performance with different combinations of textual and accounting variables. Default prediction model trained with accounting variables alone yields an AUC of 87.67. Using a combination of text embeddings along with accounting variables provides a

Table 4. Comparing the performance of hybrid models built by combining curated and contextual embeddings on the *SEC-imbalanced* dataset.

Embedding	Combining Strategy	AUC (%)	Cumulative Decile Values					
			1	2	3	4	5	6-10
AWE [22]	PCA + XGBoost	77	0.65	0.73	0.78	0.84	0.86	1
FinBERT-ASE-FF		**82**	**0.74**	**0.90**	**0.91**	**0.94**	**0.94**	1
FinBERT-Finetuned		77	0.66	0.72	0.80	0.84	0.89	1
FinBERT-SE-LSTM		74	0.63	0.69	0.75	0.79	0.81	1
SEC-BERT-ASE-FF		77	0.70	0.78	0.84	0.86	0.91	1
SEC-BERT-Finetuned		78	0.60	0.65	0.68	0.74	0.79	1
SEC-BERT-SE-LSTM		77	0.63	0.68	0.75	0.84	0.87	1
AWE [22]	FC + Sigmoid	92.92	0.63	0.81	0.91	0.96	0.99	1
FinBERT-ASE-FF		**96.08**	0.74	**0.91**	**0.96**	**0.99**	0.99	1
FinBERT-Finetuned		82.85	0.63	0.71	0.74	0.79	0.82	1
FinBERT-SE-LSTM		95.89	**0.75**	0.88	0.93	**0.99**	1	1
SEC-BERT-ASE-FF		94.80	0.69	0.87	0.94	0.96	0.99	1
SEC-BERT-Finetuned		87.80	0.69	0.79	0.84	0.85	0.88	1
SEC-BERT-SE-LSTM		95.29	0.74	0.90	0.91	0.94	0.99	1

Table 5. Comparing the performance of XGBoost model on previous years data on the *SEC-imbalanced* dataset using the best hybrid model setting.

Previous n-years data	AUC (%)	Cumulative Decile Values					
		1	2	3	4	5	6-10
1	**82**	0.74	**0.90**	0.91	0.94	0.94	**1**
2	80	0.74	0.90	0.94	0.94	0.96	1
3	81	**0.75**	0.87	0.93	0.96	0.97	1
4	78	0.74	0.88	**0.96**	**0.97**	**0.99**	1
5	80	0.74	0.85	0.94	0.96	0.97	1

major boost to the performance, with ASE-FF and SE-LSTM embeddings delivering an increment of at least 20 basis points in AUC over AWE embeddings.

Results on FS Dataset. Table 7 shows that XGBoost model with curated features outperforms deep learning models on the FS dataset. Among the deep learning models, while AUC suggests that AWE based model is the best model, it is very challenging to adjudge a clear winner between all the techniques given the nature of the FS dataset and limited number of samples available for evaluation. SEC-BERT embeddings combined with a SE-LSTM architecture is able to capture 43% of the defaults from 20% of financial statements. Due to the space limitations, the paper only presents results on the best performing setting on the real world FS dataset from all the experiments performed. However, the analysis drawn on the SEC dataset, hold true here as well.

Table 6. Comparing the performance of models built by combining Accounting (ACCG) variables and contextual embeddings with FC+Sigmoid strategy on *SEC-imbalanced*.

Model Strategy	AUC (%)	Cumulative Decile Values					
		1	2	3	4	5	6-10
ACCG	87.67	0.78	0.89	0.89	0.89	0.89	1
ACCG + AWE [22]	99.03	1	1	1	1	1	1
ACCG + FinBERT-ASE-FF	99.23	1	1	1	1	1	1
ACCG + FinBERT-Finetuned	88.69	0.89	0.89	0.89	0.89	0.89	1
ACCG + FinBERT-SE-LSTM	99.34	1	1	1	1	1	1
ACCG + SEC-BERT-ASE-FF	99.30	1	1	1	1	1	1
ACCG + SEC-BERT-Finetuned	87.11	0.78	0.89	0.89	0.89	0.89	1
ACCG + SEC-BERT-SE-LSTM	**99.73**	1	1	1	1	1	1

Table 7. Comparing the performance of XGBoost models with curated features and deep learning models evaluated on the FS dataset. XGBoost model with curated features - Dictionary (D), Sentiment (S), Theme (Th) and ASE significantly outperforms deep learning based models.

Model	AUC (%)	Cumulative Decile Values					
		1	2	3	4	5	6-10
AWE [22]	68.51	0.14	0.36	0.5	0.64	0.71	1
FinBERT-ASE-FF	45.04	0	0.21	0.29	0.36	0.5	1
FinBERT-Finetuned	56.12	0.07	0.14	0.21	0.29	0.57	1
FinBERT-SE-LSTM	54.96	0.14	0.14	0.21	0.29	0.57	1
SEC-BERT-ASE-FF	61.37	0	0.07	0.29	0.36	0.71	1
SEC-BERT-Finetuned	50.58	0.14	0.14	0.14	0.29	0.5	1
SEC-BERT-SE-LSTM	68.37	**0.21**	**0.43**	**0.5**	0.5	0.64	1
XGBoost - D/S/Th	70.18	0.14	0.21	0.36	0.57	0.71	1
XGBoost - D/S/Th/ASE	**78.00**	**0.21**	0.36	**0.5**	**0.57**	**0.79**	1

5   Conclusions

This paper demonstrates that comprehensive 360-° features from the qualitative textual data can efficiently predict the financial default as well as can boost the performance when paired with the accounting and market based quantitative features. Besides the dictionary and theme-based features, the paper also presents a novel method of encoding textual data using large language models specifically built for finance domain and three deep learning architectures for learning contextual embeddings. The document processing pipeline proposed in the paper solves a very practical real-world consideration by including companies in financial modelling that do not submit the financial statements in a structured and standard format. This makes our approach pragmatic and close to the real operational settings. Finally, the comprehensive experimental section demonstrates the efficacy of 360-° features from the textual data and the different ablation studies helps understand the influence on different factors in the financial default prediction problem.

References

1. Zhao, Z., Xu, S., Kang, B.H., et al.: Investigation and improvement of multi-layer perceptron neural networks for credit scoring. Expert Syst. Appl. **42**(7), 3508–3516 (2015)
2. Chong, E., Han, C., Park, F.C.: Deep learning networks for stock market analysis and prediction: methodology, data representations, and case studies. Expert Syst. Appl. **83**, 187–205 (2017)
3. Loughran, T., McDonald, B.: Textual Analysis in Accounting and Finance: A Survey. https://doi.org/10.2139/ssrn.2504147 (2016)
4. Hosaka, T.: Bankruptcy prediction using imaged financial ratios and convolutional neural networks. Expert Syst. Appl. **117**, 287–299 (2019)
5. Beaver, W.H.: Financial ratios as predictors of failure. J. Accounting Res., 71–111 (1966)
6. Araci, D.: FinBERT: Financial Sentiment Analysis with Pre-trained Language Models. https://arxiv.org/abs/1908.10063 (2019)
7. Huang, A.H., Wang, H., Yang, Y.: FinBERT: a large language model for extracting information from financial text. Contemporary Accounting Research (2022)
8. Loukas, L., et al.: FiNER: financial numeric entity recognition for XBRL tagging. https://arxiv.org/abs/2203.06482 (2022)
9. Shen, Z., et al.: LayoutParser: a unified toolkit for deep learning based document image analysis. In: 16th International Conference on Document Analysis and Recognition, Lausanne, Switzerland, pp. 131–146. https://doi.org/10.1007/978-3-030-86549-8_9 (2021)
10. Li, J., Xu, Y., Lv, T., Cui, L., Zhang, C., Wei, F.: DiT: self-supervised pre-training for document image transformer. In: Proceedings of the 30th ACM International Conference on Multimedia (2022)
11. Huang, Y., Lv, T., Cui, L., Lu, Y., Wei, F.: LayoutLMv3: pre-training for Document AI with Unified Text and Image Masking. arXiv:2204.08387 (2022)

12. Lombardo, G., Pellegrino, M., Adosoglou, G., Cagnoni, S., Pardalos, P.M., Poggi, A.: Machine learning for bankruptcy prediction in the American stock market: dataset and benchmarks. Future Internet. **14**(8), 244. https://doi.org/10.3390/fi14080244(2022)
13. Lombardo, G., Pellegrino, M., Adosoglou, G., Cagnoni, S., Pardalos, P.M., Poggi, A.: Deep Learning with Multi-Head Recurrent Neural Networks for Bankruptcy Prediction with Time Series Accounting Data. Available at SSRN: https://ssrn.com/abstract=4191839 (2022)
14. Edward, I.: Altman: financial ratios, discriminant analysis and the prediction of corporate bankruptcy. J. Financ. **23**(4), 589–609 (1968)
15. Shin, K.S., Lee, T.S., Kim, H.: An application of Support Vector Machines in bankruptcy prediction model. Expert Syst. Appl. **28**(1), 127–135 (2005)
16. Nanni, L., Lumini, A.: An experimental comparison of ensemble of classifiers for bankruptcy prediction and credit scoring. Expert Syst. Appl. **36**(2), 3028–3033 (2009)
17. Kim, S.Y., Upneja, A.: Predicting restaurant financial distress using decision tree and AdaBoosted decision tree models. Econ. Model. **36**, 354–362 (2014)
18. Atiya, A.F.: Bankruptcy prediction for credit risk using neural networks: a survey and new results. IEEE Trans. Neural Networks **12**(4), 929–935 (2001)
19. Tsai, C.F., Wu, J.W.: Using neural network ensembles for bankruptcy prediction and credit scoring. Expert Syst. Appl. **34**(4), 2639–2649 (2008)
20. Yoshihara, A., Fujikawa, K., Seki, K., Uehara, K.: Predicting stock market trends by recurrent deep neural networks. In: Pham, D.-N., Park, S.-B. (eds.) PRICAI 2014. LNCS (LNAI), vol. 8862, pp. 759–769. Springer, Cham (2014). https://doi.org/10.1007/978-3-319-13560-1_60
21. Lin, X.: Header and footer extraction by page association. In: Proceedings SPIE 5010, Document Recognition and Retrieval X, 13 January 2003. https://doi.org/10.1117/12.472833
22. Mai, F., Tian, S., Lee, C., et al.: Deep learning models for bankruptcy prediction using textual disclosures. Eur. J. Oper. Res. **274**(2), 743–758 (2019)
23. Ohlson, J.A.: Financial ratios and the probabilistic prediction of bankruptcy. J. Account. Res. **18**, 109–131 (1980)
24. U.S. Securities and Exchange Comission. https://www.sec.gov/edgar/search-and-access. Accessed 15 Jan 2023
25. Why a global recession is inevitable in 2023?. https://www.economist.com/the-world-ahead/2022/11/18/why-a-global-recession-is-inevitable-in-2023?gclid=CjwKCAiA5Y6eBhAbEiwA_2ZWIT-e4RQK695FLW-F_YuXnMT0Tx4w3Qcx4BdMXPv0P8A_S8guWgh0bRoCKsUQAvD_BwE&gclsrc=aw.ds. Accessed 15 Jan 2023
26. An Introduction to XBRL. https://www.xbrl.org/guidance/xbrl-glossary/. Accessed 10 Feb 2023
27. Karl Pearson F.R.S., 1901. LIII. On lines and planes of closest fit to systems of points in space. The London, Edinburgh, and Dublin Philosophical Magazine and Journal of Science, 2(11), pp. 559–572
28. Chen, T., Guestrin, C.: XGBoost: a scalable tree boosting system. In: Proceedings of the 22nd ACM SIGKDD International Conference on Knowledge Discovery and Data Mining, pp. 785–794. ACM, New York (2016). https://doi.org/10.1145/2939672.2939785
29. Wolf, T., et al.: Transformers: state-of-the-art natural language processing. In: Proceedings of the 2020 Conference on Empirical Methods in Natural Language Pro-

cessing: System Demonstrations. Association for Computational Linguistics (2020). https://www.aclweb.org/anthology/2020.emnlp-demos.6

30. Loukas, L., Fergadiotis, M., Androutsopoulos, I., Malakasiotis, P.: EDGAR-CORPUS: billions of tokens make the world go round. In: The Proceedings of the Workshop on Economics and Natural Language Processing - co-located with EMNLP (2021)

31. LoPucki, L.M.: UCLA-LoPucki Bankruptcy Research Database. UCLA School of Law. Print, Los Angeles, California (2005)

32. Peng, B., Chersoni, E., Hsu, Y.-Y., Huang, C.-R.: Is domain adaptation worth your investment? comparing BERT and FinBERT on financial tasks. In: Proceedings of the Third Workshop on Economics and Natural Language Processing, pp. 37–44, Punta Cana, Dominican Republic. Association for Computational Linguistics (2021)

33. Decile. In: The Concise Encyclopedia of Statistics. Springer, New York, NY. https://doi.org/10.1007/978-0-387-32833-1_99 (2008)

34. Fawcett, T.: An introduction to ROC analysis. Pattern Recogn. Lett. **27**(8), 861–874 (2006). ISSN 0167–8655, https://doi.org/10.1016/j.patrec.2005.10.010

35. Devlin, J., Chang, M.-W., Lee, K., Toutanova, K.: BERT: pre-training of deep bidirectional transformers for language understanding. In: Proceedings of the 2019 Conference of the North American Chapter of the Association for Computational Linguistics: Human Language Technologies, Volume 1 (Long and Short Papers), pp. 4171–4186, Minneapolis, Minnesota. Association for Computational Linguistics (2019)

36. Radford, A.: Improving language understanding with unsupervised learning (2018)

RealCQA: Scientific Chart Question Answering as a Test-Bed for First-Order Logic

Saleem Ahmed[(✉)] [iD], Bhavin Jawade, Shubham Pandey, Srirangaraj Setlur[iD], and Venu Govindaraju[iD]

University at Buffalo, Buffalo, USA
{sahmed9,bjawade,spandey8,setlur,govind}@buffalo.edu

Abstract. We present a comprehensive study of chart visual question-answering(QA) task, to address the challenges faced in comprehending and extracting data from chart visualizations within documents. Despite efforts to tackle this problem using synthetic charts, solutions are limited by the shortage of annotated real-world data. To fill this gap, we introduce a benchmark and dataset for chart visual QA on real-world charts, offering a systematic analysis of the task and a novel taxonomy for template-based chart question creation. Our contribution includes the introduction of a new answer type, 'list', with both ranked and unranked variations. Our study is conducted on a real-world chart dataset from scientific literature, showcasing higher visual complexity compared to other works. Our focus is on template-based QA and how it can serve as a standard for evaluating the first-order logic capabilities of models. The results of our experiments, conducted on a real-world out-of-distribution dataset, provide a robust evaluation of large-scale pre-trained models and advance the field of chart visual QA and formal logic verification for neural networks in general. Our code and dataset is publicly available (https://github.com/cse-ai-lab/RealCQA).

Keyword: Charts and Document Understanding and Reasoning

1 Introduction

The chart question-answering[QA] task has recently received attention from a wider community [8,17,20,24,26,30]. While generic multi-modal QA tasks have been studied widely, the chart-based QA task is still in its developmental phase, especially for real-world scientific document understanding applications.

Recent works have provided a structure for question type classification [17] while iteratively adding complexity in chart types [8] and answer types [26]. However, no existing work fills the gap of QA on real-world charts [11] with structured output prediction.

Two main approaches for synthetic Chart-QA are: (i) considering the whole input image as a matrix of pixels to generate output in the form of text, answer

© The Author(s), under exclusive license to Springer Nature Switzerland AG 2023
G. A. Fink et al. (Eds.): ICDAR 2023, LNCS 14189, pp. 66–83, 2023.
https://doi.org/10.1007/978-3-031-41682-8_5

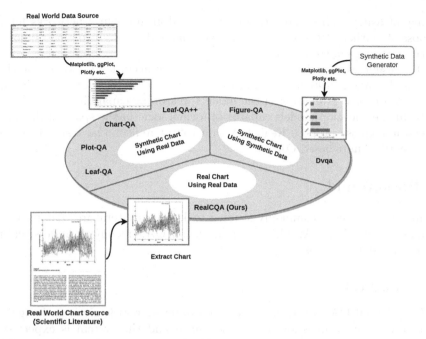

Fig. 1. Existing datasets in chart visual QA either are fully synthetic charts generated from synthetic data (Right sector in the above ellipse) or synthetic charts generated from real data (left sector). None of these datasets handle the complexity of the distribution of real-world charts found in scientific literature. We introduce the first chart QA dataset (RealCQA) in the third category (lower sector in the above figure) which consists of real-world charts extracted from scientific papers along with various categories of QA pairs. [Best viewed digitally in color].

types, etc. [8,20,30] or (ii) first extracting tabular data by identifying, classifying chart structural components and, then treating the task as a table-QA task [24].

These include either numeric answers (regression task) or single-string answers from the charts vocabulary (classification task). We further propose a structured and unstructured list answer type task, where answers can contain delimiter-spaced strings, where the order of strings might/or not matter. We also include new chart types for the scatter and box plots with curated chart-specific questions.

With the advent of representational learning over multi-modal data for document understanding [2,14,21,27,34,35], the task of knowledge representation [4] and reasoning in latent space has improved significantly from previous heuristic-driven methods used to capture propositional logic.

Recent works have paved the way for linking FOC_2 (First Order Logic with two variables and counting capability) with neural networks [3]. Tasks such as learning to reason over mathematical expressions [1,23] make a neat test bed for NSC(Neuro-Symbolic-Computing) [5].

There has been a rich history of research in building logic-based systems, for Theorem Proving, Conjecture Solving etc. Recent advances have seen the efficacy of using sophisticated transformers and graph neural networks for the purpose of mathematical reasoning over very large datasets involving millions of intermediate logical steps [18].

One recent work claims almost a 20% jump in accuracy for synthetic chart QA, just by augmenting pretraining with mathematical reasoning [22] although they lack a robust evaluation of the models' reasoning capability.

To further develop models capable of formal logic in the space of document understanding, we propose RealCQA as a robust multimodal testbed for logic and scientific chart-based QA.

2 Background

We first discuss more commonly studied tasks in the literature that provide a foundation for ChartQA. These include visual QA, document understanding, and formal logic systems.

2.1 Visual QA

VQA, or Visual QA, is a task where a computer system is given an image and a natural language question about the image and the system is expected to generate a natural language answer [19]. VQA systems aim to mimic the ability of humans to understand and reason about visual information and language and to use this understanding to generate appropriate responses to questions. Specific variations of this task include image captioning and multi-modal retrieval.

Image Captioning is the task where a computer system is given an input image and is expected to generate a natural language description of the content of the image [9] . This description should capture the main objects, actions, and events depicted in the image, as well as the relationships between them. Image captioning systems typically use machine learning algorithms to learn how to generate descriptive captions from a large dataset of images and their corresponding human-generated captions.

Multimodal Retrieval involves retrieving images and text that are related to each other based on their content [16]. This task involves the integration of computer vision and natural language processing techniques and is used in a variety of applications such as image and text search, image annotation, and automated customer service. In image-text cross-modal retrieval, a computer system is given a query in the form of either an image or a text and is expected to retrieve images or texts that are related to the query. To perform image-text cross-modal retrieval effectively, the system must be able to understand and reason about the visual and linguistic content of both images and text and to identify the relationships between them. This typically involves the use of machine learning algorithms that are trained on large datasets of images and text and their corresponding relationships.

2.2   Document Understanding

The field of document intelligence encompasses a broad range of tasks [6,31], such as localization, recognition, layout understanding, entity recognition, and linking. In this section, we describe the downstream tasks of document-QA, Table-QA, and Infographic-QA, which build up to Chart-QA

VQA for Document Understanding has been explored in works such as [36] which involve document pages comprising of tables, text and QA-pairs. The documents are sampled from financial reports and contain lots of numbers, requiring discrete reasoning capability to answer questions. Relational-VQA models use reasoning frameworks based on FOL to answer questions about visual scenes. Researchers have also explored other figure types such as Map-based QA [7]. CALM [12] proposes extending [25] with prior knowledge reasoning, and [29] proposes models for non-English document understanding through QA. Key requisites for Document-QA [DQA] include (i)*Robust feature representation:* One of the main challenges in DQA is to effectively represent the visual and semantic content of documents. The development of robust feature representations that capture the relationships between objects, properties, and concepts in documents is a key area of research in the field. (ii)*Large-scale datasets:* Another challenge in DQA is the lack of large-scale datasets that can be used to train and evaluate models. The development of large-scale datasets that include a wide variety of documents and questions is crucial for advancing the field. (iii)*Integration of prior knowledge and context:* In order to accurately answer questions about documents, models must be able to effectively integrate prior knowledge and context into their reasoning process. This requires the development of algorithms that can reason about the relationships between objects and concepts in a document, and that can incorporate prior knowledge and context into the decision-making process. (iv)*Relational reasoning:* DQA often requires reasoning about relationships between objects and concepts in a document. (v)*Multi-modal fusion:* DQA requires the integration of information from multiple modalities, including visual and semantic content. Recent works include [25,28,32,33].

Table QA is a natural language processing (NLP) task that involves answering questions about the information presented in tables. This task requires models to understand the structure and content of the table, as well as the meaning of the natural language question, in order to generate a correct answer. The table contents are provided as text input. Recent literature in the TableQA task include [13,15], which present models for generating SQL queries from natural language questions about tables.

2.3   Chart-VQA

We discuss two common approaches for this specific sub-area of IQA where the input is a chart image and a corresponding query.

Semi-Structured Information Extraction (SIE) [24] involves the follow-ing steps: **(i)** *Chart Text Analysis:* Extract the tick labels, legend, axis and chart tiles, and any other text in the image. **(ii)** *Chart Structure Analysis:* Tick associ-ation for corresponding data value interpolation of xy coordinate and the nearest tick label and legend mapping to individual data-series components labels. **(iii)** *Visual Element Detection [VED]:* Localize the chart component (line, box, point, bar) and association with x-tick and legend name. **(iv)** *Data Extraction:* Inter-polate the value represented by each data component by using the VED module and calculate the value from bounding ticks.

This reduces Chart-VQA to a Table-VQA task. However, this adds additional complexity as errors are now also introduced during the data-extraction task.

Classification-Regression [20], [26] approach has proven to be effective for chart comprehension, allowing machine learning models to accurately classify and predict the values and trends depicted in charts. In this school of thought, the input is directly treated as just pixels, usually relying on the implicit repre-sentation of chart components, plot area, visual elements, and underlying data. These features are aggregated alongside text features of the question string where the model learns their corresponding relations to predict either a classification answer(string) or a regression answer(numeric). Usually, models use visual fea-tures from a Mask-RCNN-based backbone, trained to detect chart text and structure. These are input alongside tokenized textual queries. Answer predic-tion involves predicting numeric or string type, where floats are regressed and tokens classified.

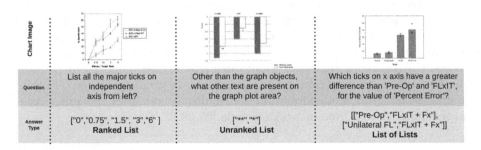

Fig. 2. List type answers have many uses-cases specifically in chart QA but has not been considered by existing CQA datasets. RealCQA introduces List type QA pairs with both (i) Ranked List (ii) Unranked List. Items in a list can consist of sets of up to 2 items.

2.4 Logic Order and Reasoning

We discuss formal logic, requirements for a testbed, and its applicability in the context of Chart-QA.

Zero-Order Logic [ZOL] is the basic unit of meaning, the atomic formula, a proposition that makes an assertion, *e.g.* answering the root level taxonomy questions: *'Is this chart of type A', 'Is there a title in the chart', 'Is the dependent axis logarithmic'* etc. Complex statements can be formed by combining atomic formulas using logical connectives such as 'and', 'or', and 'not'.

First-Order Logic [FOL] also known as predicate logic, is a type of formal logic that is used to study the relationships between objects and their properties. It allows the expression of propositions or statements that make assertions about properties and relations, and it provides a formal language for making logical inferences based on these assertions. In first-order logic, we have quantifiers $\forall$ (for all) and $\exists$ (there exist) to make statements about the entire domain of discourse by involving variables that range over the objects being discussed, *e.g.* a closed set of all tick labels in a chart. We can talk about the properties of these objects or relationships between them by using 1-place predicates or multi-place predicates, respectively, *e.g.* comparing data series values at different tick locations. These predicates can be viewed as sets where their elements are those of the domain that satisfy some property or n-tuples that satisfy some relation, *'Is the sum of the value of $\langle Y\ title \rangle$ in $\langle$ i-th x tick$\rangle$ and $\langle (i + 1)$th x tick$\rangle$ greater than the maximum value of $\langle$ title$\rangle$ across all $\langle$plural form of X title$\rangle$?'*

N-Th Order Logic uses the same quantifiers to range over predicates. This essentially allows the quantification of sets. Using the quantifiers for elements of the sets is provisional as required. These involve the 'List' type questions with structured output, *e.g.* *'Which pairs of major ticks on independent axis have a difference greater than $\langle$i-th x tick$\rangle$ and $\langle$j-th x tick$\rangle$, for the value of $\langle Y\ title \rangle$, arranged in the increasing order of difference?'*. Constraining the current scope for better evaluation, we limit this to 2nd order logic, *i.e.* we create questions with set outputs of at most 2 elements per item in the list as depicted in Fig. 2.

A Testbed for Formal Logic must satisfy specific requirements to ensure the correctness of the system being tested. The first requirement is a formal specification, which should precisely define the system's syntax, semantics, and model-checking algorithms. The second requirement is a set of test cases that covers all possible scenarios and validates the system's behavior, including its ability to handle edge cases and exceptional conditions. The third requirement is a repeatable, reliable, and easily extensible test harness that can accommodate new test cases. Finally, a verification environment must be created to host the testbed and provide necessary resources for performing the tests.

These ensure that formal logic systems are thoroughly tested and that the test results are accurate and reliable. The exact nature of the requirements and techniques used to meet them will vary depending on the system being tested and the context in which it is used. Overall, rigorous testing is crucial for establishing the correctness of logical systems and ensuring their applicability to real-world problems. Existing experimental setups for evaluating Chart-QA satisfy most of the above points, except the formal specification, which we provide for a subset of our questions. These are manually curated and verified. We will describe this in greater detail.

CQA for FOL represents our concept of utilizing the template-based chart QA task as a testbed of predicate logic. The innate structure of data which populates a scientific chart aligns naturally with the previously stated formal specification requirements. Prior research has studied VQA with charts, however, a formal testbed has not been studied.

In FOL, sentences are written in a specific syntax and structure to allow for precise and unambiguous representation of meaning. To translate a normal sentence into FOL, we need to identify the objects, individuals, and relationships described in the sentence, and express them using predicates, variables, and logical connectives. For example, the template *Is the difference between the value of ⟨ Y title ⟩ at ⟨ ith x-tick ⟩ and ⟨ jth x-tick ⟩ greater than the difference between any two ⟨ plural form of X-title ⟩ ?* can be converted to FOL as :

$$\forall i, j, p, q : (i \neq j \neq p \neq q) \rightarrow (|Y_i - Y_j| > |X_p - X_q|)$$

where ⟨...⟩ represents variables in the template and 'Y' the space of values of the dependent-variable, and 'X' for the independent variable in the chart, respectively. While curating reasoning type [17] questions, we create a subset of binary questions specifically over FOL that are valid.

In this study, our aim is to further advance the development of the chart and visual data parsing systems. Previous research has documented the limitations posed by the limited availability of annotated real-world data [11]. While absolute accuracy on a specific ChartQA dataset may not guarantee broad generalizability, this study is a step toward establishing a comprehensive understanding of this complex and evolving field. We believe that leveraging the manually curated templates and structured output generated from the semantic structure of charts presents an opportunity to effectively evaluate the multi-modal predicate logic parsing capabilities of modern neural networks, such as large-scale pre-trained language and layout models.

3   Dataset

In this section, we describe the dataset used in our study. The dataset, called RealCQA, was created by utilizing real-world chart images and annotations used in publicly conducted chart understanding challenges [11]. Figure 1 shows the current existing datasets in the CQA domain. The challenge tasks around chart understanding are shown in Fig. 3, along with the annotated data used from the publicly released train-test splits.

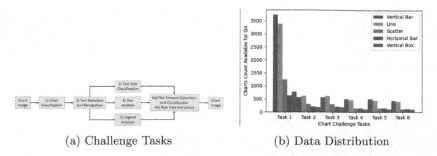

(a) Challenge Tasks   (b) Data Distribution

Fig. 3. Train and Test Structure, Retrieval, Reasoning by answer type. For List Type, we only curate reasoning questions for kth order FOL testing. String/Unranked refers to a small subset of string-type retrieval or reasoning answers where multiple equivalent conditions exist: While reading the question string, a human would expect a single answer but multiple data series have the same maximum/minimum etc. resulting in multiple correct single-string instance answers.

3.1 RealCQA

To generate question templates for RealCQA, we compiled templates from previous works [8,30], and [26]. These templates were adapted to our data and augmented with new chart-type questions, list questions, and binary FOL reasoning questions, forming a total of 240 templates.

The distribution of taxonomy and answer types for RealCQA is shown in Fig. 4. We have tried to keep the templates for different answer types with equal proportions. However, when these are used to create the actual QA pairs, the data gets skewed depending on underlying availability. Our dataset consists of a majority of 'Reasoning' type questions, as seen in previous works [17]. However, we also focus on creating binary reasoning questions that satisfy FOL. These form a major chunk of the dataset since templates with variables for i-th/j-th tick/data series are combinatorial in nature, and we create them exhaustively over the closed set of objects present in the chart.

We use the 'Structure, Retrieval, Reasoning' taxonomy proposed in previous works [17] to categorize our questions. However, we further demarcate them as Types 1, 2, 3, 4, depending on their characteristics. Type-1 refers to any questions that can be formed at the (root) level of the whole chart image, mostly ZOL. Type-2 further refers to ZOL questions for specific chart components, requiring the model to identify them. Type-3 and Type-4 are data retrieval/reasoning. Each has a further specific sub-class depending on the exact component, chart type, etc., as shown in Fig. 4.

The statistics of the dataset are shown in Fig. 5 using the previous nomenclature of 'Structural', 'Retrieval', and 'Reasoning'. For the List Type, we only curate reasoning questions for k^{th} order FOL testing. String/Unranked refers to a small subset of string-type retrieval or reasoning answers where multiple equivalent conditions exist. While reading the question string, a human would expect a single answer, but multiple data series have the same maximum/minimum, resulting in multiple correct single-string instance answers. These are generally outliers.

Overall, the RealCQA dataset offers a diverse range of questions that require various levels of chart understanding and reasoning abilities. The dataset is publicly available and can be used for further research and evaluation in the chart understanding domain.

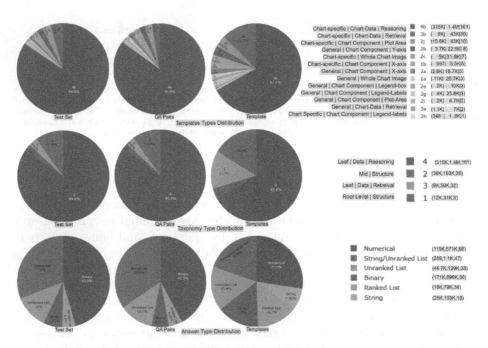

Fig. 4. Taxonomy and Type distribution: First pie is test set QA pairs, second training set QA pairs and third training set templates. Legend is in decreasing order. Best viewed digitally in color.

	Structural	Retrieval	Reasoning
Binary	63153	3069	631677
Numerical	55467	40535	474966
Ranked List	10936	0	68470
String	62318	6489	34074
String/Unranked	0	325	827
Unranked List	2394	0	197216

(a) Train Distribution

	Structural	Retrieval	Reasoning
Binary	18072	499	152561
Numerical	10646	7721	97024
Ranked List	1788	0	14601
String	17259	959	6661
String/Unranked	0	77	192
Unranked List	541	0	44161

(b) Test Distribution

Fig. 5. Train and Test Structure, Retrieval, Reasoning by answer type.

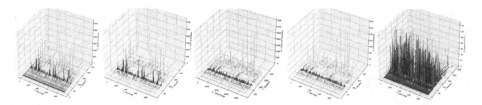

Fig. 6. Trend across different sampling strategies 1-5. X-axis represents each of the 9357 test-images, Y-axis the 240 templates each plotted with different colored bars at every 10th index, and Z axis shows the count of QA pairs.

Table 1. Total QA pairs per sampling strategy, bold shows minimum.

Answer Type	Sample 1	Sample 2	Sample 3	Sample 4	Sample 5
Total	367139	322404	276091	231356	**203735**
String	19525	3489	18548	**2512**	19046
Numeric	115391	107153	93096	**84858**	78680
Ranked	16389	13903	13357	**10871**	14228
Unranked	44702	43310	27019	25627	**25041**
Binary	171132	154549	124071	107488	**66740**

3.2   Sampling Strategies for Dataset Evaluation

For the purpose of general chart visual question answering, generating balanced and representative datasets is of paramount importance for training and evaluating models. However, when it comes to logic testbeds, the over-representation of specific templates or question types is not necessarily a disadvantage, as it allows for a more nuanced assessment of a model's logical reasoning capabilities. Nonetheless, it is still relevant to explore the impact of dataset sampling on evaluation results.

To this end, we devised five different sampling strategies and evaluated their effect on our dataset. The first strategy, exhaustive sampling, consists of including all available question-answer pairs. The remaining strategies aim to modify the distribution of questions per chart based on different criteria. Specifically, the second strategy, increasing lower bound, focuses on charts with a minimum number of questions greater than or equal to a threshold K. This strategy aims to address under-represented question types, such as root and structural questions. Conversely, the third strategy, decreasing upper bound, selects charts with a maximum number of questions less than or equal to a threshold L. This strategy is intended to address over-represented question types, typically combinatorial binary reasoning questions. The fourth strategy combines the effects of the second and third strategies, aiming to remove both under and over-represented charts. Finally, the fifth strategy, flat cap, selects a fixed number of questions per chart per template, thereby creating a more uniform dataset.

Figure 6 illustrates how the different sampling strategies affect the number of questions per chart per template, while Table 1 provides the actual number of QA-pairs per sampling. To be specific, we calculate the lower and upper 10% for the second and third strategies, respectively. For the flat cap strategy, we randomly select 150 QA-pairs for each template per chart.

By analyzing the impact of the different sampling strategies, we can gain a better understanding of how removing specific sections from the test-set affects evaluation results. These findings can be useful for modulating the training-set as required, and ultimately for developing more robust and accurate visual question-answering models.

3.3 Evaluation Metrics

In this study, we propose an evaluation metric based on the accuracy of answers. The proposed task involves four types of answers, each with its specific calculation method.

First, for numerical answers, we measure the accuracy of regression errors using L2 or L1 differences or the ER-error rate. In PlotQA-D [1], we consider a regression answer correct if it falls within ±5% tolerance from the ground truth value. Second, for single string answers, we use string-matching edit distance and count perfect matches as correct. Third, for unordered lists of strings, we use string-matching edit distance. For each of the K queries and M matches, we calculate $K \times M$ scores, and the mutually exclusive best match is aggregated per string instance normalized by K. Fourth, for ranked order lists of strings, we use the nDCG@K ranking metric, where K is the size of the ground-truth list. nDCG is a normalized version of the DCG (Discounted Cumulative Gain) metric, which is widely used to evaluate the ranking quality of information retrieval systems, such as search engines and recommendation systems. This metric assigns a relevance score to each item in a ranked list based on the user's preferences, and then discounts these scores using a logarithmic function, with items appearing lower in the ranking receiving lower scores. Lastly, for nested lists, where each item is a set, we evaluate the results invariant of set order, but list order matters in ranked lists.

4   Experiments

We benchmark multiple existing generic visual QA and chart-specific visual QA methods on RealCQA. Figure 7 shows the generic existing architecture used for CQA task. The model either learns visual and data features separately or in the same shared space and then uses some fusion model to generate the final answer. These are primarily trained on synthetic charts. Here, we evaluate multiple baseline models that have been proposed recently including ChartQA and CRCT. We present both the synthetic pre-training evaluation on RealCQA and RealCQA finetuned evaluation. Below we briefly discuss the model architecture for the baseline methods in more detail:

VLT5 [10]. VLT5 is a state-of-the-art unified framework that leverages a multimodal text conditioning language objective to perform different tasks within a unified architecture. In this framework, the model learns to generate labels in the text space based on the visual and textual inputs. In our study, we use VLT5 to perform the task of table-based question answering. Specifically, VLT5 takes

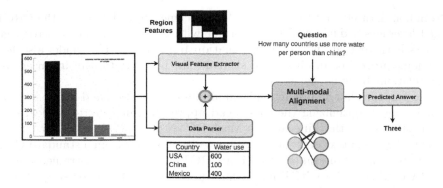

Fig. 7. A generalized framework to represent existing methods for chart QA.

as input pre-trained region-based visual features obtained from Faster-RCNN, which was pre-trained on PlotQA [26]. These visual features, along with textual tokens, are projected and fed through a unified bi-directional multi-modal encoder. Additionally, a language decoder is trained in an auto-regressive setting to perform text generation. In the textual context, we provide a pre-extracted gold standard table of the chart as concatenated input along with the query question.

We present the performance of VLT5 on the RealCQA test-set, which comprises approximately 683 charts with gold-data table annotation. The evaluation is conducted on the charts in this test-set, with a score of zero assigned to the remaining charts. The results of this evaluation are presented in Table 2 and Table 3, where Row 1 shows the performance of VLT5 segregated by Answer-Type and Question-Type, respectively.

ChartQA [24]. The authors of ChartQA introduced a large-scale benchmark dataset comprising of 9.6K human-written questions and 23.1K questions generated from human-written chart summaries. To evaluate the effectiveness of the dataset, two transformer-based multimodal architectures, namely VisionTapas and VLT5, were benchmarked on ChartQA using data tables and visual features as context.

In this study, we fine-tuned the VLT5 multi-modal encoder with ChartQA visual features that were pre-trained on PlotQA. We utilized ChartQA pre-trained Mask RCNN visual features with VLT5 multi-modal attention in both Row 1 and Row 2 of Table 2 and Table 3, respectively. Results were segregated based on Answer-Type and Question-Type, as presented in Table 2 and Table 3. Since the evaluation requires a data-table, we evaluated the model on 683 charts and assigned zero scores to the remaining QAs.

CRCT [20]. The paper proposes a novel ChartVQA approach called Classification Regression Chart Transformer (CRCT) that aims to address the limitations of existing methods in the field. The authors argue that the saturation of previous methods is due to biases, oversimplification, and classification-oriented Q&A

in common datasets and benchmarks. To overcome these challenges, the CRCT model leverages a dual-branch transformer with a chart element detector that extracts both textual and visual information from charts. The model also features joint processing of all textual elements in the chart to capture inter and intra-relations between elements.

The proposed hybrid prediction head unifies classification and regression into a single model, optimizing the end-to-end approach using multi-task learning. For visual context, they fine-tuned a Mask-RCNN on PlotQA, while for textual context, they used text detections and recognition output from a standard OCR such as tesseract. We evaluated both a CRCT model fully pre-trained on the PlotQA dataset and a CRCT model fine-tuned on RealCQA for stage 2 with pre-trained FasterRCNN. We report the performance of both models in Row 3 and Row 4 of Table 2 and Table 3, respectively.

Table 2. Performance of existing Visual Question Answering Methods on RealCQA based on Answer Type

	Total Accuracy	String	Numeric	Rank	Unranked	Binary
VLT5 (VG Pretrained)	0.2399	0.0008	0.0325	0.0106	0.0002	0.4916
VLT5 (RealCQA Finetuned)	**0.3106**	**0.3068**	0.1487	0.0246	0.0048	**0.5275**
CRCT (PlotQA Pretrained)	0.1787	0.0350	0.0412	0.0015	0.0016	0.3515
CRCT (RealCQA Finetuned)	0.1880	0.0323	**0.3158**	**0.0286**	**0.0124**	0.1807

4.1   Results

We present the quantitative results of our experiments, which are summarized in Table 2 and Table 3. We find that the VLT5 model [10], does not perform well on the RealCQA dataset when using pre-trained Mask RCNN visual features and RealCQA's gold-data table as input. The performance of 49.16% on binary-type answers is as bad as random assignment. However, by fine-tuning the VLT5 model's multi-modal alignment module on RealCQA with modified tokenization to handle list-type answers, we observe significant improvements in performance on all answer types (Table 1) and question types (Table 2). Specifically, the performance on string type answers improves from 0.008% to 30.68%, and the overall accuracy of QA pairs improves from 23.99% to 31.06%. Table 3 compares the performance of the CRCT model [20] fully pre-trained on PlotQA and VG Pretrained VLT5 on Root, Structure, and Retrieval type questions. We find that the fully pre-trained CRCT outperforms VG Pretrained VLT5 on these

Table 3. Performance of existing Visual Question Answering Methods on Real-CQA based on Question Complexity

	Root	Structure	Retrieval	Reasoning
VLT5 (VG Pretrained)	0.0764	0.1416	0.0765	0.2620
VLT5 (RealCQA Finetuned)	**0.1800**	**0.4352**	**0.5877**	**0.2937**
CRCT (PlotQA Pretrained)	0.0773	0.2338	0.1168	0.1778
CRCT (RealCQA Finetuned)	0.0115	0.1497	0.3131	0.1959

question types. However, fine-tuning the CRCT model on RealCQA leads to significant improvements in performance on Numeric, Ranked List, and Unranked List type answers, as shown in Table 2. Notably, fine-tuned CRCT achieves the best performance of 31.58% on numeric type answers. Our results highlight the importance of RealCQA as a standard test bed for evaluating chart visual QA methods, as even models that perform well on synthetic datasets such as PlotQA and FigureQA struggle to generalize to real-world chart distributions.

We present an ablation study that examines the impact of different sampling strategies on the performance of our model. The results of this study are summarized in Table 4. The study reveals that the 4th sampling strategy, which combines both the upper and lower bounds, consistently achieves the highest overall, string, and binary type accuracy across various experimental settings. On the other hand, the 5th strategy, which produces the most uniform test set, yields top accuracies for numeric and list-type answers. However, this strategy has the smallest size in terms of overall, unranked, and binary questions, and it attains the highest accuracy for unranked list type questions, which are representative of Kth Order Logic questions. It is worth noting that the 5th strategy may remove most of the challenging QA pairs, making it less desirable for our objective.

Overall, this study highlights the importance of carefully selecting the sampling strategy to obtain a test set that is representative of the distribution of real-world chart visual QA. The 4th strategy appears to be a promising choice, as it achieves high accuracy across different answer types while maintaining a sufficient number of challenging QA pairs.

Table 4. An ablation on the performance of different models based on different sampling strategies. Here, LB - Lower Bound, UB - Upper Bound, Full - No Sampling.

Data	Model	Total Accuracy	String	Numeric	Rank	Unrank	Binary
Full	VLT5 Pre-Trained	0.2399	0.0008	0.0325	0.0106	0.0002	0.4916
LB			0.0032	0.0252	0.0092	0.0001	0.5182
UB			0.0008	0.0345	0.0090	0.0002	0.4947
LB+UB			0.0044	0.0255	0.0069	0.0001	0.5334
150max			0.0008	0.0403	0.0122	0.0003	0.4349
Full	VLT5 (RealCQA Fine-Tuned)	0.3106	0.3068	0.1487	0.0246	0.0048	0.5275
LB			0.4004	0.1305	0.0201	0.0041	0.5487
UB			0.3165	0.1585	0.0246	0.0070	0.5258
LB+UB			0.5084	0.1365	0.0187	0.0061	0.5559
150max			0.3146	0.1827	0.0283	0.0085	0.4947
Full	CRCT Pretrained	0.1787	0.0350	0.0412	0.0015	0.0000	0.3515
LB			0.0224	0.0359	0.0013	0.0000	0.3473
UB			0.0367	0.0428	0.0012	0.0000	0.3742
LB+UB			0.0299	0.0363	0.0009	0.0000	0.3716
150max			0.0359	0.0499	0.0017	0.0000	0.3601
Full	CRCT (RealCQAFine-Tuned)	0.1880	0.0323	0.3158	0.0286	0.0124	0.1807
LB			0.0759	0.3196	0.0263	0.0102	0.1923
UB			0.0319	0.3316	0.0327	0.0190	0.1977
LB+UB			0.0903	0.3379	0.0309	0.0158	0.2170
150max			0.0322	0.3371	0.0329	0.0220	0.1316

5   Conclusion

In addition to our contribution of curating a novel FOL-Testbed and a dataset for the evaluation of CQA for real charts, we have also thoroughly evaluated several state-of-the-art visual question answering models on the RealCQA dataset. Our experiments reveal that while some models perform well on synthetic datasets like PlotQA and FigureQA, their performance significantly drops when tested on RealCQA, demonstrating the need for a more realistic and challenging benchmark like RealCQA. We have shown that our proposed method, CRCT, significantly outperforms previous models on several question types, especially on numeric type questions. Our ablation study further highlights the importance of sampling strategies in constructing a diverse and representative test set.

Overall, our study emphasizes the importance of multimodal learning and reasoning in visual question answering and provides insights into the limitations and opportunities of current state-of-the-art models. Future work can build on our findings by exploring more sophisticated models that integrate text, image, and reasoning more effectively, as well as developing new evaluation metrics that capture the full complexity of real-world chart questions. Additionally, expanding the dataset to cover a wider range of chart types and complexities can further improve the generalization capabilities of visual question answering models and lead to more impactful applications in areas such as data analysis and decision-making.

References

1. Ahmed, S., Davila, K., Setlur, S., Govindaraju, V.: Equation attention relationship network (earn) : a geometric deep metric framework for learning similar math expression embedding. In: 2020 25th International Conference on Pattern Recognition (ICPR), pp. 6282–6289 (2021). https://doi.org/10.1109/ICPR48806.2021.9412619
2. Appalaraju, S., Jasani, B., Kota, B.U., Xie, Y., Manmatha, R.: DocFormer: end-to-end transformer for document understanding. In: Proceedings of the IEEE/CVF International Conference on Computer Vision, pp. 993–1003 (2021)
3. Barceló, P., Kostylev, E.V., Monet, M., Pérez, J., Reutter, J., Silva, J.P.: The logical expressiveness of graph neural networks. In: 8th International Conference on Learning Representations (ICLR 2020) (2020)
4. Battaglia, P.W., et al.: Relational inductive biases, deep learning, and graph networks. arXiv preprint arXiv:1806.01261 (2018)
5. Besold, T.R., et al.: Neural-symbolic learning and reasoning: a survey and interpretation. CoRR abs/1711.03902 (2017)
6. Borchmann, Ł., et al.: Due: End-to-end document understanding benchmark. In: Thirty-fifth Conference on Neural Information Processing Systems Datasets and Benchmarks Track (Round 2) (2021)
7. Chang, S., Palzer, D., Li, J., Fosler-Lussier, E., Xiao, N.: MapQA: a dataset for question answering on choropleth maps. arXiv preprint arXiv:2211.08545 (2022)
8. Chaudhry, R., Shekhar, S., Gupta, U., Maneriker, P., Bansal, P., Joshi, A.: LeafQA: locate, encode & attend for figure question answering. In: Proceedings of the IEEE/CVF Winter Conference on Applications of Computer Vision, pp. 3512–3521 (2020)
9. Chen, C., et al.: Neural caption generation over figures. In: Adjunct Proceedings of the 2019 ACM International Joint Conference on Pervasive and Ubiquitous Computing and Proceedings of the 2019 ACM International Symposium on Wearable Computers, pp. 482–485 (2019)
10. Cho, J., Lei, J., Tan, H., Bansal, M.: Unifying vision-and-language tasks via text generation. In: International Conference on Machine Learning, pp. 1931–1942. PMLR (2021)
11. Davila, K., Xu, F., Ahmed, S., Mendoza, D.A., Setlur, S., Govindaraju, V.: ICPR 2022: challenge on harvesting raw tables from infographics (chart-infographics). In: 2022 26th International Conference on Pattern Recognition (ICPR), pp. 4995–5001 (2022). https://doi.org/10.1109/ICPR56361.2022.9956289
12. Du, Q., Wang, Q., Li, K., Tian, J., Xiao, L., Jin, Y.: CALM: commen-sense knowledge augmentation for document image understanding. In: Proceedings of the 30th ACM International Conference on Multimedia, pp. 3282–3290 (2022)
13. Eisenschlos, J.M., Gor, M., Müller, T., Cohen, W.W.: MATE: multi-view attention for table transformer efficiency. CoRR abs/2109.04312 (2021)
14. Gu, J., et al.: Unidoc: unified pretraining framework for document understanding. Adv. Neural. Inf. Process. Syst. **34**, 39–50 (2021)
15. Herzig, J., Nowak, P.K., Müller, T., Piccinno, F., Eisenschlos, J.M.: Tapas: weakly supervised table parsing via pre-training. arXiv preprint arXiv:2004.02349 (2020)
16. Jawade, B., Mohan, D.D., Ali, N.M., Setlur, S., Govindaraju, V.: NAPReg: nouns as proxies regularization for semantically aware cross-modal embeddings. In: Proceedings of the IEEE/CVF Winter Conference on Applications of Computer Vision (WACV), pp. 1135–1144 (2023)

17. Kafle, K., Price, B., Cohen, S., Kanan, C.: DVQA: understanding data visualizations via question answering. In: Proceedings of the IEEE Conference on Computer Vision and Pattern Recognition, pp. 5648–5656 (2018)

18. Kaliszyk, C., Chollet, F., Szegedy, C.: HolStep: a machine learning dataset for higher-order logic theorem proving. arXiv preprint arXiv:1703.00426 (2017)

19. Kodali, V., Berleant, D.: Recent, rapid advancement in visual question answering: a review. In: 2022 IEEE International Conference on Electro Information Technology (eIT), pp. 139–146 (2022). https://doi.org/10.1109/eIT53891.2022.9813988

20. Levy, M., Ben-Ari, R., Lischinski, D.: Classification-regression for chart comprehension. In: Avidan, S., Brostow, G., Cisse, M., Farinella, G.M., Hassner, T. (eds.) Computer Vision – ECCV 2022. ECCV 2022. Lecture Notes in Computer Science, vol. 13696, pp. 469–484. Springer, Cham (2022). https://doi.org/10.1007/978-3-031-20059-5_27

21. Li, P., et al.: SelfDoc: self-supervised document representation learning. In: Proceedings of the IEEE/CVF Conference on Computer Vision and Pattern Recognition, pp. 5652–5660 (2021)

22. Liu, F., et al.: MatCha: enhancing visual language pretraining with math reasoning and chart derendering. arXiv preprint arXiv:2212.09662 (2022)

23. Mansouri, B., Agarwal, A., Oard, D.W., Zanibbi, R.: Advancing math-aware search: the ARQMath-3 lab at CLEF 2022. In: Hagen, M., et al. (eds.) ECIR 2022. LNCS, vol. 13186, pp. 408–415. Springer, Cham (2022). https://doi.org/10.1007/978-3-030-99739-7_51

24. Masry, A., Long, D.X., Tan, J.Q., Joty, S., Hoque, E.: ChartQA: a benchmark for question answering about charts with visual and logical reasoning. arXiv preprint arXiv:2203.10244 (2022)

25. Mathew, M., Karatzas, D., Jawahar, C.: DocVQA: A dataset for VQA on document images. In: Proceedings of the IEEE/CVF Winter Conference on Applications of Computer Vision, pp. 2200–2209 (2021)

26. Methani, N., Ganguly, P., Khapra, M.M., Kumar, P.: PlotQA: reasoning over scientific plots. In: Proceedings of the IEEE/CVF Winter Conference on Applications of Computer Vision, pp. 1527–1536 (2020)

27. Powalski, R., Borchmann, Ł, Jurkiewicz, D., Dwojak, T., Pietruszka, M., Pałka, G.: Going Full-TILT Boogie on document understanding with text-image-layout transformer. In: Lladós, J., Lopresti, D., Uchida, S. (eds.) ICDAR 2021. LNCS, vol. 12822, pp. 732–747. Springer, Cham (2021). https://doi.org/10.1007/978-3-030-86331-9_47

28. Qi, L., et al.: Dureadervis: a chinese dataset for open-domain document visual question answering. In: Findings of the Association for Computational Linguistics: ACL 2022. pp. 1338–1351 (2022)

29. Ščavnická, Š., Štefánik, M., Kadlčík, M., Geletka, M., Sojka, P.: Towards general document understanding through question answering. RASLAN 2022 Recent Advances in Slavonic Natural Language Processing, p. 183 (2022)

30. Singh, H., Shekhar, S.: Stl-cqa: Structure-based transformers with localization and encoding for chart question answering. In: Proceedings of the 2020 Conference on Empirical Methods in Natural Language Processing (EMNLP), pp. 3275–3284 (2020)

31. Tanaka, R., Nishida, K., Yoshida, S.: VisualMRC: machine reading comprehension on document images. In: Proceedings of the AAAI Conference on Artificial Intelligence, vol. 35, pp. 13878–13888 (2021)

32. Tito, R., Mathew, M., Jawahar, C.V., Valveny, E., Karatzas, D.: ICDAR 2021 Competition on Document Visual Question Answering. In: Lladós, J., Lopresti, D., Uchida, S. (eds.) ICDAR 2021. LNCS, vol. 12824, pp. 635–649. Springer, Cham (2021). https://doi.org/10.1007/978-3-030-86337-1_42

33. Wu, X., et al.: A region-based document VQA. In: Proceedings of the 30th ACM International Conference on Multimedia, pp. 4909–4920 (2022)

34. Xu, Y., Li, M., Cui, L., Huang, S., Wei, F., Zhou, M.: LayoutLM: pre-training of text and layout for document image understanding. In: Proceedings of the 26th ACM SIGKDD International Conference on Knowledge Discovery & Data Mining, pp. 1192–1200 (2020)

35. Zhong, X., Tang, J., Yepes, A.J.: PubLayNet: largest dataset ever for document layout analysis. In: 2019 International Conference on Document Analysis and Recognition (ICDAR), pp. 1015–1022. IEEE (2019)

36. Zhu, F., Lei, W., Feng, F., Wang, C., Zhang, H., Chua, T.S.: Towards complex document understanding by discrete reasoning. In: Proceedings of the 30th ACM International Conference on Multimedia, pp. 4857–4866 (2022)

An Iterative Graph Learning Convolution Network for Key Information Extraction Based on the Document Inductive Bias

Jiyao Deng[1,2], Yi Zhang[3,4(✉)], Xinpeng Zhang[1], Zhi Tang[1], and Liangcai Gao[1(✉)]

[1] Wangxuan Institute of Computer Technology, Peking University, Beijing, China
`dengjiyao@stu.pku.edu.cn`, {`zhangxinpeng,tangzhi,gaoliangcai`}`@pku.edu.cn`
[2] Center for Data Science, Peking University, Beijing, China
[3] National Engineering Laboratory for Big Data Analysis and Applications, Peking University, Beijing, China
`zhangyi03@pku.edu.cn`
[4] Beijing Institute of Big Data Research, Beijing, China

Abstract. Recently, there has been growing interest in automating the extraction of key information from document images. Previous methods mainly focus on modelling the complex interactions between multimodal features(text, vision and layout) of documents to comprehend their content. However, only considering these interactions may not work well when dealing with unseen document templates. To address this issue, in this paper, we propose a novel approach that incorporates the concept of document inductive bias into the graph convolution framework. Our approach recognizes that the content of a text segment in a document is often determined by the context provided by its surrounding segments and utilizes an adjacency matrix hybrid strategy to integrate this bias into the model. As a result, the model is able to better understand the relationships between text segments even when faced with unseen templates. Besides, we employ an iterative method to perform graph convolution operation, making full use of the textual, visual, and spatial information contained within documents. Extensive experimental results on two publicly available datasets demonstrate the effectivness of our methods.

Keywords: Document Understanding · Key Information Extraction · Graph Convolutional Networks · Documents Inductive Bias

1 Introduction

Key Information Extraction (KIE) is the process of extracting structured information from documents, such as retrieving key information fields from images like invoices, receipts, and financial forms. The field of KIE has recently gained significant attention due to its numerous practical applications in areas like document analytics, rapid indexing, and efficient archiving in commercial settings.

G. A. Fink et al. (Eds.): ICDAR 2023, LNCS 14189, pp. 84–97, 2023.
https://doi.org/10.1007/978-3-031-41682-8_6

The increasing demand for KIE is driven by the need for organizations to manage and analyze large amounts of document data efficiently.

Previous research in KIE has primarily focused on how to model the complex interactions between the multimodal features of documents, including textual, visual and spatial features. LayoutLM [34] extends BERT [6] model by incorporating the 2D layout embeddings of the documents into the input embeddings during pre-training, and introduces visual information in the fine-tuning stage to get final token embeddings which are then fed into a classifier to do field type classification. LayoutLMv2 [33] and LayoutLMv3 [13] further integrate the visual information in the pre-training stage, significantly improves the image understanding ability of the model. Graph-based models, such as VRD [23] and PICK [35], model documents as graphs where the graph nodes are text segments and the edges represent spatial relationships between nodes. They utilize graph convolutional networks to extract non-local features that represent the context of the graph nodes. VRD uses a fully connected graph, where every node can attend all other nodes in the graph, which may lead to the aggregation of irrelevant and noisy node information. To overcome this limitation, PICK uses a graph learning module to reduce the noise by learning a soft adjacency matrix from encoded node embeddings. However, most previous methods neglect to introduce the inductive bias of the documents in their models. This bias plays a critical role in accurately extracting information. As shown in Fig. 1, the text segment "28.80" with the category of TOTAL can be properly understood by considering the semantics of its neighboring segments, "Received: 30.00" and "Change: 1.20". In the case of unseen document templates, utilizing this document inductive bias allows the model to understand the relationships between text segments and directly capture the structure of the information contained within the document.

Therefore, in this paper, we propose a novel approach to incorporate the document inductive bias into the existing graph convolution framework for KIE. We initialize an adjacency matrix based on the relative distance between text segments in the document, aiming to better capture the semantic relationships between text segments in the document. This matrix is then combined with another adjacency matrix learned from the node embeddings using a graph learning method to form a hybrid adjacency matrix, which leverages both the inherent structure of the document and the learned representations. Furthermore, inspired by [2], we incorporate an iterative convolution operation into our framework to iteratively optimize the hybrid adjacency matrix and avoid graph topology inconsistencies. By updating the node embeddings, edge embeddings and graph structure jointly, this operation enables our model to make full use of the textual, visual and spatial information of the document, leading to a richer contextual representation.

The main contributions of this paper are as follows:

- We propose an iterative learning convolution network based on the document inductive bias for KIE. To our best knowledge, our model is the first KIE

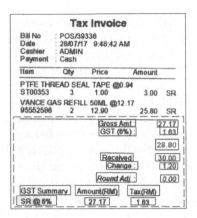

Fig. 1. The text segment "28.80" with the category of TOTAL can be properly understood by considering the semantics of its neighboring segments, "Received: 30.00" and "Change: 1.20".

approach to introduce the document inductive bias into the graph convolution framework.

- We use an iterative convolution operation to refine the graph structure, node embeddings and edge embeddings jointly to make full use of multi-modal features of documents.
- Extensive experiments show that our model achieves results comparable to the state of-the-art method.

2   Related Work

Key Information Extraction. KIE is a subfield of natural language processing that involves extracting key information from documents, such as named entities, dates, and numerical values. Conventional KIE systems [4,27,28] used rule-based approaches that required manual rule design. To overcome this, some KIE systems utilized Named Entity Recognition (NER) [19] framework and transformed documents into one-dimensional sequences, but ignoring the two-dimensional spatial layout and visual information of the documents. Recent research in KIE has focused on incorporating the visual information of rich text documents by modeling the KIE task as a computer vision task. Chargrid [15] utilized a fully convoluted encoder-decoder network and proposed a new type of text representation by encoding each document page as a 2D character grid, thus two-dimensional layout information was naturally preserved. Textual information was introduced by performing one-hot encoding on characters. VisualWord-Grid [16] replaced character-level text information with word-level Word2Vec features. BERTgrid [5] used BERT to obtain contextual text representation. However, the pre-trained parameters of BERT are fixed and do not fully play the role of the language model during model training. ViBERTgrid [22] used

CNN and BERT as backbone networks and the parameters of them could be trained jointly. The experimental results show that the combined training strategy significantly improves the performance of ViBERTgrid.

Documents could also be encoded as graphs, where the text segments in a document were encoded as nodes in the graph, the spatial relation between text segments were encoded as edges. The graph is then processed using graph convolution networks, which propagate information between nodes through the edges, resulting in node embeddings that capture the contextual information of the corresponding text segments. SDMG-R [29] proposed an end-to-end Spatial Dual-Modality Graph Reasoning method, using a learnable edge weight matrix to iteratively propagate messages between nodes. They also released a new dataset named Wildreceipt. GraphDoc [37] integrated the graph structure into a pre-training model, using graph attention to replace the original attention mechanism. This allows each input node to only attend to its neighboring nodes in the graph and learn a generic representation from less data.

Recently, researchers have been showing a growing interest in using large pre-trained models to solve information extraction tasks. LayoutLM was the first to introduce layout information in the pre-training stage. Since then, many methods have employed different strategies to represent layout information and image information in documents. TILT [26] utilized a pre-trained encoder-decoder Transformer to learn layout information, visual features, and textual semantics. LAMBERT [8] modified the Transformer Encoder architecture in a way that allows it to use layout features obtained from an OCR system. Specifically, it added layout embeddings based on the bounding boxes of tokens to the input embeddings of Transformer and added 1D and 2D relative bias when calculating the self-attention logits. DocFormer [1] adopted a novel approach to combine text, vision, and spatial features by adding shared position features to the vision and text features separately and passing them to the Transformer layer independently. BROS [12] re-focused on the combinations of text and their spatial information without relying on visual features, using a relative position encoding method to calculate the attention logits between text blocks, which improves the model's ability to perceive spatial position. The model is trained on unlabeled documents with the use of an area masking strategy. However,all these methods require large datasets (e.g., IIT-CDIP [20], which contains more than 6 million scanned documents with 11 million scanned document images.) and computational resources.

Graph Neural Network. Graph Neural Networks(GNN) have been widely used for graph classification [7,24] and node classification [18,21] because of their powerful node embedding representation ability. Graphs can model various systems, including social networks, biological networks, citation networks, and documents, as a graph is defined by a set of vertices and edges. The vertices represent entities in the graph, such as text segments in a document, and the edges represent the relationships between these entities. [7,9,31] defined convolution operation directly on node groups of neighbors in graph to propagate information between nodes according to the graph structure. However, these approaches typically use a fixed graph structure, which may not be optimal for downstream tasks. To address this limitation, Jiang et al. [14] proposed a novel Graph Learning-Convolutional Network(GLCN) that integrates both graph learning and graph convolution into a unified network architecture to learn an optimal graph structure. In this paper, we combine both graph learning to learn a graph structure and graph convolution to model the context of text segments in the document. We define a knn graph structure based on the inductive bias of the document and combine it with the learned graph structure to form a hybrid graph structure, which is iteratively updated to reach its optimal form for the downstream tasks.

3   Methodology

As illustrated in Fig. 2, our approach is composed of three modules: 1)a multi-modal feature extraction module to extract both textual features and visual features from documents; 2)a graph module for feature propagation; 3)an information extraction module for entity extraction.

3.1   Notation

Given a document image I with text segment s_i that have been detected by an off-the-shelf OCR engine. Text segment s_i contains a recognized text sentence t_i and a bounding box b_i. The recognized text sentence is represented as $t_i = \{c_1, \cdots, c_L\}$, where c_i is a character in the text sentence and L is the length of the text sentence. We label each character as $\mathbf{y}_i = \{y_1, \cdots, y_L\}$ using the IOB scheme (similar to [35]). The bounding box b_i is represented as $b_i = (p^{tl}, p^{tr}, p^{bl}, p^{br})$, which gives the coordinates of the four vertices of the bounding box.

We model documents as graphs $\mathcal{G} = (\mathcal{N}, \mathcal{E})$, node $n_i \in \mathcal{N}$, edge $e_{ij} \in \mathcal{E}$, The adjacency matrix of the graph is denoted as A.

3.2   Multi-modal Feature Extraction Module

This module extracts multi-modal features from documents images, we consider both the textual information and visual information of documents inspired by the existing work [35]. Specifically, to extract multi-modal features of each text segment in the document, we first encode the characters of the text sentence in the text segment using Word2Vec method to obtain a textual embedding. Then we use CNN to encode the visual portion of each text segment and obtain a visual embedding. Finally, we use a Transformer Encoder [30] to obtain a fusion embedding that serves as the node embedding for the graph module.

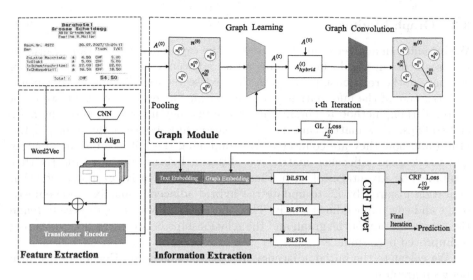

Fig. 2. Overview of our approach

Given a text sentence $t_i = \{c_1, \cdots, c_L\}$ in the text segment, we encode each character c_i using the Word2Vec method and define the textual embedding of the text segment s_i as follows:

$$T_i = [\mathbf{W_v}c_1; \ldots; \mathbf{W_v}c_L], \tag{1}$$

where $T_i \in \mathbb{R}^{L \times D}$, D is the dimension of the model, $\mathbf{W_v}$ is a learnable embedding matrix.

For the visual embedding, we use ResNet50 [11] as our CNN backbone. First, we input the document image I into the ResNet50 to obtain the feature map, and then use the RoI Align [10] with the bounding box b_i on the feature map to obtain the visual embedding of the text segment s_i.

$$V_i = \mathrm{RoI}_{b_i}(\mathrm{ResNet}(I)) \tag{2}$$

where $V_i \in \mathbb{R}^D$. We expand V_i as $\mathbb{R}^{L \times D}$ and combine it with T_i through element-wise addition operation, which serves as the input to the Transformer Encoder. This allows the visual and textual features to complement each other, resulting in non-local fusion features for the text segment s_i.

$$F_i = \mathrm{Transformer}(V_i + T_i) \tag{3}$$

where $F_i \in \mathbb{R}^{L \times D}$ is the non-local fusion feature for the text segment s_i. By performing average pooling operation on F_i, we obtain $X_i \in \mathbb{R}^D$ as the input for the Graph Module.

3.3 Graph Module

We initialize an adjacency matrix based on the topk Euclidean distance between text segments in the document. This initial adjacency matrix provides a rough estimate of the relationships between the text segments, allowing us to capture the structure of the information contained within the document. Next, we use a graph learning method to obtain a learned adjacency matrix based on the node embeddings. This matrix captures the relationships between the nodes in the graph and how they are connected. By combining the initial adjacency matrix and the learned adjacency matrix, we obtain a hybrid adjacency matrix that combines the advantages of both. Finally we use an iterative graph convolution operation inspired by [2] to update the hybrid adjacency matrix, node embeddings and edge embeddings jointly. This operation allows us to iteratively refine the representation of the graph and the relationships between the nodes, leading to improved performance.

We calculate the initial adjacency matrix $\mathbf{A}^{(0)}$ based on Euclidean distance between graph nodes.

$$\mathbf{A}_{ij}^{(0)} = \begin{cases} 1, j \in \mathcal{N}(i) \\ 0, j \notin \mathcal{N}(i) \end{cases} \tag{4}$$

where $\mathcal{N}(i)$ is the neighbors of node i. We select the topk nodes of node i as its neighbors based on the Euclidean distance.

Given node embeddings $(n_1^{(t)}, \ldots, n_N^{(t)})$ at t-th iteration, $n_i^{(0)} = X_i \in \mathbb{R}^D$, N is the number of nodes in the graph. We learn the adjacency matrix $\mathbf{A}^{(t+1)}$ from current node embeddings inspired by [31] as follows:

$$\begin{cases} \mathbf{A}_{ij}^{(t+1)} = softmax_j(a_{ij}) \\ a_{ij}^p = LeakRelu(M_1(n_i^{(t)}) + M_2(n_j^{(t)})), a_{ij} = \frac{1}{C} \sum_{p=1}^{C} a_{ij}^p \end{cases} \tag{5}$$

where M_1, M_2 are two MLP layers that map node embeddings to a vector of dimension 1. To improve the stability of the learning process, we extend it to a multi-head version(similarly to [2,30,31]). Finally, we average the results from C independent heads to obtain the final adjacency matrix.

We use the loss based on [2] to update the parameters of M_1 and M_2:

$$\mathcal{L}_G^{(t+1)} = \frac{\lambda}{N^2} \sum_{i,j} \mathbf{A}_{ij}^{(t+1)} \left\| \mathbf{n}_i^{(0)} - \mathbf{n}_j^{(0)} \right\|_2^2 + \frac{\mu}{N^2} \| \mathbf{A}^{(t+1)} \|_F^2 \tag{6}$$

where $\| \cdot \|_F$ denotes Frobenius-Norm. The hyperparameters λ and μ control the smoothness and sparsity of the learned adjacency matrix respectively.

We linearly combine two adjacency matrices $\mathbf{A}^{(0)}, \mathbf{A}^{(t+1)}$. $\mathbf{A}^{(0)}$ reflects the graph topology in real-world documents and directly expresses the relationships between text segments in the documents. On the other hand, $\mathbf{A}^{(t+1)}$ reflects the graph topology after feature transformations, which is crucial for learning node embeddings for downstream tasks.

$$\mathbf{A}_{hybrid}^{(t+1)} = \alpha \mathbf{D}^{(0)-1/2} \mathbf{A}^{(0)} \mathbf{D}^{(0)-1/2} + (1 - \alpha) \mathbf{A}^{(t+1)} \tag{7}$$

where $\mathbf{D}^{(0)} = diag(d_1, \ldots, d_N)$ is a diagonal matrix, $d_i = \sum_{j=1}^{j=N} \mathbf{A}_{ij}^{(0)}$, α is a hyperparameter to control the weight of the linear combination of $\mathbf{A}^{(0)}$ and $\mathbf{A}^{(t+1)}$.

We explore a relative 2D position encoding method [12] to encode edges e_{ij} to represent the spatial relation between node i and node j. Given a bounding box $b_i = (p^{tl}, p^{tr}, p^{bl}, p^{br})$, $p = (x, y)$, the relative 2D position encoding between node i and node j is calculated as $\overline{p}_{i,j} = \left[\mathbf{f}^{\text{sinu}} (x_i - x_j) ; \mathbf{f}^{\text{sinu}} (y_i - y_j) \right]$. $\mathbf{f}^{\text{sinu}}$ indicates a sinusoidal function [30]. Finally, edges $e_{ij}^{(0)}$ is calculated by combining the relative 2D position encoding of the four vertices of the bounding box:

$$e_{ij}^{(0)} = \boldsymbol{W}^{\text{tl}} \overline{\boldsymbol{p}}_{i,j}^{\text{tl}} + \boldsymbol{W}^{\text{tr}} \boldsymbol{p}_{i,j}^{\text{tr}} + \boldsymbol{W}^{\text{br}} \overline{\boldsymbol{p}}_{i,j}^{\text{br}} + \boldsymbol{W}^{\text{bl}} \overline{\boldsymbol{p}}_{i,j}^{\text{bl}} \tag{8}$$

where $e_{ij}^{(0)} \in \mathbb{R}^D$ is the edge embedding between node i and node j at iteration 0, $\boldsymbol{W}^{\text{tl}}, \boldsymbol{W}^{\text{tr}}, \boldsymbol{W}^{\text{br}}, \boldsymbol{W}^{\text{bl}}$ are four learnable projection matrices. In this way, we can explicitly encode the relative spatial relationship between two nodes.

Inspired by [23], we use node-edge-node triplets (n_i, e_{ij}, n_j) to perform graph convolution.

$$h_{ij}^{(t)} = \text{ReLU}(\boldsymbol{W}_i n_i^{(t)} + \boldsymbol{W}_j n_j^{(t)} + e_{ij}^{(t)}) \tag{9}$$

$$n_i^{(t+1)} = \sigma \left(\mathbf{A}_{hybrid}^{(t+1)} \mathbf{h}_i^{(t)} \boldsymbol{W}_{\text{GCN}} \right) \tag{10}$$

where W_i, W_j, W_{GCN} are layer-specific matrices that can be learned in the graph convolution layers, σ is an activation function, such as ReLU. The information of nodes at both ends of the edge i and j is aggregated in $h_{ij}^{(t)}$ based on the spatial information represented in the edge embedding e_{ij}, which provides stronger representation ability compared to a single node when performing graph convolution. Node i aggregates the information of all its neighbor nodes through the hybrid adjacency matrix obtained by Eq. (7), resulting in $n_i^{(t+1)}$ containing both global layout information and context information.

To enhance the representational power of edge embeddings, we update node embeddings using edge information and iteratively refine edge embeddings based on the updated node information. The edge embedding $e_{ij}^{(t+1)}$ is formulated as:

$$e_{ij}^{(t+1)} = MLP(h_{ij}^{(t)}) \tag{11}$$

We repeat the calculation of Eqs. (5)-(11) starting from t = 0 until $\left\| \mathbf{A}^{(t+1)} - \mathbf{A}^{(t)} \right\|_F^2 < \epsilon \left\| \mathbf{A}^{(1)} \right\|_F^2$, ϵ is a hyperparameter, and we set a maximum number of iterations to prevent an infinite loop.

3.4   Information Extraction Module

We adopt a standard BiLSTM+CRF structure as our information extraction module. For a text segment s_i, its non-local fusion features $F_i \in \mathbb{R}^{L \times D}$ and its corresponding graph node embedding $n_i^{(t)} \in \mathbb{R}^D$ at the t-th iteration in the graph module are inputted into the information extraction module.

At each iteration in the graph module, we concatenate each character embedding $f_* \in \mathbb{R}^D$ in F_i to the node embedding $n_i^{(t)}$ to form a union feature $u_i^{(t)} \in \mathbb{R}^{L \times 2D}$ that contains both multimodal features and contextual features. The parameters of the module are then optimized using the CRF loss, which at the t-th iteration is expressed as:

$$\mathcal{L}_{\mathrm{CRF}}^{(t)} = \mathrm{BiLSTM\text{-}CRF}(u_i^{(t)}; \mathbf{y_i}) \tag{12}$$

where $u_i^{(t)}$ is the union feature of the text segment s_i, and $\mathbf{y_i}$ is the ground truth label of the text segment. The final prediction of each text segment is made using the node embedding $n_i^{(T)}$, which is generated in the final iteration of the graph module.

3.5   Loss

The whole networks are trained by minimizing the following joint loss function as:

$$\mathcal{L} = \sum_{}^{T} (\mathcal{L}_G^{(t)} + \mathcal{L}_{\mathrm{CRF}}^{(t)})/T \tag{13}$$

where T is the number of the iterations.

4   Experiments

4.1   Dataset

CORD [25]. CORD is a public dataset contains 800 receipts for training, 100 receipts for validation and 100 receipts for testing. Each receipt is labelled with 30 key fields and provides several text lines with its bounding box.

WildReceipt. WildReceipt is a public dataset realised by [29], it contains 1268 receipts for training and 472 receipts for testing. These two sets have different templates, meaning that the templates in the testing set are unseen in the training set, making it a more challenging task. Each receipt is labelled with 25 categories. 12 categories are keys and 12 categories are their corresponding values, and 1 category indicates others. We evaluate WildReceipt by reporting F_1 score over 12 value categories as [29] does.

4.2   Implementation Details

Our proposed model is implemented by PyTorch and trained on 1 NVIDIA RTX 3090 GPUs with 24 GB memory. The whole network is trained from scratch using Adam optimizer [17]. We use a batch size of 2 during training. Maximum epoch is set to 40, the learning rate is set to 10^{-4} and it is decreased via 10× after 30 epochs.

Table 1. F_1 score comparison with published models on CORD and WildReceipt datasets. * refers to the results are reported by our re-implemented model

Dataset	Model	Parameters	F1
CORD	BERT [6]	110M	89.68
	BROS [12]	110M	95.73
	LiLT [32]	-	96.07
	TILT [26]	230M	95.11
	LayoutLM [34]	113M	94.72
	LayoutLMv2 [33]	200M	94.95
	LayoutLMv3 [13]	133M	96.56
	GraphDoc [37]	265M	**96.93**
	PICK [35]	-	94.18*
	Ours	66M	**95.41**
WildReceipt	LSTM-CRF [19]	-	83.20
	Chargrid [15]	-	75.39
	SDMG-R [29]	-	88.70
	TRIE [36]	-	85.99
	TRIE++ [3]	-	**90.15**
	PICK [35]	-	84.31*
	Ours	66M	**89.02**

In the multi-modal feature extraction module, the dimension of model D is set to 512. In the graph module, the value of λ and μ is set to 0.5, 0.8 respectively which control the smoothness and sparsity of the learned adjacency matrix, the value of α is set to 0.2 which control the weight of the adjacency matrix combination. We set the maximum iteration to 5 and ϵ to 0.04. The number of layer of graph convolution is set to 2. In the information extraction module, the hidden size of BiLSTM layer is set to 512.

4.3 Experimental Results

We evaluate the performance of our model based on entity-level F1 score on two public dataset: CORD and WildReceipt. The main results are shown in Table 1. In the CORD dataset, we compare our model with existing pre-trained models and achieve comparable performance, with only 1.52% lower than the state-of-the-art (SOTA) model. It's worth mentioning that the SOTA model has 4 times more parameters and was pre-trained on more data, whereas we only used the official data. Our model also outperforms a graph-based model: PICK [35] and achieves 1.23% improvement, shows the effectiveness of our proposed method in the graph framework. In the WildReceipt dataset, which is a more challenging scenario as the templates of the test set documents are unseen in the training set, we compare our model with existing lightweight SOTA architectures. Our model outperforms most models and achieves a competitive score. This result highlights the powerful ability of our model to handle unseen templates and generalize well to new, unseen scenarios.

4.4   Ablation Studies

We conduct ablation studies to verify the effectiveness of each component in our model.

The Effectiveness of Relative Spatial Features. To investigate the effect of relative spatial features, we remove the edge embeddings that represent the spatial relation between nodes when conducting graph convolution operation. The results in Table 2 show that removing the relative spatial features results in a decrease in performance compared to the full model. This suggests that the relative spatial features play an important role in capturing the spatial relation between the text segments in the documents, which helps the model to better extract key information.

Table 2. Performance comparisons on each component of our model in terms of F1 score

Model	CORD	WildReceipt
FULL Model	**95.41**	**89.02**
w/o relative spatial features	94.48	88.70
w/o initial adjacency matrix	94.69	88.08
w/o iterative convolution	94.19	88.56

The Effectiveness of the Initial Adjacency Matrix. As shown in Table 2, when the hybrid adjacency matrix is not used and only the adjacency matrix learned from graph node embeddings is utilized, the performance drops. This means that by combining the learned graph structure with the prior knowledge of the document's graph topology, the model can learn a better representation of the document. The initial adjacency matrix acts as a regularization term, guiding the model towards a more reasonable representation, which results in improved performance. This experiment highlights the importance of incorporating prior knowledge into the model to improve its performance.

Table 3. Performance comparisons of different topk on the CORD dataset

Topk	5	15	25	35	45
F1	94.43	**95.41**	95.07	94.19	94.14

The Effectiveness of Iterative Convolution. We set the number of max iteration as 1 to investigate the effect of this component. The results in Table 2 show that iterative convolution is crucial in enhancing the performance of our model. By updating the hybrid adjacency matrix in each iteration based on the updated node embeddings, the iterative convolution can resolve the graph topology inconsistencies and lead to a better representation of the nodes. The improvement in performance on both datasets highlights the effectiveness of the iterative convolution component in our model.

The Values of Topk. To investigate the impact of different values of topk when initializing the adjacency matrix, as shown in Table 3. When the topk is too small, the initial adjacency matrix can not capture the structure of the information contained within the document. As the number of topk increases, every node in the graph will aggregate useless information when conducting graph convolution, resulting in model performance degradation. In practice, we should set a task-specific number of topk depending on the layout of the documents.

5 Conclusion

In this paper, we propose a novel graph convolution framework to extract the information from document images. Through the adjacency matrix hybrid strategy, our proposed method can introduce the document inductive bias into the graph framework. And we also conduct an iterative graph convolution operation to refine graph structure and graph embeddings jointly. Our method achieves comparable results to previous methods, showing the superiority of incorporating document inductive bias into the graph convolution framework. The proposed approach provides a novel perspective on how to extract key information from document images and offers a promising direction for future research.

Acknowledgement. This work was supported by National Key R&D Program of China (No. 2021ZD0113301)

References

1. Appalaraju, S., Jasani, B., Kota, B.U., Xie, Y., Manmatha, R.: DocFormer: end-to-end transformer for document understanding. In: Proceedings of the IEEE/CVF International Conference on Computer Vision, pp. 993–1003 (2021)
2. Chen, Y., Wu, L., Zaki, M.: Iterative deep graph learning for graph neural networks: better and robust node embeddings. Adv. Neural. Inf. Process. Syst. **33**, 19314–19326 (2020)
3. Cheng, Z., et al.: TRIE++: towards end-to-end information extraction from visually rich documents. arXiv preprint arXiv:2207.06744 (2022)
4. Dengel, A.R., Klein, B.: *smartFIX*: a requirements-driven system for document analysis and understanding. In: Lopresti, D., Hu, J., Kashi, R. (eds.) DAS 2002. LNCS, vol. 2423, pp. 433–444. Springer, Heidelberg (2002). https://doi.org/10.1007/3-540-45869-7_47

5. Denk, T.I., Reisswig, C.: BERTgrid: contextualized embedding for 2D document representation and understanding. arXiv preprint arXiv:1909.04948 (2019)
6. Devlin, J., Chang, M.W., Lee, K., Toutanova, K.: BERT: pre-training of deep bidirectional transformers for language understanding. arXiv preprint arXiv:1810.04805 (2018)
7. Duvenaud, D.K., et al.: Convolutional networks on graphs for learning molecular fingerprints. Advances in Neural Information Processing Systems 28 (2015)
8. Garncarek, Ł, et al.: LAMBERT: layout-aware language modeling for information extraction. In: Lladós, J., Lopresti, D., Uchida, S. (eds.) ICDAR 2021. LNCS, vol. 12821, pp. 532–547. Springer, Cham (2021). https://doi.org/10.1007/978-3-030-86549-8_34
9. Hamilton, W., Ying, Z., Leskovec, J.: Inductive representation learning on large graphs. Advances in Neural Information Processing Systems 30 (2017)
10. He, K., Gkioxari, G., Dollár, P., Girshick, R.: Mask R-CNN. In: Proceedings of the IEEE International Conference on Computer Vision, pp. 2961–2969 (2017)
11. He, K., Zhang, X., Ren, S., Sun, J.: Deep residual learning for image recognition. In: Proceedings of the IEEE Conference on Computer Vision and Pattern Recognition, pp. 770–778 (2016)
12. Hong, T., Kim, D., Ji, M., Hwang, W., Nam, D., Park, S.: BROS: a pre-trained language model focusing on text and layout for better key information extraction from documents. In: Proceedings of the AAAI Conference on Artificial Intelligence, vol. 36, pp. 10767–10775 (2022)
13. Huang, Y., Lv, T., Cui, L., Lu, Y., Wei, F.: LayoutLMv3: pre-training for document AI with unified text and image masking. arXiv preprint arXiv:2204.08387 (2022)
14. Jiang, B., Zhang, Z., Lin, D., Tang, J., Luo, B.: Semi-supervised learning with graph learning-convolutional networks. In: Proceedings of the IEEE/CVF Conference on Computer Vision and Pattern Recognition, pp. 11313–11320 (2019)
15. Katti, A.R., et al.: Chargrid: towards understanding 2D documents. arXiv preprint arXiv:1809.08799 (2018)
16. Kerroumi, M., Sayem, O., Shabou, A.: VisualWordGrid: information extraction from scanned documents using a multimodal approach. In: Barney Smith, E.H., Pal, U. (eds.) ICDAR 2021. LNCS, vol. 12917, pp. 389–402. Springer, Cham (2021). https://doi.org/10.1007/978-3-030-86159-9_28
17. Kingma, D.P., Ba, J.: Adam: a method for stochastic optimization. arXiv preprint arXiv:1412.6980 (2014)
18. Kipf, T.N., Welling, M.: Semi-supervised classification with graph convolutional networks. arXiv preprint arXiv:1609.02907 (2016)
19. Lample, G., Ballesteros, M., Subramanian, S., Kawakami, K., Dyer, C.: Neural architectures for named entity recognition. arXiv preprint arXiv:1603.01360 (2016)
20. Lewis, D., Agam, G., Argamon, S., Frieder, O., Grossman, D., Heard, J.: Building a test collection for complex document information processing. In: Proceedings of the 29th Annual International ACM SIGIR Conference on Research and Development in Information Retrieval, pp. 665–666 (2006)
21. Li, Y., Tarlow, D., Brockschmidt, M., Zemel, R.: Gated graph sequence neural networks. arXiv preprint arXiv:1511.05493 (2015)
22. Lin, W., et al.: ViBERTgrid: a jointly trained multi-modal 2D document representation for key information extraction from documents. In: Llados, J., Lopresti, D., Uchida, S. (eds.) Document Analysis and Recognition – ICDAR 2021. ICDAR 2021. Lecture Notes in Computer Science, vol. 12821. Springer, Cham (2021). https://doi.org/10.1007/978-3-030-86549-8_35

23. Liu, X., Gao, F., Zhang, Q., Zhao, H.: Graph convolution for multimodal information extraction from visually rich documents. arXiv preprint arXiv:1903.11279 (2019)
24. Ma, Y., Wang, S., Aggarwal, C.C., Tang, J.: Graph convolutional networks with eigenpooling. In: Proceedings of the 25th ACM SIGKDD International Conference on Knowledge Discovery & Data Mining, pp. 723–731 (2019)
25. Park, S., et al.: CORD: a consolidated receipt dataset for post-OCR parsing. In: Workshop on Document Intelligence at NeurIPS 2019 (2019)
26. Powalski, R., Borchmann, Ł., Jurkiewicz, D., Dwojak, T., Pietruszka, M., Pałka, G.: Going full-TILT Boogie on document understanding with text-image-layout transformer. In: Lladós, J., Lopresti, D., Uchida, S. (eds.) ICDAR 2021. LNCS, vol. 12822, pp. 732–747. Springer, Cham (2021). https://doi.org/10.1007/978-3-030-86331-9_47
27. Riloff, E., et al.: Automatically constructing a dictionary for information extraction tasks. In: AAAI, vol. 1, pp. 2–1. CiteSeer (1993)
28. Schuster, D., Muthmann, K., et al.: Intellix-end-user trained information extraction for document archiving. In: 2013 12th International Conference on Document Analysis and Recognition, pp. 101–105. IEEE (2013)
29. Sun, H., Kuang, Z., Yue, X., Lin, C., Zhang, W.: Spatial dual-modality graph reasoning for key information extraction. arXiv preprint arXiv:2103.14470 (2021)
30. Vaswani, A., et al.: Attention is all you need. Advances in Neural Information Processing Systems 30 (2017)
31. Veličković, P., Cucurull, G., Casanova, A., Romero, A., Lio, P., Bengio, Y.: Graph attention networks. arXiv preprint arXiv:1710.10903 (2017)
32. Wang, J., Jin, L., Ding, K.: LiLT: a simple yet effective language-independent layout transformer for structured document understanding. arXiv preprint arXiv:2202.13669 (2022)
33. Xu, Y., et al.: LayoutLMv2: multi-modal pre-training for visually-rich document understanding. arXiv preprint arXiv:2012.14740 (2020)
34. Xu, Y., Li, M., Cui, L., Huang, S., Wei, F., Zhou, M.: LayoutLM: pre-training of text and layout for document image understanding. In: Proceedings of the 26th ACM SIGKDD International Conference on Knowledge Discovery & Data Mining, pp. 1192–1200 (2020)
35. Yu, W., Lu, N., Qi, X., Gong, P., Xiao, R.: Pick: processing key information extraction from documents using improved graph learning-convolutional networks. In: 2020 25th International Conference on Pattern Recognition (ICPR), pp. 4363–4370. IEEE (2021)
36. Zhang, P., et al.: TRIE: end-to-end text reading and information extraction for document understanding. In: Proceedings of the 28th ACM International Conference on Multimedia, pp. 1413–1422 (2020)
37. Zhang, Z., Ma, J., Du, J., Wang, L., Zhang, J.: Multimodal pre-training based on graph attention network for document understanding. arXiv preprint arXiv:2203.13530 (2022)

QuOTeS: Query-Oriented Technical Summarization

Juan Ramirez-Orta[1]([✉]), Eduardo Xamena[2,3], Ana Maguitman[5,6],
Axel J. Soto[5,6], Flavia P. Zanoto[4], and Evangelos Milios[1]

[1] Department of Computer Science, Dalhousie University, Halifax, Canada
juan.ramirez.orta@dal.ca
[2] Institute of Research in Social Sciences and Humanities (ICSOH), Salta, Argentina
[3] Universidad Nacional de Salta - CONICET, Salta, Argentina
[4] Escrever Ciência, São Paulo, Brazil
[5] Institute for Computer Science and Engineering, UNS - CONICET, Bahía Blanca, Argentina
[6] Department of Computer Science and Engineering, Universidad Nacional del Sur, Bahía Blanca, Argentina

Abstract. When writing an academic paper, researchers often spend considerable time reviewing and summarizing papers to extract relevant citations and data to compose the *Introduction* and *Related Work* sections. To address this problem, we propose *QuOTeS*, an interactive system designed to retrieve sentences related to a summary of the research from a collection of potential references and hence assist in the composition of new papers. *QuOTeS* integrates techniques from Query-Focused Extractive Summarization and High-Recall Information Retrieval to provide Interactive Query-Focused Summarization of scientific documents. To measure the performance of our system, we carried out a comprehensive user study where participants uploaded papers related to their research and evaluated the system in terms of its usability and the quality of the summaries it produces. The results show that *QuOTeS* provides a positive user experience and consistently provides query-focused summaries that are relevant, concise, and complete. We share the code of our system and the novel Query-Focused Summarization dataset collected during our experiments at https://github.com/jarobyte91/quotes.

1 Introduction

When writing an academic paper, researchers often spend substantial time reviewing and summarizing papers to shape the *Introduction* and *Related Work* sections of their upcoming research. Given the ever-increasing number of academic publications available every year, this task has become very difficult and time-consuming, even for experienced researchers. A solution to this problem is to use Automatic Summarization systems, which take a long document or a collection of documents as input and produce a shorter text that conveys the same information.

© The Author(s), under exclusive license to Springer Nature Switzerland AG 2023
G. A. Fink et al. (Eds.): ICDAR 2023, LNCS 14189, pp. 98–114, 2023.
https://doi.org/10.1007/978-3-031-41682-8_7

The summaries produced by such systems are evaluated by measuring their fluency, coherence, conciseness, and completeness. To this end, Automatic Summarization systems can be divided into two categories, depending on their output. In Extractive Summarization, the purpose of the system is to highlight or extract passages present in the original text, so the summaries are usually more coherent and complete. On the other hand, in Abstractive Summarization, the system generates the summary by introducing words that are not necessarily in the original text. Hence, the summaries are usually more fluent and concise. Although there have been significant advances recently [1], these complementary approaches share the same weakness: it is very hard for users to evaluate the quality of an automatic summary because it means that they have to go back to the original documents and verify that the system extracted the correct information.

Since evaluating summarization systems by hand is very difficult, several automatic metrics have been created with this purpose: BLEU [2], ROUGE [3], and METEOR [4] all aim to measure the quality of the summary produced by the system by comparing it with a reference summary via the distribution of its word n-grams. Despite being very convenient and popular, all these automatic metrics have a significant drawback: since they only look at the differences in the distribution of words between the system's summary and the reference summary, they are not useful when the two summaries are worded differently, which is not necessarily a sign that the system is performing poorly.

Therefore, although Automatic Summarization systems display high performance when evaluated on benchmark datasets [5], they often cannot satisfy their users' needs, given the inherent difficulty and ambiguity of the task [6]. An alternative approach to make systems more user-centric is Query-Focused Summarization [6], in which the users submit a query into the system to guide the summarization process and tailor it to their needs. Another alternative approach to this end is Interactive Summarization [7], in which the system produces an iteratively improved summary. Both of these approaches, and several others, take into account that the *correct* summary given a document collection depends on both the users and what they are looking for.

In this paper, we introduce *QuOTeS*, an interactive system designed to retrieve sentences relevant to a paragraph from a collection of academic articles to assist in the composition of new papers. *QuOTeS* integrates techniques from Query-Focused Extractive Summarization [6] and High-Recall Information Retrieval [8] to provide Interactive Query-Focused Summarization of scientific documents. An overview of how *QuOTeS* works and its components is shown in Fig. 1.

The main difficulty when creating a system like *QuOTeS* in a supervised manner is the lack of training data: gathering enough training examples would require having expert scientists carefully read several academic papers and manually label each one of their sentences concerning their relevance to the query, which would take substantial human effort. Therefore, we propose *QuOTeS* as a self-service tool: the users supply their academic papers (usually as PDFs), and

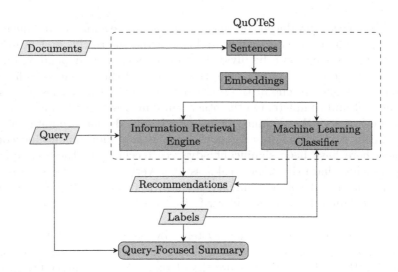

Fig. 1. Overview of how *QuOTeS* works. First, the user inputs their documents into the system, which then extracts the text present in them. Next, the system splits the text into sentences and computes an embedding for each one of them. After that, the user inputs their query, which is a short paragraph describing their research, and the system retrieves the most relevant sentences using the traditional *Vector Space Model*. The user then labels the recommendations and trains the system using techniques from High-Recall Information Retrieval to retrieve more relevant sentences until he or she is satisfied. Finally, the sentences labeled as relevant are returned to the user as the Query-Focused Summary of the collection.

QuOTeS provides an end-to-end service to aid them in the retrieval process. This paper includes the following contributions:

– A novel Interactive Query-Focused Summarization system that receives a short paragraph (called query) and a collection of academic documents as input and returns the sentences related to the query from the documents in the collection. The system extracts the text directly from the academic documents provided by the user at runtime, minimizing the effort needed to perform complex queries on the text present in the documents. Finally, the system features techniques from High-Recall Information Retrieval to maximize the number of relevant sentences retrieved.

– A novel dataset composed of *(Query, Document Collection)* pairs for the task of Query-Focused Summarization of Scientific Documents, each one with five documents and hundreds of sentences, along with the relevance labels produced by real users.

– A comprehensive analysis of the data collected during a user study of the system, where the system was evaluated using the System Usability Scale [9] and custom questionnaires to measure its usability and the quality of the summaries it produces.

2   Related Work

2.1   Query-Focused Summarization

The task of Query-Focused Summarization (QFS) was introduced in the 2005 Document Understanding Conference (DUC 2005) [6]. The focus of the conference was to develop new evaluation methods that take into account the variation of summaries produced by humans. Therefore, DUC 2005 had a single, user-oriented, question-focused summarization task that allowed the community to put some time and effort into helping with the new evaluation framework. The summarization task was to synthesize a well-organized and fluent answer to a complex question from a set of 25 to 50 documents. The relatively generous allowance of 250 words for each answer revealed how difficult it was for the systems to produce good multi-document summaries. The two subsequent editions of the conference (DUC 2006 [10] and DUC 2007 [11]) further enhanced the dataset produced in the first conference and have become the reference benchmark in the field.

Surprisingly, state-of-the-art algorithms designed for QFS do not significantly improve upon generic summarization methods when evaluated on traditional QFS datasets, as was shown in [12]. The authors hypothesized that this lack of success stems from the nature of the datasets, so they defined a novel method to quantify their Topic Concentration. Using their method, which is based on the ratio of sentences within the dataset that are already related to the query, they observed that the DUC datasets suffer from very high Topic Concentration. Therefore, they introduced TD-QFS, a new QFS dataset with controlled levels of Topic Concentration, and compared competitive baseline algorithms on it, reporting a solid improvement in performance for algorithms that model query relevance instead of generic summarizers. Finally, they presented three novel QFS algorithms (RelSum, ThresholdSum, and TFIDF-KLSum) that outperform, by a large margin, state-of-the-art QFS algorithms on the TD-QFS dataset.

A novel, unsupervised query-focused summarization method based on random walks over the graph of sentences in a document was introduced in [13]. First, word importance scores for each target document are computed using a word-level random walk. Next, they use a siamese neural network to optimize localized sentence representations obtained as the weighted average of word embeddings, where the word importance scores determine the weights. Finally, they conducted a sentence-level query-biased random walk to select a sentence to be used as a summary. In their experiments, they constructed a small evaluation dataset for QFS of scientific documents and showed that their method achieves competitive performance compared to other embeddings.

2.2   High-Recall Information Retrieval

A novel evaluation toolkit that simulates a human reviewer in the loop was introduced in [8]. The work compared the effectiveness of three Machine Learning protocols for Technology-Assisted Review (TAR) used in document review for legal proceedings. It also addressed a central question in the deployment of TAR: should the initial training documents be selected randomly, or should they be selected using one or more deterministic methods, such as Keyword Search? To answer this question, they measured Recall as a function of human review effort on eight tasks. Their results showed that the best strategy to minimize the human effort is to use keywords to select the initial documents in conjunction with deterministic methods to train the classifier.

Continuous Active Learning achieves high Recall for TAR, not only for an overall information need but also for various facets of that information, whether explicit or implicit, as shown in [14]. Through simulations using Cormack and Grossman's Technology-Assisted Review Evaluation Toolkit [8], the authors showed that Continuous Active Learning, applied to a multi-faceted topic, efficiently achieves high Recall for each facet of the topic. Their results also showed that Continuous Active Learning may achieve high overall Recall without sacrificing identifiable categories of relevant information.

A scalable version of the Continuous Active Learning protocol (S-CAL) was introduced in [15]. This novel variation requires $O(log(N))$ labeling effort and $O(Nlog(N))$ computational effort - where N is the number of unlabeled training examples - to construct a classifier whose effectiveness for a given labeling cost compares favorably with previously reported methods. At the same time, S-CAL offers calibrated estimates of Class Prevalence, Recall, and Precision, facilitating both threshold setting and determination of the adequacy of the classifier.

2.3   Interactive Query-Focused Summarization

A novel system that provides summaries for Computer Science publications was introduced in [16]. Through a qualitative user study, the authors identified the most valuable scenarios for discovering, exploring, and understanding scientific documents. Based on these findings, they built a system that retrieves and summarizes scientific documents for a given information need, either in the form of a free-text query or by choosing categorized values such as scientific tasks, datasets, and more. The system processed 270,000 papers to train its summarization module, which aims to generate concise yet detailed summaries. Finally, they validated their approach with human experts.

A novel framework to incorporate users' feedback using a social robotics platform was introduced in [17]. Using the *Nao* robot (a programmable humanoid robot) as the interacting agent, they captured the user's expressions and eye movements and used it to train their system via Reinforcement Learning. The whole approach was then evaluated in terms of its adaptability and interactivity.

A novel approach that exploits the user's opinion in two stages was introduced in [18]. First, the query is refined by user-selected keywords, key phrases, and sentences extracted from the document collection. Then, it expands the query using a Genetic Algorithm, which ranks the final set of sentences using Maximal Marginal Relevance. To assess the performance of the proposed system, 45 graduate students in the field of Artificial Intelligence filled out a questionnaire after using the system on papers retrieved from the Artificial Intelligence category of The Web of Science. Finally, the quality of the final summaries was measured in terms of the user's perspective and redundancy, obtaining favorable results.

3   Design Goals

As shown in the previous section, there is a clear research gap in the literature: on the one hand, there exist effective systems for QFS, but on the other hand, none of them includes the user's feedback about the relevance of each sentence present in the summary. On top of that, the task of QFS of scientific documents remains a fairly unexplored discipline, given the difficulty of extracting the text present in academic documents and the human effort required to evaluate such systems, as shown by [13]. Considering these limitations and the guidelines obtained from an expert consultant in scientific writing from our team, we state the following design goals behind the development of *QuOTeS*:

1. **Receive a paragraph query and a collection of academic documents as input and return the sentences relevant to the query from the documents in the collection.** Unlike previous works, *QuOTeS* is designed as an assistant in the task of writing *Introduction* and *Related Work* sections of papers in the making. To this end, the query inputted into the system is a short paragraph describing the upcoming work, which is a much more complex query than the one used in previous systems.
2. **Include the user in the retrieval loop.** As shown by previous works, summarization systems benefit from being interactive. Since it is difficult to express all the information need in a single query, the system needs to have some form of adaptation to the user, either by requiring more information about the user's need (by some form of query expansion) or by incorporating the relevance labeling in the retrieval process.
3. **Provide a full end-to-end user experience in the sentence extraction process.** So far, query-focused summarization systems have been mainly evaluated on data from the DUC conferences. A usable system should be able to extract the text from various documents provided by the user, which can only be determined at runtime. Since the main form to distribute academic documents is PDF files, the system needs to be well adapted to extract the text in the different layouts in academic publications.
4. **Maximize Recall in the retrieval process.** Since the purpose of the system is to help the user retrieve the (possibly very) few relevant sentences from the hundreds of sentences in the collection, Recall is the most critical metric when using a system like *QuOTeS*, as users can always refine the output summary to adapt it to their needs. Therefore, we use Continuous Active Learning [8] as the training procedure for the classifier inside *QuOTeS*.

4   System Design

QuOTeS is a browser-based interactive system built with *Python*, mainly using the *Dash* package [19]. The methodology of the system is organized into seven steps that allow the users to upload, search and explore their documents. An overview of how the steps relate to each other is shown in Fig. 2.

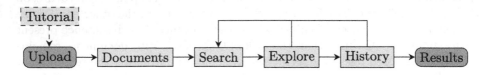

Fig. 2. Methodology of the system and its workflow.

4.1   Tutorial

In this step, the user can watch a 5-minute video[1] explaining the task that *QuOTeS* was made for and an overview of how to use the system. The main part of the video explains the different parts of the system and how they are linked together. It also explains the effect of the different retrieval options and how to download the results from the system to keep analyzing them. Since users will not necessarily need to watch the video every time they use the system, the first step they see when they access the website is the *Upload*, described below.

4.2   Upload

In this step, the users can upload their documents and get the system ready to start interacting with them via a file upload form. Once the text from all the documents has been extracted, they can click on *Process Documents* to prepare the system for the retrieval process. After that, they can select the options for the system in the *Settings* screen, which contains two drop-down menus. In the *Embeddings* menu, the user can choose how the system represents the query and the documents from three options: TFIDF embeddings based on word unigrams, TFIDF embeddings based on character trigrams and Sentence-BERT embeddings [20]. In the *Classifier* menu, the user can choose which Supervised Machine Learning algorithm to use as the backbone for the system from three options: Logistic Regression, Random Forest, and Support Vector Machine.

4.3   Documents

In this step, the user can browse the text extracted from the documents. The sentences from the papers are shown in the order they were found so that the user can verify that the text was extracted correctly. The user can select which

---

[1] Uploaded at the following anonymous link: https://streamable.com/edhej9

documents to browse from the drop-down menu at the top, which displays all the documents that have been uploaded to the system. Later on, when the user starts labeling the sentences with respect to the query, they are colored accordingly: green (for relevant) or pink (for irrelevant).

4.4   Search

This is the first main step of the system. In the text box, users can write their query. After clicking on *Search*, the system retrieves the most relevant sentences using the classical *Vector Space Model* from Information Retrieval.

The sentences below are the best matches according to the query and the representation the user picked in the *Upload* step. The user can label them by clicking on them, which are colored accordingly: green (for relevant) or pink (for irrelevant). Once the users label the sentences, they can click on *Submit Labels*, after which the system records them and shows a new batch of recommendations.

4.5   Explore

This is the second main step of the system. Here, the system trains its classifier using the labels the user submits to improve its understanding of the query. Two plots at the top show the distribution of the recommendation score and how it breaks down by document to help the user better understand the collection. The sentences below work exactly like in *Search*, allowing the user to label them by clicking on them and submitting them into the system by clicking on *Submit Labels*. Users can label the collection as much as they want, but the recommended criterion is to stop when the system has not recommended anything relevant in three consecutive turns, shown in the colored box at the top right.

4.6   History

In this step, users can review what they have labeled and where to find it in the papers. The sentences are shown in the order they were presented to the user, along with the document they came from and their sentence number to make it easier to find them. Like before, the user can click on a sentence to relabel it if necessary, which makes it change color accordingly. There are two buttons at the top: *Clear* allows the user to restart the labeling process, and *Download .csv* downloads the labeling history as a CSV file for further analysis.

4.7   Results

In the last step of *QuOTeS*, the user can assess the results. There are two plots at the top that show the label counts and how they break down by document, while the bottom part displays the query and the sentences labeled as relevant. The query along these sentences make up the final output of the system, which is the Query-Focused Summary of the collection. The user can download this summary as a *.txt* file or the whole state of the system as a JSON file for further analysis.

5   Evaluation

To evaluate the effectiveness of *QuOTeS*, we performed a user study where each participant uploaded up to five documents into the system and labeled the sentences in them for a maximum of one hour. The user study was implemented as a website written using the *Flask* package [21], where the participants went through eight screens to obtain their consent, explain the task to them and fill out a questionnaire about their perception of the difficulty of the task and the performance of *QuOTeS*. An overview of the user study is shown in Fig. 3.

Fig. 3. Overview of the user study.

5.1   Methodology

In the *Welcome Screen*, the participants were shown a quick overview of the whole user study and its duration. In the *Screening Questionnaire*, they filled out a short questionnaire indicating their education level and the frequency they read academic papers. In the *Consent Form* screen, they read a copy of the consent form and agreed to participate by clicking on a checkbox at the end. In the *Video Tutorial* screen, they watched a five-minute video about the task and how to use *QuOTeS*. In the *Results Upload* screen, they were redirected to the website of *QuOTeS* and after using the system for a maximum of one hour, they uploaded the JSON file containing the state of the system at the end of their interaction. In the *Questionnaire* screen, they filled in a three-part questionnaire to evaluate the usability of *QuOTeS*, its features and the quality of the summaries. In the *Compensation Form*, they provided their name and email to be able to receive the compensation for their participation. Finally, the *End Screen* indicated that the study was over and they could close their browser.

5.2   Participants

To recruit participants, we sent a general email call to our faculty, explaining the recruiting process and the compensation. To verify that participants were fit for our study, they filled out a screening questionnaire with only two questions, with the purpose of knowing their research experience and the frequency they normally read academic papers. The requirements to participate were to have completed at least an undergraduate degree in a university and to read academic papers at least once a month. The results of the screening questionnaire for the participants who completed the full study are shown in Table 1, while the full results of the screening questionnaire can be found in the code repository.

Table 1. Responses of the Screening Questionnaire from the participants that completed the study.

Paper Reading Frequency	Education	
	Undergraduate	Graduate
Every day	1	4
At least once a week	2	3
At least once every two weeks	0	1
At least once a month	3	1

5.3   Research Instrument

During the user study, the participants filled out a questionnaire composed of thirty questions divided into three parts: *Usability, Features*, and *Summary Quality*. In the *Usability* part, they filled out the questionnaire from the standard *System Usability Scale* [9], which is a quick and simple way to obtain a rough measure of the perceived usability of the system in the context of the task it is being used for. In the *Features* part, they answered sixteen questions about how difficult the task was and the usefulness of the different components of the system. In the *Summary Quality* part, they answered four questions about the relevance of the sentences in the system and the conciseness, redundancy, and completeness of the summaries produced. Finally, the participants submitted their opinions about the system and the user study in a free-text field. The full questionnaire presented to the participants can be found in the code repository.

5.4   Experimental Results

The frequency tables of the responses for the *System Usability Scale* questionnaire, the *Features* questionnaire, and the Summary Quality questionnaire can be found in the code repository. To make it easier to understand the responses from the questionnaires, we computed a score for the Features and Summary Quality parts in the same fashion as for the System Usability Scale: the questions with positive wording have a value from 0 to 4, depending on their position on the scale. In contrast, the questions with negative wording have a value from 4 to 0, again depending on their position on the scale. The distribution of the scores obtained during the user study is shown in Fig. 4.

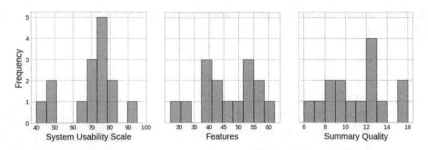

Fig. 4. Distribution of the questionnaire scores obtained during the user study. The possible range for each one of the scores is the following: *System Usability Scale* ranges from 0 to 100, with a mean of 69.67 and a median of 75; the *Features* score ranges from 0 to 64 with a mean of 45.87 and a median of 45; and the *Summary Quality* ranges from 0 to 16 with a mean of 10.67 and a median of 11. These results show that the users perceived the system as useful and well-designed and that the summaries it produces are adequate for the task.

6   Discussion

6.1   Questionnaire Responses

Overall, *QuOTeS* received a positive response across users, as the questionnaires show that the system seems to fulfill its purpose. Most of the time, the participants reported that the sentences recommended by the system seemed relevant and that the summaries appeared succinct, concise, and complete. Participants felt they understood the system's task and how it works. Furthermore, they felt that the components of the system were useful. Nonetheless, the system can be improved in the following ways:

- As shown by the last question of the *System Usability Scale* questionnaire, participants felt that they needed to learn many things before using the system. This is understandable, as *QuOTeS* is based on several concepts which are very specific to Natural Language Processing and Information Retrieval: the task of Query-Focused Summarization itself, the concept of embedding documents as points in space, and the concept of training a Machine Learning classifier on the fly to adapt it to the needs of the user. Nonetheless, knowledge of these concepts is not strictly required to obtain useful insights from the system.
- As shown by the *Features* questionnaire, the system can still be improved in terms of speed. Also, the users felt it was unclear what the different settings do and how to interpret the information in the plots. This may be improved with a better deployment and a better introductory tutorial that provides use cases for each one of the options in the settings: giving the user some guidance about when it is best to use word unigrams, character trigrams, and Sentence-BERT embeddings would facilitate picking the correct options.

The relationship between the different scores computed from the responses of the user study is shown in Fig. 5. All the scores show a clear, positive relationship with each other, with some outliers. The relationships found here are expected because all these scores are subjective and measure similar aspects of the system. Of all of them, the relationship between the System Usability Scale and the Summary Quality is the most interesting: it shows two subgroups, one in which the usability remains constant and the summary quality varies wildly, and another in which they both grow together. This may suggest that for some users, the query is so different from the collection that, although the system feels useful, they are dissatisfied with the results.

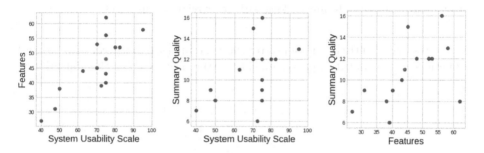

Fig. 5. Relationship between the scores computed from the questionnaires.

6.2   Analysis of the Labels Collected During the User Study

To further evaluate the performance of *QuOTeS*, we estimated the Precision and Topic Concentration using the data labeled by the users. To compute the Precision, we divided the number of sentences labeled as relevant over the total number of sentences shown to the user. To compute the Topic Concentration, we followed the approach from [12], using the Kullback-Leibler Divergence [22] between the unigram-based vocabulary of the document collection and the unigram-based vocabulary of the query-focused summary produced.

The distributions of the Precision and KL-Divergence, along with their relationship, are shown in Fig. 6. The relationship between the two metrics is noisy, but it is somewhat negative, suggesting that as the KL-Divergence decreases, the Precision increases. This result makes sense because the KL-Divergence measures how much the query deviates from the contents of the document collection.

On the other hand, Precision is displayed as a function of the Labeling Effort for each one of the participants in the user study in Fig. 7. We computed the Labeling Effort as the fraction of sentences reviewed by the user. The system displays a stable average Precision of 0.39, which means that, on average, two out of five recommendations from the system are relevant. There appear to be two classes of users: in the first class, the system starts displaying a lot of relevant sentences, and the Precision drops as the system retrieves them; in the second

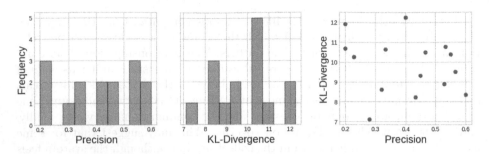

Fig. 6. Distributions of the Precision of the system (left) and the Kullback-Leibler Divergence between the word unigram distribution of the document collections and the summaries produced (center), along with their relationship (right).

class, the story is entirely the opposite: the system starts with very few correct recommendations, but it improves quickly as the user explores the collection.

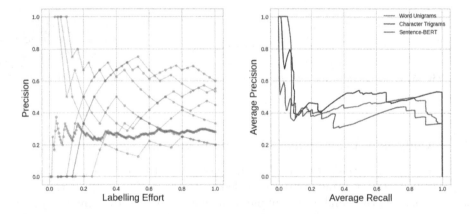

Fig. 7. Precision of the system. Precision as a function of the Labeling Effort for each one of the participants in the user study (left). Average Precision-Recall Curve of the different embeddings after removing the interactive component of *QuOTeS* (right).

The relationships between the Precision and the scores obtained from the questionnaires in the user study are shown in Fig. 8. Precision is well correlated with all the other scores, which is expected since it is the first metric perceived by the user, even before answering the questionnaires. An outlier is very interesting: one of the users gave the system low scores in terms of the questionnaires, despite having the highest Precision of the dataset. The labels produced by this user display a lower Divergence than usual, which means that his query was much closer to the collection than most users, as shown in Fig. 6. This could mean that he/she could already have excellent previous knowledge about the document collection. Therefore, although the system was retrieving relevant sentences, it was not giving the user any new knowledge.

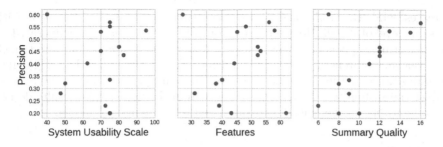

Fig. 8. Relation between the Precision of the system and the questionnaire scores.

The relationship between the Divergence and the scores is shown in Fig. 9. The relationship shown is noisier than the ones involving Precision. Although the System Usability Scale and Features scores show a positive relationship with the Divergence, this is not the case with the Summary Quality. This suggests that to have a high-quality summary, it is necessary to start with a collection close to the query. Another interesting point is that these relationships suggest that the system is perceived as more useful and better designed as the query deviates from the document collection.

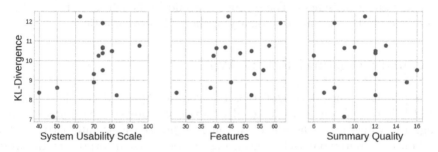

Fig. 9. Relationship between the Kullback-Leibler Divergence between the word unigram distribution of the document collection and produced summaries versus the questionnaire scores obtained in the user study.

To finalize our evaluation of *QuOTeS*, we measured its performance using the *(Query, Document Collection)* pairs collected during the user study. As a baseline, we used the traditional *Vector Space Model*, which is equivalent to disabling the *Machine Learning Classifier* component of *QuOTeS* (as shown in Fig. 1). We evaluated the three variations of the baseline system as they appear inside *QuOTeS*. The performance obtained by this baseline is shown in Fig. 7.

Even when using Sentence-BERT embeddings, the performance of the baseline system is markedly inferior compared to that of *QuOTeS*, as shown in Fig. 7. Although the Sentence-BERT embeddings start with a much higher Precision than the traditional embeddings, they quickly deteriorate as the score threshold increases, while the traditional embeddings catch up in terms of Precision with the same level of Recall. However, since none of these models obtained a satisfactory performance, it is clear that using *QuOTeS* enabled the users to find much more relevant sentences than they could have found otherwise. This highlights the importance of the Continuous Active Learning protocol in *QuOTeS*, as it enables the system to leverage the feedback from the user, so the results do not depend entirely on the embeddings produced by the language model.

6.3 Limitations

Although our experimental results are promising, the system we propose has two main limitations, given the complexity of the task and the amount of resources needed to produce benchmarks for this topic:

- First, the purpose of *QuOTeS* is not to provide fully automatic summaries since it is hard to guarantee that all the relevant sentences were retrieved in the process. Instead, its purpose is to point users in the right direction so that they can find the relevant information in the original documents.
- And second, the summaries produced by the system can still be improved using traditional techniques from Automatic Summarization. For example, their sentences in the summary could be reordered or removed to improve fluency and conciseness. These aspects would be beneficial if the goal is to produce a fully-automatic summary of the collection of articles.

7 Conclusions and Future Work

In this paper, we introduce *QuOTeS*, a system for Query-Focused Summarization of Scientific Documents designed to retrieve sentences relevant to a short paragraph, which takes the role of the query. *QuOTeS* is an interactive system based on the Continuous Active Learning protocol that incorporates the user's feedback in the retrieval process to adapt itself to the user's query.

After a comprehensive analysis of the questionnaires and labeled data obtained through a user study, we found that *QuOTeS* provides a positive user experience and fulfills its purpose. Also, the experimental results show that including both the user's information need and feedback in the retrieval process leads to better results that cannot be obtained with the current non-interactive methods.

For future work, we would like to conduct a more comprehensive user study where users read the whole papers and label the sentences manually, after which they could use *QuOTeS* and compare the summaries produced. Another interesting future direction would be to compare the system heads-on with the main non-interactive methods from the literature on a large, standardized dataset.

Acknowledgements. We thank the Digital Research Alliance of Canada (https://alliancecan.ca/en), CIUNSa (Project B 2825), CONICET (PUE 22920160100056CO, PIBAA 2872021010 1236CO), MinCyT (PICT PRH-2017-0007), UNS (PGI 24/N051) and the Natural Sciences and Engineering Research Council of Canada (NSERC) for the resources provided to enable this research.

References

1. Zhang, J., Zhao, Y., Saleh, M., Liu., P.J.: PEGASUS: pre-training with extracted gap-sentences for abstractive summarization. In: Proceedings of the 37th International Conference on Machine Learning, ICML'20. JMLR.org (2020)
2. Papineni, K., Roukos, S., Ward, T., Zhu, W.-J.: Bleu: a method for automatic evaluation of machine translation. In: Proceedings of the 40th Annual Meeting of the Association for Computational Linguistics, pages 311–318, Philadelphia, Pennsylvania, USA (2002). Association for Computational Linguistics
3. Lin, C.-Y.: ROUGE: a package for automatic evaluation of summaries. In: Text Summarization Branches Out, pp. 74–81, Barcelona, Spain (2004). Association for Computational Linguistics
4. Banerjee, S., Lavie, A.: METEOR: an automatic metric for MT evaluation with improved correlation with human judgments. In: Proceedings of the ACL Workshop on Intrinsic and Extrinsic Evaluation Measures for Machine Translation and/or Summarization, pp. 65–72, Ann Arbor, Michigan (2005). Association for Computational Linguistics
5. Rush, A.M., Chopra, S., Weston, J.: A neural attention model for abstractive sentence summarization. In: Proceedings of the 2015 Conference on Empirical Methods in Natural Language Processing, pp. 379–389, Lisbon, Portugal (2015). Association for Computational Linguistics
6. Dang, H.T.: Overview of DUC 2005. In: Proceedings of the Document Understanding Conference. vol. 2005, pp. 1–12 (2005)
7. Leuski, A., Lin, C.-Y., Hovy, E.: iNeATS: interactive multi-document summarization. In: The Companion Volume to the Proceedings of 41st Annual Meeting of the Association for Computational Linguistics, pp. 125–128, Sapporo, Japan (2003). Association for Computational Linguistics
8. Cormack, C.V., Grossman, M.R.: Evaluation of machine-learning protocols for technology-assisted review in electronic discovery. In: Proceedings of the 37th International ACM SIGIR Conference on Research and Development in Information Retrieval, SIGIR '14, pp. 153–162, New York, NY, USA (2014). Association for Computing Machinery
9. Brooke, J.: SUS - a quick and dirty usability scale. Usability Eval. Ind. **189**(194), 4–7 (1996)
10. Dang, H.T.: Overview of DUC 2006. In: Proceedings of the Document Understanding Conference. vol. 2006, pp. 1–10 (2006)
11. Dang, H.T.: Overview of DUC 2007. In Proceedings of the Document Understanding Conference. vol. 2007, pp. 1–53 (2007)
12. Baumel, T., Cohen, R., Elhadad, M.: Topic concentration in query focused summarization datasets. In: Proceedings of the Thirtieth AAAI Conference on Artificial Intelligence, vol. 30 (2016)
13. Shinoda, K., Aizawa, A.: Query-focused scientific paper summarization with localized sentence representation. In: BIRNDL@ SIGIR (2018)

14. Cormack, G.V., Grossman, M.R.: Multi-faceted recall of continuous active learning for technology-assisted review. In: Proceedings of the 38th International ACM SIGIR Conference on Research and Development in Information Retrieval, SIGIR '15, pp. 763–766, New York, NY, USA (2015). Association for Computing Machinery

15. Cormack, G.V., Grossman, M.R.: Scalability of continuous active learning for reliable high-recall text classification. In: Proceedings of the 25th ACM International on Conference on Information and Knowledge Management, CIKM '16, pp. 1039–1048, New York, NY, USA (2016). Association for Computing Machinery

16. Erera, S., et al.: A summarization system for scientific documents. In: Proceedings of the 2019 Conference on Empirical Methods in Natural Language Processing and the 9th International Joint Conference on Natural Language Processing (EMNLP-IJCNLP): System Demonstrations, pp. 211–216, Hong Kong, China (2019). Association for Computational Linguistics

17. Zarinbal, M., et al.: A New Social Robot for Interactive Query-Based Summarization: Scientific Document Summarization. In: Ronzhin, A., Rigoll, G., Meshcheryakov, R. (eds.) ICR 2019. LNCS (LNAI), vol. 11659, pp. 330–340. Springer, Cham (2019). https://doi.org/10.1007/978-3-030-26118-4_32

18. Bayatmakou, F., Mohebi, A., Ahmadi, A.: An interactive query-based approach for summarizing scientific documents. Inf. Discovery Delivery 50(2), 176–191 (2021)

19. Plotly. Dash. Python package, https://plotly.com/dash, 2013. Visited on August 30, 2022

20. Reimers, N., Gurevych, I.: Sentence-BERT: sentence embeddings using siamese bert-networks. In: Proceedings of the 2019 Conference on Empirical Methods in Natural Language Processing. Association for Computational Linguistics, vol. 11 (2019)

21. Pallets Projects. Flask: web development, one drop at a time. Python package, https://flask.palletsprojects.com, 2010. Visited on August 30, 2022

22. Kullback, S., Leibler, R.A.: On information and sufficiency. Ann. Math. Stat. 22(1), 79–86 (1951)

A Benchmark of Nested Named Entity Recognition Approaches in Historical Structured Documents

Solenn Tual[1]($\boxtimes$)⬥, Nathalie Abadie[1]⬥, Joseph Chazalon[2]⬥,
Bertrand Duménieu[3]⬥, and Edwin Carlinet[2]⬥

[1] LASTIG, Univ. Gustave Eiffel, IGN-ENSG, 94160 Saint-Mandé, France
{solenn.tual,nathalie-f.abadie}@ign.fr
[2] EPITA Research Laboratory (LRE), Le Kremlin-Bicêtre, France
{edwin.carlinet,joseph.chazalon}@lre.epita.fr
[3] CRH-EHESS, Paris, France
bertrand.dumenieu@ehess.fr

Abstract. Named Entity Recognition (NER) is a key step in the creation of structured data from digitised historical documents. Traditional NER approaches deal with flat named entities, whereas entities are often nested. For example, a postal address might contain a street name and a number. This work compares three nested NER approaches, including two state-of-the-art approaches using Transformer-based architectures. We introduce a new Transformer-based approach based on joint labelling and semantic weighting of errors, evaluated on a collection of 19th-century Paris trade directories. We evaluate approaches regarding the impact of supervised fine-tuning, unsupervised pre-training with noisy texts, and variation of IOB tagging formats. Our results show that while nested NER approaches enable extracting structured data directly, they do not benefit from the extra knowledge provided during training and reach a performance similar to the base approach on flat entities. Even though all 3 approaches perform well in terms of F1-scores, joint labelling is most suitable for hierarchically structured data. Finally, our experiments reveal the superiority of the IO tagging format on such data.

Keywords: Natural Language Processing · Nested Name Entity Recognition · Pre-trained language models · NER on noisy texts

1 Introduction

Named entity recognition (NER) is a classic natural language processing (NLP) task used to extract information from textual documents. A named entity is an entity with a specific meaning, such as a person, an organisation, or a place. Entities are often nested [12]. State-of-the-art approaches are divided into flat NER approaches, which associate a span with zero or one entity type, and nested

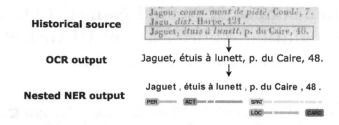

Fig. 1. Overview of NER in Historical Documents, where multiple entities can be assigned to the same span of text. English: *Jaguet, glasses cases seller, 48 Cairo Square.*

named entity recognition approaches, which can associate a given span with zero, one, two or more entity types. In other words, a nested entity is a part of one or more other entities. Base entities can be associated to create more complex entities. Nested named entities are also referred to as tree-structured entities [8], structured entities [26] or hierarchical named entities [18].

We focus on the case of nested named entities in historical documents with an application to the 19th century trade directories of Paris, France. Each directory contains tens of thousands of entries organised into lists in alphabetical order. Each entry is a highly structured non-verbal short text (see Fig. 1). It refers to the name of a company or a person, followed by a description of varying complexity that often includes details on their professional activities, and ends with one or several addresses. Additional information like military or professional awards are sometimes given. The directories were published by multiple editors between the late 18th century and the second half of the 19th century.

Similar work, aiming to extract information from historical directories, use flat NER approaches completed by a post-processing stage to create hierarchical entities. Bells et al. [6] used directories published between 1936 and 1990 to build a database of gas stations in the city of Providence, Rhode Island, United States. The pipeline consists of the following steps: an image processing block, an Optical Character Recognition (OCR) block, and a NER block performed using a rule-based approach. To limit the impact of OCR errors, the authors clean the texts before the NER stage. Structured entities, such as addresses, are built up in post-processing. Albers and Perks [4] developed a similar pipeline to extract data from Berlin directories from 1880, and performed statistical analysis about Berlin population. Abadie et al. [2] introduce a deep learning approach in their pipeline to perform NER on noisy entries extracted from Paris trade directories. Structured postal addresses are also built in post-processing before geocoding.

We focus on nested NER in historical sources to extract structured entities without rule-based post-processing. It requires taking into account the main characteristics of the books. First, entries contain two levels of entities. Top-level entities are built using bottom-level entities. For example, an address consists of two or three other types of entities: a street name, a number and sometimes a geographical feature type (e.g. *depôt* - warehouse). All of these spans are not formally linked as an address with flat NER approaches, while they are linked with nested NER approaches. These approaches provide a more complex semantic view of directories, which improves searchability. Second, directory entries do not have the same structure throughout the corpus: the entity enumeration pattern varies. Finally, directory entries are extracted from digitised documents using OCR models. This requires dealing with noisy texts at the NER stage. The amount of noise changes depending on the quality of the original print, the conservation status of historical sources and the resolution of the digitized images. The OCR engine is a source of errors such as missing or extra characters. The multiple possible entity patterns and the noisy entries encountered in the collection of directories make the use of rule-based methods inappropriate.

We aims to evaluate several nested named entity recognition approaches. The main contributions are new nested named entity datasets using Paris trade directories built on an existing flat NER dataset, and a benchmark of three nested named entity recognition approaches using transfer-learning models fine-tuned on a ground-truth dataset and on a noisy dataset. We use two pre-trained BERT-based models: a state-of-the-art French model and this same model pre-trained on domain-specific noisy examples to assess the impact of unsupervised training on the performances. We also compare two tagging formats to evaluate their influence on NER performance in highly structured texts.

This paper is organised as follows: (i) a state of the art of nested NER approaches, OCR noise, and tagging strategies on NER tasks; (ii) a presentation of the evaluated approaches; (iii) an introduction of our datasets, experiments and metrics; (iv) an evaluation of our approaches from both quantitative and qualitative points of view regarding the effects of the tagging strategy and unsupervised pre-training on transfer-learning-based models.

2   Related Works

We want to produce searchable structured data using historical directories. This paper aims to deal with the properties of such historical sources: short and highly structured texts, including OCR noise, which contained nested named entities.

2.1   Nested Named Entity Recognition Approaches

The survey presented by Wang et al. [28] identifies groups of nested NER approaches, among them are rule-based approaches, layered-based approaches, and region-based approaches. While the first one mostly relieves early methods, the others often fall under supervised learning, which includes traditional machine learning and deep learning approaches.

Handcrafted rule-based approaches aim to recognise patterns in documents. These methods require a high level of data and linguistic expertise to identify and design extraction rules [25]. In most cases, they involve searching for vocabulary in gazetteers or dictionaries as in flat NER approaches [20]. Layered approaches handle the nested NER task using cascades of flat NER layers. Each layer is trained to recognise entities of a given group according to their depth level in the dataset. Jia et al. [13] suggest stacking predictions made by independent layers for each entity level. Models often use n-level outputs to improve the recognition of n+1 level entities with a Layered BiLSTM+CRF model [14] or with a BERT+CRF model [27]. Region-based approaches use two-step models to detect candidates and classify them. There are multiple entity listing strategies: enumeration strategies (such as n-gram) or boundary-based strategies that aim to identify the first and last tokens of potential entities [29]. Furthermore, Wang et al. [28] introduce various methods which have in common the use of a unique tag created by concatenating multiple labels from the nested entities. Agrawal et al. [3] propose an intuitive approach using this unique multilevel label to fine-tune a flat NER BERT-based model. Deep learning methods outperform state-of-the-art approaches most of the time [17,28].

2.2   Named Entity Recognition on Noisy Inputs

With the rise of the digitisation of historical documents, the need to extract and structure the information they contain has increased dramatically. NER is a useful way to produce a high-level semantic view of these documents. Performing this task on historical documents raises specific challenges as stated by Ehrmann et al. [10]. Most of them are processed with automatic tools such as OCR or Handwritten Text Recognition (HTR), which are likely to misread some characters. Thus, it involves dealing with noisy inputs in the NER stage. Dinarelli and Rosset [9] suggest pre-processing texts before a NER stage. They process a dataset of French newspapers from the 19th century and underline the impact of OCR errors on the recognition of tree-structured named entities. Abadie et al. [2] study the effects of OCR noise on flat NER with transfer learning models in 19th century directory entries. According to Li et al. [17], they did an unsupervised pre-training of the French BERT-based model CamemBERT [19] with domain-specific texts provided by OCR and fine-tuned this model with noisy annotated examples. It improved the tolerance of the NER model to noise, and its performances have been improved compared to a non-specialised model.

2.3   Labels for NER on Highly Structured Documents

In NER task, each word (or token) of an entity is associated with a label, also called a tag. Labels follow specific writing formats. The most widely used format is the Inside-Outside-Beginning (IOB) model introduced by Ramshaw and Marcus [23], and its variants: IO, IOB2, IOE, IOBES, BI, IE, or BIES. In practice, these formats define prefixes in front of each class name to specify the position of the token in the entity. Table 1 presents tags associated with a directory entry in the IO and IOB2 formats. These tags are required for the training stage. The choice of tag format has an impact on the performance of NER models. Alshammari and Alanazi [5] show that IO outperforms the other tagging standards on fully developed and non-noisy texts among seven IOB-like formats.

Table 1. IO and IOB2 labels example for *Aubery jr., Quincampoix street, passage Beaufort* entry (translated in English from Cambon almgene, 1841). Labels are defined in Table 3.

Token	IO	IOB2
Aubery	I-PER	**B-PER**
je	I-PER	I-PER
.	I-PER	I-PER
r	I-LOC	**B-LOC**
.	I-LOC	I-LOC
Quincamp	I-LOC	I-LOC
.	I-LOC	I-LOC
pass	I-LOC	**B-LOC**
.	I-LOC	I-LOC
Beaufort	I-LOC	I-LOC
.	O	O

2.4 Conclusion

The following benchmark concentrates on the three main properties of our dataset: nested entities, noisy inputs, and highly-structured texts. We compare three nested named entity recognition approaches based on deep learning models to recognise complex entities in directory entries. We measure the impact of two tagging formats on entity recognition in short texts with repetitive entity patterns. We evaluate the effects of domain adaptation of BERT-based models on noisy inputs. Finally, we aim to evaluate the contribution of training with nested entities on flat named entity recognition.

3 Considered Nested NER Approaches

In this section, we present the three evaluated approaches. We focus on intuitive, low-complexity and high-performance approaches using transfer learning models. We implement two state-of-the-art proposals and introduce a new approach that incorporates hierarchical information into training.

3.1 State-of-the-Art Nested NER Approaches

Independent NER Layers (Abbreviated as [M1]). Jia et al. [13] propose an approach to fine-tune a pre-trained BERT-based model for each level of entities in the dataset (see Table 2). Each model is called a layer and is completely independent of the others. The predictions made by each layer are merged to obtain nested entities. This approach reduces the risk of error propagation: a level 2 entity can be detected even if level 1 is wrong. However, the independence of the NER layers does not allow us to control the combination of the predictions made by each layer. The assignment of a class to a span at level N does not favour or disfavour the assignment of another class at level N+1.

Table 2. IOB2 tags used to train M1 and M2. The entry used as an example is *Dufour (Gabriel), bookseller, Vaugirard Street, 7.* Labels are defined in Table 3.

Token	Level-1 [M1]	Level-2 [M1]	Joint label [M2]
Dufour	B-PER	O	B-PER+O
(	I-PER	O	I-PER+O
Gabriel	I-PER	O	I-PER+O
)	I-PER	O	I-PER+O
,	O	O	O+O
libraire	B-ACT	O	B-ACT+O
,	O	O	O+O
r	B-SPAT	B-LOC	B-SPAT+B-LOC
.	I-SPAT	I-LOC	I-SPAT+I-LOC
de	I-SPAT	I-LOC	I-SPAT+I-LOC
Vaugirard	I-SPAT	I-LOC	I-SPAT+I-LOC
,	I-SPAT	O	I-SPAT+O
7	I-SPAT	B-CARDINAL	I-SPAT+B-CARDINAL

Transfer Learning Approach Using Joint Labelling [M2]. Agrawal et al. [3] propose a BERT-based transfer learning approach using joint labelling. For each token, the authors create a complex label composed of the labels of each depth level of the nested dataset, as described in Table 2. This label is called *joint label*. A single pre-trained BERT-based model is fine-tuned with these annotated data. The nested layers of entities are processed simultaneously as a single one. This implies an increase in the number of labels used for training.

3.2   A Hierarchical BERT-Based Transfer-Learning Approach Using Joint Labelling [M3]

We propose a new version of the BERT-based joint labelling approach of Agrawal et al. [3], inspired by computer vision work on segmentation and classification. Bertinetto et al. [7] note that the loss function used for classification tasks considers all errors with the same weight, regardless of the semantic proximity between classes. Therefore, they propose to take into account the semantic distance between classes in the training and evaluation process of the models by modifying the loss function. We replace the Categorical Cross Entropy Loss implemented in the original CamemBERT model with the Hierarchical Cross Entropy Loss [7]. This function incorporates the semantic distance calculated using a tree that defines the class hierarchy. Each joint label corresponds to a leaf of the tree, as shown in Fig. 2. The errors are weighted according to the distance between the expected class and the predicted class in the tree. For example, if the model predicts a *SPAT+FT* entity instead of a *SPAT+LOC* entity, it is a *better error* than if it predicts a *DESC+O* entity because the labels are semantically closer in the first case than in the second according to the tree. Using IOB2

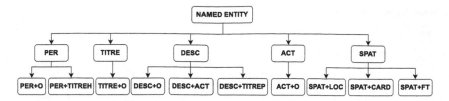

Fig. 2. Multi-level tree representation of IO tags used by the Hierarchical Loss Function called in M3 to compute the semantic distance between labels.

tags, a tree level is added to distinguish the positional prefixes that make up the labels of each class. The loss function of Bertinetto et al. [7] allows weighting the branches of the tree to increase the impact of some classification errors. This last possibility was not considered in our work.

4 Evaluation

In this section, we describe our datasets, our experiments and their associated parameters including the tagging formats, the pre-trained models and the hyper-parameters used to fine-tune them. Finally, we define our metrics.

4.1 Datasets

We have created two annotated datasets using Parisian trade directories: the ground-truth dataset with non-noisy entries and a real-world dataset with the corresponding noisy OCR outputs. The use of these two datasets allows us to compare the performance of the approaches on both clean and noisy texts.

Source Documents: Paris Trade Directories from 1798 to 1854. Our aim is to detect nested entities in entries published during the 19th century. They were printed over a long period of time by several editors using different printing techniques. The layout, content, length, and typography (font, case, abbreviations, etc.) of the entries vary greatly, as illustrated in Fig. 3. Some directories organise the information into several lists, sorted by name, activity, or street name. This means that there are many different types of entities to find. Some types of entities are common, while others are rare. Several independent organisations have digitised directories, with varying levels of quality. This results in more or less clean OCR results.

Directories Datasets. We produced two new NER datasets from the public dataset of Abadie et al. [1]. Both datasets are created from the same 8765 entries from 78 pages chosen in 18 different directories published from 1798 to 1861. The text is extracted using Pero-OCR [15,16]. The first dataset is created by first correcting the text predicted by Pero-OCR and then annotating the ground-truth named entities. To create the second dataset, also known as the *real world* dataset, we first align the text produced by Pero-OCR with the ground-truth text by means

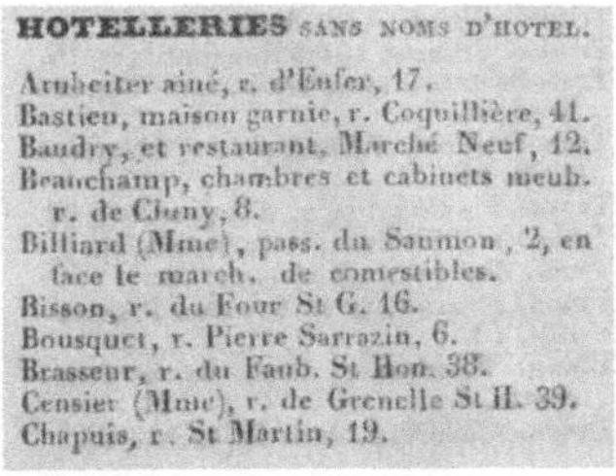

HOTELLERIES SANS NOMS D'HOTEL.

Arnbeiter aîné, r. d'Enfer, 17.
Bastien, maison garnie, r. Coquillière, 41.
Baudry, et restaurant, Marché Neuf, 12.
Beauchamp, chambres et cabinets meub. r. de Cluny, 8.
Billiard (Mme), pass. du Saumon, 2, en face le march. de comestibles.
Bisson, r. du Four St G. 16.
Bousquet, r. Pierre Sarrazin, 6.
Brasseur, r. du Faub. St Hon. 38.
Censier (Mme), r. de Grenelle St H. 39.
Chapuis, r. St Martin, 19.

(a) List ranked by activity

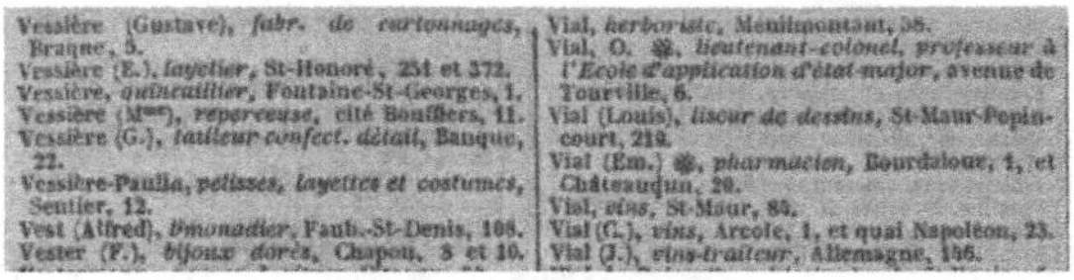

Vessière (Gustave), *fabr. de cartonnages*, Braque, 5.
Vessière (E.), *layetier*, St-Honoré, 251 et 372.
Vessière, *quincaillier*, Fontaine-St-Georges, 1.
Vessière (Mme), *repperceuse*, cité Boufflers, 11.
Vessière (G.), *tailleur confect. détail*, Banque, 22.
Vessière-Paulia, *pelisses, layettes et costumes*, Sentier, 12.
Vest (Alfred), *limonadier*, Faub.-St-Denis, 108.
Vester (F.), *bijoux dorés*, Chapon, 3 et 10.
Vial, *herboriste*, Ménilmontant, 58.
Vial, O. ✻, *lieutenant-colonel, professeur à l'Ecole d'application d'état-major*, avenue de Tourville, 6.
Vial (Louis), *liseur de dessins*, St-Maur-Popincourt, 219.
Vial (Em.) ✻, *pharmacien*, Bourdaloue, 1, et Châteaudun, 20.
Vial, *vins*, St-Maur, 84.
Vial (C.), *vins*, Arcole, 1, et quai Napoléon, 23.
Vial (J.), *vins-traiteur*, Allemagne, 156.

(b) List ranked by name

Fig. 3. Lists examples from Cambon Almgène (3a, 1839) and Didot-Bottin (3b, 1874) directories. Source: Bibliothèque Nationale de France

Table 3. Named entity types, description, and count in the ground-truth directories dataset (8445 entries).

Entity	Level	Description	Count
PER	1	Person(s) or business name	8441
ACT	1 or 2	Person or company's activities	6176
DESC	1	Complete description	371
SPAT	1	Address	8651
TITREH	2	Military or civil title relative to company's owner	301
TITREP	2	Professional rewards	94
TITRE	1	Other title	13
LOC	2	Street name	9417
CARDINAL	2	Street number	8416
FT	2	Kind of geographic feature	76

of tools implemented as part of Stephen V. Rice's thesis [22,24]. Finally, the named entity spans and their associated labels are projected from the first dataset text to the Pero-OCR text. Entries for which no named entities could be projected are also removed from the real-world dataset, which ends up with 8445 entries. For our experiments, we reduce the ground-truth dataset to the 8445 entries that are also present in the noisy dataset. The ground-truth dataset is used to evaluate approaches without the impact of the OCR and to compare its result with one of the models trained on noisy data. Entries have been randomly selected from several directories to represent the wide variety of typographic and pattern types to be learned. There are 10 types of entities described in Table 3. The maximum level of entities is 2. The entity types (Fig. 4) are structured as a *Part-Of* hierarchy: combined level-2 entities form a level-1 entity.

4.2 Experiments Summary

Nested NER on Noisy Texts. The first axis of our work focuses on the comparison of nested NER approaches on directory entries. We evaluate the ability of these approaches to detect all entities regardless of hierarchy and, in contrast, to associate the correct type of level 1 entity with any level 2 entity. We use two pre-trained

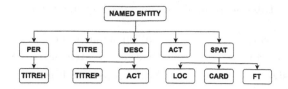

Fig. 4. *Part-of* hierarchy describing the nested entities.

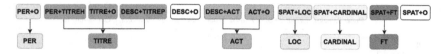

Fig. 5. Mapping between nested NER entity types and flat NER entity types.

transformer models and two tagging formats in these experiments. The robustness of the approaches to OCR noise is given particular attention.

Flat NER Vs. Nested NER. The second axis of this paper is to assess the impact of complex labels on the recognition of flat named entities. We created flat NER datasets (clean and noisy) using our tree-structured annotations. We map the joint labels to the flat labels described by Abadie et al. [2] (see Fig. 5) and re-implement their flat NER experiment. We do not use the dataset proposed by Abadie et al. [1] due to the changes we made during our tree-structured annotation phase. We also map our predicted nested entities to the corresponding flat entity types during the nested NER experiments to compare F1-score values.

4.3 Tagging Formats

The tagging formats retained for this benchmark are IO and IOB2. The IO format is the least restrictive IOB-like tagging model. If a token is part of an entity, it is tagged *I-class*, otherwise it is tagged *O*. However, it does not allow to distinguish between two entities of the same type that follow each other without separation (see Table 1). As this pattern sometimes appears in our historical records, we also use the IOB2 format. This involves creating two tags for each entity type: *B-class* and *I-class*, but we think that tagging each first token of an entity can help to divide a sequence of entities of the same type (as two successive *Addresses*). We compare the two tagging strategies by evaluating the performance on the instance segmentation task for all annotation types.

4.4 Pre-trained BERT-Based Models

We fine-tuned two BERT-based pre-trained models to evaluate the impact of unsupervised domain pre-training on performances. We choose CamemBERT [19], a French BERT-based pre-trained model that has been fine-tuned for NER (CamemBERT NER) and a domain-specific version of this model pre-trained with OCR outputs from Paris trade directories [2]. All the experiments are carried out with the hyperparameters presented in Table 4.

Table 4. Hyperparameters used for fine-tuning

Learning rate	1e–4	Callback patience	5
Weight decay	1e–5	Evaluation strategy	Steps
Batch size	16	Max steps number	5000
Optimizer	AdamW	Metric for best model	F1-Score
Seed	Run number		

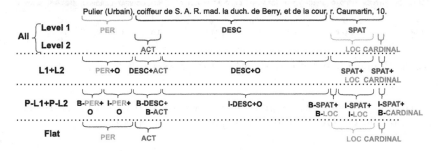

Fig. 6. Entities used for each evaluation metric. Example entry: *Pullier (Urbain), hairdresser of H.R.H Duchess of Berry and to the court, Caumartin Street, 10.*

4.5   Metrics

We use the *seqeval* [21] library to evaluate the performance of each approach. It is well suited to the evaluation of NLP tasks, including sequence labelling, and supports the IO and IOB2 tag formats. The metrics used for the evaluation are precision, recall, and F1-score. *Seqeval* first gathers tokens of the same class that follow each other, to form the predicted entities. They are then aligned with the ground-truth. Any difference in the boundary of the entity or its class, at any level, will be considered as an error.

We evaluate approaches by calculating the F1-score for the spans illustrated in Fig. 6. We can split them in two groups. First, we evaluate the ability of approaches to recognise entities independently of their structure. **All** is the global measure on entities of both levels, **Level 1 and Level 2** are the performance measure on entities of each entity level. Second, we assess how the models deal with the hierarchy of entities. **L1+L2** evaluates the fact that a Level-2 entity is part of a correctly-predicted Level-1 entity. **P-L1+P-L2** includes the evaluation of positional prefixes to the previous definition of *L1+L2*. For the IO tagging format, *P-L1+P-L2* values are equal to the *L1+L2* score and for the IOB2 tagging format, the *P-L1+P-L2* score compares expected and predicted entity types and IOB-like prefixes. Finally, the value **Flat** is the F1-score computed on the flat NER types mapped from *L1+L2* predictions (see Fig. 5).

5   Results

In this last section, we present the results of our experiments from a quantitative and a qualitative point of view.

5.1   Performance

We run the three nested NER approaches and a state-of-the-art flat NER approach with two main goals: (i) compare nested NER approaches regarding the impact of tagging formats and unsupervised domain-specific pre-training of BERT-based models and (ii) evaluate the performance of nested NER approaches regarding flat NER ones. F1-Score values are measured on the 1685 entries of the test subset and are the mean of 5 runs with fixed seeds.

Nested NER Experiments Results. Table 5 presents the F1-score measured for real-world (noisy OCR) and ground-truth datasets. Most of the time, the IO tagging format gives the best results on both datasets. Unsupervised pre-training of the models increases the results in most cases and benefits the noisy dataset. The three approaches produce close results. On the *All* entities F1-score, values are included between 95.6% and 96.6% in the ground-truth dataset and between 93.7% and 94.3% in the noisy dataset. The independent NER layers approache [M1] using IO tags outperforms all tests on both datasets on *All* and *L1* entities, while the joint labelling approach with hierarchical loss [M3] gives the best scores on *L1+L2* and *P-L1+P-L2* spans. *L2* best-score is reached with M1 on the ground-truth dataset and M3 on the noisy dataset.

M1 is trained to recognise the entities of each independent layer. Our results show that it optimises the scores of non-hierarchical features whereas M3, which is trained to maximise *P-L1+P-L2* span recognition, outperforms all approaches on hierarchical entity detection. We focus on *L1+L2* and *P-L1+P-L2* values which should be favoured to produce structured data. Thus, to recognise nested named entities in noisy inputs, our proposed M3 approach with domain-adapted BERT-based model fine-tuned with noisy annotated examples and tagged with IO labels, is the best model to use. Note that the total number of parameters trained during the fine-tuning of the M1 approach is twice as large as for M2 and M3 for a lower number of learned tags. The cost of the extra labels associated with IOB2 format (which enables instance segmentation) in terms of training requirements is not compensated by the gain provided by the ability to discriminate two touching entities (instance segmentation instead of plain semantic segmentation). IO tags give equivalents or better results in most of cases.

F1-score values measured for each entity type (see Tables 6 and 7) unsurprisingly show a drop in performance on the less represented classes (as *DESC*, *TITREH* or *FT*). Although some classes are affected very little by noise (such as *PER*, *ACT* or *LOC*), the *CARDINAL* entity type has F1-score values that drop at least by 15% points from the ground-truth dataset to the noisy dataset. This can be explained by the similarity of CARDINAL entities with OCR mistakes: the OCR engine often confuses noisy letters with numbers.

Table 5. F1 score measured for each approach, dataset, pre-trained models and tag formats (mean of 5 runs).

		Model and tags	All	L1+L2	L1	L2	P-L1+P-L2	Flat
Ground-truth	M1	CmBERT IO	96.5	95.7	96.0	**97.0**	95.7	97.0
		CmBERT IOB2	96.2	95.6	95.8	96.8	95.7	96.1
		CmBERT+ptrn IO	**96.6**	95.9	**96.3**	**97.0**	95.9	**97.3**
		CmBERT+ptrn IOB2	96.0	95.2	95.5	96.6	95.3	95.3
	M2	CmBERT IO	96.2	96.2	95.6	96.9	96.2	96.7
		CmBERT IOB2	96.0	96.0	95.3	96.8	96.0	96.7
		CmBERT+ptrn IO	96.3	96.1	96.0	96.7	96.1	96.9
		CmBERT+ptrn IOB2	96.1	96.0	95.8	96.4	96.1	96.9
	M3	CmBERT IO	96.3	96.2	95.8	96.9	96.2	96.8
		CmBERT IOB2	96.1	96.1	95.6	96.7	96.1	96.8
		CmBERT+ptrn IO	95.8	95.8	95.2	96.7	95.8	96.4
		CmBERT+ptrn IOB2	96.3	**96.3**	96.0	96.7	**96.3**	97.0
OCR	M1	CmBERT IO	93.8	93.4	93.1	94.6	93.4	94.2
		CmBERT IOB2	93.5	92.9	93.1	94.0	93.1	92.7
		CmBERT+ptrn IO	**94.3**	93.8	**94.1**	94.5	93.8	94.4
		CmBERT+ptrn IOB2	94.1	93.5	93.7	94.5	93.7	94.5
	M2	CmBERT IO	93.8	94.1	93.3	94.4	94.1	94.5
		CmBERT IOB2	93.8	94.2	93.2	94.5	94.3	94.7
		CmBERT+ptrn IO	93.9	94.1	93.4	94.4	94.1	94.6
		CmBERT+ptrn IOB2	93.7	94.1	93.1	94.5	94.2	94.8
	M3	CmBERT IO	94.1	**94.4**	93.5	**94.8**	**94.4**	94.8
		CmBERT IOB2	93.5	93.9	92.9	94.3	94.0	94.6
		CmBERT+ptrn IO	94.1	**94.4**	93.6	**94.8**	**94.4**	**94.9**
		CmBERT+ptrn IOB2	93.7	94.0	93.1	94.4	94.2	94.8

Flat NER Experiments Results. We fine-tune CamemBERT and pre-trained CamemBERT on our flat NER datasets (derived from our nested NER datasets) using only IO tags. Results are given in Table 8. As for nested NER approaches, we observe that performances with the ground-truth dataset are better than those obtained with the noisy dataset. This result shows the impact of noise on the NER task. Unsupervised pre-training has no significant effect on the results of the model fine-tuned on the ground-truth dataset, but it gives better scores on the noisy dataset. Comparison between true flat NER results and flat NER equivalent results provided by nested NER approaches shows that fine-tuning models with structured entities does not increase F1-scores.

5.2 Qualitative Analysis

In this section, we give a qualitative analysis of our results. We look at the classification errors listing the most common types of errors made by our fine-tuned models. We identify the potential sources of errors. Finally, we underline the main pros and cons of each approach

Table 6. F1 score measured for each approach, pre-trained model and tag format (mean of 5 runs) on the ground-truth dataset for each entity type.

	Model & tags	PER	ACT	DESC	TITREH	TITREP	SPAT	LOC	CARD	FT
M1	CmBERT IO	91.8	94.8	49.6	11.7	97.4	97.5	**97.9**	98.1	36.2
	CmBERT IOB2	90.4	94.0	43.3	22.5	97.3	97.7	97.6	96.5	51.5
	CmBERT+ptrn IO	**92.6**	**95.7**	**53.5**	50.5	97.2	97.6	97.6	**98.4**	53.4
	CmBERT+ptrn IOB2	89.8	93.6	44.4	41.5	97.2	97.7	97.4	97.2	50.4
M2	CmBERT IO	90.6	94.5	47.1	39.9	97.5	97.3	97.2	97.5	58.7
	CmBERT IOB2	90.3	93.8	36.8	39.8	97.3	97.4	97.4	97.9	56.0
	CmBERT+ptrn IO	90.1	94.6	47.7	**58.1**	**97.6**	98.2	97.3	**98.4**	57.3
	CmBERT+ptrn IOB2	90.1	94.4	42.9	38.6	97.1	**98.4**	97.3	97.9	62.3
M3	CmBERT IO	91.1	94.6	49.4	43.5	97.4	97.7	97.2	97.2	50.9
	CmBERT IOB2	90.8	94.3	44.8	40.7	97.3	97.3	97.4	95.8	51.8
	CmBERT+ptrn IO	90.0	94.1	36.1	45.1	97.5	96.8	97.4	96.2	42.9
	CmBERT+ptrn IOB2	90.4	94.6	48.3	44.8	97.5	**98.4**	97.4	98.1	**64.1**

Table 7. F1 score measured for each approach, pre-trained model and tag format (mean of 5 runs) on the noisy dataset for each entity type.

	Model & tags	PER	ACT	DESC	TITREH	TITREP	SPAT	LOC	CARD	FT
M1	CmBERT IO	**90.0**	92.7	49.3	20.6	94.5	94.3	94.4	69.5	48.1
	CmBERT IOB2	88.7	92.8	42.2	27.3	94.3	94.8	94.2	79.5	33.7
	CmBERT+ptrn IO	89.3	93.3	50.9	39.7	**95.1**	**96.0**	**95.1**	77.0	50.3
	CmBERT+ptrn IOB2	89.0	92.6	48.2	37.8	94.8	**96.0**	94.8	79.4	42.3
M2	CmBERT IO	89.5	93.1	47.4	39.9	94.7	95.0	94.4	74.0	41.9
	CmBERT IOB2	89.0	92.7	49.2	45.8	94.6	95.0	94.2	80.7	45.6
	CmBERT+ptrn IO	89.0	**93.4**	48.1	**57.0**	95.0	94.8	94.9	75.6	50.7
	CmBERT+ptrn IOB2	87.8	91.9	39.2	40.8	95.0	95.5	94.8	78.5	**56.5**
M3	CmBERT IO	89.6	93.0	**52.3**	54.6	**95.1**	95.0	94.5	79.9	53.9
	CmBERT IOB2	88.2	91.9	43.9	38.3	94.5	95.1	94.1	76.8	42.9
	CmBERT+ptrn IO	88.7	92.9	46.3	55.6	95.4	95.4	95.0	75.3	56.1
	CmBERT+ptrn IOB2	87.6	91.8	38.3	43.7	94.8	95.7	94.6	**82.7**	52.2

There are no hierarchy constraints on nested entity types with M1. It produces unauthorised structured entities (as *DESC/SPAT* or *SPAT/ TITREP* or *ACT/ACT*). This is a significant drawback to producing structured data. The joint labelling based approaches (M2/M3) are a great solution to avoid this issue: hierarchy constraints are defined by the joint labels list set for the training.

We focus on the classification errors made by the M2 and M3 results. They are illustrated by Fig. 7. There is a lot of confusion between the *DESC* and *ACT* entities. They contain similar information. There also are confusions between first-level *ACT* and second-level *ACT* entities which give the same kind of information whatever their entity level, as seen in Fig. 8. This first group of confusion appears in the top-left corner of Fig. 7. A first element of explanation is that infor-

Table 8. F1-score measured on flat NER datasets.

Dataset	Test	F1-Score (in %)
Ground-truth	CmBERT	96.8
	CmBERT+ptrn	96.8
OCR	CmBERT	94.7
	CmBERT+ptrn	94.9

	Predictions										
	ACT+O	DESC+O	DESC+ACT	DESC+TITREP	PER+O	PER+TITREH	SPAT+O	SPAT+CARD	SPAT+FT	SPAT+LOC	O+O
ACT+O	89.8	5.2	3.1	0	0.6	0	0	0	0	0.8	0.4
DESC+O	15.4	45	34.9	0.7	1.8	0	0	0	0	1.7	0.5
DESC+ACT	19.2	9.5	71.3	0	0	0	0	0	0	0	0
DESC+TITREP	0	38.5	4.4	57.1	0	0	0	0	0	0	0
PER+O	0.6	0.4	0	0	98	0	0	0	0	0.2	0.7
PER+TITREH	0	0	0	0	1.3	86.7	0	0	0	0	12
SPAT+O	0	0.4	0	0	0	0	95.1	1.2	0	1.8	1.4
SPAT+CARD	0	0	0	0	0	0	0.4	98.3	0	0.3	1
SPAT+FT	17.8	31.1	0	0	0	0	2.2	4.4	24.4	0	20
SPAT+LOC	0.1	0.2	0	0	0.1	0	0.3	0	0	99.1	0.2
O+O	43.6	2	1.9	0.3	5.9	0.2	0.3	2.5	0	17.4	25.9

(Gold labels on the vertical axis)

Fig. 7. Confusion matrix representing the percentage of tokens associated with each existing joint-label. Values are normalised by row and predictions have been provided by [M3] with fine-tuned CamemBERT+pre-trained model and IO tags.

mation carried by these classes are really close from a semantic point of view which favours such kind of mistakes. The *DESC* entity type is also represented by few examples in the datasets. The model often failed to recognise *TITREP* in *DESC* and *TITREH* in *PER*. There is a small number of these entities in the training dataset, which could explain these errors. Note that both *TITREP* and *TITREH* entity types always contain Unicode characters which aren't included in the CamemBERT model vocabulary. Confusion also can affect unrelated types of entities. The most often observed errors are *ACT* entities that mention a warehouse of merchandise, which are confused with a *FT* mentioned in an address. The model often misses *FT* type entities in text. For both datasets, we observed that punctuation between entities at both levels is often misclassified (see the bottom line of Fig. 7 and highlighted tokens in Fig. 8). They are included in their right-side or left-side entity. When using the IOB2 tagging model, we also

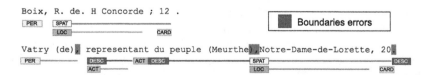

Fig. 8. Two noisy entries classified with the M3 CamemBERT+ptrn IO model: the top entry (*Boix, Concorde Street, 12*) doesn't contain mistakes whereas the bottom example (*Vatry (de), people's representative (Meurthe), Notre-Dame-de-Lorette, 20*) contains boundary errors (highlighted) and *DESC/ACT* confusions.

observed token confusions. Both the training and running time of the first approach are multiplied by the number of levels of entities in the dataset. It's not the case for M2 and M3. The first approach requires a post-processing step to merge predictions made by each flat NER layer which is time-consuming. M2 and M3 provide full confidence in the hierarchy of entities, unlike M1.

6   Conclusion

In this paper, we compare three Nested Named Entity Recognition approaches on noisy and highly structured texts extracted from historical sources: the Paris trade directories from the 19th century. The aim of this work is to produce a high-level semantic view of this collection. We do a survey of transfer-learning based approaches and select two intuitive ones: the independent NER layers approach [M1] developed by Jia et al. [13] and the joint-labelling approach [M2] developed by Agrawal et al. [3]. We suggest an update of this last approach [M3] replacing the loss function of the initial transformer model with a hierarchical loss developed by Bertinetto et al. [7]. It includes the semantic distance between expected and predicted labels. We compare the performance of each approach using a CamemBERT model and a CamemBERT model pre-trained with noisy entries. We used our new nested NER datasets with clean and noisy directories entries to fine-tune these models using IO and IOB2 tagging formats. The performance of the three approaches achieves high F1-score values with both datasets. M3 outperforms results on hierarchical entities ($L1+L2$) which are the most adapted to produce structured data. Fine-tuning a pre-trained BERT-based model increases performances on noisy entries. Finally, we conclude that the IO tagging format achieves the best performance in most cases. IOB2 tagging format does not improve our results on noisy outputs. In future works, we aim to extend our benchmark on non-highly structured historical texts written in different languages using existing datasets such as CLEF-HIPE 2020 and 2022 [11] which include nested named entities. This will give us the opportunity to extend our approach to more generic use cases based on multilingual models.

Acknowledgments. This work is supported by the French National Research Agency (ANR), as part of the SODUCO project (grant ANR-18-CE38-0013) and by the French Ministry of the Armed Forces - Defence Innovation Agency (AID).

Our datasets (images and their associated text transcription), code and models are available on Zenodo (https://doi.org/10.5281/zenodo.7864174, https://doi.org/10.5281/zenodo.7867008) and HuggingFace (https://huggingface.co/nlpso).

References

1. Abadie, N., Bacciochi, S., Carlinet, E., Chazalon, J., Cristofoli, P., Duménieu, B., Perret, J.: A dataset of French trade directories from the 19th century (FTD) (2022). https://doi.org/10.5281/zenodo.6394464
2. Abadie, N., Carlinet, E., Chazalon, J., Duménieu, B.: A benchmark of named entity recognition approaches in historical documents application to 19th century french directories. In: Document Analysis Systems: 15th IAPR International Workshop, DAS 2022, Proceedings. La Rochelle, France (2022). https://doi.org/10.1007/978-3-031-06555-2_30
3. Agrawal, A., Tripathi, S., Vardhan, M., Sihag, V., Choudhary, G., Dragoni, N.: BERT-Based Transfer-Learning Approach for Nested Named-Entity Recognition Using Joint Labeling. Appl. Sci. **12**(3), 976 (2022). https://doi.org/10.3390/app12030976
4. Albers, T., Kappner, K.: Perks and pitfalls of city directories as a micro-geographic data source. Collaborative Research Center Transregio 190 Discussion Paper No. 315 (2022). https://doi.org/10.5282/ubm/epub.90748
5. Alshammari, N., Alanazi, S.: The impact of using different annotation schemes on named entity recognition. Egypt. Inform. J. **22**(3), 295–302 (2021). https://doi.org/10.1016/j.eij.2020.10.004
6. Bell, C., et al.: Automated data extraction from historical city directories: the rise and fall of mid-century gas stations in Providence, RI. PLoS ONE **15**(8), e0220219 (2020). https://doi.org/10.1371/journal.pone.0220219
7. Bertinetto, L., Mueller, R., Tertikas, K., Samangooei, S., Lord, N.A.: Making better mistakes: Leveraging class hierarchies with deep networks. In: Proceedings of the IEEE/CVF Conference on Computer Vision and Pattern Recognition (CVPR) (2020)
8. Dinarelli, M., Rosset, S.: Models cascade for tree-structured named entity detection. In: Proceedings of 5th International Joint Conference on Natural Language Processing, pp. 1269–1278. Asian Federation of Natural Language Processing, Chiang Mai, Thailand (2011)
9. Dinarelli, M., Rosset, S.: Tree-structured named entity recognition on OCR data: analysis, processing and results. In: Proceedings of the Eighth International Conference on Language Resources and Evaluation (LREC 2012), pp. 1266–1272. European Language Resources Association (ELRA), Istanbul, Turkey (2012)
10. Ehrmann, M., Hamdi, A., Linhares Pontes, E., Romanello, M., Doucet, A.: Named entity recognition and classification in historical documents: a survey. ACM Computing Survey (2021). https://infoscience.epfl.ch/record/297355
11. Ehrmann, M., Romanello, M., Doucet, A., Clematide, S.: Hipe-2022 shared task named entity datasets (2022). https://doi.org/10.5281/zenodo.6375600
12. Finkel, J.R., Manning, C.D.: Nested named entity recognition. In: Proceedings of the 2009 Conference on Empirical Methods in Natural Language Processing, pp. 141–150. Association for Computational Linguistics, Singapore (2009)
13. Jia, L., Liu, S., Wei, F., Kong, B., Wang, G.: Nested named entity recognition via an independent-layered pretrained model. IEEE Access **9**, 109693–109703 (2021). https://doi.org/10.1109/ACCESS.2021.3102685
14. Ju, M., Miwa, M., Ananiadou, S.: A neural layered model for nested named entity recognition. In: Proceedings of the 2018 Conference of the North American Chapter of the Association for Computational Linguistics: Human Language Technologies, Vol. 1 (Long Papers), pp. 1446–1459. Association for Computational Linguistics, New Orleans, Louisiana (2018). https://doi.org/10.18653/v1/N18-1131

15. Kišš, M., Beneš, K., Hradiš, M.: AT-ST: self-training adaptation strategy for OCR in domains with limited transcriptions. In: Lladós, J., Lopresti, D., Uchida, S. (eds.) ICDAR 2021. LNCS, vol. 12824, pp. 463–477. Springer, Cham (2021). https://doi.org/10.1007/978-3-030-86337-1_31

16. Kohút, J., Hradiš, M.: TS-Net: OCR trained to switch between text transcription styles. In: Lladós, J., Lopresti, D., Uchida, S. (eds.) ICDAR 2021. LNCS, vol. 12824, pp. 478–493. Springer, Cham (2021). https://doi.org/10.1007/978-3-030-86337-1_32

17. Li, J., Sun, A., Han, J., Li, C.: A survey on deep learning for named entity recognition. IEEE Trans. Knowl. Data Eng. **34**(1), 50–70 (2022). https://doi.org/10.1109/TKDE.2020.2981314

18. Marinho, Z., Mendes, A., Miranda, S., Nogueira, D.: Hierarchical nested named entity recognition. In: Proceedings of the 2nd Clinical Natural Language Processing Workshop, pp. 28–34. Association for Computational Linguistics, Minneapolis, Minnesota, USA (2019). https://doi.org/10.18653/v1/W19-1904

19. Martin, L., et al.: CamemBERT: a tasty French language model. In: Proceedings of the 58th Annual Meeting of the Association for Computational Linguistics, pp. 7203–7219 (2020). https://doi.org/10.18653/v1/2020.acl-main.645

20. Nadeau, D., Sekine, S.: A survey of named entity recognition and classification. Lingvisticae Investigationes **30** (2007). https://doi.org/10.1075/li.30.1.03nad

21. Nakayama, H.: seqeval: a python framework for sequence labeling evaluation (2018). https://github.com/chakki-works/seqeval

22. Neudecker, C., Baierer, K., Gerber, M., Christian, C., Apostolos, A., Stefan, P.: A survey of OCR evaluation tools and metrics. In: The 6th International workshop on Workshop on Historical Document Imaging and Processing, pp. 13–18 (2021)

23. Ramshaw, L., Marcus, M.: Text chunking using transformation-based learning. In: Third Workshop on Very Large Corpora (1995), https://aclanthology.org/W95-0107

24. Santos, E.A.: OCR evaluation tools for the 21st century. In: Proceedings of the Workshop on Computational Methods for Endangered Languages. vol. 1 (2019). https://doi.org/10.33011/computel.v1i.345

25. Shen, D., Zhang, J., Zhou, G., Su, J., Tan, C.L.: Effective adaptation of hidden Markov model-based named entity recognizer for biomedical domain. In: Proceedings of the ACL 2003 Workshop on Natural Language Processing in Biomedicine, pp. 49–56. Association for Computational Linguistics, Sapporo, Japan (2003). https://doi.org/10.3115/1118958.1118965

26. Wajsbürt, P.: Extraction and normalization of simple and structured entities in medical documents. Ph.D. thesis, Sorbonne Université (2021)

27. Wajsbürt, P., Taillé, Y., Tannier, X.: Effect of depth order on iterative nested named entity recognition models. In: Tucker, A., Henriques Abreu, P., Cardoso, J., Pereira Rodrigues, P., Riaño, D. (eds.) AIME 2021. LNCS (LNAI), vol. 12721, pp. 428–432. Springer, Cham (2021). https://doi.org/10.1007/978-3-030-77211-6_50

28. Wang, Y., Tong, H., Zhu, Z., Li, Y.: Nested named entity recognition: a survey. ACM Trans. Knowl. Discov. Data **16**(6), 108:1–108:29 (2022). https://doi.org/10.1145/3522593

29. Zheng, C., Cai, Y., Xu, J., Leung, H.f., Xu, G.: A Boundary-aware neural model for nested named entity recognition. In: Proceedings of the 2019 Conference on Empirical Methods in Natural Language Processing and the 9th International Joint Conference on Natural Language Processing (EMNLP-IJCNLP), pp. 357–366. Association for Computational Linguistics, Hong Kong, China (2019). https://doi.org/10.18653/v1/D19-1034

"Explain Thyself Bully": Sentiment Aided Cyberbullying Detection with Explanation

Krishanu Maity[1]([✉]), Prince Jha[1], Raghav Jain[1], Sriparna Saha[1],
and Pushpak Bhattacharyya[2]

[1] Department of Computer Science and Engineering, Indian Institute of Technology Patna,
Patna, India
{krishanu_2021cs19,princekumar_1901cs42}@iitp.ac.in
[2] Department of Computer Science and Engineering, Indian Institute of Technology Bombay,
Bombay, India

Abstract. Cyberbullying has become a big issue with the popularity of different social media networks and online communication apps. While plenty of research is going on to develop better models for cyberbullying detection in monolingual language, there is very little research on the code-mixed languages and explainability aspect of cyberbullying. Recent laws like "right to explanations" of General Data Protection Regulation, have spurred research in developing interpretable models rather than focusing on performance. Motivated by this we develop the first interpretable multi-task model called *mExCB* for automatic cyberbullying detection from code-mixed languages which can simultaneously solve several tasks, cyberbullying detection, explanation/rationale identification, target group detection and sentiment analysis. We have introduced *BullyExplain*, the first benchmark dataset for explainable cyberbullying detection in code-mixed language. Each post in *BullyExplain* dataset is annotated with four labels, i.e., *bully label, sentiment label, target and rationales (explainability)*, i.e., which phrases are being responsible for annotating the post as a bully. The proposed multitask framework (mExCB) based on CNN and GRU with word and sub-sentence (SS) level attention is able to outperform several baselines and state of the art models when applied on *BullyExplain* dataset.

Disclaimer: The article contains offensive text and profanity. This is owing to the nature of the work, and do not reflect any opinion or stand of the authors.(The code and dataset are available at https://github.com/MaityKrishanu/BullyExplain-ICDAR.)

Keywords: Cyberbullying · Sentiment · Code-Mixed · Multi-task · Explainability

1 Introduction

Cyberbullying [35] is described as an aggressive, deliberate act committed against people using computers, mobile phones, and other electronic gadgets by one individual or a

---

Krishanu Maity, Prince Jha: Both authors contributed equally to this research.
K. Maity and P. Jha–Both authors contributed equally to this research.

© The Author(s), under exclusive license to Springer Nature Switzerland AG 2023
G. A. Fink et al. (Eds.): ICDAR 2023, LNCS 14189, pp. 132–148, 2023.
https://doi.org/10.1007/978-3-031-41682-8_9

group of individuals. According to the Pew Research Center, 40% of social media users have experienced some sort of cyberbullying[1]. Victims of cyberbullying may endure sadness, anxiety, low self-esteem, transient fear and suicidal thinking [36]. According to the report of the National Records Crime Bureau, [26] "A total of 50,035 cases were registered under Cyber Crimes, showing an increase of 11.8% in registration over 2019 (44,735 cases)". The objective of minimizing these harmful consequences emphasizes the necessity of developing techniques for detecting, interpreting, and preventing cyberbullying.

Over the last decade, most of the research on cyberbullying detection has been conducted on monolingual social media data using traditional machine learning [8, 11, 31] and deep learning models [1, 3, 38]. The contingency of code-mixing is increasing very rapidly. Over 50M tweets were analyzed by [32], in which 3.5% of tweets are code-mixed. Hence this should be our primary focus at this point in time. Code-mixing is a linguistic marvel in which two or more languages are employed in speech alternately [25]. Recently, people have started working on offensive post-detection in code-mixed languages like aggression detection [19], hate-speech detection [5], and cyberbullying detection [22,23]. However, those researchers mostly concentrated on improving the performance of cyberbullying detection using various models, without giving any insight or analysis into the explainability. With the introduction of explainable artificial intelligence (AI) [15], it is now a requirement to provide explanations/interpretations for any decision taken by a machine learning algorithm. This helps in building trust and confidence when putting AI models into production. Moreover, in Europe, legislation such as the General Data Protection Regulation (GDPR) [30] has just introduced a "right to explanation" law. Hence there is an urge in developing interpretable models rather than only focusing on improving performance by increasing the complexity of the models.

In this paper, we have developed an explainable cyberbullying dataset and a hierarchical attention-based multitask model to solve four tasks simultaneously, i.e., Cyberbullying Detection (CD), Sentiment Analysis (SA), Target Identification (TI), and Detection of Rationales (RD). To develop an explainable code-mixed cyberbullying dataset, we have re-annotated the existing *BullySent* dataset [22] with the target class (Religion, Sexual-Orientation, Attacking-Relatives-and-Friends, Organization, Community, Profession and Miscellaneous) and highlighted parts of the text that could justify the classification decision. If the post is non-bully, the rationales are not marked and the target class is selected as NA (Not Applicable). This study focuses on applying rationales, which are fragments of text from a source text that justify a classification decision. Commonsense explanations [29], e-SNLI [6], and various other tasks [10] have all employed such rationales. If these rationales are valid explanations for decisions, models that are trained to follow them might become more human-like in their decision-making and thus are more trustworthy, transparent, and reliable.

A person's sentiment heavily influences the intended content [20]. In a multitask (MT) paradigm, the task of sentiment analysis (SA) has often been considered as an auxiliary task to increase the performance of primary tasks (such as cyberbullying detection (CD) [22], Complaint Identification [34] and tweet act classification

---

[1] https://www.pewresearch.org/internet/2017/07/11/online-harassment-2017/.

(TAC) [33]). In the current set-up, sentiment analysis and rationale identification are treated as secondary tasks as the presence of rationale with sentiment information of a post certainly helps identify bully samples more accurately. Our proposed multitask model incorporates a Bi-directional Gated Recurrent Unit (Bi-GRU), Convolutional Neural Network (CNN), and Self-attention in both word level and sub-sentence (SS) level representation of input sentences. CNNs perform well for extracting local and position-invariant features [40]. In contrast, RNNs are better for long-range semantic dependency-based tasks (machine translation, language modeling) than some local key phrases. The intuition behind the usage of CNN and Bi-GRUs in both word level and sub-sentence level is to efficiently handle code-mixed data, which is noisier (spelling variation, abbreviation, no specific grammatical rules) than monolingual data.

The following are the major contributions of the current paper:

- We investigate two new issues: (i) explainable cyberbullying detection in code-mixed text and (ii) detection of their targets.
- *BullyExplain,* a new benchmark dataset for explainable cyberbullying detection with target identification in the code-mixed scenario, has been developed.
- To simultaneously solve four tasks (Bully, Sentiment, Rationales, Target), an end-to-end deep multitask framework (mExCB) based on word and subsentence label attention has been proposed.
- Experimental results illustrate that the usages of rationales and sentiment information significantly enhance the performance of the main task, i.e., cyberbullying detection.

2   Related Works

Text mining and NLP paradigms have been used to investigate numerous subjects linked to cyberbullying detection, including identifying online sexual predators, vandalism detection, and detection of internet abuse and cyberterrorism. Although the associated research described below inspires cyberbullying detection, their methods do not consider the explainability part, which is very much needed for any AI/ML task.

2.1   Works on Monolingual Data

Dinakar et al. [11] investigated cyberbullying detection using a corpus of 4500 YouTube comments and various binary and multiclass classifiers. The SVM classifier attained an overall accuracy of 66.70%, while the Naive Bayes classifier attained an accuracy of 63%. Authors in [31] developed a Cyberbullying dataset by collecting data from Formspring.me and finally achieved 78.5% accuracy by applying C4.5 decision tree algorithm using Weka tool kit. In 2020, Balakrishnan et al. [4] developed a model based on different machine learning approaches and psychological features of twitter users for cyberbullying detection. They found that combining personality and sentiment characteristics with baseline features (text, user, and network) enhances cyberbullying detection accuracy. CyberBERT, a BERT based framework developed by [27] achieved the state-of-the-art results on Formspring (12k posts), Twitter (16k posts), and Wikipedia (100k posts) dataset.

2.2 Works on Code-mixed Data

In [21], the authors attained 79.28% accuracy in cyberbullying detection by applying CNN, BERT, GRU, and Capsule Networks on their introduced code-mixed Indian language dataset. For identifying hate speech from Hindi-English code mixed data, the deep learning-based domain-specific word embedding model in [16] outperforms the base model by 12% in terms of F1 score. Authors in [19] created an aggression-annotated corpus that included 21k Facebook comments and 18k tweets written in Hindi-English code-mixed language.

2.3 Works on Rationales:

Zaidan et al. [41] proposed the concept of rationales, in which human annotators underlined a section of text that supported their tagging decision. Authors have examined that the usages of these rationales certainly improved sentiment classification performance. Mathew et al. [24] introduced HateXplain, a benchmark dataset for hate speech detection. They found that models that are trained using human rationales performed better at decreasing inadvertent bias against target communities. Karim et al. [17] developed an explainable hate speech detection approach (DeepHateExplainer) in Bengali based on different variants of transformer architectures (BERT-base, mMERT, XLM-RoBERTa). They have provided explainability by highlighting the most important words for which the sentence is labeled as hate speech.

2.4 Works on Sentiment Aware Multitasking

There are some works in the literature where sentiment analysis is treated as an auxiliary task to boost the performance of the main task. Saha et al. [33] proposed a multi-modal tweet act classification (TAC) framework based on an ensemble adversarial learning strategy where sentiment analysis acts as a secondary task. [34] developed a multi-task model based on affective space as a commonsense knowledge. They achieved a good accuracy of 83.73% and 69.01% for the complaint identification and sentiment analysis tasks, respectively. Maity et al. [22] created a Hindi-English code-mixed dataset for cyberbullying detection. Based on BERT and VecMap embeddings, they developed an attention-based deep multitask framework and achieved state-of-the-art performance for sentiment and cyberbullying detection tasks.

After performing an in-depth literature review, it can be concluded that there is no work on explainable cyberbullying detection in the code-mixed setting. In this paper, we attempt to fill this research gap.

3 *BullyExplain Dataset Development*

This section details the data set created for developing the explainable cyberbullying detection technique in a code-mixed setting.

3.1 Data Collection

To begin, we reviewed the literature for the existing code-mixed cyberbullying datasets. We found two cyberbullying datasets [21,22] in Hindi-English code-mixed tweets. We selected the BullySent [22] dataset for further annotation with rationales and target class as it was previously annotated with the bully and sentiment labels.

3.2 Annotation Training

The annotation was led by three Ph.D. scholars with adequate knowledge and expertise in cyberbullying, hate speech, and offensive content and performed by three undergraduate students with proficiency in both Hindi and English. First, a group of undergraduate computer science students were voluntarily hired through the department email list and compensated through gift vouchers and honorarium. Previously each post in BullySent [22] dataset was annotated with bully class (Bully/non-bully) and sentiment class (Positive/Neutral/Negative). For rationales and target annotation, we have considered only the bully tweets. For annotation training, we required gold standard samples annotated with rationale and Target labels. Our expert annotators randomly selected 300 memes and highlighted the words (rationales) for the textual explanation and tag a suitable target class. For rational annotation, we have followed the same strategy as mentioned in [24]. Each word in a tweet was marked with either 0 or 1, where 1 means rationale. We have considered seven target classes (Religion, Sexual-Orientation, Attacking-Relatives-and-Friends, Organization, Community, Profession and Miscellaneous) as mentioned in [24,28]. Later expert annotators discussed each other and resolved the differences to create 300 gold standard samples with rationale and target annotations. We divide these 300 annotated examples into three sets, 100 rationale annotations each, to carry out three-phase training. After the completion of every phase, expert annotators met with novice annotators to correct the wrong annotations, and simultaneously annotation guidelines were also renewed. After completing the third round of training, the top three annotators were selected to annotate the entire dataset.

3.3 Main Annotation

We used the open-source platform Docanno[2] deployed on a Heroku instance for main annotation where each qualified annotator was provided with a secure account to annotate and track their progress exclusively. We initiated our main annotation process with a small batch of 100 memes and later raised it to 500 memes as the annotators became well-experienced with the tasks. We tried to maintain the annotators' agreement by correcting some errors they made in the previous batch. On completion of each set of annotations, final rationale labels were decided by the majority voting method. If the selections of three annotators vary, we enlist the help of an expert annotator to break the tie. We also directed annotators to annotate the posts without regard for any particular demography, religion, or other factors. We use the Fleiss' Kappa score [12] to calculate the inter-annotator agreement (IAA) to affirm the annotation quality. IAA obtained

---

[2] https://github.com/doccano/doccano.

scores of 0.74 and 0.71 for the rationales detection (RD) and Target Identification (TI) tasks, respectively, signifying the dataset being of acceptable quality. Some samples from the *BullyExplain* dataset are shown in Table 1.

Ethics Note. Repetitive consumption of online abuse could distress mental health conditions [39]. Therefore, we advised annotators to take periodic breaks and not do the annotations in one sitting. Besides, we had weekly meetings with them to ensure the annotations did not have any adverse effect on their mental health.

Table 1. Some samples from annotated *BullyExplain* dataset; The green highlights mark rationales tokens. AFR - "Attacking-Relatives-and-Friends"

Tweet	Bully Label	Sentiment Label	Target
T1: Larkyaaan toh jaisyyy bht hi phalwaan hoti. Ak chipkali ko dekh kr tm logon ka sans rukk jataa Translation: Yes, I know the girls are brave . Even a tiny lizard can stop their heart	Bully	Negative	Sexual-Orientation
T2: My friend called me moti and I instantly replied with "tere baap ka khati hoon" Translation: My friend called me fatso and my instant reply was "does your father bear my cost?"	Bully	Negative	AFR
T3: Laal phool gulaab phool shahrukh bhaiya beautiful Translation: Red flowers are roses, brother Shahrukh is beautiful.	Non-bully	Positive	NA

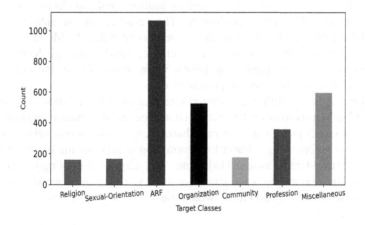

Fig. 1. Statistics of target class in our developed dataset.

3.4   Dataset Statistics

In the *BullyExplain* dataset, the average number of highlighted words per post is 4.97, with an average of 23.15 words per tweet. *maa* (mother), *randi* (whore), and *gandu* (asshole) are the top three content words for bully highlights, appearing in 33.23 percent of all bully posts. The total number of samples in the *BullyExplain* dataset is 6084, where 3034 samples belong to the non-bully class and the remaining 3050 samples are marked as bully. The number of tweets with positive and neutral sentiments are 1,536 and 1,327, respectively, while the remaining tweets have negative sentiments. Figure 1 shows the statistics of the Target class in *BullyExplain* dataset. From Fig. 1, we can observe that approximately one-third of total bully samples (3050) belong to the Attacking-Relatives-and-Friends (ARF) category (1067). This statistic reveals the nature of cyberbullying problem, where the victim's relatives and friends are the targets most of the time.

4   Explainable Cyberbullying Detection

This section presents our proposed "mExCB" model, shown in Fig. 2, for explainable cyberbullying detection. We utilized CNN and bi-GRU in both word and sub-sentence levels, along with self-attention, to make a robust end-to-end multitask deep learning model.

4.1   Text Embedding Generation

To generate the embedding of input sentence S (say) containing N number of tokens, we have experimented with BERT and VecMap.

(i) **BERT** [9] is a language model based on bidirectional transformer encoder with a multi-head self-attention mechanism. The sentences in our dataset are written in Hindi-English code-mixed form, so we choose mBERT (Multilingual BERT) pre-trained in 104 different languages, including Hindi and English. We have considered the sequence output from BERT, where each word of the input sentence has a 768-dimensional vector representation.

(ii) **VecMap** [2] is a multilingual word embedding mapping method. The main idea behind VecMap is to consider pretrained source and destination embeddings separately as inputs and align them in a shared vector space where related words are clustered together using a linear transformation matrix. As the inputs of VecMap, we have considered Fasttext [14] Hindi and English monolingual embeddings because FastText employs the character level in represented words into the vectors, unlike word2vec and Glove, which use word-level representations.

4.2   Feature Extraction

Bi-GRU and CNN have been employed to extract the hidden features from the input word embedding. (i) **Bi-GRU** [7] learns long term context-dependent semantic features into hidden states by sequentially encoding the embedding vectors, e, as

$$\overrightarrow{h}_t = \overrightarrow{GRU}_{fd}(e_t, h_{t-1}), \overleftarrow{h}_t = \overleftarrow{GRU}_{bd}(e_t, h_{t+1}) \tag{1}$$

where $\overrightarrow{h}_t$ and $\overleftarrow{h}_t$ are the forward and backward hidden states, respectively. The final hidden state representation for the input sentence is obtained as, $H_e = [h_1, h_2, h_3, ...h_N]$
where $h_t = \overrightarrow{h}_t, \overleftarrow{h}_t$ and $H_e \in \mathbb{R}^{N \times 2D_h}$. The number of hidden units in GRU is D_h.

(ii) **CNN** [18] effectively captures abstract representations that reflect semantic meaning at various positions in a text. To obtain N-grams feature map, $\mathbf{c} \in \mathbb{R}^{n-k_1+1}$ using filter $F \in \mathbb{R}^{k_1 \times d}$, we perform convolution operation, an element-wise dot product over each possible word-window, $W_{j:j+k_1-1}$. Each element c_j of feature map $\mathbf{c}$ is generated after convolution by

$$c_j = f(w_{j:j+k_1-1} * F_a + b) \tag{2}$$

Where f is a non linear activation function and b is the bias. Then we perform max pooling operation on $\mathbf{c}$. After applying F distinct filters of the same N-gram size, F feature maps will be generated, which can then be rearranged as

$$\mathbf{C} = [\mathbf{c_1}, \mathbf{c_2}, \mathbf{c_3}, ...\mathbf{c_F}]$$

Self-attention [37] has been employed to determine the impact of other words on the current word by mapping a query and a set of key-value pairs to an output. Outputs from Bi-GRU are passed through three fully connected (FC) layers, namely queries(Q), keys(K), and values(V) of dimension D_{sa}. Self-attention scores (SAT) are computed as follows:

$$SAT_i = softmax(Q_i K_i^T)V_i \tag{3}$$

where $SAT_i \in \mathbb{R}^{N \times D_{sa}}$.

4.3   mExCB :Multitask Framework for Explainable Cyberbullying Detection

We delineate the end-to-end process of the mExCB model as follows:

1. We initially compute the embedding for each word in a sentence, $S = \{s_1, s_2, ...s_N\}$. $E_{N \times D_e}^w = Embedding_{BERT/VecMap}(S)$ where D_e is the embedding dimension (for BERT $D_e = 768$ and for VecMap $D_e = 300$).
2. Word Level Representation: $E_{N \times D_e}^w$ has been passed through both Bi-GRU followed by Self-Attention and CNN layers to get word-level hidden features.
   $H_{N \times 2D_h}^w = GRU(E)$; where $D_h = 128$.
   $A_{N \times D_{sa}}^w = SAT(H)$; where $D_{sa} = 200$.
   Next, we take the mean of self-attenuated N vectors generated by GRU+SAT at the interval of l, where l is the number of words to generate a sub-sentence (SS).
   $G_{P \times D_{sa}}^w = AVG_l(A)$; where $P = N/l$.
   $C_{P \times F}^w = CNN(E)$; where $F = 200$
3. Sub-sentence Level Representation: We have added G^w and C^w to get the sub-sentence level embedding, $E_{P \times D_{sa}}^{ss}$. We followed similar steps as mentioned during the word-level representation to generate the sub-sentence level convoluted features, $C_{1 \times F}^{ss}$, and attenuated recurrent (GRU+SAT) features, $G_{1 \times D_{sa}}^{ss}$.
4. G^{ss} and C^{ss} are concatenated to get the final representation, E^s, of the given input, S. Up to this, all the layers are common for all the tasks, which basically helps in sharing task-specific information.

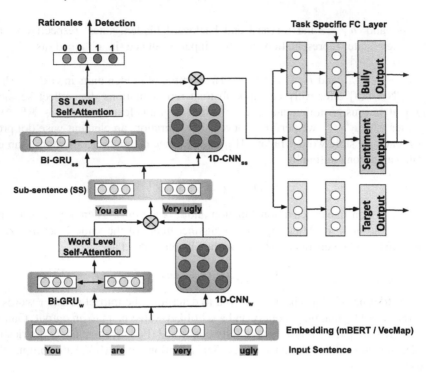

Fig. 2. Proposed Multitask Framework for Explainable Cyberbullying Detection, *mExCB*, architecture.

5. Task Specific Layers: We have three task-specific fully connected (FC) layers $(FC_1 - FC_2\text{-Softmax})$ followed by the corresponding output layer to simultaneously solve three tasks (Bully, Sentiment, and Target). For Rationales identification, we have fed G^{ss} into a sigmoid output layer which returns a binary encoding vector as an output. Each FC layer has 100 neurons. Further, outputs generated by the rationales identification and sentiment channels have been added to the last FC layer (FC_2) of the bully channel to examine how sentiment and rationales information helps in boosting the performance of cyberbullying detection.

4.4 Loss Function

Categorical cross-entropy [42] has been employed as an individual loss function for all the tasks. The final loss function, $Loss_f$ is dependent on M task-specific individual losses, $Loss_s^k$, as follows:

$$Loss_f = \sum_{k=1}^{M} \beta_k Loss_s^k \qquad (4)$$

The variable β is a hyperparameter which ranges from 0 to 1, defines the loss weights that characterise the per task loss-share to the total loss.

5 Experimental Results and Analysis

This section describes the outcomes of various baseline models and our proposed model, tested on the *BullyExplain* dataset. The experiments are intended to address the following research questions:

- **RQ1**: How does multi-tasking help in enhancing the performance of CD task?
- **RQ2**: What is the effect of different task combinations in our framework?
- **RQ3**: What is the motivation for keeping both CNN and GRU in our proposed framework?
- **RQ4**: Why do we use only the embeddings from pre-trained mBERT, but do not fine-tune the model itself for the multi-task set-up?
- **RQ5**: To handle noisy code-mixed data, which embedding is better, BERT or VecMap?

We performed stratified 10-fold cross-validation on our dataset and reported the mean metrics scores as done in [22]. During validation, we experimented with different network configurations and obtained optimal performance with batch size = 32, activation function=ReLu, dropout rate= 0.25, learning rate= 1e-4, epoch= 20. We used Adam optimizer with a weight_decay=1e-3 (for avoiding overfitting) for training. We set the value of β for the bully, rationales, sentiment, and target tasks for all the multi-task experiments by 1, 0.75, 0.66, and 0.50, respectively. All our experiments are performed on a hybrid cluster of multiple GPUs comprised of RTX 2080Ti.

5.1 Baselines Setup

We have experimented with different standard baseline techniques like CNN-GRU, BiRNN, BiRNN-Attention, and BERT-finetune, as mentioned in [24]. To investigate why both CNN and GRU are important in our model, we have performed an ablation study by adding other baselines (5 and 6) as follows:

1. **CNN-GRU**: The input is sent through a 1D CNN with window sizes of 2, 3, and 4, each with 100 filters. We employ the GRU layer for the RNN portion and then max-pool the output representation from the GRU architecture's hidden layers. This hidden layer is processed via a fully connected layer to output the prediction logits.
2. **BiRNN**: We pass the input features to BiRNN. The obtained hidden representation from BiRNN is sent to fully connected layers to obtain output.
3. **BiRNN-Attention**: This model differs from BiRNN model only in terms of the attention layer.
4. **BERT-finetune**: BERT's pooled output with dimension 768 was fed to a softmax output layer.
5. **mExCB$_{CNN}$**: In this architecture, we keep two 1D CNNs for word level and sub-sentence level encoding one after the other. The contextual hidden representation obtained from CNN is passed through multitasking channels.
6. **mExCB$_{GRU}$**: In this baseline, input features are passed through two Bi-GRU+SAT layers. Outputs from sub-sentence level representation are passed through multitasking channels.

There are four multitask variants based on how many tasks we want to solve simultaneously. As we have four tasks and our main objective is explainable cyberbullying detection, CD and RD are kept common for any multitask variants. So we have four multitask variants, i.e., CD+RD, CD+RD+SA, CD+Rd+TI, and CD+RD+SA+TI. For the CD, TI, and SA tasks, accuracy and macro-F1 metrics are used to evaluate predictive performance. For quantitative evaluation od RD task, we used a token-based, edit distance-based and sequence-based measure in the form of Jaccard Similarity (JS), Hamming distance (HD), and Ratcliff-Obershelp Similarity (ROS) metrics, respectively, as mentioned in [13].

5.2 Findings from Experiments

Table 2 and 3 illustrate the results of our proposed model, mExCB, and the other two variants of mExCB for Bully, Target, and Rationales detection tasks, respectively. Other baseline results are shown in Table 4. From the tables containing results we can conclude the following:

(RQ1) The proposed *mExCB* model outperforms all the baselines significantly for the CD task, improving 2.88% accuracy over the best baseline, BERT-finetune. For the rationale detection task, the improvement is 4.24% in terms of ROS. This improvement in performance reveals the importance of utilizing some auxiliary tasks, sentiment analysis, rationale detection, and target identification in boosting the performance of the main task (CD).

(RQ2) Unlike the four task variants (all tasks), Bully+Rationales+Sentiment (three tasks) settings attain the best results for CD and RD tasks. This finding established that increasing the number of tasks in a multitask framework does not always improve the performance of the main task compared to some less number task combinations. This decrease in the performance of the four tasks variant could be due to the task TI, which does not perform well and has a negative effect on other tasks.

Table 2. Experimental results of different multitask variants with BERT and VecMap embeddings for Bully and Target tasks. SA: Sentiment Analysis, CD: Cyberbully Detection, RD: Rationales Detection, TI: Target Identification

Embedding	Model	Bully+Rationales		Bully + Rationales +Sentiment		Bully + Rationales+Target				Bully + Rationales+Sentiment+Target			
		Bully		Bully		Bully		Target		Bully		Target	
		Acc	macro F1	Acc	macro F1	Acc	macro F1	Acc	macro F1	Acc	macro F1	Acc	macro F1
VecMap	mExCB$_{CNN}$	79.93	79.54	80.76	80.74	80.43	80.23	50.86	45.16	80.37	80.31	51.97	46.13
	mExCB$_{GRU}$	79.52	79.46	80.43	80.46	80.26	80.28	51.12	45.67	80.26	80.25	51.40	45.83
	mExCB	80.76	80.66	81.17	81.15	80.59	80.36	52.22	44.37	80.67	80.53	52.55	49.05
mBERT	mExCB$_{CNN}$	80.87	80.82	81.51	81.45	81.09	80.91	50.93	44.38	80.84	80.58	52.31	47.87
	mExCB$_{GRU}$	80.67	80.64	81.25	81.26	81.99	81.89	51.47	46.22	79.93	79.93	51.81	45.68
	mExCB	82.24	82.31	**83.31**	**83.24**	82.07	82.11	53.11	49.08	82.24	82.19	**54.54**	**50.20**

For the Target Identification (TI) task, the maximum accuracy and F1 score of 54.54 and 50.20, respectively, are attained. This low accuracy in the TI task could be due to the imbalanced nature of the Target class.

Table 3. Experimental results of different multitask variants with BERT and VecMap embeddings for RD task. RD: Rationales Detection

Embedding	Model	Bully+Rationales			Bully + Rationales +Sentiment			Bully + Rationales +Target			Bully + Rationale +Target+Sentiment		
		Rationales			Rationales			Rationales			Rationales		
		JS	HD	ROS	JS	HD	ROS	JS	HD	ROS	JS	HD	ROS
VecMap	mExCB$_{CNN}$	44.11	43.22	44.18	44.58	43.41	44.62	44.31	43.15	45.34	44.68	43.57	45.17
	mExCB$_{GRU}$	44.74	43.45	45.86	44.93	43.69	46.87	44.41	43.98	46.23	44.87	43.69	46.34
	mExCB	45.68	43.98	47.86	45.71	44.15	48.64	45.11	44.19	46.66	44.73	43.89	46.47
mBERT	mExCB$_{CNN}$	45.34	43.28	47.58	45.28	43.78	47.62	45.44	43.72	47.22	45.31	43.56	47.23
	mExCB$_{GRU}$	45.37	43.57	47.89	45.62	43.96	48.23	45.53	43.71	47.75	45.49	43.88	47.71
	mExCB	45.83	44.26	48.78	**46.39**	**44.65**	**49.42**	45.74	43.87	47.67	45.74	43.67	47.92

Table 4. Results of different baseline methods evaluated on *BullyExplain* data

Model	Bully		Bully+Rationales				
			Bully		Rationales		
	Acc	macro F1	Acc	macro F1	JS	HD	ROS
CNN GRU	78.78	78.37	79.28	79.14	43.78	42.81	43.77
BiRNN	78.62	78.55	79.44	79.51	42.11	42.53	43.12
BiRNN$_{attn}$	79.52	79.45	79.93	80.01	43.15	43.02	44.38
BERT$_{fine}$	80.18	80.08	**80.43**	**80.35**	**44.53**	**43.72**	**45.18**

(RQ3) *mExCB* has consistently performed better than mExCB$_{CNN}$ and mExCB $_{GRU}$ in both embedding strategies (BERT / VecMap). Like, in (CD+RD+SA) multi-task setting, *mExCB* performs better than mExCB$_{CNN}$ and mExCB$_{GRU}$ with improvements in the F1 score of 1.79% and 1.98%, respectively for the CD task. From Table 2, we can notice that mExCB$_{CNN}$ performs better than mExCB$_{GRU}$ for the CD task most of the time, and the reverse scenario occurs for the RD task. That is why for the RD task, only the self-attenuated sub-sentence level Bi-GRU features have been sent to the RD output layer. This finding supports the idea of using both GRU and CNN so that model can learn both long-range semantic features as well as local key phrase-based information.

(RQ4) We have already experimented with fine-tuning BERT [24] (Baseline-4) and achieved 80.18% and 80.43% (see Table 4) accuracy values in single task (CD) and multi-task (CD+RD) settings, respectively, for the CD task. On the other hand, without fine-tuning the BERT followed by CNN and GRU, we have obtained an accuracy of 83.31% for the CD task with (CD+RD+SA) multi-task setting. The BERT fine-tuning version of our proposed model achieved the highest accuracy of 81.59% for the CD task, which underperforms the non-fine-tuning version of our model. One of the possible reasons why fine-tuning is not performing well could be the less number of samples in our dataset.

(RQ5) In Table 2, we can observe that all the models performed better when BERT was used for embedding generation instead of VecMap. This result again highlights the superiority of the BERT model in different types of NLP tasks. We have not included the VecMap embedded results in Table 4 containing baselines as it performs poorly.

We have conducted a statistical t-test on the results of ten different runs of our proposed model and other baselines and obtained a p-value less than 0.05.

5.3 Comparison with SOTA

Table 5 shows the results of the existing state-of-the-art approach for CD and SA tasks on the Hindi-English code-mixed dataset. We keep the original two tasks, SA and CD, in our proposed model, *mExCB* (CD+SA) for a fair comparison between SOTA. Table 5 shows that mExCB (CD+SA) also outperforms the SOTA, illustrating that our proposed model can perform better without the introduced annotations (RD and TI). Furthermore, when we add a new task RD, the three task combination variant, *mExCB* (CD+SA+RD) outperforms both SOTA (MT-BERT+VecMap) and *mExCB*(CD+SA) with improvements in the accuracy value of 2.19% and 1.36%, respectively, for the CD task. This improvement illustrates the importance of incorporating the RD task in enhancing the performance of the main task, i.e., CD. Hence, our proposed approach is beneficial from both aspects, (i) Enhanced the performance of the CD task and (ii) Generates a human-like explanation to support the model's decision, which is vital in any AI-based task.

Table 5. Results of state-of-the-art model and the proposed model; ST: Single Task, MT: Multi-Task

Model	Bully		Sentiment	
	Acc	F1	Acc	F1
SOTA				
BERT_finetune [24]	80.18	80.08	76.10	75.62
ST- BERT+VecMap [22]	79.97	80.13	75.53	75.38
MT-BERT+VecMap [22]	81.12	81.50	77.46	76.95
Ours				
mExCB (CD+SA)	81.95	82.04	77.55	77.12
mExCB (CD+SA+RD)	**83.31**	**83.24**	**78.54**	**78.13**
Improvements	2.19	1.74	1.08	1.18

6 Error Analysis

We have manually checked some samples from the test set to examine how machine-generated rationales and bully labels differ from the human annotator's decision. Table 6 shows the predicted rationales and bully labels of a few test samples obtained by different baselines (CNN-GRU, BiRNN-Attn) and our proposed models (mExCB).

Table 6. In comparison to human annotators, rationales identified by several models are shown. Green highlights indicate agreements between the human annotator and the model. Orange highlighted tokens are predicted by models, not by human annotators. Yellow highlighted tokens are predicted by models but are not present in the original text.

Model	Text	Bully Label
Human annotator (T1)	Semi final tak usi bnde ne pahochaya hai jisko tu gandu bol raha .	Non-Bully
Translation	**The person you are calling ass*ole is the one that helped us to reach the semi finals.**	
CNN-GRU	Semi final tak usi bnde ne pahochaya hai jisko tu gandu bol raha .	Bully
BiRNN-Attn	Semi final tak usi bnde ne pahochaya hai jisko tu gandu bol raha .	Bully
mExCB$_{CNN}$	Semi final tak usi bnde ne pahochaya hai jisko tu gandu bol raha .	Bully
mExCB$_{GRU}$	Semi final tak usi bnde ne pahochaya hai jisko tu gandu bol raha .	Bully
mExCB	Semi final tak usi bnde ne pahochaya hai jisko tu gandu bol raha .	Bully
Human annotator (T2)	Abey mc gb road r ki pehle customer ki najayaz auladte	Bully
Translation	**You are the illegitimate children of first customer of GB road.**	
CNN-GRU	Abey mc gb road r ki pehle customer ki najayaz auladte	Bully
BiRNN -Attn	Abey mc gb road r ki pehle customer ki najayaz auladte	Bully
mExCB$_{CNN}$	Abey mc gb road r ki pehle customer ki najayaz auladte	Bully
mExCB$_{GRU}$	Abey mc gb road r ki pehle customer ki najayaz auladte	Bully
mExCB	Abey mc gb road r ki pehle customer ki najayaz auladte	Bully

1. It can be observed that the human annotator labeled the T1 tweet as Non-Bully. In contrast, all the models (both baselines and *GenEx* models) predicted the label as Bully highlighting the offensive word *gandu* (Asshole), supporting their predictions. This shows that the model cannot comprehend the context of this offensive word as it is not directed at anyone and has been used more in a sarcastic manner, highlighting the model's limitation in understanding indirect and sarcastic statements.

2. All the models (both baselines and our proposed models) predicted the correct label for tweet T2. But if we see the rationales predicted (highlighted part), none of the baseline models performed well compared to human decisions. Both *mExCB$_{GRU}$* and *mExCB* can predict all the words present in the ground truth rationale, but it also predicts other phrases as the rationale. In this case, we can also notice that *mExCB$_{GRU}$* predicts some more tokens which are not rationales compared to *mExCB* model. Our proposed model performs slightly better than others in the RD task. Models that excel in classification may not always be able to give reasonable and accurate rationales for their decisions as classification task needs sentence-level features [24]. In contrast, rationales detection mainly focuses on the token-level feature. This observation (low performance in RD task) indicates that more research is needed on explainability. We believe our developed dataset will help future research on explainable cyberbullying detection.

7   Conclusion and Future Work

In this paper we have made an attempt in solving the cyberbullying detection task in code-mixed setting keeping the explainability aspect in mind. As explainable AI systems help in improving trustworthiness and confidence while deployed in real-time and cyberbully detection systems are required to be installed in different social media sites for online monitoring, generating rationales behind taken decisions is a must. The contributions of the current work are two fold: (a) developing the first explainable cyberbully detection dataset in code-mixed language where rationales/phrases used for decision making are annotated along with bully label, sentiment label and target label. (b) A multitask framework *mExCB* based on word and sub-sentence label attention has been proposed to solve four tasks (Bully, Sentiment, Rationales, Target) simultaneously. Our proposed model outperforms all the baselines and beats state-of-the-art with an accuracy score of 2.19% for cyberbullying detection.

In future we would like to work on explainable cyberbully detection from code-mixed data considering image and text modalities.

Acknowledgement. Dr. Sriparna Saha gratefully acknowledges the Young Faculty Research Fellowship (YFRF) Award, supported by Visvesvaraya Ph.D. Scheme for Electronics and IT, Ministry of Electronics and Information Technology (MeitY), Government of India, being implemented by Digital India Corporation (formerly Media Lab Asia) for carrying out this research. The Authors would also like to acknowledge the support of Ministry of Home Affairs (MHA), India, for conducting this research.

References

1. Agrawal, S., Awekar, A.: Deep learning for detecting cyberbullying across multiple social media platforms. In: Pasi, G., Piwowarski, B., Azzopardi, L., Hanbury, A. (eds.) ECIR 2018. LNCS, vol. 10772, pp. 141–153. Springer, Cham (2018). https://doi.org/10.1007/978-3-319-76941-7_11
2. Artetxe, M., Labaka, G., Agirre, E.: Learning bilingual word embeddings with (almost) no bilingual data. In: Proceedings of the 55th Annual Meeting of the Association for Computational Linguistics (Volume 1: Long Papers), pp. 451–462 (2017)
3. Badjatiya, P., Gupta, S., Gupta, M., Varma, V.: Deep learning for hate speech detection in tweets. In: Proceedings of the 26th International Conference on World Wide Web Companion, pp. 759–760 (2017)
4. Balakrishnan, V., Khan, S., Arabnia, H.R.: Improving cyberbullying detection using twitter users' psychological features and machine learning. Comput. Secur. **90**, 101710 (2020)
5. Bohra, A., Vijay, D., Singh, V., Akhtar, S.S., Shrivastava, M.: A dataset of Hindi-English code-mixed social media text for hate speech detection. In: Proceedings of the Second Workshop on Computational Modeling of People's Opinions, Personality, and Emotions in Social Media, pp. 36–41 (2018)
6. Camburu, O.M., Rocktäschel, T., Lukasiewicz, T., Blunsom, P.: e-snli: Natural language inference with natural language explanations. Adv. Neural Inf. Process. Syst. **31** (2018)
7. Cho, K., Van Merriënboer, B., Bahdanau, D., Bengio, Y.: On the properties of neural machine translation: encoder-decoder approaches. arXiv preprint arXiv:1409.1259 (2014)
8. Dadvar, M., Trieschnigg, D., de Jong, F.: Experts and machines against bullies: a hybrid approach to detect cyberbullies. In: Sokolova, M., van Beek, P. (eds.) AI 2014. LNCS (LNAI), vol. 8436, pp. 275–281. Springer, Cham (2014). https://doi.org/10.1007/978-3-319-06483-3_25
9. Devlin, J., Chang, M.W., Lee, K., Toutanova, K.: Bert: pre-training of deep bidirectional transformers for language understanding. arXiv preprint arXiv:1810.04805 (2018)
10. DeYoung, J., et al.: Eraser: a benchmark to evaluate rationalized nlp models. arXiv preprint arXiv:1911.03429 (2019)

11. Dinakar, K., Reichart, R., Lieberman, H.: Modeling the detection of textual cyberbullying. In: Proceedings of the International Conference on Weblog and Social Media 2011, Citeseer (2011)
12. Fleiss, J.L.: Measuring nominal scale agreement among many raters. Psychol. Bull. **76**(5), 378 (1971)
13. Ghosh, S., Roy, S., Ekbal, A., Bhattacharyya, P.: CARES: CAuse recognition for emotion in suicide notes. In: Hagen, M., et al. (eds.) ECIR 2022. LNCS, vol. 13186, pp. 128–136. Springer, Cham (2022). https://doi.org/10.1007/978-3-030-99739-7_15
14. Grave, E., Bojanowski, P., Gupta, P., Joulin, A., Mikolov, T.: Learning word vectors for 157 languages. arXiv preprint arXiv:1802.06893 (2018)
15. Gunning, D., Stefik, M., Choi, J., Miller, T., Stumpf, S., Yang, G.Z.: Xai-explainable artificial intelligence. Sci. Robot. **4**(37), eaay7120 (2019)
16. Kamble, S., Joshi, A.: Hate speech detection from code-mixed Hindi-English tweets using deep learning models. arXiv preprint arXiv:1811.05145 (2018)
17. Karim, M.R., et al.: Deephateexplainer: explainable hate speech detection in under-resourced Bengali language. In: 2021 IEEE 8th International Conference on Data Science and Advanced Analytics (DSAA), pp. 1–10. IEEE (2021)
18. Kim, Y.: Convolutional neural networks for sentence classification. arXiv preprint arXiv:1408.5882 (2014)
19. Kumar, R., Reganti, A.N., Bhatia, A., Maheshwari, T.: Aggression-annotated corpus of Hindi-English code-mixed data. arXiv preprint arXiv:1803.09402 (2018)
20. Lewis, M., Haviland-Jones, J.M., Barrett, L.F.: Handbook of emotions. Guilford Press, New York (2010)
21. Maity, K., Saha, S.: BERT-capsule model for cyberbullying detection in code-mixed Indian languages. In: Métais, E., Meziane, F., Horacek, H., Kapetanios, E. (eds.) NLDB 2021. LNCS, vol. 12801, pp. 147–155. Springer, Cham (2021). https://doi.org/10.1007/978-3-030-80599-9_13
22. Maity, K., Saha, S.: A multi-task model for sentiment aided cyberbullying detection in code-mixed Indian languages. In: Mantoro, T., Lee, M., Ayu, M.A., Wong, K.W., Hidayanto, A.N. (eds.) ICONIP 2021. LNCS, vol. 13111, pp. 440–451. Springer, Cham (2021). https://doi.org/10.1007/978-3-030-92273-3_36
23. Maity, K., Sen, T., Saha, S., Bhattacharyya, P.: Mtbullygnn: a graph neural network-based multitask framework for cyberbullying detection. IEEE Trans. Comput. Soc. Syst. (2022)
24. Mathew, B., Saha, P., Yimam, S.M., Biemann, C., Goyal, P., Mukherjee, A.: Hatexplain: a benchmark dataset for explainable hate speech detection. arXiv preprint arXiv:2012.10289 (2020)
25. Myers-Scotton, C.: Duelling languages: Grammatical structure in codeswitching. Oxford University Press, Oxford (1997)
26. NCRB: Crime in india - 2020. National Crime Records Bureau (2020)
27. Paul, Sayanta, Saha, Sriparna: CyberBERT: BERT for cyberbullying identification. Multimedia Syst. 1–8 (2020). https://doi.org/10.1007/s00530-020-00710-4
28. Pramanick, S., et al.: Detecting harmful memes and their targets. arXiv preprint arXiv:2110.00413 (2021)
29. Rajani, N.F., McCann, B., Xiong, C., Socher, R.: Explain yourself! leveraging language models for commonsense reasoning. arXiv preprint arXiv:1906.02361 (2019)
30. Regulation, P.: Regulation (EU) 2016/679 of the European parliament and of the council. Regulation (EU) **679**, 2016 (2016)
31. Reynolds, K., Kontostathis, A., Edwards, L.: Using machine learning to detect cyberbullying. In: 2011 10th International Conference on Machine Learning and Applications and Workshops, vol. 2, pp. 241–244. IEEE (2011)

32. Rijhwani, S., Sequiera, R., Choudhury, M., Bali, K., Maddila, C.S.: Estimating code-switching on twitter with a novel generalized word-level language detection technique. In: Proceedings of the 55th Annual Meeting of the Association for Computational Linguistics (volume 1: long papers), pp. 1971–1982 (2017)
33. Saha, T., Upadhyaya, A., Saha, S., Bhattacharyya, P.: A multitask multimodal ensemble model for sentiment-and emotion-aided tweet act classification. IEEE Trans. Comput. Soc. Syst. (2021)
34. Singh, A., Saha, S., Hasanuzzaman, M., Dey, K.: Multitask learning for complaint identification and sentiment analysis. Cogn. Comput. **14**(1), 212–227 (2021). https://doi.org/10.1007/s12559-021-09844-7
35. Smith, P.K., Mahdavi, J., Carvalho, M., Fisher, S., Russell, S., Tippett, N.: Cyberbullying: its nature and impact in secondary school pupils. J. Child Psychol. Psychiatry **49**(4), 376–385 (2008)
36. Sticca, F., Ruggieri, S., Alsaker, F., Perren, S.: Longitudinal risk factors for cyberbullying in adolescence. J. Community Appl. Soc. Psychol. **23**(1), 52–67 (2013)
37. Vaswani, A., et al.: Attention is all you need. In: Advances in Neural Information Processing Systems, pp. 5998–6008 (2017)
38. Waseem, Z., Hovy, D.: Hateful symbols or hateful people? predictive features for hate speech detection on twitter. In: Proceedings of the NAACL Student Research Workshop, pp. 88–93 (2016)
39. Ybarra, M.L., Mitchell, K.J., Wolak, J., Finkelhor, D.: Examining characteristics and associated distress related to internet harassment: findings from the second youth internet safety survey. Pediatrics **118**(4), e1169–e1177 (2006)
40. Yin, W., Kann, K., Yu, M., Schütze, H.: Comparative study of CNN and RNN for natural language processing. arXiv preprint arXiv:1702.01923 (2017)
41. Zaidan, O., Eisner, J., Piatko, C.: Using "Annotator rationales" to improve machine learning for text categorization. In: Human Language Technologies 2007: The Conference of the North American Chapter of the Association for Computational Linguistics; Proceedings of the Main Conference, pp. 260–267 (2007)
42. Zhang, Z., Sabuncu, M.: Generalized cross entropy loss for training deep neural networks with noisy labels. Adv. Neural Inf. Process. Syst. **31** (2018)

LayoutGCN: A Lightweight Architecture for Visually Rich Document Understanding

Dengliang Shi[✉][iD], Siliang Liu[iD], Jintao Du[iD], and Huijia Zhu[iD]

Tiansuan Lab, Ant Group, Shanghai, China
{dengliang.sdl,liusiliang.lsl,lingke.djt,huijia.zhj}@antgroup.com

Abstract. Visually rich document understanding techniques have numerous practical application scenarios in the real world. However, existing researches tend to focus on large-scale models, especially pre-trained ones, training which is resource-consuming. Additionally, these models are unfeasible for low-resource situations, like edge applications. This paper proposes the LayoutGCN, a novel, lightweight architecture, to classify, extract, and structuralize information from visually rich documents. It treats a document as a graph of text blocks and employs convolution neural networks to encode all features from different modalities. Rich representations for text blocks containing textual, layout, and visual information are generated by a graph convolution network whose adjacency matrix is carefully designed to fully use the relative position information between text blocks. Extensive experiments on five benchmarks show the applicability of LayoutGCN for various downstream tasks and its comparable performance to existing large-scale models.

Keywords: Visually Rich Document Understanding · Lightweight Architecture · Deep Learning

1 Introduction

Visually Rich Document (VRD) refers to documents with non-trivial layout and formatting, like receipts, invoices, resumes, forms and et al. Visually Rich Document Understanding (VRDU) aims to classify, extract, and structuralize information from scanned or digital documents, which is usually regarded as a downstream task of Optical Character Recognition (OCR). Only a few works attempt to solve these two tasks in an end-to-end framework [3,34].

In addition to text, layout and visual information also plays a critical role in VRDU tasks. Recent researches on VRDU propose various approaches to efficiently fusing text, layout, and image into a single framework. These methods are primarily divided into two directions: pre-training methods and non-pre-training methods. The pre-training methods, also the most popular, are to pre-train text, layout, and image in a multi-modal model generally based on BERT-like architecture [1,5,14,20]. In the other direction, researchers treat text, layout,

G. A. Fink et al. (Eds.): ICDAR 2023, LNCS 14189, pp. 149–165, 2023.
https://doi.org/10.1007/978-3-031-41682-8_10

and image as three modalities and leverage these multi-modal features by multi-modal fusion techniques [29] or Graph Neural Networks (GNN) [23,33] without any document-specific pre-training.

Although existing approaches have achieved great success in various VRDU tasks, they all employ large-scale models, especially pre-trained models, which require a large amount of data, time, and computation power during training. As public awareness grows on climate and social responsibility issues, increasing attention has been paid to the environmental impact of training large-scale models. According to the study by Strubell et al. [28], carbon emissions of training BERT on GPU are roughly equivalent to a trans-American flight. What's more, in practice, documents should be dealt with on edge devices in some cases, which requires edge exportable models. Finally, most current research adopts pre-training techniques to jointly pre-train text, layout, and image in a single framework. These document pre-training solutions take 2D layout data as an extension of 1D positions, resulting in the underutilization of layout information. Intuitively, layout information is a shallow feature easily learned by the model for a specific type of document. That leaves us with a natural question: are those deep semantics features learned by data-hungry pre-training necessary for all VRDU tasks? The above issues motivate us to explore possible lightweight, edge-friendly, and non-pre-training methods for VRDU problems.

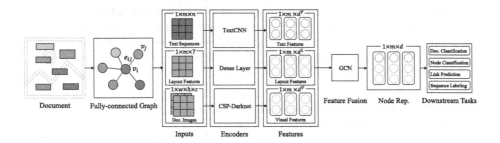

Fig. 1. Illustration of a sample flow of LayoutGCN.

In this article, we propose LayoutGCN, a novel, lightweight and practical approach for VRDU. Figure 1 provides an overview of the workflow of Layout-GCN. Firstly, we model a document as a fully-connected graph of text blocks. The text of blocks and the document image are encoded by Convolution Neural Network (CNN), and layout features are projected into hyperspace with a dense layer. Graph Convolution Network (GCN) with a carefully-designed adjacency matrix fuses textual, layout, and visual features to yield richer representation for downstream tasks. Various downstream tasks can be processed with Layout-GCN flexibly, like document classification, node classification, link prediction, and sequence labeling. We evaluate our model on various public benchmarks, and the experiment results show the effectiveness of LayoutGCN on multiple VRDU tasks. The main contributions of this paper are summarized as follows:

- We introduce a lightweight and effective model architecture for VRDU that can be applied to low-resource applications freely, like edge applications or application scenarios lacking a large amount of data.
- We propose a novel mechanism to generate weights of adjacency matrix, by which the relative position information between every two text blocks is introduced into the model as prior knowledge.
- We redefine VRDU tasks as sequence labeling, document classification, node classification, and link prediction, which can all be solved flexibly with LayoutGCN.
- Extensive experiments are conducted on various benchmarks to illustrate the practicability and effectiveness of LayoutGCN.

2   Related Work

2.1   Pre-training Methods

The recent researches on VRDU mainly focus on pre-trained approaches. LayoutLM [32] first attempt to pre-train text and layout jointly in a single framework and introduces a new document-level pre-training task called Masked Visual-Language Model (MVLM). LayoutLMv2 [31] integrates image information into the pre-training stage and proposes a spatial-aware self-attention mechanism to capture relative position information between token pairs. StructralLM [20] uses segment-level positional embeddings instead of word-level ones and designs a new pre-training objective, Cell Position Prediction (CPP). DocFormer [1] shares the learned spatial embeddings across modalities to make the model aware of the relationship between text and visual tokens. LAMBERT [5] simplifies the pre-training architecture by only injecting layout information into the model using layout embedding and relative bias based on RoBERTa [24]. LayoutLMv3 [14] proposes two pre-training tasks: Masked Image Modeling (MIM) and Word-Path Alignment (WPA).

Pre-trained methods achieve the SOTA results on most benchmarks. However, in addition to taking a large number of resources, document pre-training technology still has three major disadvantages. Firstly, unlike text pre-training, documents are not easy to access, and it is challenging to build a representative dataset for document pre-training. Moreover, pre-trained models all inherit BERT-like architecture, which is designed for sequential data and unsuitable for spatially structured data, like layout information. Finally, layout information is easy to learn in a specific VRDU task. Deep semantics features from text and images learned during pre-training are unnecessary for all VRDU tasks.

2.2   Non-pre-training Methods

DocStruct [29] is a multi-modal method to extract and structuralize information from VRD. It takes visual features as shifting features and uses an attention-based influence gate to control the influence of visual information. The main disadvantages of this method are that features from different modalities are encoded separately, and the relative position information between text blocks is ignored. The most related works to our approach are Liu et al. [23], and Yu et al. [33]. They both get richer representations containing features from multiple modalities using GNN. Liu et al. [23] build a fully-connected graph from a document and define convolution operation on node-edge-node triplets to get node representations. Nevertheless, the layout information of text blocks, like the position and size of blocks, is not considered. Meanwhile, by simply concatenating edge embedding, which primarily represents layout information, to node embedding, the graph convolution can hardly distinguish the correlation of every two nodes. Compared to Liu et al. [23], PICK [33] adds image information into the model and introduces a graph learning module to learn a soft adjacency matrix for graph convolution. However, it also underutilizes layout information, and the model can be simplified by pre-defining the weights of the adjacency matrix using the relative position information between text blocks. Inspired by the above works, we design a lightweight architecture for VRDU and focus on optimizing the mechanism of using layout features.

3 Approach

3.1 Document Modeling

A visually rich document is comprised of a document image $\mathcal{D}$ whose size is $h \times w$ and a set of text blocks. We build a fully-connected graph $\mathcal{G}$ by taking text blocks as nodes $\mathcal{V} = \{v_1, v_2, \cdots, v_m\}$ and adding edges $\mathcal{E} = \{e_{ij}, \ i = 1, 2, \cdots, m; \ j = 1, 2, \cdots, m\}$ between every two blocks, where m is the number of blocks. Generally, node v_i contains a textual sequence $\mathcal{T}_i = \{w_1^i, w_2^i, \cdots, w_n^i\}$, where n is the number of tokens, and coordinates of bounding box $[x_1^i, y_1^i, x_2^i, y_2^i]$.

3.2 Model Architecture

The whole architecture of our proposed model is shown in Fig. 2. The four components of the architecture are the textual encoder, visual encoder, layout encoder, and graph module.

Textual Encoder. In VRDU tasks, especially Information Extraction (IE) tasks, the important information is usually represented by keywords like title, the header of a table, the label of a form, or string with obvious patterns, such as named entities, date and et al. So deep semantic features from the text are unnecessary, and n-gram textual features are enough in most cases. We adopt TextCNN [17] as the textual encoder and use the same convolution operation to keep the input and output sequence the same length. Let the sequence output for $\mathcal{T}_i$ be $[s_1^i, s_2^i, \cdots, s_n^i]$, $s_t^i \in \mathbb{R}^{d_T}$, where d_T is the dimension of sequence output. Then, textual features $r_i^T \in \mathbb{R}^{d_T}$ for node v_i are gained from sequence output using 1D max pooling, like

$$r_i^T = \mathrm{MaxPooling1D}([s_1^i, s_2^i, \cdots, s_n^i])$$

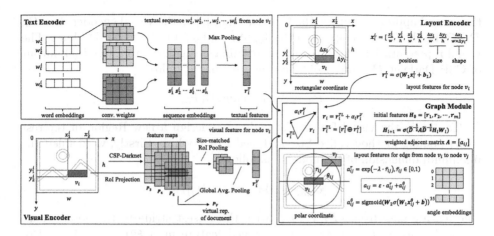

Fig. 2. Model Architecture of LayoutGCN.

Finally, to improve the stability and robustness of the model, we add layer normalization [2] and dropout [27] on the textual features.

Layout Encoder. With the coordinates of the bounding box, we can get each node's position, size, and shape. Intuitively, these layout features are all important for understanding documents. For node v_i, its layout features are represented as

$$\boldsymbol{x}_i^L = [\frac{x_1^i}{w}, \frac{y_1^i}{h}, \frac{x_2^i}{w}, \frac{y_2^i}{h}, \frac{\Delta x_i}{w}, \frac{\Delta y_i}{h}, \frac{\Delta x_i}{w \times \Delta y_i}]$$

where $\Delta x_i = x_2^i - x_1^i$ and $\Delta y_i = y_2^i - y_1^i$. The layout features are projected into a hyperspace using a fully connected layer.

$$\boldsymbol{r}_i^L = \sigma(\boldsymbol{W}_1\boldsymbol{x}_i^L + \boldsymbol{b}_1)$$

where $\sigma(\cdot)$ denotes an activation function, $\boldsymbol{W}_1 \in \mathbb{R}^{d_L \times 7}$ and $\boldsymbol{b}_1 \in \mathbb{R}^{d_L}$ are trainable weight and bias. d_L is the dimension of layout features. Following this fully connected layer, we also add layer normalization and dropout.

Visual Encoder. The image of a document also provides valuable information for VRDU sometimes. For example, in some cases, critical information is represented by bold or italic text or with a remarkable background. There are two main ways to deal with a document image. Firstly, according to the coordinates of each node's bounding box, cropping the document image into multiple image segments, and these segments are encoded separately as visual features for nodes. Nevertheless, the visual information around text blocks losses in this way. The other method is to encode the whole document image first and get the visual features of nodes using RoI pooling. The document image's feature maps can also be used for downstream tasks, like document classification. We follow the second approach and use the CSP-Darknet [6] as the visual encoder. Let P_3, P_4, P_5 be the feature maps from the 3_{th}, 4_{th} and 5_{th} block of CSP-Darknet. The visual representation P_V of the document image is gained from the final output of CSP-Darknet P_5 using global average pooling as follows:

$$P_V = \text{AvgPooling2D}(P_5)$$

To obtain the visual features of nodes, we design a new Region of Interest (RoI) pooling method called Size-matched RoI Pooling. Firstly, we defined the maximum dimension of the node v_i as follows:

$$d^i_{max} = \text{Max}([\Delta x_i, \Delta y_i])$$

The feature map from $\{P_3, P_4, P_5\}$ with the closest stride to the node's maximum dimension d^i_{max} is taken as matched feature map $P^i_k (k \in [3,4,5])$. RoI pooling layer [12] takes the matched feature maps as inputs, like

$$X^V_i = \text{RoI-Pooling}(P^i_k, [x^i_1, y^i_1, \Delta x_i, \Delta y_i])$$

With the outputs of the RoI pooling layer, we gain the visual features for node v_i using a global average pooling layer and a dense layer, as follows

$$r^V_i = \sigma(W_2 \text{AvgPooling2D}(X^V_i) + b_2)$$

where $W_2 \in \mathbb{R}^{d_V \times c_k}$ and $b_2 \in \mathbb{R}^{d_V}$ are weight matrix and bias vector, c_k is the channel of feature maps. $r^V_i \in \mathbb{R}^{d_V}$ is the visual feature for node v_i and $d_V = d_T + d_L$. Dropout and layer normalization are also applied to visual features here.

Graph Module. In the graph module, GCN (Graph Convolution Network) [30] is introduced to catch the latent relationships between nodes. Firstly, we leverage textual, layout, and visual features from the above encoders to construct initial node features. According to Wang et al. [29], the textual and layout features are reliable and can be concatenated directly as base features, like

$$r^{TL}_i = [r^T_i \oplus r^L_i]$$

where $\oplus$ indicates the concatenation operator.

The visual features should be treated as shifting features and fused with base features using the following attention-based influence gate mechanism:

$$r_i = r^{TL}_i + \alpha_i \cdot r^V_i$$
$$\alpha_i = \text{Sigmoid}(W_3[r^{TL}_i \oplus r^V_i] + b_3)$$

where $W_3 \in \mathbb{R}^{1 \times (d_T + d_L + d_V)}$ and $b_3 \in \mathbb{R}$ are weight matrix and bias, α_i is gate weight. r_i is the joint features for node v_i. Then the initial features of all nodes can be represented as $H_0 = [r_1, r_2, \cdots, r_m]$.

The richer node representations are computed by multiple GCN layers as follows:

$$H_{l+1} = \sigma(D^{-\frac{1}{2}} A D^{-\frac{1}{2}} H_l W_l)$$

where $A = [a_{ij}] \in \mathbb{R}^{m \times m}$ is the weighted adjacency matrix. $D = [d_{ij}] \in \mathbb{R}^{m \times m}$ is a diagonal matrix and $d_{ii} = \sum_{j=1}^{m} a_{ij}$. $H_l \in \mathbb{R}^{m \times d}$ is the hidden features of the l_{th} layer, $d = d_T + d_L$ when $l = 0$ otherwise $d = d_h$. d_h is the hidden size of the model. $W_l \in \mathbb{R}^{d \times d_h}$ is a layer-specific trainable weight matrix.

To catch the relative position information between text blocks, we carefully design the weights of the adjacent matrix. We model the relative position between text blocks in a polar coordinate. Intuitively, the closer the radical distance r_{ij} between node v_i and v_j is, the larger weight a_{ij} should be. The angle θ_{ij} also significantly impacts the weight a_{ij}. For example, keys usually appear directly on the left or top of values in a document. The weight a_{ij} is split into two parts:

$$a_{ij} = \varepsilon \cdot a_{ij}^r + a_{ij}^\theta$$

where a_{ij}^r and a_{ij}^θ are weights related to r_{ij} and θ_{ij} respectively. $\varepsilon \in \mathbb{R}$ is a trainable parameter.

The weight a_{ij}^r is computed by

$$a_{ij}^r = e^{-\lambda \cdot \bar{r}_{ij}}, \bar{r}_{ij} = r_{ij} / \sqrt{w^2 + h^2}$$

where λ is a positive constant to control the decay speed of weight along with radical distance.

Since it is difficult to build the relationship between weight a_{ij}^θ and angle θ_{ij} explicitly, we let the model learn this relationship itself. Inspired by position embedding [7], we discretize the angle θ_{ij} with an interval δ as follows:

$$\bar{\theta}_{ij} = \text{Round}(\theta_{ij}/\delta)$$

$\bar{\theta}_{ij}$ is mapped into a fixed embedding $x_{ij}^\theta \in \mathbb{R}^{d_\theta}$ which is initialized randomly. The relationship between a_{ij}^θ and x_{ij}^θ is represented by two linear layers:

$$a_{ij}^\theta = \text{Sigmoid}(W_5 \cdot \text{ReLU}(W_4 x_{ij}^\theta + b_4))$$

where $W_4 \in \mathbb{R}^{d_h \times d_\theta}$, $W_5 \in \mathbb{R}^{1 \times d_h}$ and $b_4 \in \mathbb{R}^{d_h}$ are learnable weight matrix and bias vector.

Finally, we concatenate the outputs $H = [h_i]$ from the last layer of GCN with the initial node features as final node representations. The final representation of node v_i can be represented as

$$\tilde{r}_i = [r_i \oplus h_i]$$

Node representation $\tilde{r}_i$ fully integrates information from text, layout, and image and can significantly benefit multiple downstream tasks.

3.3   Downstream Tasks

Sequence Labeling. Extracting information from a document is usually treated as a sequence labeling task. To keep the boundary information of text blocks, we treat text from each block as an independent sequence. Take block v_i as an example, we concatenate the node representation $\tilde{r}_i$ to the sequence output from the textual encoder as the inputs of the decoder for sequence labeling. The decoder for the sequence labeling task consists of a linear projection layer and a CRF [19] layer, which can be represented as

$$z_t^i = W_6[s_2 \oplus \tilde{r}_i] + b_5, \ t = 1, 2, \cdots, n$$
$$y_i^s = \text{CRF}([z_1^i, z_2^i, \cdots, z_n^i])$$

where W_6 and b_5 is weight matrix and bias vector respectively. y_i^s is predicted label sequence.

Node Classification. When a complete text block represents the target information, the information extraction task can also be defined as the problem of node classification. The node representation $\tilde{r}_i$ is taken as the input of classifier:

$$y_i^n = \text{Softmax}(W_7\tilde{r}_i)$$

where W_7 is trainable weight matrix and y_i^n is the probability of labels.

Link Prediction. In the case of rebuilding the structure or partial structure of a document, like a form understanding task, the structure can be defined as links between text blocks so that this kind of task can be converted as link prediction. For link prediction tasks, we first apply a linear projection on node representations as follows:

$$\hat{r}_i = \sigma(W_8\tilde{r}_i + b_6)$$

where W_8 and b_6 is weight matrix and bias. The link between node v_i and v_j is predicted as follows:

$$y_{ij}^p = \text{Sigmoid}(W_9(\hat{r}_i \cdot \hat{r}_j))$$

where W_9 is a trainable weight matrix and y_{ij}^p is the predicted probability of link between node v_i and node v_j.

Document Classification. In our model architecture, document classification is equivalent to graph classification. An attention-based graph pooling module is employed to get the representation of the document graph.

$$P_G = \text{Softmax}(\frac{QK^T}{\sqrt{d}})V$$

where Q, K and V are all the node representations $[\tilde{r}_1, \tilde{r}_2, \cdots, \tilde{r}_m]$. In addition, the visual representation of document image P_V also helps to classify a document. Both graph representation $\boldsymbol{P}_G$ and $\boldsymbol{P}_V$ are taken as the classifier's input.

$$\boldsymbol{y}^d = \text{Softmax}(\boldsymbol{W}_{10}[\boldsymbol{P}_G \oplus \boldsymbol{P}_V])$$

where $\boldsymbol{W}_{10}$ is a trainable matrix and $\boldsymbol{y}^d$ represents probability of labels.

4 Experiments

In this section, we evaluate LayoutGCN on five public benchmarks, FUNSD [16], SROIE [15], CORD [25], Train-Ticket [33], and RVL-CDIP [11]. The involved downstream tasks include sequence labeling, node classification, link prediction, and document classification. We compare the experimental results with fifteen strong baseline models to verify the performance of our approach.

4.1 Datasets Description

FUNSD can be used for entity detection and relation extraction tasks. It contains 199 real, fully annotated, scanned forms, 50 of which are used for the test set. This article redefined entity detection as node classification and relation extraction as link prediction. The entity-level F1 score is used as the evaluation metric of the entity detection task and accuracy for the relation extraction task.

SROIE dataset (Task 3) is to extract text for a number of key fields from given scanned receipts. There are 626 training samples and 347 testing samples in this dataset. We deal with this task as sequence labeling and correct the OCR errors in both training and testing samples. The evaluation metric for this task is entity-level F1 score too.

CORD is also a dataset for information extraction on receipts. It comprises 800 receipts for training, 100 for validation, and 100 for testing. We treat this task as node classification. The model's performance is also evaluated by entity-level F1 score on this dataset.

Train-Ticket refers to the dataset created by Yu et al. [33] for extracting information from train tickets. It contains 400 real images and 1,530 synthetic images sampled from the data provided by Guo et al. [10]. Eighty real samples are used for testing, and the rest are for training. We define this task as sequence labeling. The evaluation metric is the entity-level F1 score.

RVL-CDIP is designed for image classification, and the number of samples is up to 400,000. The images are categorized into 16 classes, with 25,000 images per class. The training set contains 320,000 samples, while the validation and testing set each have 40,000 samples. We employ the public OCR toolkit Tesseract[1] to get the text blocks from each image. The overall classification accuracy is reported for evaluation.

---

[1] https://github.com/tesseract-ocr/tesseract.

4.2   Settings

All the experiments are performed with the same settings. The hyper-parameters of our model are listed in Table 1. Token embeddings in the textual encoder are character-level. LayoutGCN-Lite refers to the model without a visual encoder.

Table 1. Parameters of Model in All Experiments.

Parameters	Value	Parameters	Value
batch size	32	filter size of TextCNN	2,3,4,5
number of epochs	200	decay speed λ	50
learning rate	0.001	angle interval θ	10
keep prob. of dropout	0.4	angle embedding dim. d_θ	128
hidden size d_h	256	number of GCN layers	2
token embedding dim	128	size of CSP-Darknet	small
filter number of TextCNN	32,64,128,256	RoI polling size	7

For the CORD and RVL-CDIP datasets, we do not use the validation set. We train the model on all five datasets for 200 epochs without any early stopping on one NVIDIA Tesla V100 32 GB GPU. Our model is trained from scratch using Adam [18] as the optimizer to minimize the training loss without any pre-training. To ensure the statistical stability of the model performance evaluation, we repeat every experiment five times to eliminate random fluctuations, and the average score is reported.

4.3   Results

We compare LayoutGCN with existing published strong baseline models and categorize them into the base and large models by their number of training parameters. These baseline models include single text modal models [4,24], document pre-training models [1,5,14,31] and non-pre-training model [33]. PICK is the only non-pre-training approach with competitive performance.

The experiment results of our model on the five benchmarks and the comparison with the baseline models are listed in Tabel 2. LayoutGCN achieves competitive results on the benchmarks compared to existing baseline models and even surpasses some large models. Although there is a certain gap with SOTA results, our approach is still competitive, considering the much smaller model size and no need for pre-training. For FUNSD, the result of LayoutGCN is close to the baseline models except for LayoutLMv3. The performance of our approach is difficult to improve because the training data is much tiny, with only 149 samples. There is an apparent gap between LayoutGCN and several baseline models on dataset SROIE. In addition to the lack of training samples, we will later illustrate another reason in the case study. The training samples for CORD and Train-Ticket datasets are relatively more, and our model achieves almost the same score as SOTA results. In practice, LayoutGCN can achieve applicable performance with around a thousand training samples in various applications.

The size of the RVL-CDIP training set is enormous, but LayoutGCN still has a gap with the baseline models. This phenomenon is partly because the Tesseract OCR does not perform as well as industrial OCR services used by the baseline works. It also reflects a disadvantage of LayoutGCN. It is unsuitable for documents with widely different layouts. The RVL-CDIP dataset contains 16 types of document images, and the layout of various types of documents is quite different, which is too complicated for our model. In addition, our method's deep semantic representation ability for text and image is not that strong. These factors result in LayoutGCN hardly achieving results comparable to the baseline models on the RVL-CDIP dataset.

Table 2. Experiment Results and Performance Comparisons on FUNSD, SROIE, CORD, Train-Ticket, RVL-CDIP Datasets, where **ED** denotes entity detection, and **RE** denotes relation extraction. Parameters denoted with * are estimated from the model architecture.

Model	Param	FUNSD		SROIE	CORD	Ticket	RVL-CDIP
		ED	RE				
BERT(base) [4]	110M	60.26	27.65	90.99	89.68	–	89.81
RoBERTa(base) [24]	125M	66.48	-	91.07	93.54	–	90.06
LayoutLM(base) [32]	160M	78.66	45.86	94.38	94.72	–	94.42
PICK [33]	90M*	–	–	96.10	–	**98.60**	–
LayoutLMv2(base) [31]	200M	82.76	42.91	96.25	94.95	–	95.25
BROS(base) [13]	110M	83.05	**71.46**	96.28	96.50	–	–
DocFormer(base) [1]	183M	83.34	–	–	96.33	–	**96.17**
TILT(base) [26]	230M	–	–	97.65	95.11	–	95.25
SelfDoc(base) [21]	137M	83.36	–	–	–	–	92.81
UDoc(base) [8]	272M	87.93	–	–	–	–	95.05
StrucTexT [22]	107M	83.09	44.10	96.88	–	–	–
LAMBERT [5]	125M	–	–	**98.17**	94.41	–	–
XYLayoutLM(base) [9]	160M*	83.35	–	–	–	–	–
LayoutLMv3(base) [14]	133M	**90.29**	–	–	**96.56**	–	95.44
BERT(large) [4]	340M	65.63	29.11	92.00	90.25	–	89.92
RoBERTa(large) [24]	355M	70.72	–	92.80	93.80	–	90.11
LayoutLM(large) [32]	343M	78.95	42.83	95.24	94.93	–	95.25
LayoutLMv2(large) [31]	426M	84.20	70.57	97.81	96.01	–	95.64
BROS(large) [13]	340M	84.52	**77.01**	96.62	97.28	–	–
DocFormer(large) [1]	536M	84.55	–	–	96.99	–	95.50
TILT(large) [26]	780M	–	–	**98.10**	96.33	–	95.52
StructralLM(large) [20]	355M	85.14	–	–	–	–	**96.08**
LayoutLMv3(large) [14]	368M	**92.08**	–	–	97.46	–	95.93
LayoutGCN-Lite	0.8M	81.64	36.54	93.85	**96.06**	98.24	85.64
LayoutGCN	4.5M	**82.06**	**37.08**	**94.29**	95.68	**98.52**	**89.76**

In summary, LayoutGCN shows comparable performance to most large-scale models on four public datasets and even achieves results close to SOTA on two with dramatically fewer parameters. The gap with the baseline models is expected to be eliminated when adding a bit more training samples. The LayoutGCN-Lite can get competitive results only with n-gram features from text and layout features, indicating that deep semantic features are not necessary for all VRDU tasks. Finally, our approach may have few advantages in VRDU tasks for documents with widely different layouts.

4.4   Case Study

To verify the effectiveness of our designed mechanism for generating weights of adjacency matrix, we analyze the weights of adjacency matrix for specific samples. Since the relationship between nodes is easier to distinguish in documents from the SROIE dataset, we randomly pick an example from the test set. Figure shows the adjacency matrix weights for the node "170.00" and node "02/01/2019 2:47:14 PM". The node "170.00" is the value for the key field total, and the content of its neighbor nodes with high weights appears around the total frequently in this dataset. For node "02/01/2019 2:47:14 PM" which contains the value of key field date, the neighbor node with the highest weight has the keyword "Date" which is a strong signal for detecting text for key field date. These facts prove that our model can efficiently learn the layout information about the relative position between nodes (Fig. 3).

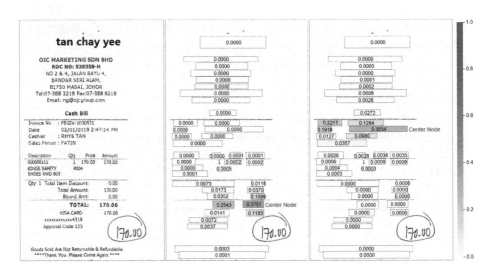

Fig. 3. Illustration of adjacency matrix weights for sample X00016469671 from test set of SROIE dataset. "Center Node" indicates the node for which the weights are showed.

Another observation is that two adjacent nodes may have similar adjacency matrix weights in some situations, making it impossible for the model to distinguish between the two nodes if they also have similar text. As shown in Figure , the nodes "RM83.00" and "RM100.00" have very similar weights with the other nodes, especially since they share the same weight with node "TOTAL" which has essential information for determining the value of field total. On the other hand, the node "RM83.00" has similar weights with nodes "CHECKS PAID" and "TOTAL", while the weight between node "RM100.00" and nodes "TOTAL", "CASH" is very close. LayoutGCN does not learn the accurate representations for nodes "RM83.00" and "RM100.00". Due to these issues, the model fails to recognize the value for the field total in this case. We attribute this problem to the inappropriateness of the angle discretization scheme. The angle between two nodes is discretized uniformly with a fixed interval in our method. If using a large interval, the angular resolution will be insufficient. In the case shown in Figure , the angle between node "RM100.00" and nodes "TOTAL" and "CASH" are both zero after discretized. On the contrary, a small angle interval will decrease the robustness of the model. A better method for discretizing angle may be non-uniform, and we will study it in future works (Fig. 4).

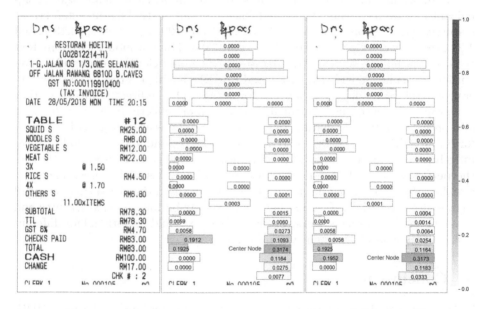

Fig. 4. Illustration of adjacency matrix weights for sample X51006647933 from test set of SROIE dataset. "Center Node" indicates the node for which the weights are showed.

4.5 Ablation Study

As mentioned above, the contribution of features from different modalities is uneven in various tasks. In this study, we enable the encoders for text, layout, and image step by step to explore their contribution to different tasks, especially the layout. We will further verify the effectiveness of the model modules for layout information in this section.

Table 3. Experiment results for Multimodal Features, where **ED** denotes entity detection and **RE** denotes relation extraction.

Model			FUNSD		SROIE	CORD	Ticket	RVL-CDIP
Text	Layout	Image	ED	RE				
√			77.14	10.40	72.67	88.16	86.08	82.17
√	√		81.64	36.54	93.85	**96.06**	98.24	85.64
√	√	√	**82.06**	**37.08**	**94.29**	95.68	**98.52**	**89.76**

As shown in Tabel 3, textual features are most informative in all tasks. A significant improvement can be observed when merging textual features and layout features, and the score increases by 5.83%, 251.35%, 29.15%, 8.98%, 14.13%, and 4.2%, respectively. Compared to BERT and RoBERTa, our model with only text achieves much lower scores on the datasets SROIE and CORD. After adding layout features, our model beats BERT and RoBERTa and gets results close to several pre-trained models with layout information. This indirectly illustrates that our proposed method uses layout information more effectively. Virtual features also provide helpful information, but their contributions are relatively low except for RVL-CDIP, 0.51%, 1.48%, 0.47%, -0.40%, 0.29%, and 4.81%. Visual features introduce more noise than helpful information to the model if used directly, and the performance of the model decrease after adding visual features. The Size-matched RoI Pooling and influence gated mechanism in the visual encoder avoid such issues successfully. Considering the low contribution of visual features and dramatically increasing model size, virtual features could be optional for different tasks, and LayoutGCN-Lite is recommended.

5 Conclusion

This paper introduces LayoutGCN, a novel and lightweight model for VRDU. LayoutGCN solves VRDU tasks using shallow semantic features from text, layout, and image. For layout, we model the relationship between the relative position of the text blocks and adjacency matrix weights, and introduce layout information to the model as prior knowledge, enhancing layout information utilization. We conduct extensive experiments on five publicly available benchmarks. The results demonstrate that LayoutGCN is practical for various downstream tasks and can achieve comparative performance compared to existing large-scale models. The tiny model size and no need for pre-training make the cost of LayoutGCN in practice much lower than existing large-scale models. LayoutGCN provides a feasible and effective method for the VRDU applications of low-resource scenarios. Moreover, we demonstrate the feasibility of dealing with VRDU tasks using only shallow semantics from the three modalities and providing a new perspective to solve the problems in the VRDU field.

Acknowledgements. This research work is sponsored by Ant Group Security and Risk Management Fund.

References

1. Appalaraju, S., Jasani, B., Kota, B.U., Xie, Y., Manmatha, R.: Docformer: end-to-end transformer for document understanding. In: Proceedings of the IEEE/CVF International Conference on Computer Vision, pp. 993–1003 (2021)
2. Ba, J.L., Kiros, J.R., Hinton, G.E.: Layer normalization. arXiv preprint arXiv:1607.06450 (2016)
3. Cheng, Z., et al.: Trie++: towards end-to-end information extraction from visually rich documents. arXiv preprint arXiv:2207.06744 (2022)
4. Devlin, J., Chang, M.W., Lee, K., Toutanova, K.: Bert: pre-training of deep bidirectional transformers for language understanding. arXiv preprint arXiv:1810.04805 (2018)
5. Garncarek, Ł, et al.: LAMBERT: layout-aware language modeling for information extraction. In: Lladós, J., Lopresti, D., Uchida, S. (eds.) ICDAR 2021. LNCS, vol. 12821, pp. 532–547. Springer, Cham (2021). https://doi.org/10.1007/978-3-030-86549-8_34
6. Ge, Z., Liu, S., Wang, F., Li, Z., Sun, J.: Yolox: exceeding yolo series in 2021. arXiv preprint arXiv:2107.08430 (2021)
7. Gehring, J., Auli, M., Grangier, D., Yarats, D., Dauphin, Y.N.: Convolutional sequence to sequence learning. In: International Conference on Machine Learning, pp. 1243–1252. PMLR (2017)
8. Gu, J., et al.: Unidoc: unified pretraining framework for document understanding. Adv. Neural Inf. Process. Syst. **34**, 39–50 (2021)
9. Gu, Z., et al.: Xylayoutlm: towards layout-aware multimodal networks for visually-rich document understanding. In: Proceedings of the IEEE/CVF Conference on Computer Vision and Pattern Recognition, pp. 4583–4592 (2022)
10. Guo, H., Qin, X., Liu, J., Han, J., Liu, J., Ding, E.: Eaten: entity-aware attention for single shot visual text extraction. In: 2019 International Conference on Document Analysis and Recognition (ICDAR), pp. 254–259. IEEE (2019)
11. Harley, A.W., Ufkes, A., Derpanis, K.G.: Evaluation of deep convolutional nets for document image classification and retrieval (2015)
12. He, K., Gkioxari, G., Dollár, P., Girshick, R.: Mask r-cnn. In: Proceedings of the IEEE International Conference on Computer Vision, pp. 2961–2969 (2017)
13. Hong, T., Kim, D., Ji, M., Hwang, W., Nam, D., Park, S.: Bros: a pre-trained language model focusing on text and layout for better key information extraction from documents. In: Proceedings of the AAAI Conference on Artificial Intelligence, vol. 36, pp. 10767–10775 (2022)
14. Huang, Y., Lv, T., Cui, L., Lu, Y., Wei, F.: Layoutlmv3: pre-training for document AI with unified text and image masking. arXiv preprint arXiv:2204.08387 (2022)

15. Huang, Z., et al.: Icdar 2019 competition on scanned receipt ocr and information extraction. In: 2019 International Conference on Document Analysis and Recognition (ICDAR), pp. 1516–1520. IEEE (2019)
16. Jaume, G., Ekenel, H.K., Thiran, J.P.: Funsd: a dataset for form understanding in noisy scanned documents. In: 2019 International Conference on Document Analysis and Recognition Workshops (ICDARW) (2019)
17. Kim, Y.: Convolutional neural networks for sentence classification. arXiv preprint arXiv:1408.5882v2 (2014)
18. Kingma, D.P., Ba, J.: Adam: a method for stochastic optimization. arXiv preprint arXiv:1412.6980 (2014)
19. Lafferty, J., McCallum, A., Pereira, F.C.: Conditional random fields: probabilistic models for segmenting and labeling sequence data (2001)
20. Li, C., et al.: Structurallm: structural pre-training for form understanding. arXiv preprint arXiv:2105.11210 (2021)
21. Li, P., et al.: selfdoc: self-supervised document representation learning. In: Proceedings of the IEEE/CVF Conference on Computer Vision and Pattern Recognition, pp. 5652–5660 (2021)
22. Li, Y., et al.: Structext: structured text understanding with multi-modal transformers. In: Proceedings of the 29th ACM International Conference on Multimedia, pp. 1912–1920 (2021)
23. Liu, X., Gao, F., Zhang, Q., Zhao, H.: Graph convolution for multimodal information extraction from visually rich documents. arXiv preprint arXiv:1903.11279 (2019)
24. Liu, Y., et al.: Roberta: a robustly optimized bert pretraining approach. arXiv preprint arXiv:1907.11692 (2019)
25. Park, S., et al.: Cord: a consolidated receipt dataset for post-ocr parsing. In: Workshop on Document Intelligence at NeurIPS 2019 (2019)
26. Powalski, R., Borchmann, Ł, Jurkiewicz, D., Dwojak, T., Pietruszka, M., Pałka, G.: Going full-TILT boogie on document understanding with text-image-layout transformer. In: Lladós, J., Lopresti, D., Uchida, S. (eds.) ICDAR 2021. LNCS, vol. 12822, pp. 732–747. Springer, Cham (2021). https://doi.org/10.1007/978-3-030-86331-9_47
27. Srivastava, N., Hinton, G., Krizhevsky, A., Sutskever, I., Salakhutdinov, R.: Dropout: a simple way to prevent neural networks from overfitting. J. Mach. Learn. Res. 15(1), 1929–1958 (2014)
28. Strubell, E., Ganesh, A., McCallum, A.: Energy and policy considerations for deep learning in nlp. arXiv preprint arXiv:1906.02243 (2019)
29. Wang, Z., Zhan, M., Liu, X., Liang, D.: Docstruct: a multimodal method to extract hierarchy structure in document for general form understanding. arXiv preprint arXiv:2010.11685 (2020)
30. Welling, M., Kipf, T.N.: Semi-supervised classification with graph convolutional networks. In: J. International Conference on Learning Representations (ICLR 2017) (2016)
31. Xu, Y., et al.: Layoutlmv2: multi-modal pre-training for visually-rich document understanding. arXiv preprint arXiv:2012.14740 (2022)
32. Xu, Y., Li, M., Cui, L., Huang, S., Wei, F., Zhou, M.: Layoutlm: pre-training of text and layout for document image understanding. In: Proceedings of the 26th ACM SIGKDD International Conference on Knowledge Discovery & Data Mining, pp. 1192–1200 (2020)

33. Yu, W., Lu, N., Qi, X., Gong, P., Xiao, R.: Pick: processing key information extraction from documents using improved graph learning-convolutional networks. In: 2020 25th International Conference on Pattern Recognition (ICPR), pp. 4363–4370. IEEE (2021)
34. Zhang, P., et al.: Trie: end-to-end text reading and information extraction for document understanding. In: Proceedings of the 28th ACM International Conference on Multimedia, pp. 1413–1422 (2020)

Topic Shift Detection in Chinese Dialogues: Corpus and Benchmark

Jiangyi Lin[1], Yaxin Fan[1], Feng Jiang[2], Xiaomin Chu[1], and Peifeng Li[1(✉)]

[1] School of Computer Science and Technology, Soochow University, Suzhou, China
{jylin,yxfansuda}@stu.suda.edu.cn, {xmchu,pfli}@suda.edu.cn
[2] School of Data Science, The Chinese University of Hong Kong, Shenzhen, China
jeffreyjiang@cuhk.edu.cn

Abstract. Dialogue topic shift detection is to detect whether an ongoing topic has shifted or should shift in a dialogue, which can be divided into two categories, i.e., response-known task and response-unknown task. Currently, only a few investigated the latter, because it is still a challenge to predict the topic shift without the response information. In this paper, we first annotate a Chinese Natural Topic Dialogue (CNTD) corpus consisting of 1308 dialogues to fill the gap in the Chinese natural conversation topic corpus. And then we focus on the response-unknown task and propose a teacher-student framework based on hierarchical contrastive learning to predict the topic shift without the response. Specifically, the response at high-level teacher-student is introduced to build the contrastive learning between the response and the context, while the label contrastive learning is constructed at low-level student. The experimental results on our Chinese CNTD and English TIAGE show the effectiveness of our proposed model.

Keywords: Dialogue topic shift detection · Hierarchical contrastive learning · Chinese natural topic dialogues corpus

1 Introduction

Dialogue topic shift detection is to detect whether a dialogue's utterance has shifted in the topic, which can help the dialog system to change the topic and guide the dialogue actively. Although dialog topic shift detection is a new task, it has become a hotspot due to its remarkable benefit to many downstream tasks, such as response generation [1] and reading comprehension [2,3], and can help those real-time applications produce on-topic or topic-shift responses which perform well in dialogue scenarios [4–6].

The task of dialogue topic shift detection can be divided into two lines, i.e., response-known task and response-unknown task, as shown in Fig. 1. The former can gain the response information and obtain a better result, while the latter is the opposite. Moreover, both of them are not accessible to future information. This is the biggest difference from the task of text topic segmentation, in which

G. A. Fink et al. (Eds.): ICDAR 2023, LNCS 14189, pp. 166–183, 2023.
https://doi.org/10.1007/978-3-031-41682-8_11

all the basic utterances are visible to each other. That is, those existing topic segmentation models cannot be applied to dialogue topic shift detection since it depends on the response and its subsequent utterances heavily. Therefore, it is more difficult to discern differences between utterances in the task of dialogue topic shift detection. Due to the absence of future utterances, dialogue topic shift detection is still a challenging task.

In this paper, we focus on the response-unknown task of topic shift detection in Chinese dialogues. There are two issues in the response-unknown task of topic shift detection in Chinese dialogues, i.e., lack of annotated corpus in Chinese and how to predict the response.

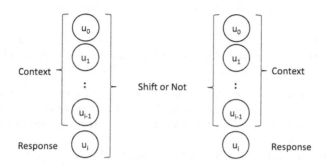

Fig. 1. Two lines of dialogue topic shift detection tasks to detect whether it exists topic shift between the utterances u_{i-1} and u_i, where the response-known task (left) can use the response u_i, while the response-unknown task (right) can be regarded as topic shift prediction without the response u_i.

There are only a few publicly dialogue topic shift corpus available and most of them are provided for the segmentation task, which does not satisfy natural conversation. Xie et al. [7] provided a detailed definition of the dialogue topic shift detection task, and annotated an English dialogue topics corpus TIAGE. Although it can fill the gap in the corpus of English conversation topics, its scale is still too small. In Chinese, Xu et al. [8] annotated a Chinese dialogue topic corpus. However, due to its small size and poor quality, this is detrimental to the further research and development of Chinese dialogue topic shift tasks. To fill the gap in the Chinese natural dialogue topic corpus, we first annotated a Chinese Natural Topic Dialogue (CNTD) corpus which consists of 1308 dialogues with high quality.

Xie et al. [7] also established a benchmark for this response-unknown task based on the T5 model [9] and this benchmark only used the context to predict topic shift and performed poorly due to the lack of the response information. Thus, it is more challenging to predict the topic shift in natural dialogue without useful response information.

The teacher-student framework has been used widely to obtain information that is not available to the model [1]. To solve the issue of the lack of response information, we propose a teacher-student framework to introduce the response information. The teacher can obtain the response information, and the student can learn the response information from the teacher through knowledge distillation. To facilitate knowledge transfer, the student mimics the teacher on every layer instead of just the top layer, which alleviates the delayed supervised signal problem using hierarchical semantic information in the teacher [10].

Besides, we construct hierarchical contrastive learning in which we consider the teacher-student as high-level and the student as low-level. At high-level, we build an information simulation loss between the context and the response to improve the semantic information of the student model with more reliable predictive information. At low-level, we design a semantic coherence-aware loss to better distinguish the different shift cases and produce more reliable prediction results.

Finally, the experimental results on our Chinese CNTD and the English TIAGE show that our proposed model outperforms the baselines. The contributions of this paper are as follows.

- We manually annotate a corpus with 1308 dialogues based on NaturalConv to fill the gaps in the Chinese natural dialogues topic corpus.
- We propose a teacher-student framework to learn the response information for the topic shift detection task.
- We introduce hierarchical contrastive learning to further improve performance.
- The experimental results both on the CNTD and TIAGE datasets show that our model outperforms the baselines.

2   Related Work

2.1   Corpus

Previous studies explored the dialogue topic tasks and published the annotated topic dialogue corpus. For English, Xie et al. [7] annotated the TIAGE consisting of 500 dialogues with 7861 turns based on PersonaChat [11]. Xu et al. [8] built a dataset including 711 dialogues by joining dialogues from existing multi-turn dialogue datasets: MultiWOZ Corpus [12], and Stanford Dialog Dataset [13]. Both corpora are either small or limited to a particular domain, and neither applies to the study of the natural dialogue domain.

For Chinese, Xu et al. [8] annotated a dataset including 505 phone records of customer service on banking consultation. However, this corpus is likewise restricted to a few specialized domains while natural dialogues are more complicated. Natural dialogues have a range of topic shift scenarios, unrestricted topics, and more free colloquialisms in the utterances. The above corpus is insufficient to fill the gap in the Chinese natural dialogue topic corpus.

2.2   Topic Shift Detection in Dialogues

The task of dialogue topic shift detection is also in its initial stage and only a few studies focused on this task. As we mentioned in Introduction, topic segmentation is a similar task. Hence, we first introduce the related work of topic segmentation in dialogues. Due to the lack of training data on dialogue, early approaches of dialogue topic segmentation usually adopted an unsupervised approach using word co-occurrence statistics [14] or sentence topic distributions [15] to measure sentence similarity between turns to achieve detection of thematic or semantic changes. Recently, with the availability of large-scale corpora sampled from Wikipedia, supervised methods for monologic topic segmentation have grown rapidly by using partial tokens as ground-truth segmentation boundaries, especially neural-based methods [16–18]. These supervised solutions are favored by researchers due to their more robust performance and efficiency.

Dialogue topic shift detection is strongly different from dialogue topic segmentation. For the dialogue topic detection task, Xie et al. [7] proposed a detailed definition with two lines: the response-known task considering both the context and the response, and the response-unknown task considering the context only. However, methods based solely on the context are still scarce. Only Xie et al. [7] predict the topic shift or not based on the T5 model. Sun et al. [19] introduce structural and semantic information to help the model detect topic shifts in online discussions, which is similar to response-known task. It is imperative to address the dialogue topic shift detection. In general, the dialogue topic shift detection task is still a challenge, as it can only rely on the context information of the dialogue. In this paper, we solve the lack of response information by utilizing knowledge distillation and hierarchical contrastive learning.

3   Corpus

The existing corpus of Chinese dialogue topic detection [8] is small and does not satisfy natural conversation. Although the English dialogue topic corpora can be converted into Chinese by machine translation, they lack natural conversation colloquiality and are small in size. Therefore, we annotate a Chinese dialogue topic detection corpus CNTD based on NaturalConv dataset [20].

In this section, we show our annotation guidelines and outline the reasons for our selection of corpus sources, as well as the manual annotation procedure and data statistics. We also analyze the topic shift distribution in CNTD.

3.1   Strengths

Each dialogue in our corpus has a piece of news as a base document, which is not available in other corpus and can be used as additional information for further research and expansion. The news is from six domains, which brings our conversations closer to natural dialogue. Besides, the speakers in our corpus are not restricted in any way, which also makes it closer to natural dialogues. In addition, we annotated the fine-grained dialogues topics, refer to Sect. 3.2. Fine-grained labels are beneficial to promote further research on dialogue topics.

Compared with the existing Chinese topic corpus annotated by Xu et al. [8], the dialogues in our corpus do not have meaningless and repetitive turns. Also, the corpus is more than twice the size of the other corpus. In addition, the news in the corpus can be studied as additional information for the dialogues.

3.2   Annotation Guidelines

Following the annotation guidelines in TIAGE [7], we distinguish each dialogue
turns whether changed the topic compared with the context. The response of a
speaker to the dialogue context usually falls into one of the following cases in
dialogues where the examples can be found in Table 1.

Table 1. Different scenarios of response in dialogues.

B "Recently playing handheld games, "dunking master" to look back on the period of watching anime."
A "I see this game so many platforms are pushing ah, but I have not played."
B "Yeah, it's pretty fun, you can go play it when the time comes. There are anime episodes."→ not a topic shift
(a) Commenting on the previous context.
A "What grade is your child in?"
B "He's a freshman." → not a topic shift
(b) Question Answering.
A "The Laval Cup is about to start, and the European team has two kings, Federer and Nadal."
B "I know Federer, he is one of the best in the tennis world."→ not a topic shift
(c) Developing The Conversation to Sub-topics
B "Haha, so what do you usually like to do sports ah? Do you usually go out for a run?"
A "Rarely, usually lying at home watching TV, running and so on is to see their fat can not pretend to look."
B "I also, then what TV are you watching lately? Have you been watching "Elite Lawyers"?"→ topic shift
(d) Introducing A Relevant But Different Topic.
B "This movie I saw crying, Iron Man died, really moved."
A "Yes, the special effects of this movie are very good."
B "Who is your favorite actor?"→topic shift
(e) Completely Changing The Topic.

– Commenting on the previous context: The response is a comment on what is said by the speaker previously;
– Question answering: The response is an answer to the question that comes from the speaker previously;
– Developing the dialogue to sub-topics: The response develops to a sub-topic compared to the context;
– Introducing a relevant but different topic: The response introduces a relevant but different topic compared to the context;
– Completely changing the topic: The response completely changes the topic compared to the context.

Among them, we uniformly identify the two cases of greeting and farewell specific to CNTD as the topic shift.

3.3 Data Source

We chose the NaturalConv dataset [20] as the source corpus, which contains about 400K utterances and 19.9K dialogues in multiple domains. It is designed to collect a multi-turn document grounded dialogue dataset with scenario and naturalness properties of dialogue.

We consider NaturalConv as a promising dataset for dialogue topic detection for the following reasons: 1) NaturalConv is much closer to human-like dialogue with the natural property, including a full and natural setting such as scenario assumption, free topic extension, greetings, etc.; 2) NaturalConv contains about 400K utterances and 19.9K dialogues in multiple domains; 3) The average turn number of this corpus is 20, and longer dialogue contexts tend to exhibit a flow with more topics; 4) The corpus has almost no restrictions or assumptions about the speakers, e.g., no explicit goal is proposed [21].

3.4 Annotation Process

We have three annotators for coarse-grained annotations and two for fine-grained annotations. Both annotations are divided into three stages as follows.
Co-annotation Stage. First, for coarse-grained annotations, we draw a total of 100 dialogues from each domain of the NaturalConv dataset proportionally for a total of 2014 dialogue turns. In this stage, three annotators are asked to discuss every 20 dialogues they annotated, and each annotator is asked to give a reason for the annotation during the discussion. Finally, the Kappa value of all annotators for coarse-grained annotations at this stage is 0.7426. In addition, we annotated the fine-grained information based on the results of the complete coarse-grained annotations. Two annotators annotated the same 150 dialogues and discussed them several times for consistency. Finally, the kappa value of all annotators for fine-grained annotations at this stage is 0.9032. These kappa values confirm that our annotators already have sufficient annotation capabilities for independent annotation, as well as the high quality of our corpus.

Independent-annotation Stage. We ensured the quality of each annotator's annotation and judging criteria before starting the second phase of annotation. For both granularity annotations, we randomly assign the dialogues drawn from each domain to each annotator for independent annotation. At this stage, we annotate 1208 dialogues for coarse-grained annotations and 1158 dialogues for fine-grained annotations.

Semi-automatic Rechecking Stage. Finally, we use a semi-automatic rechecking process to ensure that the corpus is still of high quality. On the one hand, we automatically format the dialogues with annotations to detect formatting problems caused by manual annotation. On the other hand, we automatically match the related news to each dialogue and check that the topic attributes are consistent with the dialogue to rule out any possible errors.

Table 2. Category and proportion of the corpus.

Category	Train	Val.	Test	Sum.
Health(8%)	85	11	11	107
Education(16%)	167	22	21	210
Technology(17%)	176	22	22	220
Sports(33%)	347	45	46	438
Games(8%)	86	11	11	108
Entertainment(17%)	180	23	22	225
Total	1041	134	133	1308

Table 3. Details of CNTD.

	Min.	Max.	Avg.
Dialogue Turns	20	26	20.1
Utterance Words	1	141	21.0
Dialogue Words	194	888	421.7
Dialogue Topics	2	9	5.2
Topic Turns	1	17	4.2

3.5 Annotation Results

Due to the limited time, we randomly select 1308 dialogues from the Natural-Conv dataset and annotate them with four annotators. Finally, we construct a Chinese natural topic dialogues corpus containing 26K dialogue turns.

As shown in Table 2, we randomly split them into 1041 train, 134 validation, and 133 test dialogues respectively, according to the percentage of different categories. In addition, we show the details of CNTD in Table 3, which shows that our corpus has enough topics and long turns which is suitable for dialogue topic detection. Finally, there are the statistics of our fine-grained labels, as shown in Table 4.

We count the number of dialogues with different numbers of topics, as shown in Fig. 2. On another side, we count the distribution of topic shift signals in dialogues, shown in Fig. 3. We can see there are a total of 21 turns and three peaks of topic shift signals, which occur in 2nd, 4th, and 18th turns, respectively. The reason is that the dialogue in our corpus usually starts with a greeting and ends with a farewell, which leads to more topic shifts at the beginning and end of the dialogues. In addition, the NaturalConv corpus gives a piece of news as the base document of the dialogue, so there are more frequent transitions from news to derived topics, leading to the third highest peak in 4th turn. However, we think this is consistent with a natural dialogue scenario because people often talk about recent news after daily greetings.

Table 4. Statistics for fine-grained labels.

Fine-grained labels	Count
Commenting on the previous context	15091
Question answering	3505
Developing the dialogue to sub-topics	857
Introducing a relevant but different topic	3106
Completely changing the topic	2439

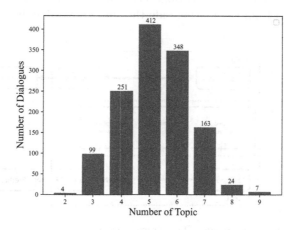

Fig. 2. Number of dialogues with different numbers of topics.

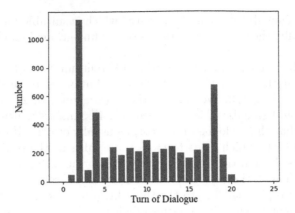

Fig. 3. Topic shift distribution of CNTD.

4   Model

The framework of our model is shown in Fig. 4. We propose a teacher-student framework based on Hierarchical Contrastive Learning, which contains two parts: knowledge distillation and hierarchical contrastive learning which consists of two different contrastive learning.

4.1   Knowledge Distillation

Existing studies cannot effectively predict topic shifts due to the lack of future information. To address this problem, we introduced a teacher-student framework for dialogue topic shift detection. The student side learns an implicit way of topic prediction from the teacher side through knowledge distillation.

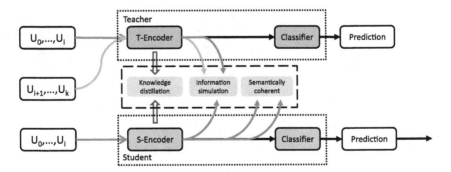

Fig. 4. Architecture of the model. The figure contains a teacher encoder, which can obtain the context and the response, and a student encoder, which has only context as input, where u_{i+1} denotes the response in the current dialogue sample and k denotes the length of the dialogue. In addition, the student is prompted to learn from the teacher through knowledge distillation and hierarchical contrastive learning which contains information simulation loss, and semantic coherence-aware loss.

In this framework, we employ a pre-trained model as an encoder on both sides to obtain semantic representations of the dialogues. Besides, we share the encoder weights on the teacher when encoding the context and the response information. For instance, the representation of $[CLS]$ of the last hidden layer state is taken as a dialogue-level representation H as follows.

$$H_{C,i}^T = E_T\left(X_{C,i}\right) \tag{1}$$

$$H_{P,i}^S = E_S\left(X_{P,i}\right) \tag{2}$$

where $H \in R^{N \times d}, i \in I = \{1, 2, ..., N\}$ denotes the index of samples in the batch, $C \in \{P, F, A\}$ where P represents the context, F represents the response, A represents the full dialog. E represents the encoder, T represents the teacher, and S represents the student.

On the teacher side, we connect the dialogue-level representations and then feed them into the linear layer to obtain the final detection results while on the student as follows.

$$Z_i^T = W_T[H_{P,i}; H_{F,i}; H_{A,i}] \tag{3}$$

$$Z_i^S = W_S H_{P,i} \tag{4}$$

where W_T and W_S denote the linear layers on the teacher and student sides, respectively.

In addition to calculating the cross-entropy loss of the final detection results, we establish the mean squared error loss between each hidden layer of the teacher and student encoders.

4.2 Hierarchical Contrastive Learning

We consider that the response information learned by knowledge distillation solely is insufficient. And the unbalanced proportion of shift or not in the dialogue corpus makes the model perform poorly in distinguishing different shift cases. Therefore, we propose hierarchical contrast learning, consisting of information simulation loss at the high level and semantic coherence-aware loss at the low level, respectively. Both losses are based on contrast learning, but the former is to strengthen the learning of features and the latter is to alleviate the imbalance of labels.

For information simulation loss, this loss enables active learning for each context representation. And this effectiveness has been demonstrated by several works [1,22,23], the incorporation of global information permits local information representation with some predictive information. This helps our encoder obtain a representation with more reliable predictive information.

To alleviate the unbalanced proportion of labels in the dialogues, we propose a semantic coherence-aware loss based on supervised contrast learning (SCL). The main concept of SCL [24,25] is to regard the samples of different categories as positive and negative samples from each other to address the issue of significant quantitative imbalance in the dataset [26]. This loss effectively alleviates the imbalance of shift cases and helps the model further distinguish between different shift cases.

High-level Information Simulation Loss. We build the information simulation loss on both sides so that the context representations can be mapped to the same high-dimensional space. Thus, it is easier to learn the response information to improve final detection results. The following equation can be used to describe this loss.

$$L_{ISL}^M = \sum_{i \in I} \log \frac{exp\left(H_{P,i}^M \cdot H_{A,i}^T\right)}{\sum_{j \in A(i)} exp\left(H_{P,i}^M \cdot H_{A,j}^T\right)} \tag{5}$$

where $A(i) = I - \{i\}$ denotes the samples in the current batch other than itself, P denotes the context, A denotes the full dialog, and $M \in \{T, S\}$ where T denotes the teacher, S denotes the student.

Low-level Semantic Conherent-aware Loss. For a batch with N training samples, a copy of the dialogue's last hidden state H is made to obtain $\overline{H}$ that is considered as the positive, and its gradient is detached. This results in $2N$ samples, then the semantic coherence-aware loss of all samples in a batch can be expressed as follows.

$$U = [H; \overline{H}] \tag{6}$$

$$L_{SCL} = \sum_{i \in I} \frac{-1}{|P(i)|} \sum_{p \in P(i)} \log \frac{exp\left(U_i \cdot U_p\right)}{\sum_{a \in A(i)} exp\left(U_i \cdot U_a\right)} \tag{7}$$

where $U \in R^{2N \times d}, i \in I = \{1, 2, ..., 2N\}$ denotes the index of samples in a batch, and $P(i) = I_{j=i} - \{i\}$ denotes samples of the same category as i but not itself.

4.3   Model Training

We train our model in two steps. The teacher is trained first, and its loss consists of two parts: cross-entropy loss between predictions and manual annotation labels, named L_{NCE}. And the information simulation loss for learning response representation is named L_{ISL}^T. The overall training loss is as follows.

$$L^T = L_{NCE} + L_{ISL}^T \tag{8}$$

$$L^S = L_{NCE} + L_{KD} + L_{ISL}^S + L_{SCL} \tag{9}$$

Then, we train the student, which consists of four parts: cross-entropy loss between predictions and manual annotation labels named L_{NCE}, knowledge distillation between each hidden layer of both sides named L_{KD}, information simulation loss for the dialogue context named L_{ISL}^S, and the semantic coherence-aware loss for different shift cases named L_{SCL}. The above equation represents its overall training loss. As it proves to be arduous to fine-tune the weights assigned to the four losses, we ultimately opt for equal weighting across all of them.

5 Experiments and Analysis

5.1 Experimental Settings

Based on the train/validation/test dataset of CNTD we partitioned in Table 2 and previous work on TIAGE [7], we extract (context, response) pairs from each dialogue as input and the label of response as a target for the response-unknown task. In our experiments, every utterance except the first utterance of the dialogue can be considered as a response. As for evaluation, we report Precision (P), Recall (R), and Micro-F1 scores.

We use BERT as an encoder and fine-tune it during training. For both the TIAGE and CNTD corpus, all pre-trained model parameters are set to default values. We conduct our experiments on NVIDIA GeForce GTX 1080 Ti and NVIDIA GeForce GTX 3090 with batch sizes of 2 and 6 for both CNTD and TIAGE, with the initial learning rates of 2e–5. And we set the epochs of training to 20, and the dropout to 0.5.

For the pre-trained models in the experiment, we apply BERT-base-Chinese and MT5-base to obtain the semantic representation of the dialogues in CNTD, and we apply BERT-base-uncased and T5-base to obtain the semantic representation of the dialogues in TIAGE.

5.2 Experimental Results

Dialogue topic shift detection is a new task and there is no complex model available, besides a simple T5 [7] that can be considered as the SOTA model. Since we employ BERT as our encoder and the T5 model is used in TIAGE, we use the pre-trained models of T5 [9] and BERT [27] as baselines. For BERT, we connect the utterances in the context and separate the last utterance with $[SEP]$. For T5, we also connect utterances in the context and classify the undecidable predicted results to the 'not a topic shift' category.

Table 5. Comparison of our model and three baselines on CNTD.

Model	P	R	F1
T5	27.1	46.8	34.3
BERT	55.5	43.8	48.9
TS	52.7	47.4	49.9
Ours	**56.0**	**52.0**	**53.9**

Table 5 shows the performance comparison between our model and the baselines, in which TS denotes our teacher-student model without the hierarchical comparative learning (HCL) and Ours denotes our final model, i.e., the addition of SCL on the student side based on the addition of ISL on both the teacher and student sides.

It can be found that on CNTD, our model achieves a good improvement and improves both precision and recall in comparison with the baselines. Although T5 does not perform poorly on recall, its precision is inadequate in comparison with BERT, and it is clear that T5 is not effective in predicting topics. In contrast, TS improved by 1.0 in Micro-F1 in comparison with BERT, which confirms that the teacher-student framework is effective in introducing response information. As well, Ours improved by 4.0 in micro-F1 in comparison with TS, and also showed significant improvement in P and R, which fully demonstrates that our HCL can improve the model's ability to discriminate between different topic situations. In particular, our model improves on CNTD by 5.0 in comparison with the best baseline BERT, which shows the effectiveness of our proposed model.

5.3   Ablation Study

To verify the effectiveness of the components used in our model, we conduct ablation studies on CTND, and the experimental results are shown in Table 6.

If we remove ISL on the teacher side $(-ISL_S)$ or the student side $(-ISL_T)$, the performance of the model decreased by 1.5 and 1.3 on the Micro-F1 value, respectively, with the largest decrease after removing the ISL on the student side. Although $-ISL_T$ has the highest precision in predicting topics and lower error probability than $Ours$ and $-ISL_S$. However, it can be seen that adding ISL at both the teacher and student sides can better improve the correct prediction rate. Moreover, if we remove ISL both on the teacher and student side $(-ISL_{TS})$, it achieves a similar performance on Micro-F1, in comparison with $-ISL_S$ and $-ISL_T$. However, it achieves the highest precision (58.8%).

If we remove SCL (-SCL) or HCL (-HCL) from our model, the Micro-F1 value of the models -SCL and -HCL drop from 53.9 to 52.4 (−1.5) and 49.9 (−4.0), respectively. These results show that our Semantic Conherent-aware Loss(SCL), and Hierarchical Contrastive Learning(HCL) are effective for this task, especially HCL.

Table 6. Results of our model and its variants on CNTD where ISL, SCL, and HCL refer to the information simulation Loss, semantic conherent-aware Loss, and hierarchical contrastive Learning, respectively, and T, S, and TS refer to teacher side, student side and both of them.

	P	R	F1
Ours	56.0	52.0	**53.9**
-ISL$_S$	53.6	50.8	52.2
-ISL$_T$	56.1	49.3	52.6
-ISL$_{TS}$	**58.8**	47.0	52.3
-SCL	51.6	**53.1**	52.4
-HCL	52.7	47.4	49.9

Table 7. Performance on dialogues with the different number of topics.

Topic Number	Our Model(F1)	BERT(F1)
2	100	66.7
3	49.0	42.6
4	64.2	53.7
5	58.5	46.3
6	50.3	50.7
7	44.4	47.5
8	44.4	40.0
9	76.9	54.5

5.4   Analysis on Different Angles of Performance

In addition, we explore the performance of the dialogues with different numbers of topics to analyze our model in comparison with BERT, as shown in Table 7. It can be found that our model has a better performance than BERT on dialogues with fewer topics. Our model gets at least a 6% improvement in topic shift prediction on dialogues with 2 to 5 topics and obtains above-average performance. And when the number of topics increases to 9, the performance improves because the conversation length is still about 20 and the topics shift more significantly.

In Table 8, we also investigate the recall of the topic shift detection for various topic turns. Our model is improved for varying degrees across topic turns, with the most significant improvements in turns 7–9. Even in long topic shift cases, our model can obtain an effective boost. However, the performance of our model inevitably decreases compared to short topic shift cases. When there are fewer topic turns, the topic shift situation is simpler, so it is easier to determine. When the length of turns becomes longer and the situation becomes complicated, the topic of long turns has more information so it is easier to identify.

5.5   Results on English TIAGE

As shown in Table 9, it can be found that our model also achieves a good improvement on English TIAGE. Although our model is not the best on precision, we

Table 8. Performances of topic shift with different turn lengths.

Topic Turns	Our Model(Recall)	BERT(Recall)
1–3	56.6	53.8
4–6	30.3	21.4
7–9	28.8	11.9
10–12	40.0	32.0
13–17	40.0	30.0

Table 9. Results on TIAGE.

	TIAGE		
	P	R	F1
T5	**34.0**	17.0	22.0
BERT	28.1	17.9	21.7
TS	26.9	20.1	22.9
Ours	27.4	**28.3**	**27.8**

obtain the best performance on both recall and Micro-F1 values, especially on micro-F1 with a 5.8% improvement over T5. This proves that our model achieves the best performance both in English and Chinese.

5.6 Case Study and Error Analysis

We also conducted a case study. The prediction made by our model, the BERT model on the instance, and the manual labels are shown in Table 10. Compared with the BERT model, it is obvious that our model can accurately anticipate the change of topic in the instances corresponding to the utterances "Yes, that's right.", "It is, indeed, should pay attention to it." etc., belonging to the question-answering scenario. However, if you respond "Well, the policy has been implemented in place this time." and "And now we are promoting the development of children's creative and practical skills." etc. belonging to the commenting on the previous context scenario, our model or BERT cannot accurately predict the topic shift in this scenario. This shows that detecting the topic shifts in natural dialogue is still challenging.

We further analyze the errors of the prediction produced in our experiments. Specifically, we analyzed the example to explore whether the error in the results of this example is prevalent in other dialogues. From Table 10, we can find that the wrong predictions at 14th and 18th turn. We predict "The teaching equipment must be updated, right?" as 'not a topic shift' and "Well, thanks to the government!" as 'topic shift'.

We counted the appearance of many errors, and the errors are mainly divided into two categories. One is for the "Introducing a relevant but different topic" type of utterance. It was predicted that no topic shift occurred due to the lack of information about the future of the conversation. The other is the "commenting on the previous context" category. Since this type of response does not affect the integrity of the previous topic, it is mostly predicted to be a topic shift.

Table 10. The results of BERT, Ours, and Human of different turns where "1" indicates that a topic shift has occurred and "0" indicates the opposite. We omit the lines with all 0.

Turns	BERT	Ours	Human
3	0	1	1
5	0	1	1
11	0	1	1
13	1	0	0
14	0	0	1
16	1	0	0
17	0	1	1
18	1	1	0
19	0	1	1

6 Conclusion

Based on the NaturalConv dataset, we create the CNTD dataset with manual annotations, which fill the gap in the Chinese natural dialogues topic corpus. And we propose a teacher-student model based on hierarchical contrastive learning to solve the lack of response information. We introduced response information through a teacher-student framework and constructed information simulation learning in high-level teachers and students and semantic conherent-aware learning in low-level students. The experiment results demonstrate that our model can perform better in dialogue with few topics. However, detecting the long turns topics or the dialogues with more topics remains a complex problem. Our future work will focus on how to better use response information and news information to detect topic shifts in real-time.

Acknowledgements. The authors would like to thank the three anonymous reviewers for their comments on this paper. This research was supported by the National Natural Science Foundation of China (Nos. 62276177, 61836007 and 62006167), and Project Funded by the Priority Academic Program Development of Jiangsu Higher Education Institutions (PAPD).

References

1. Dai, S., Wang, G., Park, S., Lee, S.: Dialogue response generation via contrastive latent representation learning. In: Proceedings of the 3rd Workshop on Natural Language Processing for Conversational AI, pp. 189–197 (2021)
2. Li, J., et al.: Dadgraph: a discourse-aware dialogue graph neural network for multi-party dialogue machine reading comprehension. In: 2021 International Joint Conference on Neural Networks (IJCNN), pp. 1–8. IEEE (2021)

3. Li, Y., Zhao, H.: Self-and pseudo-self-supervised prediction of speaker and key-utterance for multi-party dialogue reading comprehension. Find. Assoc. Comput. Linguist. EMNLP **2021**, 2053–2063 (2021)
4. Ghandeharioun, A., et al.: Approximating interactive human evaluation with self-play for open-domain dialog systems. Adv. Neural Inf. Process. Syst. **32**, 13658–13669 (2019)
5. Einolghozati, A., Gupta, S., Mohit, M., Shah, R.: Improving robustness of task oriented dialog systems. arXiv preprint arXiv:1911.05153 (2019)
6. Liu, B., Tur, G., Hakkani-Tur, D., Shah, P., Heck, L.: Dialogue learning with human teaching and feedback in end-to-end trainable task-oriented dialogue systems. In: Proceedings of NAACL-HLT, pp. 2060–2069 (2018)
7. Xie, H., Liu, Z., Xiong, C., Liu, Z., Copestake, A.: Tiage: a benchmark for topic-shift aware dialog modeling. In: Findings of the Association for Computational Linguistics: EMNLP, vol. 2021, pp. 1684–1690 (2021)
8. Yi, X., Zhao, H., Zhang, Z.: Topic-aware multi-turn dialogue modeling. In: Proceedings of the AAAI Conference on Artificial Intelligence, vol. 35, pp. 14176–14184 (2021)
9. Raffel, C., et al.: Exploring the limits of transfer learning with a unified text-to-text transformer. J. Mach. Learn. Res. **21**(140), 1–67 (2020)
10. Li, Z., et al.: Hint-based training for non-autoregressive machine translation. In: Proceedings of the 2019 Conference on Empirical Methods in Natural Language Processing and the 9th International Joint Conference on Natural Language Processing (EMNLP-IJCNLP), pp. 5708–5713 (2019)
11. Zhang, S., Dinan, E., Urbanek, J., Szlam, A., Kiela, D., Weston, J.: Personalizing dialogue agents: I have a dog, do you have pets too? In: Proceedings of the 56th Annual Meeting of the Association for Computational Linguistics (Volume 1: Long Papers), pp. 2204–2213 (2018)
12. Budzianowski, P., et al: Multiwoz-a large-scale multi-domain wizard-of-oz dataset for task-oriented dialogue modelling. In: Proceedings of the 2018 Conference on Empirical Methods in Natural Language Processing, pp. 5016–5026 (2018)
13. Eric, M., Krishnan, L., Charette, F., Manning, C.D.: Key-value retrieval networks for task-oriented dialogue. In: Proceedings of the 18th Annual SIGdial Meeting on Discourse and Dialogue, pp. 37–49 (2017)
14. Eisenstein, J., Barzilay, R.: Bayesian unsupervised topic segmentation. In: Proceedings of the 2008 Conference on Empirical Methods in Natural Language Processing, pp. 334–343 (2008)
15. Du, L., Buntine, W., Johnson, M.: Topic segmentation with a structured topic model. In: Proceedings of the 2013 conference of the North American Chapter of the Association for Computational Linguistics: Human language technologies, pp. 190–200 (2013)
16. Koshorek, O., Cohen, A., Mor, N., Rotman, M., Berant, J.: Text segmentation as a supervised learning task. In: Proceedings of the 2018 Conference of the North American Chapter of the Association for Computational Linguistics: Human Language Technologies, Volume 2 (Short Papers), pp. 469–473 (2018)
17. Badjatiya, Pinkesh, Kurisinkel, Litton J.., Gupta, Manish, Varma, Vasudeva: Attention-based neural text segmentation. In: Pasi, Gabriella, Piwowarski, Benjamin, Azzopardi, Leif, Hanbury, Allan (eds.) ECIR 2018. LNCS, vol. 10772, pp. 180–193. Springer, Cham (2018). https://doi.org/10.1007/978-3-319-76941-7_14
18. Arnold, S., Schneider R., Cudré-Mauroux, P., Gers, F.A., Alexander Löser. Sector: A neural model for coherent topic segmentation and classification. Trans. Assoc. Comput. Linguist **7**, 169–184, 2019

19. Yingcheng Sun and Kenneth Loparo. Topic shift detection in online discussions using structural context. In 2019 IEEE 43rd Annual Computer Software and Applications Conference (COMPSAC), volume 1, pages 948–949. IEEE, 2019
20. Wang, X., Li, C., Zhao, J., Dong, Yu.: Naturalconv: A chinese dialogue dataset towards multi-turn topic-driven conversation. In Proceedings of the AAAI Conference on Artificial Intelligence **35**, 14006–14014 (2021)
21. Wenquan Wu, Zhen Guo, Xiangyang Zhou, Hua Wu, Xiyuan Zhang, Rongzhong Lian, and Haifeng Wang. Proactive human-machine conversation with explicit conversation goal. In Proceedings of the 57th Annual Meeting of the Association for Computational Linguistics, pages 3794–3804, 2019
22. Aaron van den Oord, Yazhe Li, and Oriol Vinyals. Representation learning with contrastive predictive coding. arXiv preprint arXiv:1807.03748, 2018
23. Shaoxiong Feng, Xuancheng Ren, Hongshen Chen, Bin Sun, Kan Li, and Xu Sun. Regularizing dialogue generation by imitating implicit scenarios. In Proceedings of the 2020 Conference on Empirical Methods in Natural Language Processing (EMNLP), pages 6592–6604, 2020
24. Li, S., Yan, H., Qiu, X.: Contrast and generation make bart a good dialogue emotion recognizer. In Proceedings of the AAAI Conference on Artificial Intelligence **36**, 11002–11010 (2022)
25. Beliz Gunel, Jingfei Du, Alexis Conneau, and Veselin Stoyanov. Supervised contrastive learning for pre-trained language model fine-tuning. In International Conference on Learning Representations
26. Yanran Li, Hui Su, Xiaoyu Shen, Wenjie Li, Ziqiang Cao, and Shuzi Niu. Dailydialog: A manually labelled multi-turn dialogue dataset. In Proceedings of the Eighth International Joint Conference on Natural Language Processing (Volume 1: Long Papers), pages 986–995, 2017
27. Jacob Devlin Ming-Wei Chang Kenton and Lee Kristina Toutanova. Bert: Pretraining of deep bidirectional transformers for language understanding. In Proceedings of NAACL-HLT, pages 4171–4186, 2019

Detecting Forged Receipts with Domain-Specific Ontology-Based Entities & Relations

Beatriz Martínez Tornés[(✉)] [iD], Emanuela Boros [iD], Antoine Doucet [iD],
Petra Gomez-Krämer [iD], and Jean-Marc Ogier [iD]

University of La Rochelle, L3i, 17000 La Rochelle, France
{beatriz.martinez_tornes,Emanuela.Boros,Antoine.Doucet,
Petra.Gomez-Kramer,Jean-Marc.Ogier}@univ-lr.fr

Abstract. In this paper, we tackle the task of document fraud detection. We consider that this task can be addressed with natural language processing techniques. We treat it as a regression-based approach, by taking advantage of a pre-trained language model in order to represent the textual content, and by enriching the representation with domain-specific ontology-based entities and relations. We emulate an entity-based approach by comparing different types of input: raw text, extracted entities and a triple-based reformulation of the document content. For our experimental setup, we utilize the single freely available dataset of forged receipts, and we provide a deep analysis of our results in regard to the efficiency of our methods. Our findings show interesting correlations between the types of ontology relations (e.g., has_address, amounts_to), types of entities (product, company, etc.) and the performance of a regression-based language model that could help to study the transfer learning from natural language processing (NLP) methods to boost the performance of existing fraud detection systems.

Keywords: Fraud detection · Language models · Ontology

1 Introduction

Document forgery is a widespread problem, while document digitization allows for easier exchange for companies and administrations. Coupled with the availability of image processing and document editing software as well as cost-effective scanners and printers, documents face many risks to be tampered with or counterfeited [21], where counterfeiting is the production of a genuine document from scratch by imitation and forgery is the alteration (tampering) of one or more elements of an authentic document.

First, one of the main challenges of document fraud detection is the lack of freely available annotated data, as many studies around fraud do not consider the actual documents and focus on the transactions (such as credit card fraud, insurance fraud, or even financial fraud) [6,27,39]. Collecting real forged

G. A. Fink et al. (Eds.): ICDAR 2023, LNCS 14189, pp. 184–199, 2023.
https://doi.org/10.1007/978-3-031-41682-8_12

documents is also difficult because real fraudsters would not share their work, and companies or administrations are reluctant to reveal their security breaches and cannot share sensitive information [34,42,46]. Moreover, the challenge of working with a corpus of potentially fraudulent administrative documents is the scarcity of fraud as well as the human expertise required to spot the fraudulent documents [7,12,30]. Taking an interest in real documents actually exchanged by companies or administrations is important for the fraud detection methods developed to be usable in real contexts and for the consistency of authentic documents to be ensured. However, this type of administrative document contains sensitive private information and is usually not made available for research [6].

Second, most of the research in document forensics is focused on the analysis of images of documents, as most of these are scanned and exchanged as images by companies and administrations. Document forgery detection is thus often defined as a tampering detection computer vision (CV) task [9,13,15,20]. A document image can be tampered with in different ways with the help of image editing software. The modification can be done in the original digital document or in the printed and digitized version of the document, which is usually a scanned document, as the mobile-captured document contains too many distortions. Thus, the document can then be printed and digitized again to hide the traces of the fraud [18,24].

In these regards, the *Find it!* competition [4] was, to the best of our knowledge, the only attempt to encourage both CV and natural language processing (NLP) methods to be used for document forgery detection, by providing a freely-available parallel (image/text) forged receipt corpora. However, the number of participants was low (five submissions) and only one of them incorporated content features in the form of rule-based check modules (i.e., looking at inconsistencies in article prices and the total to pay), which proved to be rather effective (an F1 of 0.638).

We, thus, consider that NLP and knowledge engineering (KE) could be used to improve the performance of fraudulent document detection by addressing the inconsistencies of the forgery itself [4]. Hence, while CV methods rely on finding imperfections, by either aiming to detect irregularities that might have occurred during the modification process [8] or by focusing on printer identification, in order to verify if the document has been printed by the original printer [19,33], NLP methods could bridge the gap between image and textual inconsistencies [43]. We experiment with a pre-trained language model regression-based approach while also tailoring the textual input by generating ontology-based entities and relations in documents in order to provide more semantic content of a forged French receipt dataset [3,4]. Our findings show interesting correlations between the types of ontology relations (e.g., has_contactDetail, has_address), the types of entities (product, company, etc.) and the performance of our approach that could foster further research and help to study the transfer learning from NLP methods to boost the performance of existing fraud detection systems.

The paper is organized as follows. Section 2 presents the state of the art with CV-based fraud detection methods, as well as in NLP. Section 3 introduces the forgery detection receipt dataset we used in this study. Our semantic-aware approach is described in Sect. 4, focusing on our alternative textual ontology-based inputs. We then present the experiments and results in Sect. 5. Finally, Sect. 6 states the conclusions and future work.

2 Related Work

Computer Vision-based Fraud Detection. Most of the research in document forensics is part of the field of computer vision (CV) [9,13,15,20]. As a result, most fraud detection datasets are not focused on the semantic content and its alteration. Therefore, they are not usable by a textual approach. Indeed, most datasets are synthetically curated to evaluate a particular CV approach [6,34,39]. For a concrete example, the automatically generated documents in [8] were also automatically tampered with (in terms of noise and change in size, inclination, or position of some characters). The payslip corpus was created by randomly completing the various fields required for this type of document [42]. Datasets used for source scanner (or printer) identification consist of the same documents scanned (or printed) by different machines, without any actual content modification [38]. These datasets are suitable for image-based approaches [14,16,17], but are not relevant for content analysis approaches, as the forged documents are as inconsistent as the authentic ones. Some works focus on the detection of graphical indices of the document modification such as slope, size, and alignment variations of a character with respect to the others [8], font or spacing variations of characters or in a word [9], the variation of geometric distortions of characters introduced by the printer [41], the text-line rotation and alignment [44] or an analysis of the document texture [17]. The authors of [2] use distortions in the varying parts of the documents (not the template ones) through pair-wise document alignment to detect forgery. Hence, the methods need several samples of a class (template). A block-based method for copy and move forgery detection was also proposed which is based on the detection of similar characters using Hu and Zernike moments, as well as PCA and kernel PCA combined with a background analysis [1]. The principle of detecting characters is similar to that of [8] using Hu moments. The method of [17] is the more generic, as it is not related to a certain type of tampering. It is based on an analysis of LBP textures to detect discontinuities in the background and residuals of the image tampering. Due to the difficulty of the task and the lack of generality of these methods, only a few works have been proposed for this task.

Natural Language Processing-based Fraud Detection. Since most of the research in document forensics is focused on the analysis of images of documents, NLP-based fraud detection suffers from a lack of previous work. However, existing fraud detection approaches, in a broader sense than document forgery, mainly focus on supervised machine learning (e.g., neural networks, bagging ensemble methods, support vector machine, and random forests) based on manual feature engineering [6,25,27,30,34]. Knowledge graph embeddings-based approaches have also been proposed to tackle content-based fraud detection in these types of documents [40, 43]. However, this approach congregated all the documents in order to learn a representation of the different extracted semantic relations and used graph-based methods in order to add data from external sources. Moreover, the approach did not prove to be efficient, compared to CV state-of-the-art results. Recently, language models based on BERT [22, 28] have been developed and proven to outperform state-of-the-art results in anomaly detection in system logs, and records of events generated by computer systems. However, the methods are sequential, and cannot be applied in receipt fraud detection where segments of text can be erased (e.g., the removal of a purchased product and its price).

3   Forged Receipt Dataset

The freely available dataset [3, 4] that we utilize is composed of 998 images of French receipts and their associated optical character recognition (OCR) results. It was collected to provide an image/text parallel corpus and a benchmark to evaluate our text-based methods for fraud detection. The forged receipts are the result of tampering workshops, in which participants were given a standard computer with several image editing software to manually alter both images and associated OCR results of the receipts. Thus, the dataset contains realistic forgeries, consistent with real-world situations such as fraudulent refund claims made by modifying the price of an item as shown in Fig. 1 (a), its label, the means of payment, etc. The forgery can also target an undue extension of warranty by modifying the date (however, unless the date is implausible, as, in the example

Fig. 1. Price forgery.   **Fig. 2.** Date forgery.

Fig. 3. Address forgery.

shown in Fig. 2 (b), there is no semantic inconsistency). Other forgeries can involve the issuing company with the aim of money laundering, as in the example in Fig. 3 (c) which produces a false invoice to a false company.

The receipts were collected locally in the research laboratory they were developed, which results in a high frequency of stores in the vicinity. Although this can be seen as a bias, we consider it remains close to a real application case, in which a company stores the documents/invoices it emits. Given the quality of most receipts, in terms of ink, paper and care to avoid crumpling, the automated OCR results were not usable. They were thus manually corrected, both automatically (to tackle recurring errors such as "€" symbols at the end of lines) and manually. The manual correction was performed participatively[1]. The dataset of 998 documents is split into 498 documents for training and 500 for testing, each with 30 forged documents. Thus, the data is imbalanced, according to a realistic distribution of the data. Indeed, there is typically less than 5% of forged documents in document flows, a distribution similar to outliers [4,36].

4 Language Model Regression-based Approach

We base our fraud detection model on the pre-trained model CamemBERT [32] which is a state-of-the-art pre-trained language model for French based on the RoBERTa model [31].

4.1 Model Description

CamemBERT [32] is a stack of Transformer [45] layers, where a Transformer block (encoder) is a deep learning architecture based on multi-head attention mechanisms with sinusoidal position embeddings. In detail, let $\{x_i\}_{i=1}^{l}$ be a token input sequence consisting of l words, denoted as $\{x_i\}_{i=1}^{l} = \{x_1, x_2, \ldots x_i, \ldots x_l\}$, where $x_i (1 \leq x_i \leq l)$ refers to the i-th token in the sequence of length l. CamemBERT, similarly to other language models, expect the input data in a specific format: a special token, [SEP], to mark the end of a sentence or the separation between two sentences, and [CLS], at the beginning of a text, used for classification or regression tasks. We chose CamemBERT's [CLS] token output vector $[CLS]$, denoted by $CamemBERT_{[CLS]}$, as the input of the model and then, apply $CamemBERT$ for further fine-tuning: $f(\{x_i\}_{i=1}^{l}) = CamemBERT_{[CLS]}W_t$ where $W_t \in R^{d_{model} \times 1}$ are the learnable parameters of the linear projection layer and d_{model} is the hidden state dimension of Camem-BERT.

As previously mentioned, we treat the fraud detection task as a regression task and thus a numeric score $s_x \in [0, 1]$ is assigned to the input example $\{x_i\}_{i=1}^{l}$ for quantifying its forging level, which is defined as $s_x = \sigma(f(\{x_i\}))$ where σ is the sigmoid function $\sigma(z) = \frac{1}{1+e^{-z}}$ that returns a numeric score $s_x \in [0, 1]$. Finally, the predicted values are thresholded at 0.5.

---

[1] The platform is available at https://receipts.univ-lr.fr/.

4.2 Domain-specific Forged Receipt Input

In order to better explore the semi-structured specific nature of the receipts, we experimented with four main types of input:

1. **Text**: the raw text of a receipt without any pre-processing;
2. **Entities**: we detect the present entities based on a receipt ontology and concatenate them with a space separator (e.g. "Carrefour") as described below;
3. **Text + Entities**: we augment the receipt *Text* by introducing special markers for each type of entity (e.g., company, product) and replace each entity in the text with its text surrounded by the entity type markers [10];
4. **Knowledge-base Triples**: based on the same ontology but extracting also the semantic relations.

We, first, present the ontology, and then, we detail the detection and the usage of the entities and relations (triples).

Table 1. Receipt ontology object properties. Object properties connect two individuals (a subject and object) and can have a defined *domain* class to specify the class membership of the individuals serving as subjects, and an *image* class to define the class membership of the individuals serving as objects. The table does not list the *type* relation, associating every entity with its type. These data properties are the following: *has_ date, has_ time, amounts_ to, has_ total_ price, has_ number_ of_ items, has_ full_ payment_ of, weights, has_ price_ per_ kg, has_ unit_ price, has_ quantity,* and *has_ return_ money_ amount.*

Domain	Object Property	Inverse Property	Image
City	has_zipCode	is_zipCode_of	ZipCode
Company	has_contactDetail	is_contactDetail_of	ContactDetail
Company	has_address	is_address_of	Address
Company	has_email_address	is_email_address_of	EmailAdress
Company	has_fax	is_fax_of	FaxNumber
Company	has_website	is_website_of	Website
Company	has_phone_number	is_phone_number_of	PhoneNumber
Company	issued	is_issued_by	Receipt
Product	has_expansion	is_expansion_of	Expansion
Company	has_registration	is_registration_of	Registration
Receipt	has_intermediate_payment		IntermediatePayment
Receipt	concerns_purchase		Product
Receipt	contains	is_written_on	Product, Registration, ContactDetail
SIREN	includes	is_component_of	SIRET, RCS, TVA IntraCommunity
City, ZipCode	part_of		Address
Company	is_located_at		City
Company	sells	is_sold_by	Product

Receipt Ontology. The receipt ontology was built by [5] and it is domain-specific, being curated to account for all the information present in the receipts (e.g., a concept for the receipts, instantiated by their IDs, a concept for the issuing company, another for its contact information, etc.). The authors of the dataset chose to represent its semantic contents with an ontological model in order to explicitly represent what humans can imply through their understanding of an invoice. Indeed, much implicit information lies in the understanding of the layout, the format, the content and their combinations. For example, an address written under a company name corresponds to the address of the company. We can also note that a company can have several addresses, or that an address can be shared by several companies in the case of an office building. Moreover, the ontological model allows maintaining a certain flexibility (new classes can be added for other types of business documents, other types of registrations relevant to other countries, etc.) and enable the use of reasoners. However, we utilized the ontology as a starting point in our approach as we were not interested in the description, the inferences and the reasoning per se, but wanted to explore a less formal semantic enrichment of the content of the documents in order to propose a more generalizable approach. Therefore, we focused on the domain-specific entities and relations described and populated in the ontology to build our experiments. The entity detection and the knowledge-base triple detection are presented hereunder.

The ontology is originally written in French, consequently, all labels in this article have been translated, and it was automatically populated with manually-crafted regular expressions based on the regularities of a receipt document. For instance, the products and their prices were extracted from the lines of the document finished by the "€" symbol, or using it as a decimal separator, excluding the lines that report the total or the payment. The extraction process was performed as a finite state machine to adapt to more varied structures, such as prices and products not being aligned. The ontology was populated dynamically using the Python library Owlready2[2]. Table 1 lists the object properties that link the information present in the receipts in order to provide an exhaustive list of the extracted information. We note that the receipt is an entity itself, represented by the label of its ID (a numerical value).

Entity Detection. We kept all domain-specific entity types, even in the cases where they produce redundant information, in order to maintain the granularity of the semantic annotation. For instance, when an address is correctly extracted, it is represented through several entity types: its full address, its city, and its postcode. Each entity subjected to any type of alteration (removal, addition, or modification) was counted. The modifications were not counted in themselves, only the entities altered were: for instance, a date "11/02/2017" altered to "10/02/2016" counts as one modified entity, even if it has suffered two graphic modifications. We grouped the entity types into four categories: company information (name, address, phone number, etc.), product information (label, price,

---

[2] https://owlready2.readthedocs.io/en/v0.37/.

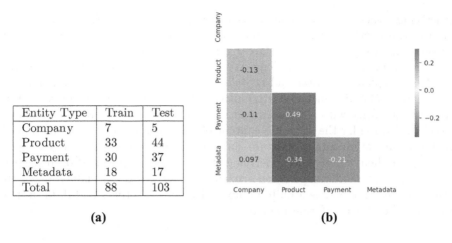

Entity Type	Train	Test
Company	7	5
Product	33	44
Payment	30	37
Metadata	18	17
Total	88	103

(a)   (b)

Fig. 4. (a) Modified entities in the data splits. (b) Entity type correlation matrix.

quantity, or weight), payment information (total, paid amount, discount total), and receipt metadata (date, time, etc.). The number of modified entities per data split and entity types are presented in Fig. 4 (a).

Most of the altered entities involve amounts of money (product and payment entities), even if those are not always consistently modified. Figure 4 (b) shows how the forged receipts are correlated to the entity types. We notice a slightly high coefficient value (0.49) for payment and product entity types, proving a strong correlation between these two. This correlation proves the realistic nature of the forgeries, as an effort has been made to maintain the coherence of the receipt. Indeed, if the price of one or more products is modified, but the total amount remains as is, the forgery becomes easily detected by a human calculating the sum of the amounts. As these entities are not independent, we have considered these types of fraud as inconsistent, and consider them easier to detect by a context-based approach than forgeries involving the receipt metadata (i.e., date and time of purchase).

Finally, in order to take advantage of the semantic details of these entities (entity types), we modify the initial $\{x_i\}_{i=1}^l$ token sequence to give: $x = [x_0, x_1, \ldots, [ENTITY_{start}]x_i[ENTITY_{end}], \ldots, x_n]$ where n is the length of the sequence and $ENTITY$ is the entity type of $x_i \in [payment, company, etc.]$. We, afterward, feed this token sequence into $CamemBERT_{[CLS]}$ instead of $\{x_i\}_{i=1}^l$ [3].

Knowledge-base Triple Detection. In order to go beyond the extracted entities and provide more information about the relations between the entities, we chose to incorporate the domain-specific relationships present in the ontology curated by the authors of the dataset [5]. Our goal was to bring the

---

[3] This strategy has been previously explored in research for different NLP tasks [10, 11,35].

underlying structure of the documents to the forefront by explicitly stating the relations between entities. Those relations include object properties, relations between entities such as *has_ address* and *type* relations, that associate each entity with its class in the ontology. We made sure to remove inverse relations, e.g., *has_ telephone_ number* and *is_ telephone_ number_ of* by keeping only one of each pair. We also included attributive relations, i.e., data properties, that associate an entity with a value (numeric, date, or time). We use the extracted knowledge to render the semantic content of the receipts more explicit. Indeed, as the document's content does not have a syntactic structure, the extracted relations can help convey the underlying structure of the information present in the receipts. The knowledge-based triples serve as a text normalization of the content of the receipts, as a finite number of relations (object and data properties) describe all the information extracted.

5   Experiments

We compare our model with two baseline methods.

First, we consider a *numerical inconsistency checker*, by simulating a checker that assigns the forged class to any document in which there is a simple numerical inconsistency, not relying on any external knowledge. The numeric inconsistency checker accounts for forgeries that a human with a calculator could spot. We consider a simple numerical inconsistency any discrepancy between the total and the sum of the prices, between the total and the total paid, or between the quantity, the unitary price, and the price of the product. However, if the only numeric inconsistency lies in a tax estimation, we consider that our checker does not have the tools to notice the inconsistency, as it requires equation-solving skills.

Second, we consider a support vector machine (SVM) regression classifier with default hyperparameters as our baseline model applied on the term frequency-inverse document frequency (TF-IDF) representation of the unigrams and bigrams extracted from lowercased receipts.

We also compare our results to two CV methods proposed for the dataset. The Verdoliva [15,16] architecture is also based on an SVM and combines three different approaches: a copy-move forgery detection module, based on [14], a camera-signature extraction (and comparison) module [15,16], and a forgery detection based on local image features, originally proposed as a steganalysis method [13]. We also report the results proposed by Fabre [4], that fed the preprocessed images (discrete wavelet transform and grayscale) to a pre-trained model Resnet152 [23] for classification.

5.1   Hyperparameters

We experimented with an SVM with default hyperparameters (C of 1.0 and ϵ of 0.1). In the CamemBERT experiments, we use AdamW [26] with a learning rate of 1×10^{-5} for 2 epochs with mean squared error (MSE) loss. We also considered a maximum sentence length of 256 (no receipt is longer than this). We experiment with a CamemBERT endpoint (CamemBERT-base, with 110M parameters). The evaluation is performed in terms of precision (P), recall (R), and F1.

5.2 Results

Table 2 details the classification results. As the classification is very imbalanced, we report only the results for the "Forged" class.

Table 2. Evaluation results for forged receipt detection.

Method	P	R	F1
Numeric inconsistency checker	100.0	46.67	63.34
CV Approaches			
Fabre [4]	36.4	93.3	52.3
Verdoliva [4]	90.6	96.7	93.5
Baselines			
SVM (text)	7.73	53.33	13.50
SVM (entities)	5.24	33.33	9.05
SVM (text + entities)	5.77	40.00	10.08
SVM (triples)	**29.41**	**100.0**	**45.45**
CamemBERT Approaches			
CamemBERT (text)	6.61	50.0	11.67
CamemBERT (entities)	8.76	73.33	15.66
CamemBERT (text + entities)	7.39	63.33	13.24
CamemBERT (triples)	93.75	**100.0**	**96.77**

We notice how the methods using *Triples* as their input outperform the others, even in their TF-IDF representation, recall is equal to one, meaning all forged receipts are successfully retrieved. Figure 5 presents the area under the receiver operating characteristic (ROC) curve (AUC) for our CamemBERT-based experiments. Not surprisingly, we observe that the *Triples* approach has an AUC near to 1 which means it has a good measure of separability, while the others are closer to 0.

In the case of *Triples*, we observed only two mislabelled true receipts. In one of them, receipt 211, the total price is rather blurry in the image, so it has been manually corrected in the OCR output. However, the value of the total uses "," instead of "." as a decimal separator. As we can see in Fig. 1, the usual separator is ".". This manually induced irregularity could explain this error. The other mislabelled true receipt shows no salient irregularity. The only thing we have noted is that the total amount is expensive (over 87 euros).

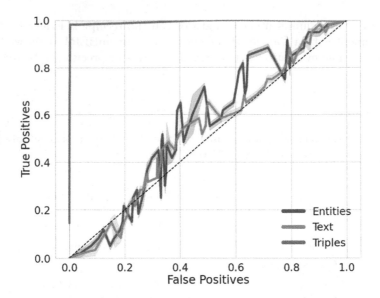

Fig. 5. ROC curve.

Concerning the comparison with the numeric inconsistency checker, we noted that our approach performs better, even for receipts without numerical inconsistency. However, it is important to take notice of the strict definition of inconsistency we have used. Indeed, we only consider inconsistencies in the interaction of the entities themselves, as it provides a stable way to annotate. However, most of the "consistent" forgeries that we consider the most difficult to detect (and the numeric inconsistency checker misses) are implausible. For instance, many of the receipts in which only the date has been modified are actually assigned to an impossible date as in receipt number 334, where the month is numbered 17 or receipt number 662, where the year (2018) is actually after the data collection stopped.

Results per Entity Types. We analysed the results in terms of their count of modified entities and found that it has no statistical impact. We performed an independent t-test in order to compare the number of altered entities in correctly

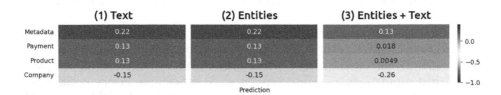

Fig. 6. Entity types correlating with the fraud predictions.

detected forged receipts with the number of altered entities in undetected forged receipts with the results of our approaches based on *Text*, *Entities* and *Text + Entities*. We did not find any statistically significant difference in the means of the two groups, whether we looked into the count of total modified entities, the count of product-related entities, company-related, payment-related, or the receipt metadata entities (p-value > 5%). There was no use in analysing this kind of error for the triples approach (a recall of 100%).

Figure 6 presents the correlation coefficients between the predictions of each model and the modified entity types. We observe for all approaches a rather weak positive correlation between the existence of forged Metadata, Payment, and Product entities, and an even weaker negative correlation with the Company-related entities. This allows us to understand that, while there is an influence regarding which entities are forged, generally, the Metadata, Payment, and Product entities could be correlated with the performance of our fraud detection methods.

Results per Ontology Relation Types. We also analysed the results of forged receipt detection using only one relation type at a time, as shown in Table 3. The relation *contains* exists between the receipt ID and any registration or contact detail of the company present in the receipt (phone, email, fax, address, etc.). Keeping the *contains* relation allows to keep and structure the information related to the company-specific entities. The model trained with such input mislabels uncommon receipts, such as a highway toll or an hourly parking ticket, whose structure is very different from supermarket receipts. Company information is not among the most modified (only seven entities in the train set and five in the test), which leads to the belief that the data may have other biases. For instance, up to almost 50% of the forged receipts were emitted by the same company (Carrefour). Carrefour is indeed the most common company, but it represents only 30% of all receipts. Moreover, the ID associated with each receipt is not entirely random, as receipts are at least sorted by their emitting company.

Table 3. Evaluation results for forged receipt detection.

Method	P	R	F1
Per triple type			
amounts_to	63.83	100.0	77.92
contains	44.12	100.0	61.22

Furthermore, *amounts_to* is the relation to the value of the amount of the intermediate payment. When there is only one mean of payment, it is equivalent to the total amount, however, when two or more means of payment are used, *amounts_to* projects the relation to those amounts. As we can see in Table 3, keeping only this relation still yields very effective results. This triple type considers exclusively information related to general numeric values (totals and payment information). A certain bias is to be expected in the modified numerical values, such as Benford's law [37] which describes the non-normal distribution of naturally occurring numerical data, and it has been used in accounting fraud detection. In real-life occurring numbers (such as prices, population numbers, etc.), the first digit is likely to be small. Indeed, the authors of the dataset report that using Benford's law to look for anomalous numerical data results in a recall of 70% [5]. These results, taken as input only the triples *amounts_ to*, are very encouraging for the ability of our approach to leverage statistical information, even on numerical values, to detect forgery.

6 Conclusions and Future Work

This paper proves that content-based methods are up to the challenge of document fraud detection on the same level as image-based methods. Our initial goal was to build a baseline and to encourage future work in the NLP domain to address document forgery detection, and the results exceeded our expectations. Our semantic-aware approach based on the CamemBERT pre-trained model projecting the relations between entities to represent the content of the receipts achieves high recall values by efficiently leveraging the information extracted from the documents in the form of triples. Ideally, we would like to test our approach on other realistic forgery datasets in order to experiment with other document types and more complex use-cases, however we know of no other publicly available forgery detection corpus. Moreover, as administrative documents are often exchanged as their scanned images, as future work, we propose to continue this line of work by using multimodal approaches [29,47,48].

Acknowledgements. This work was supported by the French defence innovation agency (AID), the VERINDOC project funded by the Nouvelle-Aquitaine Region.

References

1. Abramova, S., et al.: Detecting copy-move forgeries in scanned text documents. Electron. Imaging **2016**(8), 1–9 (2016)
2. Ahmed, A.G.H., Shafait, F.: Forgery detection based on intrinsic document contents. In: 2014 11th IAPR International Workshop on Document Analysis Systems, pp. 252–256 (2014)
3. Artaud, C., Doucet, A., Ogier, J.M., d'Andecy, V.P.: Receipt dataset for fraud detection. In: First International Workshop on Computational Document Forensics (2017)
4. Artaud, C., Sidère, N., Doucet, A., Ogier, J.M., Yooz, V.P.D.: Find it! fraud detection contest report. In: 2018 24th International Conference on Pattern Recognition (ICPR), pp. 13–18 (2018)
5. Artaud, C.: Détection des fraudes : de l'image à la sémantique du contenu. Application à la vérification des informations extraites d'un corpus de tickets de caisse, PhD Thesis, University of La Rochelle (2019)

6. Behera, T.K., Panigrahi, S.: Credit card fraud detection: a hybrid approach using fuzzy clustering & neural network. In: 2015 Second International Conference on Advances in Computing and Communication Engineering (2015)

7. Benchaji, I., Douzi, S., El Ouahidi, B.: Using genetic algorithm to improve classification of imbalanced datasets for credit card fraud detection. In: International Conference on Advanced Information Technology, Services and Systems (2018)

8. Bertrand, R., Gomez-Krämer, P., Terrades, O.R., Franco, P., Ogier, J.M.: A system based on intrinsic features for fraudulent document detection. In: 2013 12th International Conference on Document Analysis and Recognition, pp. 106–110. Washington, DC (2013)

9. Bertrand, R., Terrades, O.R., Gomez-Krämer, P., Franco, P., Ogier, J.M.: A conditional random field model for font forgery detection. In: 2015 13th International Conference on Document Analysis and Recognition (ICDAR), pp. 576–580 (2015)

10. Boros, E., Moreno, J., Doucet, A.: Event detection with entity markers. In: European Conference on Information Retrieval, pp. 233–240 (2021)

11. Boros, E., Moreno, J.G., Doucet, A.: Exploring entities in event detection as question answering. In: Hagen, M., et al. (eds.) ECIR 2022. LNCS, vol. 13185, pp. 65–79. Springer, Cham (2022). https://doi.org/10.1007/978-3-030-99736-6_5

12. Carta, S., Fenu, G., Recupero, D.R., Saia, R.: Fraud detection for e-commerce transactions by employing a prudential multiple consensus model. J. Inf. Secur. Appl. **46**, 13–22 (2019)

13. Cozzolino, D., Gragnaniello, D., Verdoliva, L.: Image forgery detection through residual-based local descriptors and block-matching. In: 2014 IEEE International Conference on Image Processing (ICIP) (2014)

14. Cozzolino, D., Poggi, G., Verdoliva, L.: Efficient dense-field copy-move forgery detection. IEEE Trans. Inf. Forensics Secur. **10**(11), 2284–2297 (2015)

15. Cozzolino, D., Verdoliva, L.: Camera-based image forgery localization using convolutional neural networks. In: 2018 26th European Signal Processing Conference (EUSIPCO) (2018)

16. Cozzolino, D., Verdoliva, L.: Noiseprint: A CNN-based camera model fingerprint. IEEE Trans. Inf. Forensics Secur. **15**, 144–159 (2020)

17. Cruz, F., Sidere, N., Coustaty, M., d'Andecy, V.P., Ogier, J.M.: Local binary patterns for document forgery detection. In: 2017 14th IAPR International Conference on Document Analysis and Recognition (ICDAR), vol. 1 (2017)

18. Cruz, F., Sidère, N., Coustaty, M., Poulain D'Andecy, V., Ogier, J.: Categorization of document image tampering techniques and how to identify them. In: Pattern Recognition and Information Forensics - ICPR 2018 International Workshops, CVAUI, IWCF, and MIPPSNA, Revised Selected Papers, pp. 117–124 (2018)

19. Elkasrawi, S., Shafait, F.: Printer identification using supervised learning for document forgery detection. In: 2014 11th IAPR International Workshop on Document Analysis Systems, pp. 146–150 (2014)

20. Fridrich, J., Kodovsky, J.: Rich models for steganalysis of digital images. IEEE Trans. Inf. Forensics Secur. **7**(3), 868–882 (2012)

21. Gomez-Krämer, P.: Verifying document integrity. Multimedia Security 2: Biometrics, Video Surveillance and Multimedia Encryption, pp. 59–89 (2022)

22. Guo, H., Yuan, S., Wu, X.: Logbert: log anomaly detection via bert. In: 2021 International Joint Conference on Neural Networks (IJCNN), pp. 1–8 (2021)

23. He, K., Zhang, X., Ren, S., Sun, J.: Deep residual learning for image recognition (2015). 10.48550/ARXIV.1512.03385, https://arxiv.org/abs/1512.03385

24. James, H., Gupta, O., Raviv, D.: OCR graph features for manipulation detection in documents (2020)

25. Kim, J., Kim, H.-J., Kim, H.: Fraud detection for job placement using hierarchical clusters-based deep neural networks. Appl. Intell. **49**(8), 2842–2861 (2019)
26. Kingma, D.P., Ba, J.: Adam: a method for stochastic optimization. arXiv preprint arXiv:1412.6980 (2014)
27. Kowshalya, G., Nandhini, M.: Predicting fraudulent claims in automobile insurance. In: 2018 Second International Conference on Inventive Communication and Computational Technologies (ICICCT) (2018)
28. Lee, Y., Kim, J., Kang, P.: Lanobert: system log anomaly detection based on bert masked language model. arXiv preprint arXiv:2111.09564 (2021)
29. Li, P., et al.: Selfdoc: self-supervised document representation learning. In: Proceedings of the IEEE/CVF Conference on Computer Vision and Pattern Recognition, pp. 5652–5660 (2021)
30. Li, Y., Yan, C., Liu, W., Li, M.: Research and application of random forest model in mining automobile insurance fraud. In: 2016 12th International Conference on Natural Computation, Fuzzy Systems and Knowledge Discovery (ICNC-FSKD) (2016)
31. Liu, Y., et al.: Roberta: A robustly optimized bert pretraining approach. ArXiv abs/1907.11692 (2019)
32. Martin, L., et al.: CamemBERT: a tasty French language model. In: Proceedings of the 58th Annual Meeting of the Association for Computational Linguistics, pp. 7203–7219. Association for Computational Linguistics (2020). 10.18653/v1/2020.acl-main.645, https://aclanthology.org/2020.acl-main.645
33. Mikkilineni, A.K., Chiang, P.J., Ali, G.N., Chiu, G.T., Allebach, J.P., Delp III, E.J.: Printer identification based on graylevel co-occurrence features for security and forensic applications. In: Security, Steganography, and Watermarking of Multimedia Contents VII, vol. 5681, pp. 430–440. International Society for Optics and Photonics (2005)
34. Mishra, A., Ghorpade, C.: Credit card fraud detection on the skewed data using various classification and ensemble techniques. In: 2018 IEEE International Students' Conference on Electrical, Electronics and Computer Science (SCEECS) (2018)
35. Moreno, J.G., Boros, E., Doucet, A.: TLR at the NTCIR-15 FinNum-2 task: improving text classifiers for numeral attachment in financial social data. In: Proceedings of the 15th NTCIR Conference on Evaluation of Information Access Technologies, Tokyo Japan, pp. 8–11 (2020)
36. Nadim, A.H., Sayem, I.M., Mutsuddy, A., Chowdhury, M.S.: Analysis of machine learning techniques for credit card fraud detection. In: 2019 International Conference on Machine Learning and Data Engineering (iCMLDE), pp. 42–47 (2019)
37. Nigrini, M.J.: Benford's Law: Applications for Forensic Accounting, Auditing, and Fraud Detection, vol. 586. Wiley (2012)
38. Rabah, C.B., Coatrieux, G., Abdelfattah, R.: The supatlantique scanned documents database for digital image forensics purposes. In: 2020 IEEE International Conference on Image Processing (ICIP) (2020)
39. Rizki, A.A., Surjandari, I., Wayasti, R.A.: Data mining application to detect financial fraud in indonesia's public companies. In: 2017 3rd International Conference on Science in Information Technology (ICSITech) (2017)
40. Rossi, A., Firmani, D., Matinata, A., Merialdo, P., Barbosa, D.: Knowledge graph embedding for link prediction: a comparative analysis. ACM Trans. Knowl. Discov. Data **15**(2), 14:1-14:49 (2021)

41. Shang, S., Kong, X., You, X.: Document forgery detection using distortion mutation of geometric parameters in characters. J. Electron. Imaging **24**(2), 023008 (2015)
42. Sidere, N., Cruz, F., Coustaty, M., Ogier, J.M.: A dataset for forgery detection and spotting in document images. In: 2017 Seventh International Conference on Emerging Security Technologies (EST) (2017)
43. Tornés, B.M., Boros, E., Doucet, A., Gomez-Krämer, P., Ogier, J.M., d'Andecy, V.P.: Knowledge-based techniques for document fraud detection: a comprehensive study. In: Gelbukh, A. (eds.) Computational Linguistics and Intelligent Text Processing. CICLing 2019. Lecture Notes in Computer Science, vol. 13451, pp. 17–33. Springer, Cham (2023). https://doi.org/10.1007/978-3-031-24337-0_2
44. Van Beusekom, J., Shafait, F., Breuel, T.M.: Text-line examination for document forgery detection. Int. J. Doc. Anal. Recogn. (IJDAR) **16**(2), 189–207 (2013)
45. Vaswani, A., et al.: Attention is all you need. Advances in Neural Information Processing Systems 30 (2017)
46. Vidros, S., Kolias, C., Kambourakis, G., Akoglu, L.: Automatic detection of online recruitment frauds: characteristics, methods, and a public dataset. Future Internet **9**(1), 6 (2017)
47. Xu, Y., et al.: Layoutlmv2: multi-modal pre-training for visually-rich document understanding. In: ACL-IJCNLP 2021 (2021)
48. Xu, Y., Li, M., Cui, L., Huang, S., Wei, F., Zhou, M.: Layoutlm: pre-training of text and layout for document image understanding. In: Proceedings of the 26th ACM SIGKDD International Conference on Knowledge Discovery & Data Mining (2020)

CED: Catalog Extraction
from Documents

Tong Zhu[1], Guoliang Zhang[1], Zechang Li[2], Zijian Yu[1], Junfei Ren[1],
Mengsong Wu[1], Zhefeng Wang[2], Baoxing Huai[2], Pingfu Chao[1],
and Wenliang Chen[1(✉)]

[1] Institute of Artificial Intelligence, School of Computer Science and Technology,
Soochow University, Suzhou, China
{tzhu7,glzhang,zjyu,jfrenjfren,mswumsw,pfchao,wlchen}@suda.edu.cn
[2] Huawei Cloud, Hangzhou, China
{lizechang1,wangzhefeng,huaibaoxing}@huawei.com

Abstract. Sentence-by-sentence information extraction from long documents is an exhausting and error-prone task. As the indicator of document skeleton, catalogs naturally chunk documents into segments and provide informative cascade semantics, which can help to reduce the search space. Despite their usefulness, catalogs are hard to be extracted without the assist from external knowledge. For documents that adhere to a specific template, regular expressions are practical to extract catalogs. However, handcrafted heuristics are not applicable when processing documents from different sources with diverse formats. To address this problem, we build a large manually annotated corpus, which is the first dataset for the Catalog Extraction from Documents (CED) task. Based on this corpus, we propose a transition-based framework for parsing documents into catalog trees. The experimental results demonstrate that our proposed method outperforms baseline systems and shows a good ability to transfer. We believe the CED task could fill the gap between raw text segments and information extraction tasks on extremely long documents. Data and code are available at https://github.com/Spico197/CatalogExtraction

Keywords: Catalog Extraction · Information Extraction · Intelligent Document Processing

1 Introduction

Information in long documents is usually sparsely distributed [13,21], so a preprocessing step that distills the structure is necessary to help reduce the search space for subsequent processes. Catalogs, as the skeleton of documents, can naturally locate coarse information by searching the leading section titles. As exemplified in Fig. 1, the debt balance "474.860 billion yuan" appears in only one segment in the credit rating report that is 30 to 40 pages long. Taking the whole

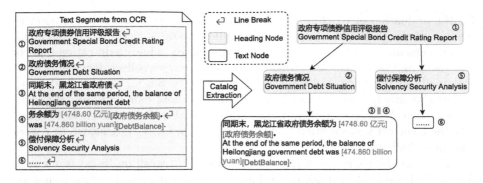

Fig. 1. An example of catalog extraction. The text segments on the left are converted to a catalog tree on the right. The third and the fourth segments are concatenated after catalog extraction.

document into Information Extraction (IE) systems is not practical in this condition. By searching the catalog tree, this entity can be located in the "Government Debt Situation" section with prior knowledge. Unfortunately, most documents are in plain text and do not contain catalogs in an easily accessible format. Thus, we propose the Catalog Extraction from Documents (CED) task as a preliminary step to any extremely long document-level IE tasks. In this manner, fine-grained entities, relations, and events can be further extracted within paragraphs instead of the entire document, which is pragmatic in document-level entity relationship extraction [12,14,15] and document-level event extraction [1].

Designing handcrafted heuristics may be a partial solution to the automatic catalog extraction problem. However, the performance is limited due to three major challenges: 1) Section titles vary across documents, and there are almost no common rules. For documents that are in the same format or inherited from the same template, the patterns of section titles are relatively fixed. Therefore, it is common to use regular expression matching to obtain the whole catalog. However, such handcrafted heuristics are not reusable when the formats of documents change, and researchers have to design new patterns from scratch, making catalog extraction laborious. 2) Catalogs have deep hierarchies with five- to six-level section headings. As the level of section headings deepens, titles become increasingly complex, and simple rule systems usually cannot handle fine-grained deep section headings well. 3) A complete sentence may be cut into multiple segments due to mistakes in data acquisition tools. For example, Optical Character Recognition (OCR) systems are commonly used for obtaining document texts. However, these systems often make mistakes, and sentences may be incorrectly cut into several segments by line breaks. These challenges increase the difficulties of using handcrafted rules.

To address the CED task, we first construct a corpus with a total of 650 manually annotated documents. The corpus includes bid announcements, financial announcements, and credit rating reports. These three types of documents vary in length and catalog complexity. This corpus is able to serve as a benchmark for the evaluation of CED systems. Among these three sources, bid announcements are the shortest in length with simple catalog structures, and financial announcements contain multifarious heading formats, while credit rating reports have deep and nested catalog structures. In addition, we collect documents from Wikipedia with catalog structures as a large-scale corpus for general model pre-training to enhance the transfer learning ability. These four types of data cover the first two challenges in catalog extraction. We also chunk sentences to simulate the incorrect segmentation problem observed in OCR systems, which covers the third challenge in CED.

Based on the constructed dataset, we design a transition-based framework for the CED task. The catalog tree is formulated as a stack and texts are encased in an input queue. These two buffers are used to help make action predictions, where each action stands for a control signal that manipulates the composition of a catalog tree. By constantly comparing the top element of the catalog stack with one text piece from the input queue, the catalog tree is constructed while action predictions are obtained. The final experimental results show that our method achieves promising results and outperforms other baseline systems. Besides, the model pre-trained on Wikipedia data is able to transfer the learned information to other domains when training data are limited.

Our contributions are summarized as follows:

- We propose a new task to extract catalogs from long documents.
- We build a manually annotated corpus for the CED task, together with a large-scale Wikipedia corpus with catalog structures for pre-training. The experimental results show the efficacy of low-resource transfer.
- We design a transition-based framework for the task. To the best of our knowledge, this is the first system that extracts catalogs from plain text segments without handcrafted patterns.

2   Related Work

Since CED is a new task that has not been widely studied, in this section, we mainly introduce approaches applied to similar tasks below.

Parsing Problems: Similar to other text-to-structure tasks, CED can be recognized as a parsing problem. A common practice to build syntactic parsers is biaffine-based frameworks with delicate decoding algorithms (e.g., CKY, Eisner, MST) to obtain global optima [4,19]. However, when the problem shifts from sentences to documents, former token-wise encoding and decoding methods become less applicable. As to documents, there are also many popular discourse parsing theories [6,8,11], which aim to extract the inner semantics among Elementary Discourse Units (EDU). However, the number of EDUs in current corpora is small. For instance, in the popular RST-DT corpus, the average number of EDU is only 55.6 per document [17]. When the number of EDUs grows larger, the transition-based method becomes a popular choice [7]. Our proposed CED task is based on naive catalog structures that are similar to syntactic structures, but some traditional parsing mechanisms are not suitable since one document may contain thousands of segments. To this end, we utilize the transition-based method to deal with the CED task.

Transition-based Applications: The transition-based method parses texts to structured trees in a bottom-up style, which is fast and applicable for extremely long documents. Despite the successful applications in syntactic and discourse parsing [5,7,20], transition-based methods are widely used in information extraction tasks with particular actions, such as Chinese word segmentation [18], discontinuous named entity recognition [3] and event extraction [16]. Considering all the characteristics of the CED task, we propose a transition-based method to parse documents into catalog trees.

3   Dataset Construction

In this section, we introduce our constructed dataset, the ChCatExt. Specifically, we first elaborate on the pre-processing, annotation and post-processing methods, then we provide detailed data statistics.

3.1   Processing & Annotation

We collect three types of documents to construct the proposed dataset, including bid announcements[1], financial announcements[2] and credit rating reports[3]. We adopt Acrobat PDF reader to convert PDF files into docx format and use Office Word to make annotations. Annotators are required to: 1) remove running titles, footers (e.g., page numbers), tables and figures; 2) annotate all headings with different outline styles; and 3) merge mis-segmented paragraphs. To reduce the annotation bias, each document is assigned to two annotators, and an expert will check the annotations and make final decisions in case of any disagreement. Due to the length and structure variations, one document may take up to twenty minutes for an annotator to label.

After the annotation process, we use pandoc[4] and additional scripts to parse these files into program-friendly JSON objects. We insert a pseudo root node before each object to ensure that every document object has only one root. In real applications, documents are usually in PDF formats, which are immutable and often image-based. Using OCR tools to extract text contents from those files is a common practice. However, the OCR tools often split a natural sentence apart when a sentence is physically cut by line breaks or page turnings in PDF, as shown in Fig. 1. To simulate real-world scenarios, we randomly sample

---

[1] http://ggzy.hebei.gov.cn/hbjyzx.
[2] http://www.cninfo.com.cn.
[3] https://www.chinaratings.com.cn and https://www.dfratings.com.
[4] https://pandoc.org.

204   T. Zhu et al.

Table 1. Data statistics. BidAnn refers to bid announcements, FinAnn is financial announcements and CreRat is credit rating reports. One node may contain multiple segments in its content, and we list the number of nodes here. Depth represents the depth of the document catalog tree (text nodes are also included). Length is obtained by counting the number of document characters.

Source	#Docs	Avg.Length	Avg.#Nodes			Avg.Depth
			Heading	Text	Total	
BidAnn	100	1,756.76	8.04	30.61	38.65	3.00
FinAnn	300	3,504.22	12.09	52.31	64.40	3.79
CreRat	250	15,003.81	27.70	81.07	108.77	4.59
Total ChCatExt	650	7,658.30	17.47	60.03	77.50	3.98
Wiki	214,989	1,960.41	11.07	19.34	30.41	3.86

some paragraphs with a probability of 50% and chunk them into segments. For heading strings, we chunk them into segments with lengths of 7 to 20 with jieba[5] assistance. This makes heading segmenting more natural, for example, "招标公告" will be split into "招标 (zhao biao)" and "公告 (gong gao)" instead of "招 (zhao)" and "标公告 (biao gong gao)". For other normal texts, we split them into random target lengths between 70 and 100. Since the workflow is rather complicated, we will open-source all the processing scripts to help further development.

In addition to the above manually annotated data, we collect 665,355 documents from Wikipedia[6] for model pre-training. Most of these documents are shallow in catalog structures and short in text lengths. We keep documents with a catalog depth from 2 to 4 to reach higher data complexity, so that these documents are more similar to the manually annotated ones. After that, 214,989 documents are obtained. We chunk these documents in the same way as the manually annotated ones to simulate OCR segmentation.

3.2 Data Statistics

Table 1 lists the statistics of the whole dataset. Among the three types, BidAnn has the shortest length and the shallowest structure, and the headings are similar to each other. FinAnn is more complex in structure than BidAnn and contains more nodes. Moreover, there are many forms of headings in FinAnn without obvious patterns, which increases the difficulty of catalog extraction. CreRat is the most sophisticated one among all types of data. Its average length is 8.5 times longer than BidAnn while the average depth is 4.59. However, it contains fewer variations in headings, which may be easier for models to locate. Compared to manually annotated domain-specific data, Wiki is easy to obtain. The structure

[5] https://github.com/fxsjy/jieba.
[6] https://dumps.wikimedia.org/zhwiki/20211220/.

Table 2. An example of transition-based catalog tree construction. Elements in **bold** represent the current stack top s, and elements in underline represent the input text q. $ means the terminal of $\mathcal{Q}$ and the finale of action prediction.

Step	Catalog Tree Stack $\mathcal{S}$	Input Queue $\mathcal{Q}$	Predicted Action
1	**Root**	Credit Rating Report, ...	Sub-Heading
2	Root [**Credit Rating Report**]	Debt Situation, ...	Sub-Heading
3	Root [Credit Rating Report [**Debt Situation**]]	The balance, ...	Sub-Text
4	Root [Credit Rating Report [Debt Situation [**The balance**]]]	was 474 billion yuan., ...	Concat
5	Root [Credit Rating Report [Debt Situation [**The balance was 474 billion yuan.**]]]	Security Analysis, ...	Reduce
6	Root [Credit Rating Report [**Debt Situation** [The balance was 474 billion yuan.]]]	Security Analysis, ...	Reduce
7	Root [**Credit Rating Report** [Debt Situation [The balance was 474 billion yuan.]]]	Security Analysis, ...	Sub-Heading
8	Root [Credit Rating Report [Debt Situation [The balance was 474 billion yuan.] **Security Analysis**]]	Texts, $	Sub-Text
9	Root [Credit Rating Report [Debt Situation [The balance was 474 billion yuan.] Security Analysis [**Texts**]]]	$	$

depth is similar to that of FinAnn while its length is 1.5k shorter. Because of the large size, Wiki is well suited for model pre-training and parameter initializing.

It is worth noting that leaf nodes can be heading or normal texts in catalog trees. Since normal texts cannot lead a section, all texts are leaf nodes in catalogs. However, headings could also be leaf nodes if the leading section has no children. Such a phenomenon appears in approximately 24% of documents. Therefore one node cannot be recognized as a text node simply by the number of children, which makes the CED task more complicated.

4   Transition-Based Catalog Extraction

In this section, we introduce details of our proposed TRAnsition-based Catalog trEe constRuction method *TRACER*. We first describe the transition-based process, and then introduce the model architecture.

4.1   Actions and Transition Process

The transition-based method is designed for parsing trees from extremely long texts. Since the average length of our CreRat documents is approximately 15k

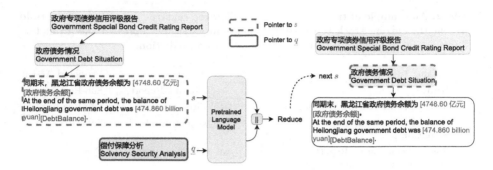

Fig. 2. Framework of the transition-based catalog extraction.

Chinese characters, popular global optimized tree algorithms are apparently too costly to be utilized here.

Action design plays an important role in our transition-based method. There are two buffers here: 1) the input queue $\mathcal{Q}$ providing one text segment q at each time; and 2) the tree buffer $\mathcal{S}$ that records the final catalog tree, where the current stack top points to s. Actions are obtained by comparing s and q continuously, which results in the buffer changing. As the comparison process continues, actions compose a control sequence to build the target catalog tree simultaneously.

To solve the mentioned challenges, actions are designed to distinguish between headings and texts. Our actions can also capture the difference between headings from adjacent depth levels. In this way, we construct the catalog tree without regard to its depth and complexity. Additionally, we propose an additional action for text segment concatenation. Based on these facts, we design 4 actions as follows:

- Sub-Heading: current input text q is a child heading node of s;
- Sub-Text: current input text q is a child text node of s;
- Concat: current input text q is the latter part of s and their contents should be concatenated;
- Reduce: the level of q is above or at the same level as s, and s should be updated to its parent node.

An example is provided in Table 2. To start the prediction, a *Root* node is given in advance. The first heading *Credit Rating Report* is regarded as a child of *Root*. Then, *Debt Situation* becomes another heading node. After that, the **Sub-Text** action suggests that *The balance* is the child node of *Debt Situation* as the body text. Action **Concat** concatenates two body text. Next, action **Reduce** leads to the second layer from the third one. We can eventually build a catalog tree with such a sequence of actions. Furthermore, we present two constraints to avoid illegal results. The first one is that the action between *Root* node and the first input q can only be Sub-Heading or Sub-Text; Another constraint restricts text nodes to be leaf nodes in the tree, and only Reduce and Concat actions are allowed when s is not a heading. If the predicted action is illegal, we take the second-best prediction as the final result.

4.2 Model Architecture

As Fig. 2 shows, the given inputs s and q are encoded via a pre-trained language model (PLM). Here, we use a light version of Chinese whole word masking RoBERTa (RBT3) [2] to obtain encoded representations $\boldsymbol{s}$ and $\boldsymbol{q}$. After concatenation, $\boldsymbol{g} = \boldsymbol{s}||\boldsymbol{q}$ is fed into Feed-Forward Networks (FFN). The FFN is composed of two linear transform layers with ReLU activation function and dropout. Finally we adopt the softmax function to obtain the predicted probabilities as shown in Eq. 1.

$$o = \text{FFN}(\boldsymbol{g}),$$
$$p(\mathcal{A}|s,q) = \text{softmax}(o), \tag{1}$$

where $\mathcal{A}$ denotes all the action candidates. In this way, we can capture the implicit semantic relationship between two nodes.

During prediction, we take the action with maximal probability p as the predicted result:

$$a_i = \underset{a \in \mathcal{A}}{\text{argmax}}\, p(\mathcal{A}|s,q),$$

where $a_i \in \mathcal{A}$ is the predicted action. As discussed in Sect. 4.1, we use two extra constraints to help force decoding legal action results. If a_i is an illegal action, we sort the predicted probabilities in reverse order, and then find the legal result with the highest probability.

As for training, we take cross entropy as the loss function to help update the model parameters:

$$\mathcal{L} = -\sum_i \mathbb{I}_{y_a = a_i} \log p(a_i|s,q),$$

where $\mathbb{I}$ is the indicator function, y_a is the gold action, and $a_i \in \mathcal{A}$ is the predicted action.

5   Experiments

5.1   Datasets

We further split the datasets into train, development, and test sets with a proportion of 8:1:1 for training. To fully utilize the scale advantage of the Wiki corpus, we use it to train the model for 40k steps and subsequently save the PLM parameters for transferring experiments.

Table 3. Main results on ChCatExt. The scores are calculated via the method described in Sect. 5.2. Please beware the overall scores are NOT the average of Heading and Text scores. Heading and Text scores are obtained from a subset of predicted tuple results, where all the node types are "Heading" or "Text". The overall scores are calculated from the universal set, so they are often lower than the Heading and Text scores. WikiBert represents the PLM that is trained on the wiki corpus in advance.

Methods	Heading			Text			Overall		
	P	R	F1	P	R	F1	P	R	F1
Pipeline	88.637	86.595	87.601	81.627	82.475	82.047	76.837	77.338	77.085
Tagging	87.456	88.241	87.846	81.079	81.611	81.344	77.746	78.800	78.269
TRACER	**90.634**	**90.341**	**90.486**	83.031	**85.673**	**84.328**	**81.017**	**83.818**	**82.390**
w/o Constraints	89.911	89.713	89.811	82.491	84.948	83.698	80.216	83.035	81.596
TRACER w/ WikiBert	88.671	89.785	89.221	**83.308**	85.025	84.156	80.820	83.357	82.063

5.2   Evaluation Metrics

We use the overall micro F1 score on predicted tree nodes to evaluate performances. Each node in a tree can be formulated as a tuple: (level, type, content), where level refers to the depth of the current node, type refers to the node type (either *Heading* or *Text*), and content refers to the string that the node carries. The F1 score can be obtained by comparing gold tuples and predicted tuples.

$$P = \frac{N_r}{N_p}, \quad R = \frac{N_r}{N_g}, \quad F1 = \frac{2PR}{P+R},$$

where N_r denotes the number of correctly matched tuples, N_g represents the number of gold tuples and N_p denotes the number of predicted tuples.

5.3   Baselines

Few studies focus on the catalog extraction task, thus we propose two baselines for objective comparisons.

1) Classification Pipeline: The catalog extraction task can be formulated in two steps: segment concatenation and tree prediction. For the first step, we take the text pairs as input and adopt the [CLS] token representation to predict the concatenation results. Suppose the depth of a tree is limited, the depth level can be regarded as a classification task with MaxHeadingDepth + 1 labels, where "1" stands for the text node label. We use PLM with TextCNN [9] to make level predictions.

2) Tagging: Inheriting the idea of two-step classification from above, the whole task can be formulated as a tagging task. The segment concatenation sub-task reflects the BIO tagging scheme, and the level depth and node type are tagging labels. We use PLM with LSTM and CRF to address this tagging task.

5.4 Experiment Settings

Experiments are conducted with an NVIDIA TitanXp GPU. We use RBT3[7], a Chinese RoBERTa variation, as the PLM. We use AdamW [10] to optimize the model with a learning rate of 2e-5. Models are trained for 10 epochs. The training batch size is 20, and the dropout rate is 0.5. We take 5 trials with different random seeds for each experiment and report average results on the test set with the best model evaluated on the development set. For the classification pipeline and the tagging baselines, we set the maximal heading depth to 8.

5.5 Main Results

Table 4. Transferring F1 results: train on the source set, evaluate on the target set. Training on Wiki is the process of obtaining the WikiBert, so results in TRACER w/ WikiBert are absent since the experiments are duplicates.

tgt ↓ src ↓	TRACER				TRACER w/ WikiBert			
	BidAnn	FinAnn	CreRat	Wiki	BidAnn	FinAnn	CreRat	Wiki
BidAnn	88.076	**25.557**	8.400	2.703	**88.200**	25.260	**11.741**	-
FinAnn	7.391	**69.249**	15.543	11.388	**8.100**	68.588	**20.174**	-
CreRat	2.361	14.420	**92.790**	14.029	**7.000**	**30.821**	92.290	-

Table 5. Transfer F1 results: train on k documents from the source set, evaluate on the whole target set.

tgt ↓ src ↓	TRACER								
	BidAnn			FinAnn			CreRat		
	k=3	5	10	k=3	5	10	k=3	5	10
BidAnn	63.033	10.969	7.242	0.713	20.798	9.164	0.000	11.490	**39.264**
FinAnn	63.460	**17.758**	10.613	0.815	28.177	11.755	0.047	11.337	48.543
CreRat	77.259	14.363	14.845	1.725	25.110	**12.636**	3.255	10.768	70.277
TRACER w/ WikiBert									
BidAnn	**66.644**	14.578	18.719	2.355	21.482	18.385	1.024	12.781	30.626
FinAnn	**67.040**	15.509	**15.467**	3.515	32.568	14.125	1.765	25.285	56.192
CreRat	79.029	16.424	14.936	4.517	27.528	12.398	18.659	19.238	67.775

From Table 3, we find that our proposed TRACER outperforms the classification pipeline and tagging baselines by 5.305% and 4.121% overall F1 scores. The pipeline method requires two separate steps to reveal catalog trees, which may accumulate errors across modules and lead to an overall performance decline. Although the tagging method is a stronger baseline than the pipeline one, it still cannot match TRACER. The reason may be the granularities that these methods

---

[7] https://huggingface.co/hfl/rbt3.

focus on. The pipeline and the tagging methods directly predict the depth level for each node, while TRACER pays attention to the structural relationships between each node pair. Besides, since the two baselines need a set of predefined node depth labels, TRACER is more flexible and can predict deeper and more complex structures.

As discussed in Sect. 4.1, we use two additional constraints to prevent TRACER from generating illegal trees. The significance of these constraints is presented in the last line of Table 3. If we remove them, the overall F1 score drops 0.794%. The decline is expected, but the variation is small, which shows the robustness of the TRACER model design.

Interestingly, the PLM trained on the Wiki corpus does not bring performance improvements as expected. This may be due to the different data distributions between Wikipedia and our manually annotated ChCatExt. The following transferring analysis section Sect. 5.6 contains more results with WikiBert.

5.6   Analysis of Transfer Ability

One of our motivations for building a model to solve the CED task is that we want to provide a general model that fits all kinds of documents. Therefore, we conduct transfer experiments under different settings to find the interplay among different sources with diverse structure complexities. The results are listed in Table 4 to 7.

Table 6. Transfer F1 results: train on the source set, further train on k target documents, evaluate on the target set.

src ↓ tgt ↓	TRACER								
	BidAnn			FinAnn			CreRat		
	k=3	5	10	k=3	5	10	k=3	5	10
BidAnn	-	87.995	74.630	**26.640**	-	29.607	**37.991**	56.658	-
FinAnn	-	87.991	75.921	24.502	-	**35.672**	**38.560**	68.287	-
CreRat	-	**88.923**	79.061	27.988	-	43.066	52.139	72.954	-
TRACER w/ WikiBert									
BidAnn	-	**91.400**	**76.626**	25.709	-	**29.729**	29.762	53.406	-
FinAnn	-	**93.608**	**76.106**	28.035	-	33.698	36.217	65.825	-
CreRat	-	88.777	**81.020**	**32.345**	-	**45.488**	**57.580**	**73.519**	-

Table 7. Transfer F1 results: concatenate the source set with k target documents, train on the merged set, evaluate on the target set.

src ↓ tgt ↓	TRACER								
	BidAnn			FinAnn			CreRat		
	k=3	5	10	k=3	5	10	k=3	5	10
BidAnn	-	**80.924**	73.703	27.237	-	29.528	**45.813**	**56.273**	-
FinAnn	-	**88.902**	76.137	24.800	-	31.989	32.173	22.583	-
CreRat	-	**88.310**	**82.551**	**31.768**	-	**45.847**	**61.107**	**73.933**	-
TRACER w/ WikiBert									
BidAnn	-	78.647	**79.606**	**28.243**	-	**33.735**	45.226	55.887	-
FinAnn	-	83.227	**76.556**	**30.823**	-	**34.878**	**36.559**	**62.070**	-
CreRat	-	83.823	76.442	29.587	-	35.086	59.217	71.713	-

We first train models on three separate source datasets and make direct predictions on target datasets. From the left part of Table 4, we can obtain a rough intuition of the data distribution. The model trained on BidAnn makes poor predictions on FinAnn & CreRat, and gets only 7.391% and 2.361% F1 scores, which also conforms with former discussions in Sect. 3.2. BidAnn is the easiest one among the three sources of datasets, so the generalization ability is less robust. FinAnn is shallower in structure, but it contains more variations. The model trained on FinAnn only obtains a 69.249% F1 score evaluated on FinAnn itself. However, it gets better results on BidAnn (25.557%) and CreRat (14.420%) than the others. The model trained on CreRat gets 92.790% on itself. However, it does not generalize well on the other two sources. We also provide the zero-shot cross-domain results from Wiki to the other three subsets. Although the results are poor under the zero-shot setting, the pre-trained WikiBert shows great transfer ability. Comparing results horizontally in Table 4, we find that the pre-trained WikiBert could provide good generalization and outperforms the vanilla TRACER among 6 out of 9 transferring data pairs. The other 3 pairs' results are very close and competitive.

To further investigate the generalization ability of pre-training on the Wiki corpus, we take an extreme condition into consideration, where only a few documents are available to train a model. In this case, as shown in Table 5, we train models with only k source documents and calculate the final evaluation results on the whole target test set. Each model is evaluated on the original source development set to select the best model and then the best model makes final predictions on the target test set. TRACER w/ WikiBert outperforms vanilla TRACER among 23 out of 27 transferring pairs. There is no obvious upward trend when increasing k from 3 to 10, which is unexpected and suggests that the model may suffer from overfitting problems on such extremely small training sets.

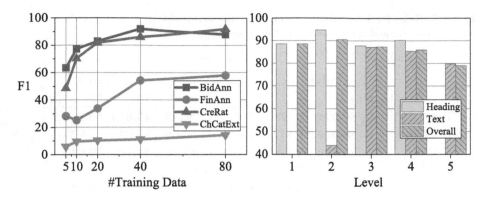

Fig. 3. F1 scores with different numbers training data scales (left) and levels (right). The scale means the number of documents participating in training. The results with different levels are evaluated on ChCatExt.

In most cases of real-world applications, a few target documents are available. Supposing we want to transfer models from source sets to target sets with k target documents available, there are two possible methods to utilize such data. The first one is to train on the source set, and then further train with k target documents; the other one is to concatenate the source set and k targets into a new train set. We conduct experiments under these two settings. The results are presented in Table 6 and 7. Comparing the vanilla TRACER model results, we find that concatenating has 10 out of 18 pairs that outperform the further training method. From k=3 to 10, there are 2, 3, and 5 pairs that show better results, indicating that the concatenation method is better as k increases. WikiBert has different effects under these two settings. In the further training method, WikiBert is more powerful (11 out of 18 pairs), while it is less useful in the concatenation method (8 out of 18 pairs).

Overall, we find that: 1) WikiBert achieves good performances, especially when the training set is small; 2) If there are k target documents available besides the source set, WikiBert is not a must, and concatenating the source set with k targets to make a new train set may produce better results.

5.7 Analysis on the Number of Training Data

The left part of Fig. 3 shows the average results on each separate dataset with different training data scales. Although BidAnn is the smallest data, the model still gets a 63.460% F1 score and surpasses the other datasets. Interestingly, a decline is observed in BidAnn when the number of training documents increases from 40 to 80. We take it as a normal fluctuation representing a performance saturation since the performance standard deviation is 4.950% when the training data scale is 40. Besides, we find that TRACER has good performance on CreRat. This indicates that TRACER performs well in datasets with deeply nested documents if the catalog heading forms are less varied. In contrast, TRACER is lower in performance on FinAnn than BidAnn and CreRat, and it is more data-hungry than other data sources. For ChCatExt, the merged dataset, performance grows slowly with the increase of training data scale, and more data are needed to be fully trained. Comparing the overall F1 performance of 82.390% on the whole ChCatExt, the small scale of the training set may lead to a bad generalization.

5.8 Analysis on Different Depth

From the right bar plot of Fig. 3, it is interesting to see the F1 scores are 0% in level 1 text and level 5 heading. This is mainly due to the golden data distribution characteristics that there are no text nodes in level 1, and there are few headings in deeper levels, leading to zero performances. The F1 score on level 2 text is only 43.938%, which is very low compared to the level 3 text result. Considering that there are only 6.092% of text nodes among all the level 2 nodes, this indicates that TRACER may be not robust enough. Combining the above factors, we find that the overall performance increases from level 1 to 2 and then decreases as the level grows deeper. To reduce the performance decline with deeper levels, additional historical information needs to be considered in future work.

6 Conclusion and Future Discussion

In this paper, we build a large dataset for automatic catalog extraction, including three domain-specific subsets with human annotations and large-scale Wikipedia documents with automatically annotated structures. Based on this dataset, we design a transition-based method to help address the task and get promising results. We pre-train our model on the Wikipedia documents and conduct experiments to evaluate the transfer learning ability. We expect that this task and new data could boost the development of Intelligent Document Processing.

We also find some imperfections from the experimental results. Due to the distribution gaps, pre-training on Wikipedia documents does not bring performance improvements on the domain-specific subsets, although it is proven to be useful under the low-resource transferring settings. Besides, the current model only compares two single nodes each time and misses the global structural histories. Better encoding strategies may need to be discovered to help the model deal with deeper structure predictions. We leave these improvements to future work.

Acknowledgments. This work was supported by the National Natural Science Foundation of China (Grant No. 61936010) and Provincial Key Laboratory for Computer Information Processing Technology, Soochow University. This work was also supported by the Priority Academic Program Development of Jiangsu Higher Education Institutions, and the joint research project of Huawei Cloud and Soochow University. We would also like to thank the anonymous reviewers for their valuable comments.

References

1. Chen, Y., Liu, S., Zhang, X., Liu, K., Zhao, J.: Automatically labeled data generation for large scale event extraction. In: Proceedings of the 55th Annual Meeting of the Association for Computational Linguistics (Volume 1: Long Papers), pp. 409–419. Association for Computational Linguistics, Vancouver (2017). 10.18653/v1/P17-1038, https://aclanthology.org/P17-1038
2. Cui, Y., Che, W., Liu, T., Qin, B., Yang, Z.: Pre-training with whole word masking for chinese BERT. IEEE/ACM Trans. Audio Speech Lang. Process. **29**, 3504–3514 (2021). https://doi.org/10.1109/TASLP.2021.3124365
3. Dai, X., Karimi, S., Hachey, B., Paris, C.: An effective transition-based model for discontinuous NER. In: Proceedings of the 58th Annual Meeting of the Association for Computational Linguistics, pp. 5860–5870. Association for Computational Linguistics (2020). 10.18653/v1/2020.acl-main.520, https://aclanthology.org/2020.acl-main.520
4. Dozat, T., Manning, C.D.: Deep biaffine attention for neural dependency parsing. In: 5th International Conference on Learning Representations, ICLR 2017, Toulon, France, April 24–26, 2017, Conference Track Proceedings. OpenReview.net (2017), https://openreview.net/forum?id=Hk95PK9le
5. Dyer, C., Ballesteros, M., Ling, W., Matthews, A., Smith, N.A.: Transition-based dependency parsing with stack long short-term memory. In: Proceedings of the 53rd Annual Meeting of the Association for Computational Linguistics and the 7th International Joint Conference on Natural Language Processing (Volume 1: Long Papers), pp. 334–343. Association for Computational Linguistics, Beijing (2015). https://doi.org/10.3115/v1/P15-1033,https://aclanthology.org/P15-1033
6. Halliday, M.A.K., Matthiessen, C.M., Halliday, M., Matthiessen, C.: An Introduction to Functional Grammar. Routledge (2014)
7. Ji, Y., Eisenstein, J.: Representation learning for text-level discourse parsing. In: Proceedings of the 52nd Annual Meeting of the Association for Computational Linguistics (Volume 1: Long Papers), pp. 13–24. Association for Computational Linguistics, Baltimore (2014). https://doi.org/10.3115/v1/P14-1002,https://aclanthology.org/P14-1002
8. Kamp, H., Reyle, U.: From Discourse to Logic: Introduction to Modeltheoretic Semantics of Natural Language, Formal Logic and Discourse Representation Theory, vol. 42. Springer (2013)
9. Kim, Y.: Convolutional neural networks for sentence classification. In: Moschitti, A., Pang, B., Daelemans, W. (eds.) Proceedings of the 2014 Conference on Empirical Methods in Natural Language Processing, EMNLP 2014, October 25–29, 2014, Doha, Qatar, A meeting of SIGDAT, a Special Interest Group of the ACL, pp. 1746–1751. ACL (2014). https://doi.org/10.3115/v1/d14-1181, https://doi.org/10.3115/v1/d14-1181
10. Loshchilov, I., Hutter, F.: Decoupled weight decay regularization. In: 7th International Conference on Learning Representations, ICLR 2019, New Orleans, LA, USA, May 6–9, 2019. OpenReview.net (2019), https://openreview.net/forum?id=Bkg6RiCqY7
11. Mann, W.C., Thompson, S.A.: Rhetorical structure theory: toward a functional theory of text organization. Interdisc. J. Study Discourse **8**(3), 243–281 (1988). https://doi.org/10.1515/text.1.1988.8.3.243

12. Peng, N., Poon, H., Quirk, C., Toutanova, K., Yih, W.t.: Cross-sentence N-ary relation extraction with graph LSTMs. Transactions of the Association for Computational Linguistics **5**, 101–115 (2017) https://doi.org/10.1162/tacl_a_00049, https://aclanthology.org/Q17-1008

13. Yang, H., Chen, Y., Liu, K., Xiao, Y., Zhao, J.: DCFEE: a document-level Chinese financial event extraction system based on automatically labeled training data. In: Proceedings of ACL 2018, System Demonstrations, pp. 50–55. Association for Computational Linguistics, Melbourne (2018). 10.18653/v1/P18-4009, https://aclanthology.org/P18-4009

14. Yao, Y., et al.: DocRED: a large-scale document-level relation extraction dataset. In: Proceedings of the 57th Annual Meeting of the Association for Computational Linguistics, pp. 764–777. Association for Computational Linguistics, Florence (2019). 10.18653/v1/P19-1074, https://aclanthology.org/P19-1074

15. Yu, D., Sun, K., Cardie, C., Yu, D.: Dialogue-based relation extraction. In: Proceedings of the 58th Annual Meeting of the Association for Computational Linguistics, pp. 4927–4940. Association for Computational Linguistics (2020). https://aclanthology.org/2020.acl-main.444

16. Zhang, J., Qin, Y., Zhang, Y., Liu, M., Ji, D.: Extracting entities and events as a single task using a transition-based neural model. In: Proceedings of the Twenty-Eighth International Joint Conference on Artificial Intelligence, IJCAI-19, pp. 5422–5428. International Joint Conferences on Artificial Intelligence Organization (2019). https://doi.org/10.24963/ijcai.2019/753

17. Zhang, L., Xing, Y., Kong, F., Li, P., Zhou, G.: A top-down neural architecture towards text-level parsing of discourse rhetorical structure. In: Proceedings of the 58th Annual Meeting of the Association for Computational Linguistics, pp. 6386–6395. Association for Computational Linguistics (2020). https://aclanthology.org/2020.acl-main.569

18. Zhang, M., Zhang, Y., Fu, G.: Transition-based neural word segmentation. In: Proceedings of the 54th Annual Meeting of the Association for Computational Linguistics (Volume 1: Long Papers), pp. 421–431 (2016)

19. Zhang, Y., Li, Z., Zhang, M.: Efficient second-order TreeCRF for neural dependency parsing. In: Proceedings of the 58th Annual Meeting of the Association for Computational Linguistics, pp. 3295–3305. Association for Computational Linguistics (2020). https://aclanthology.org/2020.acl-main.302

20. Zhu, M., Zhang, Y., Chen, W., Zhang, M., Zhu, J.: Fast and accurate shift-reduce constituent parsing. In: Proceedings of the 51st Annual Meeting of the Association for Computational Linguistics (Volume 1: Long Papers), pp. 434–443. Association for Computational Linguistics, Sofia (2013), https://aclanthology.org/P13-1043

21. Zhu, T., et al.: Efficient document-level event extraction via pseudo-trigger-aware pruned complete graph. In: Raedt, L.D. (ed.) Proceedings of the Thirty-First International Joint Conference on Artificial Intelligence, IJCAI-22, pp. 4552–4558. International Joint Conferences on Artificial Intelligence Organization (2022). https://doi.org/10.24963/ijcai.2022/632, main Track

A Character-Level Document Key Information Extraction Method with Contrastive Learning

Xinpeng Zhang[1] , Jiyao Deng[1,2], and Liangcai Gao[1(✉)]

[1] Wangxuan Institute of Computer Technology, Peking University, Beijing, China
{zhangxinpeng,gaoliangcai}@pku.edu.cn, dengjiyao@stu.pku.edu.cn
[2] Center for Data Science, Peking University, Beijing, China

Abstract. Key information extraction (KIE) from documents has become a major area of focus in the field of natural language processing. However, practical applications often involve documents that contain visual elements, such as icons, tables, and images, which complicates the process of information extraction. Many of current methods require large pre-trained language models or multi-modal data inputs, leading to demanding requirements for the quality of the data-set and extensive training times. Furthermore, KIE datasets frequently suffer from out-of-vocabulary (OOV) issues. To address these challenges, this paper proposes a document KIE method based on the encoder-decoder model. To effectively handle the OOV problem, we use a character-level CNN to encode document information. We also introduce a label feedback mechanism in the decoder to provide the label embedding back to the encoder for predicting adjacent fields. Additionally, we propose a similarity module based on contrastive learning to address the problem of content diversity. Our method requires only text inputs, has fewer parameters, but still achieves comparable results with state-of-the-art methods on the document KIE task.

Keywords: document understanding · key information extraction · natural language processing · contrastive learning

1 Introduction

Semi-structured documents exist widely in scientific and technological literature, financial statements, newspapers and magazines, etc. They are used to store and display data compactly. They contain a lot of useful information and are valuable data assets. Unlike plain text documents, semi-structured documents (such as tables, bills, posters, etc.) contain not only text information, but also layout, structure, and image information. They have richer visual elements. Therefore, they are more difficult to recognize and understand and pose new challenges to researchers in the field of natural language processing.

The information extraction task of a document is to extract, summarize and analyze the given key information fields from the document. From a practical

G. A. Fink et al. (Eds.): ICDAR 2023, LNCS 14189, pp. 216–230, 2023.
https://doi.org/10.1007/978-3-031-41682-8_14

point of view, it is of great value to automatically collect personal information, important dates, addresses and amounts from documents such as tables, bills, contracts, etc. In recent years, visual information extraction from documents has attracted the attention of many researchers.

The task of visual information extraction for semi-structured documents is a derivative of the information extraction task for plain text documents. Although layout and image features are added to semi-structured documents, they are still very similar. With the success of language modeling in text understanding tasks, these approaches have also been widely used in the field of document understanding. These methods serialize the document and then apply the sequence model such as RNN [22], LSTM [6] and Transformer [28] to it.

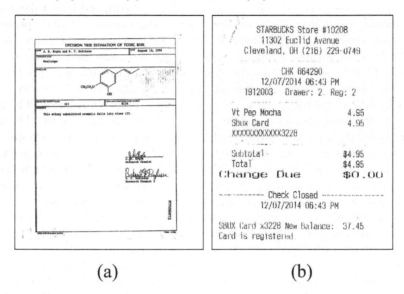

(a)   (b)

Fig. 1. Examples of two document KIE datasets, (a) from FUNSD and (b) from SROIE. The aim of KIE task is to extract texts of a number of key fields from given documents.

Therefore, in the early days of this field, the focus was on improving the language model on serialized text and applying it to semi-structured documents. As a result, several well-known language models were developed, including GPT [25], BERT [3], RoBERTa [21], and LayoutLM [32], which have achieved good results. However, these large pre-training models often consume a significant amount of computing resources and require a large amount of training data. Additionally, while these models have strong semantic understanding due to the large amount of training data, they often go beyond the needs of document information extraction, as visual information extraction usually involves documents with relatively simple semantics, such as tables, receipts, and tickets. Furthermore, these models have limited understanding of structures and spatial relationships, as some methods still convert documents into serialized text for processing without utilizing location information and image information, while others simply add rules for location information, with limited response to this information.

To address the above limitations, we propose a document information extraction method based on the encoder-decoder model. The method encodes document information using character-level Convolutional Neural Networks (CNNs) [15] to better address the Out-Of-Vocabulary (OOV) problem. Furthermore, a label feedback mechanism is introduced to the decoder to return the label embedding to the encoder for the prediction of the adjacent field. A similarity module based on contrastive learning is also proposed to address the issue of content diversity. Our method only requires text input and has fewer parameters compared to the state-of-the-art methods, but it still achieves comparable results in document information extraction tasks.

In encoding phase, due to the high frequency of abbreviations, unusual words and proper nouns in the KIE data-set, word-level or subword-level tokenization methods may cause very low word frequency in dictionary and OOV problems. In addition, the semantic information of these words is poorly interpretable and usually does not form a complete sentence, the semantic association between words is sparse, and the advantage of self-attention mechanism or pre-training is not obvious here. For this reason, we applied a character-level encoder, which takes CharLM as the backbone network to encode input information. The granularity of the character-level is finer than that of the token-level. The most commonly-used characters, such as letters, digits and punctuation, are sufficient to cover the most of tokens formed by their combinations. Instead of maintaining a vocabulary of enormous size at the token-level, we build up an embedding layer to encode each character. In this way, we can avoid most OOV problem and control the size of the embedding matrix at the appropriate scale.

In decoding phase, we apply the common BiLSTM + CRF [10] decoder and sum the CRF loss [16] and contrastive loss from similarity module together to train the model. Label feedback mechanism is applied in the decoder to return the predicted labels from the decoder to the encoder as label embedding, which is similar to the principle of "segment embedding" in BERT. Intuitively, the category to which a field belongs affects the label prediction of its adjacent fields. This approach is to make full use of label information to increase the semantic association between different words. At the same time, this model also applies to unlabeled documents or partially labeled documents, allowing the model to train on low-quality data-sets.

The similarity module is designed to mitigate the impact of content diversity on character-level encoders. Character-level embedding is content-aware and might result in dispersive encoding vectors because the characters in the text field may be various. Ideally, if the text fields corresponding to the embedding vectors generated by the encoder contain similar patterns, they should be close. However, common patterns in data-sets, such as SROIE and FUNSD, are often manifested by different label contents. The same field may have very different content in different documents, which is detrimental to our model capturing common patterns in documents. To overcome this problem, we need to map documents or fields in similar patterns to a tighter vector space to mitigate the impact of content diversity. We assume that similar fields, such as two company names, is supposed to have similar patterns. Therefore, we train the encoder module through the contrastive learning between similar or dissimilar text area.

We have noticed that many document semantics are relatively simple and contain a large number of professional terms and abbreviations, which usually leads to OOV problem, so character-level encoding is an effective method, and our experiments have also proven this. The main contributions of this article are:

1. We apply character-level encoder in our document KIE model to try to solve the OOV problem.
2. We apply a contrastive learning module to learn the similarity between documents or fields, which will help to build a more versatile model.
3. Compared with other methods, our method has significantly fewer parameters, but achieves results comparable to the state-of-the-art method.

2 Related Work

2.1 Data-Sets

Commonly used public data-sets for document KIE include SROIE [9], FUNSD [12], CORD [23] and Kleister [26]. In addition, many studies in recent years have published data-sets for specific application scenarios while proposing the method, such as Chinese VAT invoices [20], train tickets, medical prescriptions [33], taxi receipts [34], and test paper titles [30].

2.2 Pre-trained Language Model

Traditional language model, such as RNN, LSTM and BERT, which are mainly applied in serialized text, are the basic methods in document KIE. The basic process is to treat documents as text sequences after optical character recognition. Then pre-trained language models have achieved good results in document KIE. After that LayoutLM introduces layout information and image area information in the training process, making full use of the characteristics of two-dimensional documents, and achieving better results on this task. Since then, most methods have adopted different ways to represent the layout information and image information of documents.

Many NLP and deep learning methods have shown promising results when applied to serialized text, exhibiting good generalization. However, when it comes to processing two-dimensional documents, serializing them can result in the loss of information and difficulties in encoding the relative positions of different fields. This is particularly the case in tabular data where different rows have different lengths, making it challenging to align columns and represent column information in the model's input. To overcome this issue, early methods resorted to adjusting the input embedding or model structure. For example, TILT method adds the contextual vision embedding from Unet to the input of the Transformer, and adds spatial bias from the y-axis direction of the document to the attention score [24]. LAMBERT [4] adds layout embedding based on the bounding box of tokens to the input embedding of RoBERTa, and added 1-D and 2-D relative bias when calculating the self-attention score. SPADE [11] describes the information extraction task as a spatial dependency resolution question. BROS [7] further improves SPADE by proposing a new position encoding method and pre-training tasks based on area masking.

Currently, LayoutLMv2/v3 [8,31] is a widely adopted baseline approach for document KIE tasks. It integrates layout embedding and visual embedding into the input of BERT. The layout embedding is computed based on the bounding box coordinates of each word, while the visual embedding is obtained from a visual encoder with ResNeXt-FPN [18] as its backbone network. The pre-training tasks, such as masked visual-language modeling, text-image alignment, and Text-Image Matching, are more suitable for two-dimensional documents than the original pre-training tasks of BERT, allowing the model to effectively utilize multi-modal information to understand the relationships between text, layout, and images.

2.3 Multi-modal Method

It should be noted that while attention mechanism-based models, such as LayoutLM, aim to integrate multi-modality, they still primarily focus on adding layout and image representation information to the original language model, such as BERT, which is essentially an extension of text sequences.

Recently, researchers have proposed several innovative approaches to multi-modal information fusion. They have combined and optimized the models from multiple perspectives, including the document representation model, information fusion techniques, the design of end-to-end joint training architecture, the application of graph network structure, and the idea of task transformation. These advances are expected to have significant impact on future research. In the following sections, we will introduce some of the most influential methods in this field.

TRIE is a unified end-to-end text reading and information extraction network [34]. It uses multi-modal vision of text reading and text feature fusion for information extraction.The framework enables the text reading module and the information extraction module to be jointly trained. PICK obtains richer semantic representation by combining different features of the document [33]. And it inserts the graph network into the encoder decoder model for association analysis, which is a representative work of applying graph learning to solve the problem of information extraction. VIES is a unified end-to-end trainable framework for simultaneous text detection, recognition and information extraction [30]. It introduces a visual coordination mechanism and a semantic coordination mechanism to collect visual and semantic features from the text detection and recognition branches . MatchVIE, for the first time, applies the key value matching model on KIE task and proves that the method of modeling the key value relationship can effectively extract visual information [27].

In addition, Unidoc learns a generic representation by making use of three self-supervised losses, encouraging the representation to model sentences, learn similarities, and align modalities [5]. LiLT a effective language-independent layout transformer, which can be pre-trained on the structured documents of a single language and then fine-tuned on other languages [29]. DocFormer is pretrained in an unsupervised fashion, combines text, vision and spatial features using a multi-modal self-attention layer [1]. StrucText introduces a segment-token aligned encoder and designs a novel pre-training strategy with three self-supervised tasks to learn a richer representation [17]. Each of these methods claims to achieve high scores in document KIE tasks.

2.4 2D Grid-Based Method

Methods based on 2D-grid are also a way to solve the problem. This approach is based on a two-dimensional feature grid, and field values are extracted from the grid using a standard instance split model. This type of method originated in Chargrid. Chargrid introduces 2D grids as a new type of text representation [13]. VisualWordgrid combines these grid representations with 2D feature maps of document images to generate a more powerful multi-modal 2D document representation [14]. BERTgrid represents documents as a grid of context lexical feature vectors, and incorporates a BERT network into the network [2]. ViBERTgrid is proposed to overcome the disadvantage of not fully utilizing the performance advantage of language model in methods such as Chargrid and BERTgrid. Unlike BERTgrid, the parameters of BERT and CNN in multi-modal backbone network proposed by ViBERTgrid are jointly trained [19]. The experimental results show that the combined training strategy significantly improves the performance of ViBERTgrid.

3 Methodology

Our Method consists of three modules: a character level encoder module, a similarity module with contrastive learning and a decoder of BiLSTM + CRF. We apply character-level encoding and label feedback training method, and contrastive learning module is applied for distinguishing different field in a document.

As shown in Fig. 2, in the input phase, the document information is input by line, and the character level embedding of each word is obtained. After passing through a multi filter CNN network, a max over time pooling operation is applied to obtain a fixed dimensional token embedding of the word, which is added by label embedding from the decoder. Then we input the combined embedding into a BiLSTM network, and then obtain the tag information of the current sentence through a layer of CRF - for example, on SROIE data-sets, we preset five labels, respectively representing company, address, date, amount and other information.

During the first round of training, our model followed the above process. In the second and subsequent round of training, on the basis of the predicted

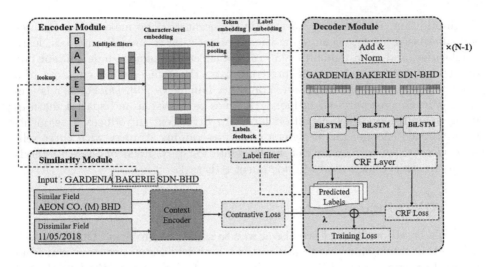

Fig. 2. The overall framework of our proposed method. It consists of three main parts: Character-level encoder module with CharLM as the backbone for document feature representation; Decoder module that consists of BiLSTM + CRF and labels feedback mechanism for predicting labels, modifying label embedding and calculating loss; Similarity module measures the similarity between different fields (or document) and generate the contrastive loss as part of the training loss. Note that the training of each sample will take up to N-1 rounds.

labels from first round, we feedback the label with the highest confidence to the training process as label embedding, add it to the token embedding from the CNN layer, and then get a new round of predicted information. Repeat this process until the preset number of training rounds is reached (in fact, for the KIE task with N labels, in theory, we can train the model for N-1 rounds for one sample. However, according to the actual effect, each round of training will feed back the label with the highest confidence as the known information to the input stage, and training for 2-3 rounds is enough. We will also discuss this topic later.)

3.1   Character Level Encoder Module

In the information extraction task, there are a large number of abbreviations, uncommon words, numeric phrases and proper nouns (such as address, person name, date, amount, etc.) in the document. The word frequency distribution is very different from the general text. If we choose common word-level or subword-level encoding, we need to construct a large dictionary, which is not conducive to the pre-training model, and also limits the migration of the model. Take SROIE or FUNSD for example, diverse bill words, such as date, address, amount and serial number, pose an obstacle in embedding matrix construction at the token-level. The unpredictable random tokens would incur the out-of-vocabulary (OOV) problem.

To deal with this problem, we alternatively utilize the char-level embedding to encode document tokens. Then we sum up all char-level embedding vectors together within a token through max-over time pooling as its token embedding vector. In this way, we can avoid most of the OOV problems and control the size of the embedding matrix at the appropriate scale.

In the encoding phase, we partially apply the character-aware CNN model in CharLM. We select 92 uppercase letters, lowercase letters, numbers, punctuation symbols and special symbols to represent the vast majority of the words that appear in the document. If there are still uncovered characters, we uniformly replace them with a special symbol.

As shown in Fig. 2, first layer performs a lookup of character embedding of dimension 4 and stacks them to form the matrix C_k (For example, the gray box in the upper left corner of the figure represents the character embedding of the word "BAKERIE"). Then convolution operations are applied between C_k and multiple filter matrices. In the example in Fig. 2, we have twelve filters-four filters of width 2, three filters of width 3, three filters of width 4 and two filters of width 5. A max-over-time pooling operation is applied to obtain a fixed-dimensional token embedding of the word:

$$TE_k[i] = maxpooling(cov(C_k, F_i)) \qquad (1)$$

where F_i is the i-th convolution filter vector and $TE_k \in R^{d_t}$, and d_t is the dimension of token embedding, which is equal to the total number of filter matrices:

$$d_t = \sum_{i \in FW} FN_i \qquad (2)$$

where FN_i is the number of filter vectors with the width i, and FW is the set of all width values. In Fig. 2, for example, the value of FW is {2,3,4,5} and FN is {4,3,3,2}, thus $d_t = 4 + 3 + 3 + 2 = 12$.

The output from the encoder will be added to the label embedding and then input into the decoder. The final embedding is :

$$E_k = TE_k + LE_k, \quad E_k, TE_k, LE_k \in R^d \qquad (3)$$

where d is equal to d_t, and the dimension of output embedding E, token embedding TE and label embedding LE are the same.

3.2   Decoder Module with Label Feedback Mechanism

Our decoder adopts the structure of BiLSTM + CRF. The output embedding matrix of the encoder is input to the BiLSTM network after a layer of residual connection and layer normalization, and the prediction result of the label is obtained in the CRF layer - this word belongs to the company, address, date, amount, or nothing. When we get the predicted label, we apply the label feedback mechanism to return the result to encoder as a label embedding. let y_k^t be the predicted labels of the k-th words in t-th rounds of training, and f_k^{t+1} be the feature map of the k-th word in (t+1)-th rounds of training. As mentioned earlier, the output embedding is the sum of token embedding and label embedding:

$$E_k^{t+1} = TE_k^{t+1} + LE_k^t \tag{4}$$

$$LE_k^t = Flt(y_k^t) \tag{5}$$

where Flt represents the filtering and pre-processing of predicted label information, which is the output of the label filter in Fig. 2:

$$Flt(y)[*, 0:d] = \begin{cases} default\ tag, & y \notin T_{max} \\ predicted\ tag, & y \in T_{max} \end{cases} \tag{6}$$

where d is the dimensions of token embedding, and T_{max} is the set of label with the highest overall confidence.

For partially labeled data, the label feedback also works. First, training is based on tagged data, whereas for unlabeled data, only the filtering method needs to be adjusted. Because we cannot know the distribution of the overall confidence level in advance, we select the label of the highest confidence level for the current sample to give feedback:

$$Flt(y)[*, 0:d] = \begin{cases} predicted\ tag, & conf(y) = max(T_c) \\ default\ tag, & otherwise \end{cases} \tag{7}$$

where T_c is the set of all possible label values, and conf(y) is the confidence of label y. On the test set, we adopt the same process to obtain the results.

In addition to the predicted label, the CRF loss output by the decoder will be weighted summed with the contrastive learning loss as the final training loss. And the CRF loss is defined as its negative log form:

$$\begin{aligned} Loss_{CRF} &= -log\frac{P_{real}}{\sum_k P_k} \\ &= -(S_{real} - log(\sum_k e^{s_k})) \end{aligned} \tag{8}$$

3.3 Similarity Module with Contrastive Learning

The similarity module is proposed to solve the new problems caused by character level embedding, which has a finer granularity than word level embedding and is content aware. Since the characters in the similar text field may be various, it will lead to dispersive encoding vectors. In addition, their encoding vectors may also vary greatly, which is not conducive to our model capturing common patterns in similar field.[41] In order to mitigate the impact of content diversity, we need to map the text field in similar patterns to a tighter vector space. We assume that similar text fields should have similar patterns. Therefore, we can assist the encoder to acquire features through contrastive learning between similar or dissimilar text fields.

In our experiments, we specify that the text lines with the same label and common words are similar (in fact, we have tried clustering algorithms, but the result is not stable). The distance between two token embedding from similar fields should be closer, and for different ones, they should be far apart. After defining similar text fields, we adopt contrast learning loss to help train the whole model. For each document field, we randomly select $|S|$ similar fields and $|D|$ dissimilar fields. Then we add the contrastive learning loss to the objective function for optimization.

In terms of contrastive learning loss calculation, first we need to select the scale of S and D, which respectively represent the token embedding of similar texts and dissimilar texts of a text field during contrastive learning. In our experiment, we set $|S| = 1$ and $|D| = 4$. And the contrastive loss is defined as:

$$Loss_{cont} = -log\frac{e^{TE \cdot TE^s}}{\sum_{TE_d \in D} e^{TE \cdot TE^d}} \qquad (9)$$

And the training loss is the weighted sum of CRF loss and contrastive loss with the weight coefficient $\lambda = 0.005$:

$$Loss_{training} = Loss_{CRF} + \lambda Loss_{cont} \qquad (10)$$

4 Experiment

4.1 Data-Sets

We selected SROIE and FUNSD as the evaluation data-set. SROIE was provided by ICDAR 2019 competition [9], short for Scanned Receipts OCR and Information Extraction (SROIE). Since then, SROIE data-set has become one of the most commonly used data-sets in the field of document KIE, on which most of the relevant work has been evaluated. The data-set has 1000 whole scanned receipt images. Each receipt image contains around about four key text fields: company, address, date and total amount. The text of documents in the data-set mainly consists of digits and English characters. For the KIE task, each image in the data-set is annotated with a text file. Since the data-set itself can also be used for OCR tasks, OCR recognition results for each image are provided for the KIE task to reduce the impact of OCR performance on the KIE model (But on the contrary, the OCR errors inherent in the data-set are a major problem for researchers).

FUNSD is short for Form Understanding in Noisy Scanned Documents, which is a data-set for text detection, optical character recognition, spatial layout analysis and form understanding [12]. The data-set contains 199 real, fully annotated, scanned forms.

4.2 Evaluation Metrics

We adopt the evaluation protocol provided by ICDAR 2019 Robust Reading Challenge: For each test receipt image, the extracted text is compared to the ground truth. An extract text is marked as correct if both submitted content and category of the extracted text matches the ground truth; Otherwise, marked as incorrect. The precision is computed over all the extracted texts of all the test receipt images. F1 score is computed based on precision and recall:

$$F1 = 2 \cdot \frac{Precision \cdot Recall}{Precision + Recall} \tag{11}$$

There are OCR errors and random prefix in SROIE data-set, so we modified the OCR mismatch in the data-set and removed the "RM" prefix that appears randomly in the total cost label to make the OCR results consistent with the picture.

4.3   Results

Table 1. F1 score comparison on SROIE and FUNSD dataset

Modality	Method	F1 score		param
		FUNSD	SROIE	
Text+Layout+Image	LayoutLMv2[a]	0.8420	0.9781	426M
	LayoutLMv3[a]	**0.9208**	-	368M
	VIES	–	0.9612	–
	PICK	–	0.9612	–
	TRIE	–	0.9618	–
	StrucTexT	–	0.9688	–
	TILT[a]	–	**0.9810**	780M
	Unidoc	0.8793	–	272M
Text+Layout	LayoutLM[a]	0.7789	0.9524	343M
	ViBERTgrid	–	0.9640	157M
	LAMBERT	–	**0.9693**	125M
	BROS[a]	0.8452	0.9548	340M
	LiLT	**0.8841**	–	–
	Our method[b]	0.8373	0.9635	**43M**
Text	RoBERTa[a]	0.7072	0.9280	355M
	BERT[a]	0.6563	0.9200	340M
	Our method	**0.7527**	**0.9610**	**35M**

[a] All listed methods are the LARGE version.
[b] The layout information is input in the same way as LayoutLM: we add the word embedding and the 2-D position embedding, which is calculated from the coordinates of the bounding box. See LayoutLM [32] for detail.

We compare our methods with PICK, LayoutLM, BERT, TRIE, ViBERT-grid, StrucTexT, TILT, BROS, Unidoc, LiLT and LAMBERT. We grouped different methods according to modality and compared their performance and parameters. The results are shown in Table 1.

Overall, more modalities mean higher scores. The result shows that our model achieves an F1 score of 0.9610 on SROIE and 0.7527 on FUNSD, which is significantly better than other methods only with text information. Moreover, our method achieves scores comparable to those of the state-of-the-art method on SROIE with a minimum number of parameters(35M), which is much less than most methods.

4.4 Ablation Study

To investigate the effectiveness of the predicted label feedback mechanism and the model performance under different network components, we further experimented with different configurations on our model(Only-text version). We eliminate the label feedback mechanism or similarity module from our model. In addition, we have modified the encoder to generate word-level embedding of the text for comparison experiments. The results are shown in Table 2.

Table 2. Performance of our method with different model structure on SROIE

Configuration	F1 score
CharLM (CNN + Highway + LSTM)	0.9029
Our method with word-level encoder	0.8843
Our method without label feedback	0.9276
Our method without similarity module	0.9315
Our method	**0.9610**

The results show that the character-level encoder significantly improves the F1 score, and label feedback and similarity module also contribute to it.

5 Conclusion

With the advancement in natural language processing, the scope of information that can be processed by natural language processing methods has expanded beyond one-dimensional sequential text. The extraction of information from 2-D documents such as forms, invoices, and receipts has become a growing area of interest among researchers.In the field of information extraction, we have noticed that many document semantics are relatively simple and contain a large number of professional terms and abbreviations, so character-level encoding is an effective method, and our experiments have also proven this.

In this paper, we propose a novel document KIE model that incorporates a character-level encoder, a similarity module based on contrastive learning, and a predicted label feedback mechanism. We find that character-level embedding can be effective for this task, especially when the data contains many rare words or abbreviations, because it can better address the OOV issue. The predicted labels can be combined with the input data to provide additional information, enabling the model to better understand the semantic relationship between words. Additionally, the contrastive learning module helps the encoder identify similar patterns.

Our model uses fewer parameters than most current methods and performs comparably to state-of-the-art approaches. Our method also achieves the highest score among single-modal methods. Unlike large pre-trained models, our approach focuses more on the relationships between different categories of fields in semi-structured documents, rather than the semantic information of the text itself, which we believe is beneficial for some simple document KIE tasks.

Acknowledgment. This work was supported by National Key R&D Program of China (No. 2021ZD0113301)

References

1. Appalaraju, S., Jasani, B., Kota, B.U., Xie, Y., Manmatha, R.: DocFormer: end-to-end transformer for document understanding. In: 2021 IEEE/CVF International Conference on Computer Vision (ICCV), pp. 973–983 (2021)
2. Denk, T.I., Reisswig, C.: BERTgrid: contextualized embedding for 2D document representation and understanding. In: Workshop on Document Intelligence at NeurIPS 2019 (2019)
3. Devlin, J., Chang, M.W., Lee, K., Toutanova, K.: BERT: Pre-training of deep bidirectional transformers for language understanding. In: Proceedings of the 2019 Conference of the North American Chapter of the Association for Computational Linguistics: Human Language Technologies, Volume 1 (Long and Short Papers), pp. 4171–4186 (2019)
4. Garncarek, Ł, et al.: LAMBERT: layout-aware language modeling for information extraction. In: Lladós, J., Lopresti, D., Uchida, S. (eds.) ICDAR 2021. LNCS, vol. 12821, pp. 532–547. Springer, Cham (2021). https://doi.org/10.1007/978-3-030-86549-8_34
5. Gu, J., et al.: Unidoc: unified pretraining framework for document understanding. Adv. Neural. Inf. Process. Syst. **34**, 39–50 (2021)
6. Hochreiter, S., Schmidhuber, J.: Long short-term memory. Neural Comput. **9**(8), 1735–1780 (1997)
7. Hong, T., Kim, D., Ji, M., Hwang, W., Nam, D., Park, S.: BROS: a pre-trained language model focusing on text and layout for better key information extraction from documents. In: Proceedings of the AAAI Conference on Artificial Intelligence, vol. 36, pp. 10767–10775 (2022)
8. Huang, Y., Lv, T., Cui, L., Lu, Y., Wei, F.: LayoutLMv3: pre-training for document ai with unified text and image masking. In: Proceedings of the 30th ACM International Conference on Multimedia, pp. 4083–4091 (2022)
9. Huang, Z., et al.: ICDAR 2019 competition on scanned receipt OCR and information extraction. In: 2019 International Conference on Document Analysis and Recognition (ICDAR), pp. 1516–1520 (2019)

10. Huang, Z., Xu, W., Yu, K.: Bidirectional LSTM-CRF models for sequence tagging. arXiv preprint arXiv:1508.01991 (2015)
11. Hwang, W., Yim, J., Park, S., Yang, S., Seo, M.: Spatial dependency parsing for semi-structured document information extraction. In: Findings of the Association for Computational Linguistics: ACL-IJCNLP 2021, pp. 330–343 (2021)
12. Jaume, G., Ekenel, H.K., Thiran, J.P.: FUNSD: a dataset for form understanding in noisy scanned documents. In: 2019 International Conference on Document Analysis and Recognition Workshops (ICDARW), vol. 2, pp. 1–6. IEEE (2019)
13. Katti, A.R., Reisswig, C., Guder, C., Brarda, S., Bickel, S., Höhne, J., Faddoul, J.B.: Chargrid: towards understanding 2D documents. In: Proceedings of the 2018 Conference on Empirical Methods in Natural Language Processing, pp. 4459–4469 (2018)
14. Kerroumi, M., Sayem, O., Shabou, A.: VisualWordGrid: information extraction from scanned documents using a multimodal approach. In: Barney Smith, E.H., Pal, U. (eds.) ICDAR 2021. LNCS, vol. 12917, pp. 389–402. Springer, Cham (2021). https://doi.org/10.1007/978-3-030-86159-9_28
15. Kim, Y., Jernite, Y., Sontag, D., Rush, A.M.: Character-aware neural language models. In: Proceedings of the Thirtieth AAAI Conference on Artificial Intelligence, pp. 2741–2749 (2016)
16. Lafferty, J., Mccallum, A., Pereira, F.: Conditional random fields: probabilistic models for segmenting and labeling sequence data. In: Proceedings of the Eighteenth International Conference on Machine Learning (ICML 2001), pp. 282–289 (2002)
17. Li, Y., et al.: StrucTexT: structured text understanding with multi-modal transformers. In: Proceedings of the 29th ACM International Conference on Multimedia, pp. 1912–1920 (2021)
18. Lin, T.Y., Dollár, P., Girshick, R., He, K., Hariharan, B., Belongie, S.: Feature pyramid networks for object detection. In: 2017 IEEE Conference on Computer Vision and Pattern Recognition (CVPR), pp. 936–944 (2017)
19. Lin, W., et al.: ViBERTgrid: a jointly trained multi-modal 2d document representation for key information extraction from documents. In: Lladós, J., Lopresti, D., Uchida, S. (eds.) ICDAR 2021. LNCS, vol. 12821, pp. 548–563. Springer, Cham (2021). https://doi.org/10.1007/978-3-030-86549-8_35
20. Liu, X., Gao, F., Zhang, Q., Zhao, H.: Graph convolution for multimodal information extraction from visually rich documents. In: Proceedings of the 2019 Conference of the North American Chapter of the Association for Computational Linguistics: Human Language Technologies, Volume 2 (Industry Papers), pp. 32–39. Association for Computational Linguistics (2019)
21. Liu, Y., et al.: RoBERTa: a robustly optimized BERT pretraining approach. arXiv preprint arXiv:1907.11692 (2019)
22. Mikolov, T., Karafiát, M., Burget, L., Cernocký, J., Khudanpur, S.: Recurrent neural network based language model. In: INTERSPEECH 2010, 11th Annual Conference of the International Speech Communication Association, pp. 1045–1048. ISCA (2010)
23. Park, S., et al.: CORD: a consolidated receipt dataset for post-OCR parsing. In: Workshop on Document Intelligence at NeurIPS 2019 (2019)
24. Powalski, R., Borchmann, Ł, Jurkiewicz, D., Dwojak, T., Pietruszka, M., Pałka, G.: Going Full-TILT Boogie on document understanding with text-image-layout transformer. In: Lladós, J., Lopresti, D., Uchida, S. (eds.) ICDAR 2021. LNCS, vol. 12822, pp. 732–747. Springer, Cham (2021). https://doi.org/10.1007/978-3-030-86331-9_47

25. Radford, A., Narasimhan, K., Salimans, T., Sutskever, I., et al.: Improving language understanding by generative pre-training. https://ww.nasa.gov/nh/pluto-the-other-red-planet (2018)
26. Stanisławek, T., et al.: Kleister: key information extraction datasets involving long documents with complex layouts. In: Lladós, J., Lopresti, D., Uchida, S. (eds.) ICDAR 2021. LNCS, vol. 12821, pp. 564–579. Springer, Cham (2021). https://doi.org/10.1007/978-3-030-86549-8_36
27. Tang, G., et al.: MatchVIE: exploiting match relevancy between entities for visual information extraction. In: Proceedings of the Thirtieth International Joint Conference on Artificial Intelligence, IJCAI-21, pp. 1039–1045. International Joint Conferences on Artificial Intelligence Organization (2021)
28. Vaswani, A., et al.: Attention is all you need. In: Advances in Neural Information Processing Systems, vol. 30, pp. 5998–6008 (2017)
29. Wang, J., Jin, L., Ding, K.: LiLT: a simple yet effective language-independent layout transformer for structured document understanding. In: Proceedings of the 60th Annual Meeting of the Association for Computational Linguistics (Volume 1: Long Papers), pp. 7747–7757 (2022)
30. Wang, J., et al.: Towards robust visual information extraction in real world: new dataset and novel solution. In: Proceedings of the AAAI Conference on Artificial Intelligence, vol. 35, pp. 2738–2745 (2021)
31. Xu, Y., et al.: LayoutLMv2: multi-modal pre-training for visually-rich document understanding. In: Proceedings of the 59th Annual Meeting of the Association for Computational Linguistics and the 11th International Joint Conference on Natural Language Processing (Volume 1: Long Papers), pp. 2579–2591 (2021)
32. Xu, Y., Li, M., Cui, L., Huang, S., Wei, F., Zhou, M.: LayoutLM: pre-training of text and layout for document image understanding. In: Proceedings of the 26th ACM SIGKDD International Conference on Knowledge Discovery & Data Mining, pp. 1192–1200 (2020)
33. Yu, W., Lu, N., Qi, X., Gong, P., Xiao, R.: Pick: processing key information extraction from documents using improved graph learning-convolutional networks. In: 2020 25th International Conference on Pattern Recognition (ICPR), pp. 4363–4370. IEEE (2021)
34. Zhang, P., et al.: TRIE: end-to-end text reading and information extraction for document understanding. In: Proceedings of the 28th ACM International Conference on Multimedia, pp. 1413–1422 (2020)

Multimodal Rumour Detection: Catching News that Never Transpired!

Raghvendra Kumar[1]([✉])(iD), Ritika Sinha[1]([✉])(iD), Sriparna Saha[1](iD),
and Adam Jatowt[2](iD)

[1] Indian Institute of Technology Patna, Dayalpur Daulatpur, India
raghvendra.kumar1004@gmail.com, ritika16sinha@gmail.com
[2] University of Innsbruck, Innsbruck, Austria
adam.jatowt@uibk.ac.at

Abstract. The growth of unverified multimodal content on microblogging sites has emerged as a challenging problem in recent times. One major roadblock to this problem is the unavailability of automated tools for rumour detection. Previous work in this field mainly involves rumour detection for textual content only. As per recent studies, the incorporation of multiple modalities (text and image) is provably useful in many tasks since it enhances the understanding of the context. This paper introduces a novel multimodal architecture for rumour detection. It consists of two attention-based BiLSTM neural networks for the generation of text and image feature representations, fused using a cross-modal fusion block and ultimately passing through the rumour detection module. To establish the efficiency of the proposed approach, we extend the existing PHEME-2016 data set by collecting available images and in case of non-availability, additionally downloading new images from the Web. Experiments show that our proposed architecture outperforms state-of-the-art results by a large margin.

Keywords: Rumour Detection · Multimodality · Deep learning · PHEME Dataset · Twitter

1   Introduction

Considering the recent developments in technology, there is still insufficient control over the proliferation and dissemination of information transmitted through untrusted online sources like micro-blogging sites [27]. This leads to the propagation of unverified news, especially in the context of breaking news, which may further unfold rumours among the masses [19]. These rumours, if left unmitigated, can influence public opinion, or corrupt the understanding of the event for journalists. Manually fact-checking the news in real-time is a tremendously

---

R. Kumar and R. Sinha—These authors contributed equally to this work.

G. A. Fink et al. (Eds.): ICDAR 2023, LNCS 14189, pp. 231–248, 2023.
https://doi.org/10.1007/978-3-031-41682-8_15

difficult task. So, there is a need to automate the process of rumour detection to promote credible information on online sources.

Additionally, information transfer in the modern era is increasingly becoming multimodal. *Oxford Reference*[1] defines multimodality as "The use of more than one semiotic mode in meaning-making, communication, and representation generally, or in a specific situation". Information on micro-blogging sites is rarely present in the textual mode only. These days, they contain images, videos, and audio, among other modalities of information.

Fig. 1. A sample Twitter thread

Twitter, a well-known micro-blogging site, lets users exchange information via brief text messages, that may have accompanying multimedia. For the task of rumour detection, it is actually more effective to utilize the entire Twitter thread. Reply tweets are useful in this regard as the users often share their opinions, suggestions, criticism, and judgements on the contents of the source tweet. Also, one gets a better understanding of the context when provided with visual cues (images). As elaborated through an example shown in Fig. 1, a user (through reply tweet), has indicated that the image associated with the source tweet has wrongly stated the crash site. Other users can also interact by adding their views. These replies on the source tweet help in understanding the situation better.

---

[1] https://www.oxfordreference.com.

Recent research on Twitter rumour detection usually uses the PHEME or SemEval datasets and it mainly involves machine learning-based models [19,28]. In these methods, authors extracted statistical or lexical features from textual data. These fail to capture dependencies between the features, and the semantics of the Twitter thread. To overcome this problem, deep learning-based methods were applied. These methods provide a more robust representation of the features. However, all these works are done on a textual content only.

In complex disciplines where a sole modality might not be able to offer sufficient information, multiple modalities could strengthen the overall system performance by combining the advantages of different modalities and offering a more thorough and reliable representation of the data. Also, multimodal systems are more natural to interact with as they mimic how humans use diverse senses to comprehend their surroundings. Just as importantly, multimodal systems provide a more engaging experience by using multiple modalities to create dynamic and interactive content. In this work, we put forward a deep learning-based approach that employs multimodality via an extended dataset that we have prepared to distinguish between reliable and unreliable content on micro-blogging websites such as Twitter. We also conduct thorough studies on the proposed model and the extended dataset. Attention-based RNN models have mainly been used for fusing feature representations of the two modalities [11]. Cheung *et al.* [5] have made significant progress towards extending the PHEME-2016 dataset to make it multimodal[2]. However, the authors have performed this extension only for those tweet threads where images had already been uploaded by the users, so the resulting dataset is only partially multimodal, with only 46% of the tweets containing images. We further extend the dataset to be fully multimodal.

The following is a summary of our work's significant contributions:

- We propose a dual-branch Cross-fusion Attention-based Multimodal Rumour Detection (CAMRD) framework for effectively capturing multimodal interdependencies between the text and image modalities for the detection of rumours on micro-blogging sites.
- The PHEME-2016 dataset comprises textual data only. We extend the dataset[3] by collecting images using the associated metadata and by the means of web scraping. We make use of cross-modal fusion of CAMRD to effectively capture the interdependencies between the two modalities.
- We perform extensive studies for selecting the best image amongst multiple images using various heuristics.

2   Related Works

Over the past few years, there have been significant research efforts focused on identifying rumours and misinformation. This section provides an overview of

---

[2] Unfortunately, that dataset was not made public.

[3] https://drive.google.com/file/d/1XR7g6UL8_4yqvo12alQn2iqmWvHb6iKr/view?usp=sharing.

previous works in this area and delves into specific studies related to cross-modal learning which is related to the proposed framework for detecting rumours.

Previous methods for detecting rumours heavily relied on statistical analysis. One of the earliest works examined the spread of rumours on Twitter by analyzing the frequency of certain words in data sets [25]. Despite being constrained to specific events and unable to generalize to new scenarios, these techniques became the basis for later developments in machine learning-based approaches to detect rumours. Kwon et al. [12] were the first to use machine learning techniques for this task including support vector machines, decision trees and random forests, accompanied with linguistic features for rumour detection.

Others have used Recurrent Neural Networks (RNNs) [15] and Convolutional Neural Networks (CNNs) [4], as such neural network-based models are able to effectively uncover and learn the underlying patterns within a data set. These techniques are often combined with pre-trained non-contextual word embeddings, such as GloVe [20], which yield a unique vector for each word without taking note of the context. Contextual word embeddings like BERT [8] have also been used as the vector representation, as the generated embeddings convey the semantic meaning of each word. In this study, we have used BERTweet [18] which is a publicly available, large-scale language model that has been pre-trained specifically for understanding and processing English tweets. It is trained based on the RoBERTa [14] pre-training procedure to generate contextual embeddings.

The ensemble graph convolutional neural network (GCNN) technique proposed in [1] uses a two-branch approach, a graph neural network (GNN) and a convolutional neural network (CNN) to process the features of a node in relation to its neighbouring nodes and to obtain feature representations from weighted word embedding vectors, respectively. The weighted combination of these two branches' features is considered the final feature vector. This method results in a poor representation of the conversation as it does not fully utilize the relationship between global and local information. The method outlined in [22] suggested using a recurrent neural network (RNN) for rumour detection in conversations but it suffered the same limitation of not considering the relative importance of each message in the conversation. To overcome this limitation, we have incorporated an attention-weighted average module in order to achieve a more precise representation of tweets and images.

Visual computing techniques are also useful for the analysis of social media contents, as shown in [3]. In recent times, processes which earlier had unimodal input and output such as emotion and opinion analysis, fake-news identification, and hate-speech detection have now expanded into the multimodal domain [9] which is the integration of multiple modalities of information. Another paradigm, namely, knowledge distillation which is a method used to transfer knowledge from a multimodal model to a single-modal model, has been applied in various fields like computer vision and speech processing [2,13,23].

3   Dataset

3.1   Original Dataset

As mentioned earlier, we used the PHEME-2016 dataset [28] for our experiments. It is a collection of Twitter threads consisting of 3,830 non-rumour and 1,972 rumour tweets posted during 5 breaking news, namely Charlie Hebdo, Ferguson, Germanwings crash, Ottawa Shooting, and Sydney Siege. It contains metadata regarding source tweets and reply tweets for each tweet thread.

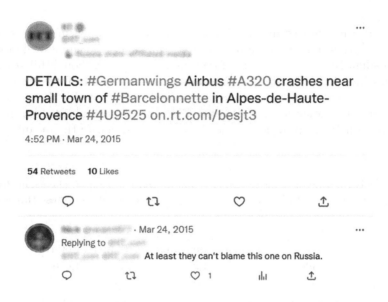

DETAILS: #Germanwings **Airbus** #A320 **crashes near small town of** #Barcelonnette **in Alpes-de-Haute-Provence** #4U9525 on.rt.com/besjt3

4:52 PM · Mar 24, 2015

54 Retweets   **10** Likes

Fig. 2. Example of a tweet thread with no image

3.2   Dataset Extension

The original PHEME-2016 dataset initially did not contain any images. So the tweet threads which already had images uploaded by the users were collected using the image information specified in the metadata, with the distribution illustrated in Table 1. For the remaining tweet threads, we augmented images through web scraping[4], as shown by an example via Fig. 2 and Fig. 3. Our criteria for downloading images included only the source tweet and not the reply tweets because reply tweets have higher chances of not being relevant to the tweet's content, relevance, and appropriateness. Also, the rationale behind making the dataset multimodal was that even though the textual data can provide valuable information, they are often limited in their ability to convey complex or detailed information. Images can provide supplementary information and help to provide

---

[4] https://github.com/Joeclinton1/google-images-download.

Fig. 3. Web scraped image for the above tweet thread showing rescuers examining the situation at the crash site

a comprehensive understanding of a situation. For example, in the case of the Ottawa shooting news topic in our dataset, the tweet "Penguins will also have the Canadian National Anthem before tonight's game. A thoughtful gesture by the Pens. Sing loud, Pittsburgh #Ottawa". was not able to fully illustrate the impact of the event, as the user may not have known that the "Penguins" here referred to a Canadian ice hockey team. However, the downloaded image of the sports team helps to provide additional context and clarify the meaning of the tweet.

Table 1. Distribution of tweet threads in rumour and non-rumour classes in PHEME-2016 Dataset, and count of images that were present originally on these threads.

News Event	Tweets		Images	
	Rumour	Non-rumour	Rumour	Non-rumour
Charlie Hebdo	458	1,621	234	1,010
Ferguson	284	859	72	414
Germanwings Crash	238	231	82	141
Ottawa Shooting	470	420	172	134
Sydney Siege	522	699	241	310

4   Problem Statement

Our research aims to create an end-to-end multimodal framework for rumour detection given an array of tweets $\{T \cup I\}$ where $T = \{T_1, T_2, ..., T_n\}$ and each T_i represents the text content (further, $T_i = \{s_i, r_{i1}, r_{i2}, ..., r_{ik}\}$ where s_i represents source tweet and reply tweets are denoted by r_{ik}), and $I = \{I_1, I_2, ..., I_n\}$ where each I_i denotes the image related to the ith tweet thread. The task is to train an end-to-end rumor detector $f : \{T \cup I\} \rightarrow \{Rumour, Non - Rumour\}$ by inter-relating the dependence of text and images.

5   Proposed Methodology

Our model primarily consists of three modules: (i) Attention-based feature extraction sub-modules of text and image modalities (ii) Cross-modal fusion module and (iii) Rumour Classification module (RC) for rumor detection, which are elaborated in detail in the following subsections. Figure 4 represents the architecture of our proposed model.

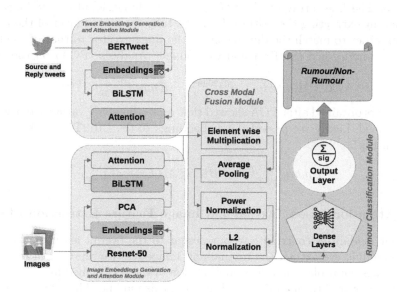

Fig. 4. Proposed dual-branch Cross-fusion Attention-based Multimodal Rumour Detection (CAMRD) model architecture

5.1   Embedding Generation

The following is the process used for the generation of embeddings across multiple modalities:

Tweet Embedding: BERTweet [18], a pre-trained language model for English tweets, was used for generating word-wise text embeddings. The *bertweet-large* model in particular (max-length = 512, Parameters = 355, pre-trained on 873M English tweets (cased)) was utilized. The normalized textual content of the source tweet s_i and reply tweets r_{ij}, where j is the number of reply tweets for a source tweet s_i were concatenated and passed through BERTweet:

$$t_i = s_i \oplus r_{ij} \tag{1}$$

$$Embed_{tweet} = BERTweet(t_i) \tag{2}$$

to get the required array of $[Embed_{tweet}]^{n*1024}$ dimensional embeddings, where n is the total count of tweet threads.

Image Embedding: The image features I_i for each tweet thread t_i are extracted using ResNet-50 [10] as indicated below,

$$Embed_{image} = ResNet50(I_i) \tag{3}$$

ResNet-50 is a convolutional neural network, 50 layers deep. A pre-trained version of the network trained on the ImageNet [7] database was used for our task. The images were first resized and normalized into a fixed dimension of 224 × 224 × 3. The final array of $[Embed_{image}]^{n*2048}$ dimensional embeddings were obtained by extracting the features from the fully connected layer of the model. Next, in order to match the dimensions of the image vector to that of the tweet vector, we performed the Principal Component Analysis (PCA) on the image vector:

$$PCA([Embed_{image}]^{n*2048}) = [Embed_{image}^{PCA}]^{n*1024} \tag{4}$$

The vectors obtained after PCA showed that more than 95% variance of the data was retained. Lastly, we passed the embeddings to the BiLSTM layer, depicted below in Eq. 5.

$$Embed_{image}^{BiLSTM} = BiLSTM(Embed_{image}^{PCA}) \tag{5}$$

5.2   Attention-based Textual and Image Feature Extraction Module

Different modalities contribute to our design to the overall feature generation. In this module, we extract the textual and image feature representations.

The most suitable semantic representation of the text embeddings (explained above) is captured through the first branch of this module. The text embeddings are fed to this branch as represented in Eq. 6. It consists of a BiLSTM layer for capturing the long-term dependencies of textual embeddings in both forward and backward directions.

$$Embed_{tweet}^{BiLSTM} = BiLSTM(Embed_{tweet}) \tag{6}$$

Some words are more important than others, contributing more toward meaningful textual representation. Hence, the output of BiLSTM is passed through an attention layer, shown in Eq. 7, for extracting those words that are crucial to the tweet's meaning, and forming the final embedding vectors out of those descriptive words [26].

$$Embed_{tweet}^{final} = Attention(Embed_{tweet}^{BiLSTM}) \tag{7}$$

The second branch of this module focuses on the critical parts of the image for understanding which aspect of the picture makes it more likely to be categorized as rumour or non-rumour. Thus, we pass the output vector of the image from the previous module to an attention layer, represented in Eq. 8.

$$Embed_{image}^{final} = Attention(Embed_{image}^{BiLSTM}) \tag{8}$$

5.3   Cross-modal Fusion Module (CMF)

We use the cross-modal fusion module to merge the final textual and image feature vectors in lieu of a plain concatenation because the CMF module magnifies the association between the vectors of both modalities. Additionally, it overcomes the necessity of choosing the appropriate lengths of the extracted vectors which poses a challenge in plain concatenation. The first step of the CMF module involves element-wise multiplication of the vectors of the two modalities which adequately encapsulates the relationship between them.

$$Embed_{mul} = Embed_{tweet}^{final} * Embed_{image}^{final} \tag{9}$$

The * in Eq. 9 denotes element-wise multiplication. Next, average pooling is performed as it retains more information often than not when compared to max or min pooling.

$$Embed_{pooled} = Avg.Pooling(Embed_{mul}) \tag{10}$$

Then power normalization is carried out to reduce the chances of over-fitting.

$$Embed_{p-norm} = sgn(Embed_{pooled}) * \sqrt{|Embed_{pooled}|} \tag{11}$$

where sgn(x) is the signum function, described in Eq. 12.

$$\begin{cases} 1 & \text{if } x > 0 \\ 0 & \text{if } x = 0 \\ -1 & \text{if } x < 0 \end{cases} \tag{12}$$

The last step in the module carries out L2 normalization, to minimize the overall error and lower the computational cost.

$$Embed_{L2-norm} = \|Embed_{p-norm}\|_2 = \sqrt{\sum(Embed_{p-norm})^2} \tag{13}$$

To recap, our CMF module aligns the tweets and image features by boosting their association and is novel compared to existing works that fixate on the plain concatenation of feature representations of different modalities. The mathematical formulation of the CMF module is explained using Eq. 9 to Eq. 13.

5.4   Rumour Classification Module

Our final rumour classification module consists of five hidden layers with rectified linear activation function as depicted by

$$ReLU(z) = max(0, z) \tag{14}$$

and an output layer with a logistic function,

$$\sigma(z) = \frac{1}{1 + e^{-z}} \tag{15}$$

as an activation function. This module intakes the L2 normalized vectors, output by the CMF module, and extrapolates them into the objective array of two classes to yield the final expectancy probability that determines whether the multimodal tweet is rumour or non-rumour. The loss function used for our model is binary cross-entropy which is calculated as

$$-\sum (y \log(p) + (1 - y) \log(1 - p)) \tag{16}$$

where y is the binary indicator (0 or 1) and p is the predicted probability observation. The optimizer used is the Adam optimizer.

6   Experiments, Results, and Analysis

In this segment, we present and analyze the various experimental configurations we have used and their respective outcomes.

6.1   Experimental Setting and Evaluation Metrics

This section describes the process of extracting embeddings, the pre-processing of tweets and images, various hyperparameters used, and all the technical implementation details. Python libraries NumPy, Pandas, Keras, and Sklearn, were used on the TensorFlow framework to execute the tests. The system's performance was assessed on the basis of parameters like accuracy, precision, recall, and F1-score.

The experimental protocol followed was random shuffling of the tweet threads, and then splitting into an 80–10–10 train-validation-test set. Fine-tuning was performed on the validation dataset. As raw tweets contain mostly unorganized and dispensable information, we preprocess the tweets using RegEx in Python by removing URLs, usernames, punctuation, special characters, numbers, and hash(#) characters in hashtags. In addition, character normalization was also done. Pre-processed tweets were then passed to the BERTweet model to generate embeddings of 1,024 dimensions. Next, a BiLSTM layer with 512 units intakes these feature vectors and passes them on to the attention module. In parallel, we re-scale images to the size of (224,224,3) and feed them to the ResNet-50 module, which produces a 2,048-dimensional feature vector that in turn is fed to PCA to reduce it to 1,024 dimensions. Next, we pass it through the attention module. Following that, both the vectors are then advanced to the CMF module which fuses them and then feeds them to the rumor classification module that has five hidden layers with 512, 256, 128, 64, and 16 units, respectively, and the output layer with 1 unit. The activation functions for the hidden layers and the output layer are the ReLU and Sigmoid, respectively. Our model is trained with a batch size of 32 and an epoch number of 100. Table 2 summarizes the details of all the hyperparameters used for training our model. The hyperparameters shown in Table 2 were concluded after performing a parameter sensitivity study in which we analyzed the effect of variations in individual parameters over a range and observed how it affected the output of the system.

Table 2. Hyperparameters utilized for training the presented model.

Parameters	Values
Tweet length	512
Image size	(224, 224, 3)
Optimizer	Adam
Learning rate	0.001
Batch Size	32
Epochs	100
Filter size	(3, 3)
Strides	(1, 1)
Padding	'same'

6.2 Results and Discussion

In addition to the augmented PHEME-2016 dataset which contains a single image per tweet thread, i.e., 5,802 tweet-image pairs in total, we also created a further expanded dataset for experimental purposes which consists of multiple images, with a majority of tweets containing two images and few of the tweets with several images, altogether, 10,035 images. Various heuristics were applied to select the best image from the multiple images, which are as follows:

- H1: Manual selection of a single image per tweet.
- H2: Selection of the first image present in the dataset per tweet.
- H3: Random selection of an image per tweet.
- H4: Selection of the image with the largest dimension present in the dataset per tweet.

The first heuristic means that a human is manually selecting the best image for a tweet thread and the instructions followed by the human to manually select the image included:

- Relevance: The image should be relevant to the content of the tweet and help to convey the message or meaning of the tweet visually.
- Coherent to the hashtags: The image should align with the hashtags and help to convey the intended message or meaning.

Table 3 illustrates that amidst the various heuristics used, manually selecting the images produces the optimal result in terms of all four evaluation parameters with our proposed model. The proposed model produces the poorest accuracy when no heuristics were applied and multiple images were used. After the establishment of the best heuristic, we conducted experiments with various combinations of image and tweet embeddings including ResNet-50, ResNet-101 [10], ResNet-152 [10], VGG-19 [21], VGG-16 [21], InceptionV3 [24], and Xception [6] and BERTweet and OpenAI, respectively. Table 4 shows that the embeddings generated by the combination of BERTweet and ResNet-50 produce the optimal outcomes in terms of accuracy, recall, and F1-score which demonstrate that it is the best-performing embedding-generating combination. The conjunction of BERTweet and ResNet-101 performs exceptionally well in terms of precision.

Table 3. Results obtained on the proposed CAMRD model using various heuristics.

Heuristic Used	Accuracy	Precision	Recall	F1-Score
H1	0.893	0.913	0.917	0.915
H2	0.877	0.911	0.906	0.909
H3	0.841	0.850	0.915	0.881
H4	0.844	0.881	0.875	0.878
None	0.832	0.858	0.897	0.877

Table 4. Results obtained after using different tweet and image embeddings on the proposed CAMRD model and the heuristics of manually selected images.

Tweet Embd.	Image Embd.	Accuracy	Precision	Recall	F1-Score
BERTweet	ResNet-50	**0.893**	0.913	**0.917**	**0.915**
BERTweet	ResNet-101	0.875	**0.914**	0.898	0.906
BERTweet	ResNet-152	0.868	0.888	**0.917**	0.902
BERTweet	VGG-19	0.860	0.897	0.890	0.894
BERTweet	VGG-16	0.862	0.887	0.908	0.897
BERTweet	InceptionV3	0.863	0.900	0.893	0.896
BERTweet	Xception	0.859	0.884	0.900	0.892
OpenAI	ResNet-50	0.886	0.897	0.891	0.894
OpenAI	ResNet-101	0.872	0.897	0.885	0.891
OpenAI	ResNet-152	0.871	0.872	0.913	0.892
OpenAI	VGG-19	0.882	0.909	0.888	0.898
OpenAI	VGG-16	0.877	0.887	0.904	0.895
OpenAI	InceptionV3	0.864	0.896	0.875	0.880
OpenAI	Xception	**0.893**	0.910	0.904	0.907

6.3   Comparison with State of the Art and Other Techniques

In this part, we compare our CAMRD model with the existing State-of-the-Art (SOTA) and various other techniques. In Multi-Task Framework To Obtain Trustworthy Summaries (MTLTS) [17], the authors have proposed an end-to-end solution that jointly verifies and summarizes large volumes of disaster-related tweets, which is the current SOTA that we use for comparison. The other techniques that we compare our model with are SMIL: Multimodal Learning with Severely Missing Modality [16], Gradient Boosting classifier which attains the best results amongst different machine learning classifiers that we experimented upon and the final technique in which we passed tweets to both the branches of our proposed model. The techniques, namely, our proposed CAMRD model and gradient boosting classifier, when used with a single image per tweet and multiple images per tweet have been represented in Table 5 with subscript SI and MI, respectively. The values of the main parameters for the Gradient Boosting technique were as follows, learning_rate = 0.1, loss = 'log_loss', min_samples_split = 2, max_depth = 3. These values were selected after parameter sensitivity testing.

Table 5 demonstrates that our proposed model when operated with a single image per tweet outperforms the SOTA and other techniques in terms of accuracy and precision, while gradient boosting classifier when used with multiple images per tweet shows superior results in terms of recall and f1-score, however, this technique seriously falters in the case of a single image per tweet.

Table 5. Comparative analysis of SOTA and other methods

Approach	Accuracy	Precision	Recall	F1-Score
$CAMRD_{SI}$	**0.893**	**0.913**	0.917	0.915
$CAMRD_{MI}$	0.832	0.858	0.897	0.877
SMIL [16]	0.815	0.823	0.835	0.829
GB_{SI}	0.741	0.765	0.873	0.816
GB_{MI}	0.884	0.884	**0.951**	**0.917**
MTLTS [17]	0.786	0.77	0.766	0.768
Text-only	0.733	0.803	0.791	0.797

6.4 Classification Examples and Error Analysis

In this section we show few examples of tweets and explain the causes for both cases, when the tweets were correctly classified, and error analysis when the tweets were wrongly classified. The first case covers the true positive and true negative scenario where tweets from each class, rumour and non-rumour were rightly categorized as depicted in Fig. 5. The left-hand side tweet thread has been correctly classified as a rumour, with the subsequent reply tweets on the source tweet questioning the authenticity of the latter. The right-hand side tweet has been correctly classified as non-rumour as the image supports the tweet.

The second case covers the false positive as shown in Fig. 6 and false negative as shown in Fig. 7 where tweets from each class, rumour and non-rumour were wrongly categorized.

Here, the image shown in Fig. 6 represents disrespectful behaviour as it shows abusive symbols in public, and the text says that it is alright to rage against what happened, so the model might be biased against public display of resentment and this may be the reason for this tweet getting misclassified. The entire Twitter thread as shown in Fig. 7 seems quite convincing, as well the image also represents a man being pointed fingers at, which is quite in line with the word 'punishable'. This may be the reason for this tweet getting misclassified.

In general, if there are discrepancies between text and image, the CAMRD model may misclassify them, which is reflected via examples (Fig. 6 and Fig. 7).

Fig. 5. Example of tweets correctly classified, left-hand side tweet thread and corresponding picture belongs to a rumour class and the right-hand side tweet and image belongs to non-rumour class

Fig. 6. Example of a non-rumour tweet wrongly classified as rumour

Fig. 7. Example of a rumour tweet wrongly classified as non-rumour

7   Conclusions and Future Works

In this paper, we have introduced a novel deep-learning based model named Cross-fusion Attention-based Multimodal Rumour Detection (CAMRD), which takes text and image(s) as input and then categorizes it as either rumour or non-rumour. Attention-based textual and visual features were extracted. Instead of a plain concatenation of both features, we have used the cross-modal fusion module which then passes it to the Multilayer Perceptron segment that classi-fies the data. Additionally, the PHEME 2016 dataset has been expanded and made fully multimodal by means of image augmentation. The diverse experi-ments conducted on the expanded dataset show that our approach is effective for multimodal rumour detection. Most social media platforms these days have seen a huge rise in textual and visual content being uploaded, in the form of short videos, commonly known as 'reels', which certainly opens our work to be extended in the direction of including videos and audio as well.

Acknowledgements. Raghvendra Kumar would like to express his heartfelt gratitude to the Technology Innovation Hub (TIH), Vishlesan I-Hub Foundation, IIT Patna for providing the Chanakya Fellowship, which has been instrumental in supporting his research endeavours. Dr. Sriparna Saha gratefully acknowledges the Young Faculty Research Fellowship (YFRF) Award, supported by Visvesvaraya Ph.D. Scheme for Electronics and IT, Ministry of Electronics and Information Technology (MeitY), Government of India, being implemented by Digital India Corporation (formerly Media Lab Asia) for carrying out this research.

Author contributions. Raghvendra Kumar, Ritika Sinha : These authors contributed equally to this work.

References

1. Bai, N., Meng, F., Rui, X., Wang, Z.: Rumour detection based on graph convolutional neural net. IEEE Access 1 (2021). https://doi.org/10.1109/ACCESS.2021.3050563

2. Bai, Y., Yi, J., Tao, J., Tian, Z., Wen, Z., Zhang, S.: Fast end-to-end speech recognition via non-autoregressive models and cross-modal knowledge transferring from bert. IEEE/ACM Trans. Audio, Speech and Lang. Proc. **29**, 1897–1911 (2021). https://doi.org/10.1109/TASLP.2021.3082299

3. Borth, D., Ji, R., Chen, T., Breuel, T., Chang, S.F.: Large-scale visual sentiment ontology and detectors using adjective noun pairs. In: Proceedings of the 21st ACM International Conference on Multimedia, MM 2013, pp. 223–232. Association for Computing Machinery, New York, NY, USA (2013). https://doi.org/10.1145/2502081.2502282

4. Chen, Y., Sui, J., Hu, L., Gong, W.: Attention-residual network with CNN for rumor detection. In: Proceedings of the 28th ACM International Conference on Information and Knowledge Management, CIKM 2019, pp. 1121–1130. Association for Computing Machinery, New York, NY, USA (2019). https://doi.org/10.1145/3357384.3357950

5. Cheung, T.H., Lam, K.M.: Transformer-graph neural network with global-local attention for multimodal rumour detection with knowledge distillation (2022). https://doi.org/10.48550/ARXIV.2206.04832, https://arxiv.org/abs/2206.04832

6. Chollet, F.: Xception: deep learning with depthwise separable convolutions. In: 2017 IEEE Conference on Computer Vision and Pattern Recognition (CVPR), pp. 1800–1807. IEEE Computer Society, Los Alamitos, CA, USA, July 2017. https://doi.org/10.1109/CVPR.2017.195, https://doi.org/ieeecomputersociety.org/10.1109/CVPR.2017.195

7. Deng, J., Dong, W., Socher, R., Li, L.J., Li, K., Fei-Fei, L.: Imagenet: a large-scale hierarchical image database. In: 2009 IEEE Conference on Computer Vision and Pattern Recognition, pp. 248–255 (2009). https://doi.org/10.1109/CVPR.2009.5206848

8. Devlin, J., Chang, M.W., Lee, K., Toutanova, K.: BERT: pre-training of deep bidirectional transformers for language understanding. In: Proceedings of the 2019 Conference of the North American Chapter of the Association for Computational Linguistics: Human Language Technologies, Volume 1 (Long and Short Papers), pp. 4171–4186. Association for Computational Linguistics, Minneapolis, Minnesota, June 2019. https://doi.org/10.18653/v1/N19-1423, https://aclanthology.org/N19-1423

9. Ghani, N.A., Hamid, S., Targio Hashem, I.A., Ahmed, E.: Social media big data analytics: a survey. Computers in Human Behavior **101**, 417–428 (2019). https://doi.org/10.1016/j.chb.2018.08.039, https://www.sciencedirect.com/science/article/pii/S074756321830414X

10. He, K., Zhang, X., Ren, S., Sun, J.: Deep residual learning for image recognition. In: Proceedings of the IEEE Conference on Computer Vision and Pattern Recognition (CVPR), June 2016

11. Jin, Z., Cao, J., Guo, H., Zhang, Y., Luo, J.: Multimodal fusion with recurrent neural networks for rumor detection on microblogs. In: Proceedings of the 25th ACM International Conference on Multimedia, MM 2017, pp. 795–816. Association for Computing Machinery, New York, NY, USA (2017). https://doi.org/10.1145/3123266.3123454, https://doi.org/10.1145/3123266.3123454

12. Kwon, S., Cha, M., Jung, K., Chen, W., Wang, Y.: Prominent features of rumor propagation in online social media. In: 2013 IEEE 13th International Conference on Data Mining, pp. 1103–1108 (2013). https://doi.org/10.1109/ICDM.2013.61

13. Liu, T., Lam, K., Zhao, R., Qiu, G.: Deep cross-modal representation learning and distillation for illumination-invariant pedestrian detection. IEEE Trans. Circ. Syst. Video Technol. **32**(1), 315–329 (2022). https://doi.org/10.1109/TCSVT.2021.3060162

14. Liu, Y., et al.: Roberta: a robustly optimized bert pretraining approach. arXiv preprint arXiv:1907.11692 (2019)

15. Ma, J., et al.: Detecting rumors from microblogs with recurrent neural networks. In: Proceedings of the Twenty-Fifth International Joint Conference on Artificial Intelligence, IJCAI 2016, pp. 3818–3824. AAAI Press (2016)

16. Ma, M., Ren, J., Zhao, L., Tulyakov, S., Wu, C., Peng, X.: Smil: multimodal learning with severely missing modality (2021). https://doi.org/10.48550/ARXIV.2103.05677, https://arxiv.org/abs/2103.05677

17. Mukherjee, R., et al.: MTLTS: a multi-task framework to obtain trustworthy summaries from crisis-related microblogs. In: Proceedings of the Fifteenth ACM International Conference on Web Search and Data Mining, ACM, February 2022. https://doi.org/10.1145/3488560.3498536

18. Nguyen, D.Q., Vu, T., Nguyen, A.T.: BERTweet: a pre-trained language model for English Tweets. In: Proceedings of the 2020 Conference on Empirical Methods in Natural Language Processing: System Demonstrations, pp. 9–14 (2020)

19. Pathak, A.R., Mahajan, A., Singh, K., Patil, A., Nair, A.: Analysis of techniques for rumor detection in social media. Procedia Comput. Sci. **167**, 2286–2296 (2020). https://doi.org/10.1016/j.procs.2020.03.281, https://www.sciencedirect.com/science/article/pii/S187705092030747X, international Conference on Computational Intelligence and Data Science

20. Pennington, J., Socher, R., Manning, C.: GloVe: global vectors for word representation. In: Proceedings of the 2014 Conference on Empirical Methods in Natural Language Processing (EMNLP), pp. 1532–1543. Association for Computational Linguistics, Doha, Qatar, October 2014. https://doi.org/10.3115/v1/D14-1162, https://aclanthology.org/D14-1162

21. Simonyan, K., Zisserman, A.: Very deep convolutional networks for large-scale image recognition. In: International Conference on Learning Representations (2015)

22. Song, C., Yang, C., Chen, H., Tu, C., Liu, Z., Sun, M.: Ced: credible early detection of social media rumors. IEEE Trans. Knowl. Data Eng. **33**(8), 3035–3047 (2021). https://doi.org/10.1109/TKDE.2019.2961675

23. Sun, S., Cheng, Y., Gan, Z., Liu, J.: Patient knowledge distillation for BERT model compression. In: Proceedings of the 2019 Conference on Empirical Methods in Natural Language Processing and the 9th International Joint Conference on Natural Language Processing (EMNLP-IJCNLP), pp. 4323–4332. Association for Computational Linguistics, Hong Kong, China, November 2019. https://doi.org/10.18653/v1/D19-1441, https://aclanthology.org/D19-1441

24. Szegedy, C., Vanhoucke, V., Ioffe, S., Shlens, J., Wojna, Z.: Rethinking the inception architecture for computer vision. In: 2016 IEEE Conference on Computer Vision and Pattern Recognition (CVPR), pp. 2818–2826 (2016). https://doi.org/10.1109/CVPR.2016.308

25. Takahashi, T., Igata, N.: Rumor detection on twitter. In: The 6th International Conference on Soft Computing and Intelligent Systems, and The 13th International Symposium on Advanced Intelligence Systems, pp. 452–457 (2012)

26. Yang, Z., Yang, D., Dyer, C., He, X., Smola, A., Hovy, E.: Hierarchical attention networks for document classification. In: Proceedings of the 2016 Conference of the North American Chapter of the Association for Computational Linguistics: Human Language Technologies, pp. 1480–1489. Association for Computational Linguistics, San Diego, California, June 2016. https://doi.org/10.18653/v1/N16-1174, https://aclanthology.org/N16-1174

27. Zubiaga, A., Aker, A., Bontcheva, K., Liakata, M., Procter, R.: Detection and resolution of rumours in social media. ACM Comput. Surv. 51(2), 1–36 (2018). https://doi.org/10.1145/3161603

28. Zubiaga, A., Liakata, M., Procter, R.: Exploiting context for rumour detection in social media. In: Ciampaglia, G.L., Mashhadi, A., Yasseri, T. (eds.) SocInfo 2017. LNCS, vol. 10539, pp. 109–123. Springer, Cham (2017). https://doi.org/10.1007/978-3-319-67217-5_8

Semantic Triple-Assisted Learning for Question Answering Passage Re-ranking

Dinesh Nagumothu$^{(\boxtimes)}$ [ID], Bahadorreza Ofoghi [ID], and Peter W. Eklund [ID]

School of Information Technology, Deakin University,
Burwood Victoria 3125, Australia
{dnagumot,b.ofoghi,peter.eklund}@deakin.edu.au

Abstract. Passage re-ranking in question answering (QA) systems is a method to reorder a set of retrieved passages, related to a given question so that answer-containing passages are ranked higher than non-answer-containing passages. With recent advances in language models, passage ranking has become more effective due to improved natural language understanding of the relationship between questions and answer passages. With neural network models, question-passage pairs are used to train a cross-encoder that predicts the semantic relevance score of the pairs and is subsequently used to rank retrieved passages. This paper reports on the use of open information extraction (OpenIE) triples in the form $< subject, verb, object >$ for questions and passages to enhance answer passage ranking in neural network models. Coverage and overlap scores of question-passage triples are studied and a novel loss function is developed using the proposed triple-based features to better learn a cross-encoder model to rerank passages. Experiments on three benchmark datasets are compared to the baseline BERT and ERNIE models using the proposed loss function demonstrating improved passage re-ranking performance.

Keywords: Passage re-ranking · Information extraction · Passage retrieval · Linked open data

1 Introduction

An open-domain question answering (QA) pipeline consists in part of a passage *ranker* that retrieves relevant passages to a question and a *reader* that answers the question from the selected passage. Previous studies [11,25] point out that answer-containing passages that are ranked more highly increase the chance of correctly answering a question, and hence improve the overall performance of the QA pipeline [14]. Passage ranking conventionally consists of a full-stage *retriever* that fetches a small set of passages to a query from a larger corpus and a *reranker*, trained to preferentially reorder the retrieved passages. Recently, pre-trained language models (PLMs), such as BERT [3] and ERNIE [23], have shown excellent natural language understanding capabilities. These models can be used for various downstream NLP tasks, including passage retrieval and re-ranking, achieving state-of-the-art performance [13].

G. A. Fink et al. (Eds.): ICDAR 2023, LNCS 14189, pp. 249–264, 2023.
https://doi.org/10.1007/978-3-031-41682-8_16

Re-ranking with a PLM can be performed with a cross-encoder architecture where the question and passage are passed to an encoder trained to identify their relevance. However, the question-passage pair contains semantic features that can help produce a better re-ranking, see e.g., [16,17]. In this paper, we analyze the use of additional information extraction, from both question and passage and its effect on re-ranking. OpenIE is used as the source of information extracted from question-passage pairs and these features are used to enhance a cross-entropy loss function that improves learning in a cross-encoder neural architecture. Information in the form of $< subject, predicate, object >$ triples are extracted for every question and its retrieved passages. Two metrics based on triple features, called *coverage* and *overlap*, are computed to be incorporated into a new cross-entropy loss function. The coverage score is defined as the proportion of question triples that are covered in the respective set of passage triples. The overlap score is defined as the average overlap of question triples over passage triples. Coverage and overlap scores are calculated for each question-passage pair with individual subject, predicate, and object scores, alongside the triple score. The intuition behind using these metrics is that passages relevant to a question, and those that are potentially answer-containing, will have a larger semantic triple coverage and overlap scores compared to less relevant passages and those that are not answer-containing. To validate this hypothesis, we conducted an exploratory analysis of coverage and overlap scores, discussed in detail in the following sections.

The extracted semantic triple features can be used to enhance the performance of the neural cross-encoder model in several ways. We have taken two approaches and experimented with: (i) the explicit late fusion of the extracted triple features into the neural model and; (ii) the elaboration of the cross-entropy loss function of the model to make use of the coverage and overlap scores computed from the analysis of the triples. The cross-encoder model consists of a PLM that takes the question-passage pairs as an input sequence and encodes them into dense vector representations. These vector representations are passed to a fully connected neural layer. The final layer contains a single neuron with a sigmoid activation function for relevance score prediction. The late fusion experiment involves the concatenation of the proposed coverage and overlap scores to the generated dense representations from the PLM before the final fully connected layer. The late fusion technique does not result in significant performance improvements to baseline models. Details of these experiments are provided in Sect. 6. The second approach, an elaboration of the cross-entropy loss function yields more promising outcomes with details provided in Sect. 5.

Binary cross-entropy loss is a commonly used loss function to train a cross-encoder model for re-ranking. The predicted relevance score of a question-passage pair from the model is compared with the actual relevance to calculate the loss. During the training of a cross-encoder model, its weights and bias are adjusted at each step to minimize the produced cross-entropy loss [19]. We propose an elaborated loss function that alters the existing cross-entropy loss formulation using the coverage and overlap scores from the triple analysis of question-passage pairs. A high coverage/overlap score and high relevance prediction, or low coverage/overlap score with low relevance prediction, will reduce the cross-entropy loss. On the other hand, a low score with high prediction, and a high score with low prediction, will result in an increased loss than the existing cross-entropy loss. When trained with the new loss function, the performance of the passage re-ranking cross-encoder model improves over the baseline PLMs, including BERT and ERNIE. The contributions of this work are therefore as follows:

- analysis of coverage and overlap metrics of semantic triples generated from OpenIE for question-passage pairs to find the distributions of coverage and overlap scores in answer-containing passages to show they are statistically different from those of irrelevant passages not containing correct answers.
- experiment with different feature fusion techniques into neural models, to find that only the development of a novel loss function based on the proposed coverage and overlap scores results in better learning of a cross-encoder model and enhance passage ranking metrics.

2   Related Work

Question answering systems conventionally consists of three stages: a passage *retriever* that fetches a small set of passages from a larger corpus, a *re-ranker* that is used to re-order the retrieved passages, and a *reader* that considers the top k-passages individually to find an answer. In the retriever stage, dense passage and query representations are generated using a dual-encoder architecture with pre-trained language models. These representations are then used to compute the similarity between the query and passage and the top k-relevant passages are then identified and passed to the next stage for re-ranking.

A re-ranker will produce a relevance score for each retrieved passage with respect to a question. monoBERT [13] was the first implementation of BERT for question-based passage re-ranking tasks that outperformed the previous state-of-the-art. It uses the BERT model with a cross-encoder architecture that has as input the question and passage as two sentences separated by a special "[SEP]" token. monoBERT limits the question to 64 tokens and the combined question-passage pair to 512 tokens. A pre-trained BERT model is fine-tuned to predict the relevance score of the question-passage pair using cross-entropy loss. In contrast, DuoBERT [14] performs multi-stage re-ranking where question-passage pairs are used to identify the top k-relevant passages. The retrieved passages are initially re-ranked using monoBERT and the top k-passages are used to perform pairwise classification, where the question and two passages at a time are passed to the cross-encoder architecture to identify the better question-passage pair according to the relevance score.

Unlike these approaches, where the retriever and re-ranker are trained individually, a unified pre-training framework was proposed in [24], where the retriever and re-ranker enhance each other using query generation and passage expansion.

The retriever usually identifies a sub-set of passages from a large collection of passages. Retrieval can be performed with high-dimensional sparse representations, like TF-IDF or BM25, or low-dimensional dense representations, with the latter producing better results [8]. Although a cross-encoder is more effective in identifying relevant documents, it is inefficient to compute for large collections. A dual-encoder architecture is used to produce dense vector representations of the passages and queries individually. The similarity of representations, between question and passage, is computed and sorted to select the top k-relevant passages from the larger collection. Dense passage retrieval (DPR) [8] uses two pre-trained BERT-base models as encoders. The dot-product is selected as the best method for measuring the similarity between the question and passage. RocketQA [20] improves the existing dual-encoder architecture using a multi-stage dual encoder, introducing de-noised hard negatives in the training set generated by the cross-encoder.

There has been prior work in using explicit features for passage re-ranking neural models. Detailed analysis of linguistic characteristics of questions and answer passages is performed in [15]. A significant difference in the distribution of linguistic features between relevant and irrelevant passages is identified. Linguistic features, like parts of speech, are combined with sentence embedding models using the feature concatenation technique in [17]. This enhances the vector representations of the questions and answer passages. A combination of Information Retrieval (IR), deep learning relevance metrics, and linguistic features are used to retrieve the passages from a large corpus. Data envelopment analysis is performed on these scores in combination to re-rank the retrieved passages [16]. To induce the domain-specific knowledge into the neural model, distilled knowledge graph embeddings using a graph encoder and text representations using the PLM are projected into a common latent space [5]. A knowledge injector is employed for interaction between the knowledge graph and text representations for improved re-ranking performance.

Similar to the passage re-ranking module, explicit features are used to enhance end-to-end QA performance as well. Fusion-in-Decoder (FiD) [7] is an end-to-end QA model containing a dense retriever and generative reader. The retrieved passages are encoded individually with passage and question text and concatenated representations in the decoder generate a final answer. KG-FiD [26] uses knowledge graphs to form semantic relationships between the retrieved passages and re-ranks them using graph neural networks resulting in improved performance. Another method is to generate unified representations [18] from heterogeneous knowledge sources by converting knowledge bases and tables into plain text and using FiD to retrieve the passage and generate an answer.

3   Methodology

The task of passage re-ranking reorders the retrieved passage set $\{p_1, p_2, ...p_k\}$ based on question q. Each passage contains a set of sentences $\{s_1, s_2, ...s_l\}$. Each sentence in the passage and the question is composed of a sequence of words

$\{pw_1, pw_2, ..., pw_n\}$ and $\{qw_1, qw_2, ..., qw_n\}$, respectively. As in the BERT [3] model, a cross-encoder architecture is created by feeding the question q and passage p_i, as sentence A and sentence B, where the two are separated by the "[SEP]" token. A single-layer neural network is applied over the "[CLS]" token from the BERT model to predict the relevance score s_i of the passage with respect to the question [13]. The predicted relevance score of each passage is sorted from high to low to obtain the re-ranked list of passages.

3.1 Information Extraction-Based Coverage and Overlap Scores

OpenIE6 [10] is used to extract triples from the questions and passages in the form of $< subject, predicate, object >$ triples. Given passage p and question q, OpenIE6 generates extractions in the form of $\{Tp_1, Tp_2, ...Tp_N\}$ and $\{Tq_1, Tq_2, ...Tq_M\}$, respectively. The generated extractions are then used to compute coverage and overlap scores. The calculation procedure of the coverage and overlap scores is shown in Algorithm 1.

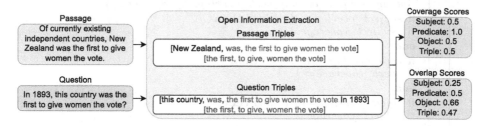

Fig. 1. Coverage and overlap score for a question-passage pair from the QUASAR-T dataset. The text in blue indicates whether the entire question triple or any of its components is covered in passage triples. The text in red indicates the overlapping part from its respective triple component. (Color figure online)

3.2 Coverage Analysis

For each passage, four coverage scores are computed, while three are from the subject, verb, and object individually. The triple coverage score is also computed based on the presence of the entire question triple in the list of passage triples. The subject coverage score of a passage defines the proportion of question subjects present in passage subjects. This score is computed by matching the subject from each question triple against all subjects present in passage triples. The average number of question subjects covered in the subjects of passage triples produces the subject coverage score for that passage. The Jaccard measurement is used to identify whether two matched subjects are the same where a high score represents high similarity. A Jaccard score of more than 0.8 is considered the threshold for classification. The same process is repeated to compute

Algorithm 1: Coverage and overlap score calculation algorithm for a passage and a question.

Input : A passage p and a question q
Output: A list of coverage and overlap scores

$T_p = \text{Open\_IE}(p)$
$T_q = \text{Open\_IE}(q)$
$[C_{subject}, C_{verb}, C_{object}, C_{triple}] = [0.0, 0.0, 0.0, 0.0]$
$[O_{subject}, O_{verb}, O_{object}, O_{triple}] = [0.0, 0.0, 0.0, 0.0]$
for t_q *in* T_q **do**
$\quad < s_q, v_q, o_q > \leftarrow T_q$
$\quad subject\_coverage = 0, verb\_coverage = 0$
$\quad object\_coverage = 0, triple\_coverage = 0$
$\quad$ **for** t_p *in* T_p **do**
$\quad\quad < s_p, v_p, o_p > \leftarrow t_p$
$\quad\quad subject\_overlap = Jaccard\_Similarity(s_q, s_p)$
$\quad\quad verb\_overlap = Jaccard\_Similarity(v_q, v_p)$
$\quad\quad object\_overlap = Jaccard\_Similarity(o_q, o_p)$
$\quad\quad triple\_overlap =$
$\quad\quad Average(subject\_overlap, verb\_overlap, object\_overlap)$
$\quad\quad$ **if** $subject\_overlap > threshold$ **then**
$\quad\quad\quad |$ $subject\_coverage = 1$
$\quad\quad$ **end**
$\quad\quad$ **if** $verb\_overlap > threshold$ **then**
$\quad\quad\quad |$ $verb\_coverage = 1$
$\quad\quad$ **end**
$\quad\quad$ **if** $object\_overlap > threshold$ **then**
$\quad\quad\quad |$ $object\_coverage = 1$
$\quad\quad$ **end**
$\quad\quad$ **if** $subject\_coverage$ *and* $verb\_coverage$ *and* $object\_coverage$ **then**
$\quad\quad\quad |$ $triple\_coverage = 1$
$\quad\quad$ **end**
$\quad$ **end**
end
$O_{subject} = Average(subject\_overlap); O_{verb} = Average(verb\_overlap)$
$O_{object} = Average(object\_overlap); O_{triple} = Average(triple\_overlap)$
$C_{subject} = Average(subject\_coverage); C_{verb} = Average(verb\_coverage)$
$C_{object} = Average(object\_coverage); C_{triple} = Average(triple\_coverage)$
return $[C_{subject}, C_{verb}, C_{object}, C_{triple}], [O_{subject}, O_{verb}, O_{object}, O_{triple}]$

---

the predicate (verb) and object coverage scores. The triple coverage score can be calculated by checking if a triple formed by the question is present in the corresponding lists of triples derived from the passage. An example of coverage score generation of a passage-question pair is given in Fig. 1. For the question "In 1893, this country was the first to give women the vote?", OpenIE6 generates two triples "this country, was, the first to give women the vote in 1893" and "the first, to give, women the vote". The question retrieves a passage "Of currently existing independent countries, New Zealand was the first to give women the vote". Triples extracted from the passage include "New Zealand, was, the first to give women the vote" and "the first, to give, women the vote". Subject coverage score is calculated by checking the number of subjects from question triples covered in the passage triples. As 1 out of 2 subjects was covered, the subject coverage score is 0.5. Predicate coverage scores and object coverage scores are calculated likewise. Both the predicates in question are present in passage triples, hence a score of 1.0. 1 out of 2 are found in passage triples so a score of 0.5 is given. The triple coverage is 0.5 as 1 complete triple out of 2 is present in the list of passage triples.

3.3 Overlap Analysis

Similar to the coverage scores, overlap scores are also computed individually for subjects, verbs, and objects of each passage, along with the entire triple overlap score. The overlap scores differ slightly from the coverage scores as they represent the actual overlap instead of the number of items overlapping. The Jaccard score is computed for every subject in question triples with every subject in the passage triples. The average of these scores determines the subject overlap score of the question-passage pair. Predicate and object overlap scores are also calculated correspondingly. From the example shown in Fig. 1, if a question and a passage generate two triples each, the subject overlap score is calculated by measuring the Jaccard score of all subject combinations forming 4 combinations in total. As one combination generated a score of 1.0 and others are 0.0, the overlap score becomes $1/4=0.25$. The predicate and object overlap scores are calculated similarly. From the example, it is evident that the object overlap is very high, and hence a score of 0.66 is generated. The triple overlap score is the average of subject, predicate, and object overlaps.

3.4 Exploratory Analysis of Coverage and Overlap Scores

We experiment with three passage ranking datasets: QUASAR-T, MS MARCO, and the TREC 2019 Deep Learning (DL) Track for evaluation purposes. Exploratory analysis is conducted on the first two datasets only as the TREC 2019 DL Track uses the same training data as MS MARCO. Each question-passage pair contains a label that determines if the question is answerable from the passage. Label '1' means the passage can answer the question (is relevant) and '0' means it cannot answer the question (is irrelevant). For exploration, the training set was segmented into relevant and irrelevant pairs. For each pair, four coverage scores and four overlap scores were computed. The detailed statistics of the coverage and overlap scores of relevant and irrelevant pairs are summarized in Tables 1 and 2. Statistical t-tests were performed by comparing the relevant and irrelevant scores to measure the difference in distributions. The null hypothesis states that the two distributions do not have significantly different means. Based on the p-values presented in Tables 1 and 2, the distributions of relevant and irrelevant pairs are significantly different and the null hypothesis is rejected.

Table 1. Comparison of mean coverage scores of relevant and irrelevant passages in the QUASAR-T and MS MARCO training sets. Note: separate analysis for MS MARCO and TREC 2019 DL was not carried out since both use the same question and passage collections.

	QUASAR-T				MS MARCO			
	Relevant	Irrelevant	Overall	p-value	Relevant	Irrelevant	Overall	p-value
Subject	0.032	0.028	0.028	1.5e-06	0.137	0.082	0.092	0.0
Predicate	0.119	0.081	0.085	2.1e-142	0.187	0.158	0.163	0.0
Object	0.012	0.004	0.005	4.2e-118	0.025	0.013	0.015	0.0
Triple	0.001	0.0	0.0	1.1e-06	0.004	0.0	0.001	0.0

Table 2. Comparison of mean overlap scores of relevant and irrelevant passages in the QUASAR-T and MS MARCO training sets. Note: separate analysis for MS MARCO and TREC 2019 DL was not carried out since both use the same question and passage collections.

	QUASAR-T				MS MARCO			
	Relevant	Irrelevant	Overall	p-value	Relevant	Irrelevant	Overall	p-value
Subject	0.057	0.053	0.053	5.0e-07	0.104	0.078	0.082	0.0
Predicate	0.097	0.07	0.074	1.0e-136	0.089	0.081	0.082	3.5e-132
Object	0.07	0.052	0.054	2.7e-246	0.044	0.039	0.04	0.0
Triple	0.075	0.058	0.06	1.3e-211	0.079	0.066	0.068	0.0

3.5 Semantic-Triple Assisted Learning Using Coverage/overlap Scores

The mean coverage and overlap scores of the subject, predicate, and objects are larger in relevant pairs compared to the irrelevant pairs in both datasets, as detailed in the previous section. These scores are therefore utilized in adjusting the loss of the cross-encoder models during training. Passage-question pairs with larger coverage/overlap scores predicted by the model as relevant or pairs with lower coverage/overlap scores predicted as irrelevant should decrease the loss. This makes the model correct its trainable parameters less aggressively since the model predictions are in line with the proposed scores. Similarly, loss should increase if the prediction is relevant and the coverage/overlap score is less, or if the prediction is irrelevant, and the scores are higher, resulting in larger adjustments to the model parameters for better predictions.

Conventionally, the binary cross-entropy loss is used when training the pointwise re-ranking models as shown in eqn. (1), where N represents the batch size, $y_i \in [0, 1]$ as "relevance" labels for the question-passage pair, and $p(y_i)$ is the log of the predicted passage-question pair.

$$\mathcal{L}_{\text{cross-entropy}} = \frac{-1}{N} \cdot \sum_{i=1}^{N} y_i \cdot \log(p(y_i)) + (1 - y_i) \cdot \log(1 - p(y_i)) \quad (1)$$

The coverage/overlap scores are combined with the existing cross-entropy loss using the linear interpolation method as shown in eqn. (2), where $\mathcal{L}_{\text{coverage/overlap}} = (p - \lambda_p) \times (\lambda_c - c)$, and α is a hyper-parameter ranging from 0 to 1 and needs to be tuned for best performance, p represents sigmoid-activated logits, λ_p is the threshold for logits probability set to 0.5, c is the average coverage/overlap score, and λ_c is the threshold for coverage/overlap scores, set to 0.06 considering all mean coverage/overlap scores.

$$\mathcal{L}_{\text{new}} = \alpha \cdot \mathcal{L}_{\text{cross-entropy}} + (1 - \alpha) \cdot \mathcal{L}_{\text{coverage/overlap}} \qquad (2)$$

The $\mathcal{L}_{\text{coverage/overlap}}$ measure is formulated in such a way that if the prediction is relevant and coverage/overlap is high, or if the prediction is irrelevant and coverage/overlap is low, a negative scalar value is generated which decreases the existing cross-entropy loss using eqn. (2). This will result in a reduced step size when correcting the model parameters as the predictions and coverage/overlap scores reflect the same intuition. Likewise, the loss increases if the prediction is irrelevant and the coverage/overlap is high, or if the prediction is relevant and the coverage is low. This condition will generate a positive value for $\mathcal{L}_{\text{coverage/overlap}}$, thus increasing the overall loss. An adjustment to the existing loss function is proposed to take advantage of the coverage/overlap scores to improve the cross-encoder model training for the passage re-ranking task.

4   Experiments

This section presents the datasets used for training, the evaluation metrics used to compare the model performances, and the model training settings.

4.1   Datasets

The performance of the proposed loss function is evaluated using three publicly available datasets, namely QUASAR-T [4], MS MARCO [12], and the TREC 2019 Deep Learning Track [2]. The QUASAR-T dataset contains open-domain trivia questions with the passages sourced from ClueWeb09 [1]. We use 300,000 question-passage pairs for training with 3,000 unique questions. For evaluation, the test set contains 3,000 questions with each question having 100 retrieved passages. MS MARCO was created for open-domain machine reading comprehension tasks where the questions are sampled from Microsoft Bing search queries. The training set contains 502,939 queries with 8.8 million passages in the corpus. MS MARCO and TREC 2019 DL use the same training data. Evaluation is performed on the MS MARCO development set containing 6,980 queries and TREC 2019 DL with 43 queries. TREC 2019 DL is a multi-class dataset with finely-graded relevance labels: "Irrelevant", "Relevant", "Highly Relevant", and "Perfectly Relevant". For the task of re-ranking, this can be transformed into binary labels by considering the "Irrelevant" label as 0 and the remainder as 1. These datasets provide 1,000 passages for each query for the re-ranking task. Additionally, we evaluated our re-ranking model on top of the BM25 retriever where the top 1,000 passages of these two datasets are retrieved from the full corpus.

4.2   Evaluation Metrics

We used the mean reciprocal rank (MRR) of the top 10 passages (MRR@10) and *recall@k* to evaluate the model performance on the MS MARCO and QUASAR-T datasets. For TREC 2019 DL, mean average precision at 10 (MAP@10) and normalized discounted cumulative gain at 10 ($NDCG$@10) were used to measure the performance. MRR@10 is based on the rank of the first relevant passage in the first 10 passages. *Recall@k* denotes the number of questions that contain a positive passage in the top-k retrieved passages. MAP@10 is the average proportion of relevant passages in the top 10 retrieved passages. While precision only considers the binary relevance of a passage, NDCG@10 is measured using the predicted relevance score and calculating the gain to evaluate the re-ranking performance of the model.

4.3   Model Training

We used two baseline PLMs, an uncased version of BERT-base [3] and ERNIE-2.0-base [23] as our cross-encoder models. Both models utilize the transformer architecture and an attention mechanism [22]. The question and passage are input as a single sequence separated by a "[SEP]" token. The "[CLS]" token in the input sequence provides dense vector representations to the question-passage pair. As the question-passage pair is passed as a single input sequence, the PLM can utilize the cross-attentions between question and passage, thus providing better representation. The "[CLS]" vector is then passed to a fully connected single neuron to predict a relevance score between 0 and 1. We use the SentenceBERT [21] implementation of cross-encoder models, which is based on the Huggingface[1] tokenization and model initialization. When training, we use the proposed loss function with Adam Optimizer and a learning rate of 1e-5 to adjust the weights and biases of the cross-encoder model. All models are trained and evaluated on a single NVIDIA Tesla A100 GPU with 48GB memory for 2 epochs with 5,000 warm-up steps and a batch size of 32. All presented evaluation results are averaged over four train/test runs. The positive-hard negative ratio is set to 1:4. MS MARCO and TREC 2019 DL are evaluated on the models trained with the MS MARCO training set. We use the Anserini-based Lucene[2] index for retrieving the top 1,000 passages for a query using the BM25 method. QUASAR-T is evaluated on a separate model trained with the QUASAR-T training data.

5   Results

The performance gains of the proposed loss function with coverage/overlap scores are measured by training the BERT and ERNIE-based cross-encoders. We train and evaluate both PLMs with the QUASAR-T and MS MARCO

---

[1] https://huggingface.co/.
[2] https://github.com/castorini/anserini.

Table 3. Model performance evaluations on the QUASAR-T dataset with the proposed semantic triple-assisted loss functions. Note: LF = loss function.

Model	MRR@10	R@1	R@2	R@3	R@4	R@5	R@10
$BERT_{base}$	46.2	36.0	46.6	53.1	57.5	60.5	67.8
+ Overlap LF	47.3	37.7	47.8	54.0	57.5	60.6	68.0
+ Coverage LF	47.8	38.1	48.9	54.6	58.1	60.9	67.8
$ERNIE_{base}$	51.3	42.8	52.0	57.1	60.6	62.9	69.2
+ Overlap LF	50.5	41.6	51.1	56.6	60.3	62.7	69.0
+ Coverage LF	**52.1**	**43.6**	**53.0**	**57.9**	**61.3**	**63.8**	**69.8**

datasets. The TREC 2019 Deep Learning Track test set is evaluated with trained MS MARCO models as both use the same training data. We also experiment with the BM25 retrieved passages for MS MARCO and TREC 2019 DL questions alongside the originally provided passages for re-ranking. The BERT and ERNIE-based cross-encoders, with the unchanged cross-entropy loss function, are considered as the baselines.

5.1 Effect of Overlap Scores

The overlap scores formulated in eqn. (2) are used to alter the existing cross-entropy loss of the baseline BERT and ERNIE models. Table 3 shows the performance of the models on the QUASAR-T test set. The overlap-based loss shows a significant improvement of 1.1% of MRR over the BERT baseline, but negatively impacted the performance of the ERNIE model. Table 4 contains the model performances on the MS MARCO development set and the TREC 2019 DL Track test set. A similar pattern is observed on MS MARCO with the overlap-based loss function improving the BERT baseline but no significant impact over the ERNIE-based models. The BERT base model with the overlap-adjusted loss function is the best-performing model over the TREC 2019 DL Track test set, with the highest MAP@10 on the provided re-ranking set and the best NDCG@10 on the BM25 retrieved set.

5.2 Effect of Coverage Scores

Using the coverage scores in place of the overlap scores results in further improvement over the baseline BERT model and a significant improvement over the ERNIE model. On the QUASAR-T test set, the MRR@10 is improved by 1.6% and 0.8% using the BERT-base and ERNIE-base models, respectively. When evaluated on the MS MARCO development set, MRR@10 and *Recall@k* show substantial improvements over the BERT baseline. On the other hand, a marginal improvement is observed in all the evaluation metrics over the ERNIE baseline. The ERNIE-based cross-encoder with coverage-adjusted loss is the best model from our experiments, outperforming all the baselines on the QUASAR-T and MS MARCO development sets.

Table 4. Model performance evaluations on the MS MARCO and TREC 2019 Deep Learning Track datasets with the proposed semantic triple-assisted overlap and coverage loss functions. Note: LF = loss function.

Method	MS MARCO-Dev							TREC 2019 DL	
	MRR@10	R@1	R@2	R@3	R@4	R@5	R@10	MAP@10	NDCG@10
BM25	18.7	10.4	17.4	23.2	27.1	30.2	39.9	11.3	50.6
$BERT_{base}$	35.5	23.8	35.8	43.3	48.2	52.1	61.4	14.3	68.2
+ Overlap LF	36.2	24.5	36.7	43.9	48.8	52.5	61.8	**14.8**	68.1
+ Coverage LF	36.3	24.5	36.7	44.1	49.0	52.4	62.2	14.7	68.2
$ERNIE_{base}$	36.5	24.4	37.2	44.4	49.3	53.0	62.4	14.6	68.2
+ Overlap LF	36.5	24.4	37.2	44.4	49.4	52.9	62.5	14.6	67.7
+ Coverage LF	36.8	24.9	37.5	44.7	49.4	53.1	62.5	14.5	67.9
BM25+$BERT_{base}$	35.9	23.7	35.9	43.6	48.9	52.8	62.9	14.1	68.9
+ Overlap LF	36.7	24.7	37.0	44.4	49.5	53.3	63.2	14.2	**69.1**
+ Coverage LF	36.8	24.9	37.1	44.2	49.2	53.0	63.0	14.3	68.9
BM25+$ERNIE_{base}$	36.7	24.3	37.1	44.9	50.1	53.8	63.9	13.6	68.5
+ Overlap LF	36.8	24.3	37.0	44.9	50.0	53.9	**64.3**	13.7	68.3
+ Coverage LF	**37.5**	**25.4**	**37.9**	**45.4**	**50.5**	**54.2**	64.1	13.7	68.1

5.3 Analyzing the Effect of α

From eqn. (2), the hyper-parameter $\alpha \in [0, 1]$ is to be fine-tuned for optimal re-ranking performances. All the model results in the previous sections were reported based on the optimal α value. We carried out an analysis on the three evaluation datasets by training the BERT and ERNIE models with α values ranging from 0.1 to 1 with an interval of 0.1. Figure. 2 shows the effect of α value on the three datasets with their respective evaluation metrics. On the QUASAR-T test set, $\alpha = 0.1$ showed a better ranking performance on all the metrics with coverage-adjusted loss. On the MS MARCO dev and TREC 2019 DL test sets, α values of 0.3 and 0.5 result in the best performance respectively. Therefore, $\alpha = 0.3$ is selected as the optimal value as a result of consistent performance gains over the baseline on all datasets.

5.4 Comparison with the State-of-the-art

Table 5 compares the performance of our proposed methods with those of the existing state-of-the-art techniques using the baseline versions of the BERT and ERNIE models. We used the MRR@10 score from the MS MARCO development set and MAP@10 and NDCG@10 from the TREC 2019 DL Track test set for comparison purposes. monoBERT [13] uses the traditional cross-encoder architecture with BERT-base PLM. Our proposed methods use the same architecture additionally with an ERNIE base backbone and were trained with the coverage and overlap-adjusted loss functions. The ERNIE base model, with the coverage-adjusted loss function, outperformed other methods when re-ranking

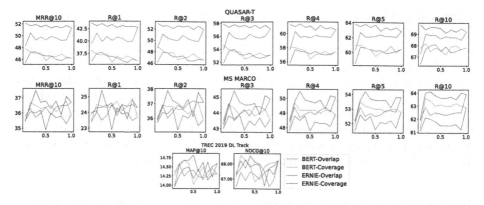

Fig. 2. The analysis of α values on different datasets and its impact on MRR@10 and Recall@k (R@1, R@2, R@3, R@4, R@5, R@10). $BERT_{base}$ and $ERNIE_{base}$ models trained with the overlap and coverage-adjusted loss functions. $\alpha = 1.0$ denotes an unchanged cross-entropy loss value, i.e. no overlap or coverage.

the BM25-retrieved top 1000 passages. BERT base, with overlap-adjusted loss, gave a better MAP on the TREC 2019 DL track.

6   Discussion and Future Work

The coverage and overlap scores computed with OpenIE-generated semantic triples have been demonstrated to act as useful features to distinguish between relevant (answer-containing) and irrelevant passages to a given natural language question. However, the utilization of these features requires careful consideration of how such features can be injected into the neural learning of a cross-encoder model.

Table 5. Performance comparison of enhanced models with proposed semantic triple-assisted loss functions with the state-of-the-art models on the MS MARCO and TREC 2019 Deep Learning Track datasets. Note: LF = loss function, '-' = no results from the literature and '†' = our implementation. All other results are directly taken from the relevant literature.

Method		MS MARCO	TREC 2019 DL	
	Model	MRR@10	MAP@10	NDCG@10
BM25	–	18.7	11.3	50.6
monoBERT [13]	$BERT_{base}$	34.7	14.30†	68.23†
ColBERT [9]	$BERT_{base}$	34.9	–	–
BERT Ranker distillation [6]	$BERT_{base}$	36.0	–	**69.23**
BM25+$ERNIE_{base}$	$ERNIE_{base}$	36.7†	13.59†	68.51†
$BERT_{base}$+**Overlap LF(Ours)**	$BERT_{base}$	36.2	**14.82**	68.13
BM25+$ERNIE_{base}$+Coverage LF(Ours)	$ERNIE_{base}$	**37.5**	13.73	68.14

The "[CLS]" token in the BERT model provides rich dense representations to the question-passage pair. On the other hand, the triple scores form a different vector representation for the same question-passage pair. To assess their usability and effect in the modeling of cross-encoders explicitly, we experimented with the late fusion of these two vectors by concatenating the "[CLS]" token with the coverage/overlap scores of the semantic triples. The fusion model was trained and evaluated with the QUASAR-T training and test datasets discussed in Sect. 4, the results of which are summarized in Table 6. However, using the scores based on these semantic triples with late fusion did not provide substantial improvements in the re-ranking performance of the baseline model.

Table 6. Model performance evaluations of BERT with the late fusion of coverage and overlap scores on the QUASAR-T dataset. Overlap scores are shown to improve the model re-ranking performance in every case.

Model	MRR@10	R@1	R@2	R@3	R@4	R@5	R@10
$BERT_{base}$	46.2	36.0	46.6	53.1	57.5	60.5	67.8
+ Overlap	46.6	36.7	47.2	53.2	57.2	60.1	67.7
+ Coverage	46.8	36.8	47.4	53.3	57.4	61.0	68.2

Although the semantic triples may not explicitly improve the modeling structure and effectiveness of re-ranking, the coverage and overlap scores were shown to play a role in improving the passage re-ranking effectiveness when used in combination with the existing cross-entropy loss function of the models. Parts of our future work will focus on the application of the coverage and overlap-adjusted loss functions in training the larger cross-encoder neural models based on the work presented in this paper. Similar performance improvements are anticipated given the consistency of performance improvement patterns observed in this work.

7   Conclusion

This paper presents the use of semantic triples for improved passage re-ranking. Coverage and overlap scores were computed from question triples and passage triples from a question-passage pair. Our findings from the exploratory analysis show that relevant, answer-containing passages to a question have higher coverage and overlap scores compared to irrelevant passages or those that do not explicitly answer the question. We experimented with two separate methods: (i) the late fusion of these explicit scores into the neural model, and (ii) adjusting the cross-entropy loss produced during model training, where the latter enhanced the re-ranking performances. The calculation of loss in the neural model was altered based on the semantic triple scores and model predictions. We evaluated our passage re-ranking performances using three benchmark datasets with an unchanged cross-entropy loss function as the baseline. The proposed loss function enhances model training and test performances of the BERT-base and ERNIE-base cross-encoder models.

Acknowledgements. This research is partly funded by the Minerals Council of Australia.

References

1. Callan, J., Hoy, M., Yoo, C., Zhao, L.: Clueweb09 data set (2009), https://lemurproject.org/clueweb09/ Accessed 28 Apr 2023
2. Craswell, N., Mitra, B., Yilmaz, E., Campos, D., Voorhees, E.M.: Overview of the TREC 2019 Deep Learning Track, https://arxiv.org/abs/2003.07820 Accessed 28 Apr 2023
3. Devlin, J., Chang, M.W., Lee, K., Toutanova, K.: Bert: Pre-training of deep bidirectional transformers for language understanding. In: Proceedings of the 2019 Conference of the North American Chapter of the Association for Computational Linguistics: Human Language Technologies, Vol 1. pp. 4171–4186 (2019)
4. Dhingra, B., Mazaitis, K., Cohen, W.W.: Quasar: Datasets for question answering by search and reading (2017), https://arxiv.org/abs/1707.03904
5. Dong, Q., et al.: Incorporating explicit knowledge in pre-trained language models for passage re-ranking. In: Proceedings of the 45th International ACM SIGIR Conference, pp. 1490–1501. SIGIR '22, ACM, New York, NY, USA (2022). https://doi.org/10.1145/3477495.3531997
6. Gao, L., Dai, Z., Callan, J.: Understanding BERT rankers under distillation. In: Proc of the 2020 ACM SIGIR on International Conference on Theory of Information Retrieval. pp. 149–152 (2020)
7. Izacard, G., Grave, E.: Leveraging passage retrieval with generative models for open domain question answering. In: Proceeding of the 16th Conference of the European Chapter of the Association for Computational Linguistics: Main Volume. pp. 874–880. Association for Computational Linguistics (2021). https://doi.org/10.18653/v1/2021.eacl-main.74
8. Karpukhin, V., et al.: Dense passage retrieval for open-domain question answering. In: Proceeding of the 2020 Conference on Empirical Methods in Natural Language Processing (EMNLP). pp. 6769–6781. Association for Computational Linguistics (2020). https://doi.org/10.18653/v1/2020.emnlp-main.550
9. Khattab, O., Zaharia, M.: ColBERT: Efficient and effective passage search via contextualized late interaction over bert. In: Proc of the 43rd International ACM SIGIR Conference on Research and Development in Information Retrieval. pp. 39–48. SIGIR '20, ACM, New York, NY, USA (2020). https://doi.org/10.1145/3397271.3401075
10. Kolluru, K., Adlakha, V., Aggarwal, S., Mausam, Chakrabarti, S.: OpenIE6: Iterative Grid Labeling and Coordination Analysis for Open Information Extraction. In: Proceeding of the 2020 Conference on Empirical Methods in Natural Language Processing (EMNLP). pp. 3748–3761. Association for Computational Linguistics, Online (Nov 2020). https://doi.org/10.18653/v1/2020.emnlp-main.306
11. Mao, Y., He, P., Liu, X., Shen, Y., Gao, J., Han, J., Chen, W.: Reader-guided passage reranking for open-domain question answering. In: Findings of the Association for Computational Linguistics: ACL-IJCNLP 2021. pp. 344–350 (2021)

12. Nguyen, T., et al.: MS MARCO: A human generated machine reading comprehension dataset. In: CoCo@ NIPs (2016)
13. Nogueira, R., Cho, K.: Passage re-ranking with BERT (2019), https://arxiv.org/abs/1901.04085
14. Nogueira, R., Yang, W., Cho, K., Lin, J.: Multi-stage document ranking with BERT (2019), https://arxiv.org/abs/1910.14424
15. Ofoghi, B.: Linguistic characterization of answer passages for fact-seeking question answering. In: Proceedings of the 37th ACM/SIGAPP Symposium on Applied Computing. p. 821–828. SAC '22, ACM, New York, NY, USA (2022). https://doi.org/10.1145/3477314.3506999
16. Ofoghi, B., Mahdiloo, M., Yearwood, J.: Data envelopment analysis of linguistic features and passage relevance for open-domain question answering. Knowl.-Based Syst. **244**, 108574 (2022). https://doi.org/10.1016/j.knosys.2022.108574
17. Ofoghi, B., Zarnegar, A.: Answer Passage Ranking Enhancement Using Shallow Linguistic Features. In: Torra, V., Narukawa, Y. (eds.) MDAI 2021. LNCS (LNAI), vol. 12898, pp. 286–298. Springer, Cham (2021). https://doi.org/10.1007/978-3-030-85529-1_23
18. Oguz, B., et al.: UniK-QA: Unified representations of structured and unstructured knowledge for open-domain question answering. In: Findings of the Association for Computational Linguistics: NAACL 2022. pp. 1535–1546. Association for Computational Linguistics, Seattle, United States (2022). https://doi.org/10.18653/v1/2022.findings-naacl.115
19. Pradeep, R., Liu, Y., Zhang, X., Li, Y., Yates, A., Lin, J.: Squeezing Water from a Stone: A Bag of Tricks for Further Improving Cross-Encoder Effectiveness for Reranking. In: Hagen, M., Verberne, S., Macdonald, C., Seifert, C., Balog, K., Nørvåg, K., Setty, V. (eds.) ECIR 2022. LNCS, vol. 13185, pp. 655–670. Springer, Cham (2022). https://doi.org/10.1007/978-3-030-99736-6_44
20. Qu, Y., et al.: RocketQA: an optimized training approach to dense passage retrieval for open-domain question answering. In: In Proceedings of NAACL (2021)
21. Reimers, N., Gurevych, I.: Making monolingual sentence embeddings multilingual using knowledge distillation. In: Proceedings of the 2020 Conference on Empirical Methods in Natural Language Processing. Association for Computational Linguistics (2020), https://arxiv.org/abs/2004.09813
22. Vaswani, A., et al.: Attention is all you need. Adv. Neural Inf. Process. Syst. **30** (2017)
23. Wang, S., et al.: Ernie 3.0 titan: exploring larger-scale knowledge enhanced pretraining for language understanding and generation (2021), https://arxiv.org/abs/2112.12731
24. Yan, M., Li, C., Bi, B., Wang, W., Huang, S.: A unified pretraining framework for passage ranking and expansion. In: Proceeding of the AAAI Conference on Artificial Intelligence. vol. 35, pp. 4555–4563 (2021)
25. Yang, W., Xie, Y., Lin, A., Li, X., Tan, L., Xiong, K., Li, M., Lin, J.: End-to-end open-domain question answering with BERTserini. In: Proceeding of the 2019 Conference of the North American Chapter of the Association for Computational Linguistics (Demonstrations). pp. 72–77. Minneapolis, Minnesota (2019). https://doi.org/10.18653/v1/N19-4013,https://aclanthology.org/N19-4013
26. Yu, D., et al.: KG-FiD: infusing knowledge graph in fusion-in-decoder for open-domain question answering. In: Proc of the 60th Annual Meeting of the Association for Computational Linguistics, Vol. 1, pp. 4961–4974. Dublin, Ireland (2022). https://doi.org/10.18653/v1/2022.acl-long.340

I-WAS: A Data Augmentation Method with GPT-2 for Simile Detection

Yongzhu Chang$^{(\boxtimes)}$, Rongsheng Zhang , and Jiashu Pu

Fuxi AI Lab, NetEase Inc., Hangzhou, China
{changyongzhu,zhangrongsheng,pujiashu}@corp.netease.com

Abstract. Simile detection is a valuable task for many natural language processing (NLP)-based applications, particularly in the field of literature. However, existing research on simile detection often relies on corpora that are limited in size and do not adequately represent the full range of simile forms. To address this issue, we propose a simile data augmentation method based on **W**ord replacement **A**nd **S**entence completion using the GPT-2 language model. Our iterative process called **I-WAS**, is designed to improve the quality of the augmented sentences. To better evaluate the performance of our method in real-world applications, we have compiled a corpus containing a more diverse set of simile forms for experimentation. Our experimental results demonstrate the effectiveness of our proposed data augmentation method for simile detection.

Keywords: GPT-2 · Simile detection · Data augmentation · Iterative

1 Introduction

Figurative language, particularly analogy, is a common feature of literature and poetry that can help to engage and inspire readers [1]. An analogy is a type of figurative language that compares two different objects in order to make a description more vivid. An analogy can take the form of a metaphor or a simile. Unlike metaphors, similes explicitly use comparative words such as *"like"*, *"as"*, and *"than"* in English, or "像" (Xiang), "如同" (Ru Tong), and "宛若" (Wan Ruo) in Chinese. Simile sentences also typically have both a TOPIC, which is the noun phrase that acts as the logical subject, and a VEHICLE, which is the logical object of the comparison and is typically a noun phrase.

The Simile Detection (SD) task is crucial for various applications, such as evaluating student essays or extracting striking sentences from literature. Previous research on SD can be broadly classified into two categories: (1) studies that focus on identifying the sentiment conveyed by a given simile sentence, such as

Supported by the Key Research and Development Program of Zhejiang Province (No. 2022C01011).

Table 1. The examples of simile sentences that contain different comparative words. The relative positions of TOPIC, VEHICLE, and comparative words are not fixed. The CW means the comparative words.

TOPIC:天气 (Weather) **VEHICLE**: 火炉 (furnace)
zh: 八月的天气 (TOPIC) 就像 (CW) 是火炉 (VEHICLE) 一样烘烤着大地。 en: The weather in August is like a furnace baking the hot earth.
zh: 八月的天气 (TOPIC) 已经炎热的如同 (CW) 火炉 (VEHICLE) 一般 en: The weather in August has been as hot as a furnace.
zh: 到了八月份，外面如火炉 (VEHICLE) 般 (CW) 的天气 (TOPIC) en: By August, the furnace like weather outside.
zh: 八月份的天气 (TOPIC) 跟火炉 (VEHICLE) 一样 (CW) en: The weather in August is the same as a furnace
zh: 这八月份的天气 (TOPIC)，俨然 (CW) 是一个大火炉 (VEHICLE) 啊 en: The August weather, as if it is a big furnace
zh: 八月的天气 (TOPIC) 宛如 (CW) 火炉 (VEHICLE) 烘烤着炎炎大地 en: The August weather is similar to a furnace baking the hot earth

irony, humor, or sarcasm; [2–4] and (2) studies that aim to determine whether a sentence is a simile or not [5,12,19]. Our paper focuses on the second type of simile detection. Many previous studies [8–10,14] on simile detection have used datasets that are limited in scope, specifically, datasets in Chinese that only contain examples with the same comparative word (e.g. "like") and in which the TOPIC always appears to the left of the comparative word and the VEHICLE always appears to the right. Nevertheless, in real-world situations, as shown in Table 1, simile sentences can involve a variety of comparative words, and the position of the TOPIC and VEHICLE relative to each other is not fixed. Additionally, existing datasets [5,13–15] for simile detection are often of limited size. To develop an industrial simile detector, a corpus containing abundant examples and diverse simile forms is necessary. However, building such a corpus can be time-consuming and labor-intensive.

To address the issue of limited data availability in simile detection, we propose a data augmentation method that leverages **W**ord replacement **A**nd **S**entence completion with the GPT-2 language model. Our approach, called **WAS** (word replacement and sentence completion), involves the following steps: (1) Word replacement: for a given simile sentence, we randomly select a comparative word from a predefined set[1] and replace the original comparative word in the sentence. (2) Sentence completion: using the replaced comparative word and the context above it as a prompt, we feed the modified simile sentence into the GPT-2 model to generate additional augmented sentences. We then use a simile detection model trained on the original corpus to rank the augmented sentences. Additionally, we investigate an iterative version of this process, called **I-WAS**,

---

[1] Comparative words collected from Chinese similes.

which aims to improve the quality of the augmented sentences by training a simile detection model on a mix of the original corpus and the augmented sentences in subsequent iterations.

To accurately evaluate the performance of simile detection models, we have compiled a corpus of simile sentences as a test set for experimentation. This corpus is more diverse than existing datasets, with a total of 606 simile sentences that cover 7 different comparative words and a range of TOPIC-VEHICLE position relations. The data is collected from the internet[2] and manually labeled. We conduct extensive experiments on both the [13] dataset and our newly collected test set. The results of these experiments demonstrate the effectiveness of our proposed data augmentation method for simile detection.

Our contributions can be summarized as follows.

- We propose a data augmentation method called **WAS** (**W**ord replacement **A**nd **S**entence completion) for generating additional simile sentences, and an iterative version of this process (**I-WAS**) to improve the quality of the augmented sentences.
- We have compiled a corpus of diverse simile forms for evaluating simile detection models, which will be made available upon publication.
- We conduct thorough experiments to demonstrate the effectiveness of our proposed data augmentation method for simile detection in real-world scenarios.

2   Related Work

2.1   Simile Detection

Previous research on simile detection has focused on two main areas. The first area involves detecting the sentiment conveyed by a simile, such as irony, humor, or sarcasm [2–4]. In this line of research, researchers have used rule-based, feature-based, and neural-based methods to identify the sentiment underlying simile sentences [3,4,16,17]. For example, [18] developed a classification module with a gate mechanism and a multi-task learning framework to capture rhetorical representation. The second area of research involves detecting whether a sentence is a simile or not [5,12,19]. Feature-based approaches have been used for this purpose, with early work focusing on datasets containing only similes with the comparative word "like" [13]. More recent studies have employed techniques such as bidirectional LSTM networks with attention mechanisms [20] and part-of-speech tags [9,22] to improve the accuracy of simile detection.

In this paper, we focus on the task of determining whether a sentence is a simile or not. As illustrated in Table 1, simile expressions come in many different forms and can be quite varied. While it is possible to extend existing methods to handle more complex and diverse simile sentences, to the best of our knowledge, the datasets provided by previous research [5,13,14] are not sufficient for

---

[2] http://www.ruiwen.com/zuowen/biyuju/.

evaluating the performance of simile detection models. Furthermore, there is currently no large-scale, annotated corpus for simile detection that is suitable for supporting data-driven approaches.

2.2 Text Data Augmentation

Variational autoencoders (VAEs) [21], generative adversarial networks (GANs) [23], and pre-trained language (PTL) generation models [24,27,43,47] are commonly used for data augmentation in sentence-level sentiment analysis and text mining tasks [28,29]. These methods typically involve encoding the input text into a latent semantic space and then generating new text from this representation. Nevertheless, these approaches are often not capable of generating high-quality text. Back translation, where text is translated from one language (e.g., English) into another (e.g., French) and then back again, has also been used for data augmentation [30]. However, this method is less controllable in terms of maintaining context consistency. Another class of data augmentation methods involves replacing words in the input text with synonyms or other related words. For example, [31,32] used WordNet [33] and Word2Vec [34] to identify words that could be replaced, while [36,44] used pre-trained models to predict the appropriate replacement words. [37] proposed inserting random punctuation marks into the original text to improve performance on text classification tasks. Recently, pre-trained language models have become popular due to their strong performance. [38] proposed a seq2seq data augmentation model for language understanding in task-based dialogue systems, and similar approaches have been used in other fields as well [39,40]. Meanwhile, in the field of aspect term extraction, many studies have also employed pre-trained models for data augmentation [41–43].

However, none of the aforementioned data augmentation methods have been applied to the simile detection task, as they tend to operate at the sentence level and may unintentionally alter the simile components of simile sentences during the augmentation process.

3   Task and Methodology

3.1   Formulation of Simile Detection

Suppose we have a training dataset of size n, denoted as $D_{train} = (x_i, y_i)_{i=1}^{n}$, where $(x_i^T, x_i^C, x_i^V) \in x_i$ represents the indices of the TOPIC, comparative word, and VEHICLE in a simile sentence, respectively, and $y_i \in \{0, 1\}$ indicates whether the sentence is a simile or not. Our data augmentation method aims to generate a new sentence $\hat{x}_i$ using the context of the original sentence x_i (specifically, the TOPIC) as a prompt while maintaining the same label as the original sentence. The generated data is then combined with the original data and used as input for training a classifier, with the corresponding labels serving as the output.

3.2   Simile Sentences Augmentation

Previous conditional data augmentation approaches, such as those proposed by [44, 47], can generate samples for datasets with unclear targets. These approaches [6, 46, 47] typically involve converting the sentences into a regular format and then using pre-trained models to generate new sentences through fine-tuning. In contrast, our proposed method, called **I**terative **W**ord replacement **A**nd **S**entence completion (I-WAS), generates TOPIC-consistent and context-relevant data samples without fine-tuning. It is important to maintain label consistency between the original and generated samples in data augmentation tasks [47]. To ensure label consistency in our approach, we first train a base simile detection model on the original corpus, replacing comparative words randomly. Then, we apply GPT-2 to generate 10 data samples for each sentence using Top-K sampling. Finally, we select the final label-consistent augmented samples using the base simile detection model to predict all of the augmented samples. We apply our augmentation method to the dataset proposed by [13]. The steps involved in our method are as follows:

Algorithm 1: Training

Input: Training dataset D_{train}; Total number of cycle M; Model parameters θ_{basic} of simile detection; Model parameter $\theta_{\text{GPT}-2}$

Output: θ

1 $\hat{D}_{\text{train}}$: the augmentation dataset ;
2 P: the dataset with simile probability ;
3 G: augmented samples dataset generated from GPT-2 ;
4 D_0: the dataset after word replacement ;
5 **select**:select augmented sample with label consistent with y_j;
6 **train**: training the model with θ ;
7 **wr**: Comparative word replace with predefined set in Table 2 ;
8 $\theta = \theta_{\text{basic}}$;
9 **for** $i \leftarrow 1$ **to** M **do**
10      $\quad D_0 \leftarrow \mathbf{wr}(D_{\text{train}})$;
11      $\quad G \leftarrow \text{Model}(\theta_{\text{GPT}-2}, D_0)$;
12      $\quad P \leftarrow \text{Model}(\theta, G)$;
13      $\quad \hat{D}_{\text{train}} \leftarrow \mathbf{select}(\text{P})$;
14      $\quad D_{\text{train}} \leftarrow D_{\text{train}} \cup \hat{D}_{\text{train}}$;
15      $\quad \theta \leftarrow \mathbf{train}(\theta, D_{\text{train}})$;
16 **end**
17 return θ

Word Replacement (WR): The original dataset only includes the comparative word "like". To add diversity to the dataset, we replace "like" with a randomly selected comparative word from a predefined set, such as "seems to" and "the same as". For example, the original sentence *"The weather in August* is *like* a furnace baking the hot earth" becomes "The weather in August is **the same as** a furnace baking the hot earth" after replacement.

Sentence Completion (SC): To generate candidate samples, we mask the context after the comparative word and use the resulting prompt as input for the GPT-2 model. For example, given the original sentence "The weather in August is same as a furnace baking the hot earth" the input for GPT-2 would be "The weather in August is same as.". We follow the same process to generate candidate samples where the VEHICLE is located to the left of the comparative word. GPT-2 generates candidate samples through sentence completion using an auto-regressive approach and the top-k sampling strategy. We set a maximum length of 50 for the generated text and a size of 10 for the candidate set.

To select a suitable sample from the candidate set with a consistent label, we follow the following procedure: first, we obtain the labels of the original sentences and the probabilities predicted by the simile detection model. Then, we select the label-consistent augmented samples by maximizing or minimizing the probability, depending on the label of the original sentence. For example, if the label is negative, we rank the probabilities of the augmented samples and choose the sentence with the minimum probability. If the minimum probability is greater than 0.5, we change the label of the augmented sample. If the label is positive for the original sentence, we choose the sample with the maximum probability. In this way, we can obtain an augmented sample for each sentence in the dataset, resulting in a new dataset $\hat{D}_{train}$.

Iterative Process: In the iterative process, we begin by obtaining the model parameters θ from the previous iteration. If it is the first iteration, θ is initialized with θ_{basic}, which is trained on D_{train}. We then replace the "like" in the original sentences with new comparative words, such as "seem". We use GPT-2 to generate the candidate set based on the replaced sentences and annotate the candidate samples with the model using the parameters θ, resulting in an augmentation dataset $\hat{D}_{train}$. Finally, we merge D_{train} and $\hat{D}_{train}$ and perform a simile detection task on the combined dataset. The details of the algorithm are shown in Algorithm 1.

4   Experiments Setup

4.1   Diverse Test Set

In real-world scenarios, simile sentences often involve a wide range of comparative words, as shown in Table 1. To address this issue, we have collected a new test set from the Internet, referred to as the Diverse Test Set (**DTS**), which consists of 1k simile sentences and covers a variety of comparative words. However, the data collected from the Internet is likely to contain noise. To obtain a high-quality validation dataset, we hire 5 professional annotators on a crowdsourcing platform[3] to annotate the collected data. The true label for a sentence is set to 1 (is) or 0 (not), where 1 indicates that the sentence is a simile. We utilize

---

[3] https://fuxi.163.com/productDetail/17.

a voting method to determine the labels for the final sentences. After filtering the data, we select 606 simile sentences covering 7 different comparative words. The number of simile sentences corresponding to different comparative words is shown in Table 2.

Meanwhile, to improve the quality of the dataset presented in this paper, we conduct a re-labeling process for the selected similes. Accordingly, we apply a set of 3 criteria to evaluate the data: (1) Creativity (C), which refers to the level of originality or novelty of the sentence; (2) Relevance (R), which measures the extent to which the VEHICLE in the sentence is related to the TOPIC; and (3) Fluency (F), which evaluates the smoothness and clarity of the sentence. These criteria are rated on a scale from 1 (not at all) to 5 (very), and each sentence is evaluated by three students. A total of 20, 15, and 10 students are used to rate the Creativity, Relevance, and Fluency of the similes, respectively. More information can be found in Table 3.

Table 2. The number of simile sentences with different comparison words in the DTS.

comparative words	像(like)	宛如(similar to)	好似(seem)	仿佛(as if)	如同(as...as)	跟...一样(same as)	好比(just like)
count	333	74	72	56	38	27	6

Table 3. Num means the number of workers employed for each labeling task and α denotes the score of the reliability proposed by [49] to measure the quality of results provided by annotators.

	Num	Average	α
Creativity (C)	20	4.4	0.86
Relevance (R)	15	3.9	0.75
Fluency (F)	10	4.1	0.81

4.2 Dataset

In this paper, we apply the dataset proposed by [13] as our training and test sets. The test set is referred to as the **Biased Test Set** (**BTS**). The training set consists of 11337 examples, including 5088 simile sentences and 6249 non-simile sentences. We follow [13] to divide the dataset into 7263 training examples and 2263 testing examples. However, the BTS contains only the comparative word "like". Thus we provide a **Diverse Test Set** (**DTS**) in this paper. In addition to the comparative word "like", our dataset includes 7 types of comparative words. The statistical analysis of the BTS and DTS datasets is shown in Table 4. It is worth noting that the BTS from [13] contains both simile and non-simile

Table 4. The statistical analysis of BTS and DTS. It is to verify the results of the experiments

	BTS	DTS
Sentences	2262	606
Simile sentences	987	606
Non-simile sentences	1275	0
Tokens	66k	16k
Comparative Words	1	7

sentences, while the DTS collected in this study only includes simile sentences. Finally, we apply two test sets to evaluate the performance of simile detection models.

4.3 Baselines

We compare the proposed augmentation approaches with the following baselines:

EDA [7]: A simple way to augment data consists of four operations: synonym replacement, random swap, random insertion, and random deletion.

BT [11]: This method involves translating a Chinese sentence into English and then translating the resulting English sentence back into Chinese.

MLM [36,47]: The sentence is randomly masked and then augmented with a pre-trained model by contextual word vectors to find top-n similar words.

To assess the effectiveness of various data augmentation techniques, we apply the Bert-base model [25][4] as the base classifier. The Bert-base model is a 12-layer transformer with 768 hidden units and 12 attention heads and has a total of 110M parameters. We conduct ablation tests to compare the performance of the data augmentation methods proposed in this paper. The specific methods and corresponding ablation tests are listed below:

I-WAS: An iterative data augmentation method with word replacement and sentence completion. More augmented data can be obtained during the iterative process. For example, I-WAS$_1$ indicates the first iteration.

I-WAS w/o wr: An augmentation approach to sentence completion with a static comparative word such as "like".

I-WAS w/o sc: A method without sentence completion that randomly replaces "like" in simile sentences with different comparative words.

It is important to note that for each method of data augmentation, we only augment a single sample for each sentence in the original training dataset, D_{train}. Additionally, the augmented dataset, $\hat{D}_{\text{train}}$, is of size I-1 times that of the D_{train} dataset in the I-WAS method. This means that the size of the augmented dataset is determined by the number of augmentation iterations, I.

---

[4] https://huggingface.co/bert-base-chinese.

Table 5. Comparison among different data augmentation methods on the biased test set and diversity test set. Bold indicates the best result, and the score means the average result of 5 runs with different seeds. The I-WAS$_1$ means the first iteration of I-WAS.

Methods	BTS		DTS
	F_1	Accuracy	Accuracy
Bert-base	87.54	88.61	66.70
+ EDA [7]	87.59	88.52	69.11
+ BT [11]	87.35	88.15	72.48
+ MLM [47]	87.79	89.08	70.20
+ I-WAS$_1$	87.63	88.53	**73.73**
+ I-WAS$_1$ w/o sc	**88.43**	**89.55**	70.30
+ I-WAS$_1$ w/o wr	87.03	88.05	68.84

4.4 Settings

Training Details: In all the experiments presented in this paper, we apply the same parameter settings for the simile detection task. We initialize the weights with the model that achieved the best performance in the previous iteration. The maximum sequence length is set to 256, and the AdamW [26] optimizer is used with a learning rate of 2e-4. The batch size is 128, and we utilize an early stop of 3 to prevent overfitting. For the sentence completion stage, we apply the GPT-2 [35] model, which is trained on a large corpus of novels and common crawl and has 5 billion parameters. The model uses a transformer architecture and consists of a 45-layer decoder with 32 attention heads in each layer. The embedding size is 3027 and the feed-forward filter size is 3072 * 4. To generate diverse similes, we obtain 10 samples using the GPT-2 model with the top-k sampling algorithm. All experiments are repeated using 5 different random seeds (16, 32, 64, 128, and 256) and the average scores are reported.

Metrics: To facilitate comparison with previous work, we apply the same evaluation metrics as [13], which are accuracy and the harmonic mean F_1 score. Since the DTS includes only simile samples, we focus on the evaluation metric of accuracy.

5 Results and Analysis

5.1 Results

The experimental results on BTS and DTS are shown in Table 5, and the following conclusions can be drawn from the results.

The Effect of the Test Set: We can observe that the accuracy of different data augmentation methods on the DTS is at least 15 points lower than on the BTS. For instance, the accuracy of the I-WAS$_1$ method on the DTS is 73.73, while it is 88.53 on the BTS. This difference is likely due to the fact that the simile sentences in the BTS only contain the comparative word "like", while the DTS includes more diverse and realistic sentence expressions with various comparative words. These results suggest that the BTS proposed by [13] may not be representative of simile detection in real-world scenarios.

I-WAS Results: Our proposed I-WAS data augmentation method achieves the highest accuracy score on the DTS compared to the other evaluated baselines (EDA, BT, MLM). Specifically, the I-WAS$_1$ method increases the accuracy score from 66.70 (using the Bert-base model) to 73.73, representing an absolute improvement of 7.03. Additionally, We also observe no significant differences in the performance of the various augmentation methods on the biased test set (BTS). These results demonstrate that the I-WAS method is particularly effective on realistic test sets.

Ablation Test Results: The results of the ablation test indicate that each of the *sc*, *wr*, and iterative process steps in the I-WAS method significantly improves the performance of simile detection on the DTS. Specifically, the gains achieved by the *sc*, *wr*, and iterative process steps were 3.43 (from 70.30 without *sc* to 73.73 with I-WAS), 4.89 (from 68.84 without *wr* to 73.73 with I-WAS$_1$), respectively. Similar to the baseline methods, the experiments in the ablation test do not show significant improvements in performance on the BTS.

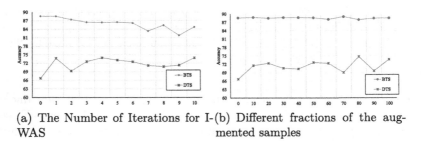

(a) The Number of Iterations for I-WAS

(b) Different fractions of the augmented samples

Fig. 1. Performance of the model trained on different conditions. Figure (a) shows the model performance under various numbers of iterations, and the training data is twice as large as the original. Meanwhile, the pre-trained model is initialized with the optimal results of the previous iteration. Figure (b) indicates the impact of the number of augmented samples on the model performance. These fractions include {10, 20, 30, 40, 50, 60, 70, 80, 90, 100}%. All the scores are the average of 5 runs with a different random seed. The y-axis shows the accuracy of the model.

Table 6. Examples of augmented simile sentences of different methods.

Source	Method	Target
父母的爱就像天上的白云，永远不会减少。 Parents' love is like the white clouds in the sky, never diminishing.	EDA	父母的爱就像天上的白云，永远不会减少。 Parents' love is ~~like~~ the ~~white clouds~~ in the sky, never ~~diminishing~~
	MLM	父母的脸就像天上的白云，永远不会减少。 Parents' face is like the white clouds in the sky, never diminishing.
	BT	父母的爱就像天上的白云，永远不会减少。 Parents' love is like the white clouds in the sky, never diminishing.
	I-WAS₁	父母的爱就好似这颗大树，永远不会枯萎。 Parents' love is the same as the tree, never wither.
书就像打开知识大门的金钥匙。 The Books are like the golden key to opening the door to knowledge.	EDA	打开就像书知识大门的金钥匙。 The opening is like the golden key to the door of book knowledge.
	MLM	书就像打开知考大门的金钥锁。 The book is like the golden key lock to opening the door to the knowledge.
	BT	这本书就像将金钥匙打开到门口。 This book is like opening the golden key to the door.
	I-WAS₁	书就仿佛一座山。 The book is like a mountain.
我大颗的泪水像细珠一样滚动着。 My large tears rolled like fine beads.	EDA	我大颗的泪水像细珠一样滚动着， My large ~~tears~~ rolled like fine beads.
	MLM	太大颗的泪都像细珠一样跳动着， Too large tears rolled like fine beads.
	BT	我的眼泪像珠子一样滚动， My tears rolled like beads.
	I-WAS₁	我大颗的泪水像断线的珠子一样不要钱的往下掉， My big tears fell like broken beads.

5.2 Analysis

Number of Iterative Processes: Fig. 1(a) shows the influence of the number of iterations on the performance of the model. We set the number of iterations to 10 and initialized the pre-trained model with the optimal model parameters from the previous iteration. As the number of iterations increases, the performance of the model on the BTS gradually decreases, while the performance on the DTS improves. This can be attributed to the data distribution drift [45]. During the continuous iterative learning process, the increasing number of augmented samples leads to a gradual shift in the distribution of the training data toward the DTS. As shown in Table 5, at the first iteration, the accuracy on the DTS improved by 7% while the accuracy on the BTS decreased by 0.1%, indicating the effectiveness of the proposed method in this paper.

Trend on Training Set Sizes: Figure 1(b) shows the performance of the model on different fractions of the augmented samples, where the largest fraction is twice the size of the original training data. These fractions include {10, 20, 30, 40, 50, 60, 70, 80, 90, 100}. It is worth noting that adding different proportions of augmented samples has varying levels of impact on the model's performance.

As we can see from the figure, the model's performance on the DTS fluctuates more dramatically than that on the BTS when different proportions of augmented samples are inserted. This is because the distribution of the original training data is similar to that of the BTS, and the various fractions of augmented samples with noise can differently impact the original distribution of the training data. When the number of augmented samples is smaller, the influence of noisy data becomes more significant. For example, if there are 10 enhanced samples and 5 mislabeled samples, the noise data accounts for 50%; when the number of enhanced samples increases to 100 and the number of mislabeled samples increases to 20, the percentage of noise data is reduced to 20%.

5.3   Augmented Sample

In this section, to evaluate the quality of the data generated by the different data augmentation approaches, we provide some examples in Table 6. We can see that, with the exception of the EDA approach, all other approaches can maintain sentence completeness, coherence between the TOPIC and VEHICLE, and label consistency. Meanwhile, the effectiveness of the BT-based sentence expansion approach depends on the complexity of the original sentence. Simple sentences are more difficult to augment using this method, as shown in the first sample in Table 6. On the other hand, complex sentences are easier to augment using this method. The MLM-based approach can change the fluency of sentences and potentially replace the TOPIC. However, the samples generated using the approach proposed in this paper increase the diversity of the VEHICLE without changing the TOPIC of the sentences, while maintaining label consistency and content fluency.

6   Discussion

Comparing the results of the BTS and DTS obtained in our experiments, we can see a significant gap, particularly in the performance of the Bert-base model. We believe that this gap may be due to the inconsistency between the training and test sets. As mentioned in Sect. 4.2, the training set used in our experiments is obtained from [13] and is consistent with the BTS. It only includes one simile pattern. However, the DTS includes several types of simile patterns, which means that a model trained on the BTS, such as the Bert-base model, may perform well on a particular pattern but not as well on others. This leads to a decrease in performance on the DTS. Additionally, the more simile patterns included in the test set, the more pronounced the decrease in model performance becomes.

In Fig. 1(a), we observe a gradual decrease in the performance of the model on the BTS during the iteration process. [45] suggests that catastrophic forgetting may be caused by changes in the data distribution or changes in the learning criterion, which are referred to as drifts. [48] propose Mean Teacher to overcome the problem that the targets change only once per epoch, and Temporal Ensembling becomes unwieldy when learning large datasets. In general, the data distribution changes over time in continual learning (or iterative learning). The decrease in performance shown in Fig. 1(a) is likely due to the changing data distribution of the original training set as the number of augmented samples increases. As the augmented samples include more patterns of similes, the model's performance in detecting these patterns improves. However, the performance on pattern-specific similes, such as those in the BTS, decreases. The problem of catastrophic forgetting in the process of continuous learning is an area for future work.

7 Conclusion and Future Work

In this paper, we introduce a data augmentation method called WAS, which uses GPT-2 to generate figurative language for simile detection. Unlike existing augmentation methods that are unaware of the target information, WAS is able to generate content-relevant and label-compatible sentences through word replacement and sentence completion. Our experimental results show that WAS performs the best on the diverse simile test set, demonstrating the effectiveness of this generative augmentation method using GPT-2. Recent studies have shown that it is more effective to incorporate constraints during the generation process through cooperative language model generation. When using a classifier to filter the generated examples, it is important to consider the impact on the performance of the base simile detection model. Future research could focus on improving the performance of the base simile detection model and exploring related topics.

Acknowledgements. This work is supported by the Key Research and Development Program of Zhejiang Province (No. 2022C01011). We would like to thank the anonymous reviewers for their excellent feedback. We are very grateful for the professional markers provided by NetEase Crowdsourcing.

References

1. Paul, A.M.: Figurative language. Philosophy Rhetoric **3**(4), 225–248 (1970). http://www.jstor.org/stable/40237206
2. Niculae, V., Danescu-Niculescu-Mizil, C.: Brighter than gold: figurative language in user generated comparisons. In: EMNLP (2014)
3. Qadir, A., Riloff, E., Walker, M.: Learning to recognize affective polarity in similes. In: EMNLP (2015)
4. Qadir, A., Riloff, E., Walker, M.: Automatically inferring implicit properties in similes. In: HLT-NAACL (2016)
5. Wei-guang, Q.: Computation of chinese simile with "xiang". J. Chinese Inf. Process. (2008)
6. Schick, T., Schutze, H.: Generating datasets with pretrained language models. ArXiv abs/2104.07540 (2021)
7. Wei, J., Zou, K.: Eda: easy data augmentation techniques for boosting performance on text classification tasks. ArXiv abs/1901.11196 (2019)
8. Ren, D., Zhang, P., Li, Q., Tao, X., Chen, J., Cai, Y.: A hybrid representation-based simile component extraction. Neural Comput. Appl. **32**(18), 14655–14665 (2020). https://doi.org/10.1007/s00521-020-04818-6
9. Zhang, P., Cai, Y., Chen, J., Chen, W.H., Song, H.: Combining part-of-speech tags and self-attention mechanism for simile recognition. IEEE Access **7**, 163864–163876 (2019)
10. Zeng, J., Song, L., Su, J., Xie, J., Song, W., Luo, J.: Neural simile recognition with cyclic multitask learning and local attention. ArXiv abs/1912.09084 (2020)
11. Yu, A.W., et al.: Fast and accurate reading comprehension by combining self-attention and convolution (2018)

12. Veale, T.: A context-sensitive, multi-faceted model of lexico-conceptual affect. In: ACL (2012)
13. Liu, L., Hu, X., Song, W., Fu, R., Liu, T., Hu, G.: Neural multitask learning for simile recognition. In: EMNLP (2018)
14. Song, W., Guo, J., Fu, R., Liu, T., Liu, L.: A knowledge graph embedding approach for metaphor processing. IEEE/ACM Trans. Audio, Speech Lang. Process. **29**, 406–420 (2021)
15. Chakrabarty, T., Saakyan, A., Ghosh, D., Muresan, S.: Flute: figurative language understanding through textual explanations (2022)
16. Hao, Y., Veale, T.: An ironic fist in a velvet glove: creative mis-representation in the construction of ironic similes. Minds Mach. **20**, 635–650 (2010)
17. Manjusha, P.D., Raseek, C.: Convolutional neural network based simile classification system. In: 2018 International Conference on Emerging Trends and Innovations in Engineering and Technological Research (ICETIETR), pp. 1–5 (2018)
18. Chen, X., Hai, Z., Li, D., Wang, S., Wang, D.: Jointly identifying rhetoric and implicit emotions via multi-task learning. In: FINDINGS (2021)
19. Miwa, M., Bansal, M.: End-to-end relation extraction using LSTMs on sequences and tree structures. ArXiv abs/1601.00770 (2016)
20. Guo, J., Song, W., Liu, X., Liu, L., Zhao, X.: Attention-based BiLSTM network for Chinese simile recognition. In: 2018 IEEE 9th International Conference on Software Engineering and Service Science (ICSESS), pp. 144–147 (2018)
21. Kingma, D.P., Welling, M.: Auto-encoding variational Bayes. CoRR abs/1312.6114 (2014)
22. Zeng, J., Song, L., Su, J., Xie, J., Song, W., Luo, J.: Neural simile recognition with cyclic multitask learning and local attention. ArXiv abs/1912.09084 (2020)
23. Goodfellow, I., et al.: Generative adversarial nets. In: NIPS (2014)
24. Anaby-Tavor, A., et al.: Do not have enough data? deep learning to the rescue! In: AAAI (2020)
25. Devlin, J., Chang, M.W., Lee, K., Toutanova, K.: Bert: pre-training of deep bidirectional transformers for language understanding. In: NAACL-HLT (2019)
26. Loshchilov, I., Hutter, F.: Decoupled weight decay regularization. In: ICLR (2019)
27. Li, Y., Caragea, C.: Target-aware data augmentation for stance detection. In: NAACL (2021)
28. Gupta, R.: Data augmentation for low resource sentiment analysis using generative adversarial networks. In: ICASSP 2019–2019 IEEE International Conference on Acoustics, Speech and Signal Processing (ICASSP), pp. 7380–7384 (2019)
29. Hu, Z., Yang, Z., Liang, X., Salakhutdinov, R., Xing, E.: Toward controlled generation of text. In: ICML (2017)
30. Edunov, S., Ott, M., Auli, M., Grangier, D.: Understanding back-translation at scale. In: EMNLP (2018)
31. Zhang, X., Zhao, J., LeCun, Y.: Character-level convolutional networks for text classification. ArXiv abs/1509.01626 (2015)
32. Wang, W.Y., Yang, D.: That's so annoying!!!: A lexical and frame-semantic embedding based data augmentation approach to automatic categorization of annoying behaviors using #petpeeve tweets. In: EMNLP (2015)
33. Miller, G.: Wordnet: a lexical database for English. Commun. ACM **38**, 39–41 (1995)
34. Mikolov, T., Sutskever, I., Chen, K., Corrado, G., Dean, J.: Distributed representations of words and phrases and their compositionality. In: NIPS (2013)
35. Radford, A., Wu, J., Child, R., Luan, D., Amodei, D., Sutskever, I.: Language models are unsupervised multitask learners (2019)

36. Kobayashi, S.: Contextual augmentation: data augmentation by words with paradigmatic relations. In: NAACL-HLT (2018)
37. Karimi, A., Rossi, L., Prati, A.: AEDA: an easier data augmentation technique for text classification. ArXiv abs/2108.13230 (2021)
38. Hou, Y., Liu, Y., Che, W., Liu, T.: Sequence-to-sequence data augmentation for dialogue language understanding. In: COLING (2018)
39. Claveau, V., Chaffin, A., Kijak, E.: generating artificial texts as substitution or complement of training data. ArXiv abs/2110.13016 (2021)
40. Papanikolaou, Y., Pierleoni, A.: DARE: data augmented relation extraction with GPT-2. ArXiv abs/2004.13845 (2020)
41. Kober, T., Weeds, J., Bertolini, L., Weir, D.J.: Data augmentation for hypernymy detection. In: EACL (2021)
42. Liu, D., et al.: Tell me how to ask again: question data augmentation with controllable rewriting in continuous space. ArXiv abs/2010.01475 (2020)
43. Li, K., Chen, C., Quan, X., Ling, Q., Song, Y.: Conditional augmentation for aspect term extraction via masked sequence-to-sequence generation. ArXiv abs/2004.14769 (2020)
44. Wu, X., Lv, S., Zang, L., Han, J., Hu, S.: Conditional BERT contextual augmentation. ArXiv abs/1812.06705 (2019)
45. Lesort, T., Caccia, M., Rish, I.: Understanding continual learning settings with data distribution drift analysis. ArXiv abs/2104.01678 (2021)
46. Ding, B., et al.: DAGA: data augmentation with a generation approach for low-resource tagging tasks. In: EMNLP (2020)
47. Kumar, V., Choudhary, A., Cho, E.: Data augmentation using pre-trained transformer models. ArXiv abs/2003.02245 (2020)
48. Tarvainen, A., Valpola, H.: Mean teachers are better role models: weight-averaged consistency targets improve semi-supervised deep learning results. In: NIPS (2017)
49. Krippendorff, K.: Computing krippendorff's alpha-reliability (2011)

Information Redundancy and Biases in Public Document Information Extraction Benchmarks

Seif Laatiri[✉], Pirashanth Ratnamogan, Joël Tang, Laurent Lam,
William Vanhuffel, and Fabien Caspani

BNP Paribas, Paris, France
{seifedinne.laatiri,pirashanth.ratnamogan,joel.tang,laurent.lam,
william.vanhuffel,fabien.caspani}@bnpparibas.com

Abstract. Advances in the Visually-rich Document Understanding (VrDU) field and particularly the Key-Information Extraction (KIE) task are marked with the emergence of efficient Transformer-based approaches such as the LayoutLM models. Despite the good performance of KIE models when fine-tuned on public benchmarks, they still struggle to generalize on complex real-life use-cases lacking sufficient document annotations. Our research highlighted that KIE standard benchmarks such as SROIE and FUNSD contain significant similarity between training and testing documents and can be adjusted to better evaluate the generalization of models.

In this work, we designed experiments to quantify the information redundancy in public benchmarks, revealing a 75% template replication in SROIE's official test set and 16% in FUNSD's. We also proposed resampling strategies to provide benchmarks more representative of the generalization ability of models. We showed that models not suited for document analysis struggle on the adjusted splits dropping on average 10,5% F1 score on SROIE and 3.5% on FUNSD compared to multi-modal models dropping only 7,5% F1 on SROIE and 0.5% F1 on FUNSD.

Keywords: Visually-rich Document Understanding · Key Information Extraction · Named Entity Recognition · Generalization Assessment

1 Introduction

Visually-rich Document Understanding (VrDU) is a field that has seen progress lately following the recent breakthroughs in Natural Language Processing and Computer Vision. This field aims to transform documents into structured data by simultaneously leveraging their textual, positional and visual attributes. Since scanned documents are often noisy, recent works addressed VrDU as a two component stream: first extracting the text with optical character recognition and then performing analysis using OCR text, document's layout and visual attributes.

Real life business documents can belong to multiple categories such as financial reports, employee contracts, forms, emails, letters, receipts, resumes and

G. A. Fink et al. (Eds.): ICDAR 2023, LNCS 14189, pp. 280–294, 2023.
https://doi.org/10.1007/978-3-031-41682-8_18

others. Thus it is challenging to create general pipelines capable of handling all types of documents. State-of-the-art models focus on document-level pre-training objectives then fine-tuning on downstream tasks. These models show promising results on public information extraction benchmarks such as FUNSD [11], SROIE [10], CORD [21], Kleister NDA [28]. However, on real world complex use cases it is difficult to replicate the same performances due to a lack of annotated samples and diversity in their templates. In this paper, we show that common benchmarks can be managed to better evaluate the generalization ability of information extraction models and thus become more viable tools for model selection for real-world business use cases.

To this end, we focus on SROIE [10] and FUNSD [11] and explore their potential to challenge the generalization power of a given model. We design experiments to measure document similarities in the official training and testing splits of these benchmarks and propose resampling strategies to render these benchmarks a better evaluation of models' performance on unseen documents. We then investigate the impact of these resampling strategies on state-of-the-art VrDU models.

2 Background

2.1 Related Work

Dataset Biases and Model Generalization in NLP. The study of dataset biases and model generalization is an important area of research that has already been conducted on several NLP tasks. In the Named Entity Recognition task, several studies have highlighted the fact that the common datasets CONLL 2003 [25] and OntoNotes 5 [34] are strongly biased by an unrealistic lexical overlap between mentions in training and test sets [3, 29]. In the co-reference task, the same lexical overlap with respect to co-reference mentions was observed in CONLL 2012 [22] and led to an overestimation of the performance of deep learning models compared to classical methods for real world applications [20]. The Natural Language Inference (NLI) task also suffers from a dataset with bias. Indeed, MultiNLI has been reported to suffer from both lexical overlap and hypothesis-only bias: the fact that hypothesis sentences contain words associated with a target label [6].

Deep Learning models are extremely sensitive to these biases. However in real world, models should be able to generalize to new out of domain data. Multiple studies experimented models memorization capability [2] and models robustness in order to perform on out of domain data in multiple tasks: translation [19], co-reference [30] or named entity recognition [29].

To the best of our knowledge, our work is the first one assessing and analyzing biases in the context of information extraction from documents.

Information Extraction Models. When performing visual analysis on documents, early work handled separately the different modalities. Preliminary work

[7,26] focused on extracting tabular data from documents by combining heuristic rules and convolutional networks to detect and recognize tabular data. Later work [27] used visual features for document layout detection by incorporating contextual information in Faster R-CNNs [24]. Follow up research [16,38] combined textual and visual information by introducing a graph convolution based model for key information extraction. This approach exhibited good results however, models are only using supervised data which is limited. In addition, these pre-training methods do not inherently combine textual and visual attributes as they are merged during fine-tuning instead.

Following the rise of Transformers, more Transformer based models were adapted for VrDU with novel pre-training objectives. LayoutLM [36] uses 2D positional embeddings in order to integrate layout information with word embeddings and is pretrained on layout understanding tasks. LayoutLMv2 [35] adds token-level visual features extracted with Convolution Neural Networks and models interactions among text, layout and image. Later work aimed to match the reconstruction pre-training objectives of masked text with a similar objective for reconstructing visual attributes such as LayoutLMv3 [9] which proposed to predict masked image areas through a set of image tokens similarly to visual Transformers [5,13]. Other recent approaches [12,15] experimented with an OCR-free setup by leveraging an encoder-decoder visual transformer.

2.2   Datasets

Multiple datasets exist to benchmark a variety of document understanding tasks. For instance RVL-CDIP [8] is an image-centric dataset for document image classification, FUNSD [11], CORD [21], SROIE [10] and Kleister-NDA [28] are datasets for key information extraction respectively from forms, receipts and contracts whilst DocVQA [18] is a benchmark for visual question answering.

In this work, we focus on the task of information extraction while investigating template similarities in the current documents distribution within the datasets. Documents having the same template are documents sharing the same layout that can be read in the same way. We decided to primarily work with SROIE and FUNSD as they are common benchmarks displaying two different types of documents. In more details, the SROIE dataset for Scanned Receipt OCR and Information Extraction was presented in the 2019 edition of the ICDAR conference. It represents processes of recognizing text from scanned restaurant receipts and extracting key entities from them. The dataset contains 1000 annotated scanned restaurant receipts split into train/test splits with the ratio 650/350. Three tasks were set up for the competition: Scanned Receipt Text Localisation, Scanned Receipt Optical Character Recognition and Key Information Extraction from Scanned Receipts.

Second, FUNSD is a dataset for Form Understanding in Noisy Scanned Documents that have been a constant benchmark in recent document information extraction work. It contains noisy scanned forms and aims at extracting and structuring their textual contents. The dataset provides annotations for text recognition, entity linking and semantic entity labeling. In this work, FUNSD refers to the revised version of the dataset released by the authors [31] containing the same documents with cleaned annotations.

2.3   Problem Statement

We formalize key information extraction as a token classification task on the tokenized text extracted from the document.

Let us denote by $T = t_{0<i\leq n}$ the sequence of text tokens t_i extracted from a document $\mathcal{D}$. Let $\mathcal{I}$ be the image of the document $\mathcal{D}$ and m the number of entity types (since we perform Inside-Outside-Beginning tagging [23] similarly to named entity recognition tasks, the number of entity classes is $2m + 1$). We aim to build a classifier $\mathcal{F}$ such that for every token t_i in the sequence:

$$\mathcal{F}(t_i|T,\mathcal{I}) = c \tag{1}$$

with $c \in \{1, \ldots, 2m + 1\}$ the target IOB class of token t_i.

In IOB tagging, an entity spans over multiple adjacent tokens $(t_i)_{i_{start}\leq i\leq i_{end}}$ and is only correctly predicted if all of its tokens are correctly predicted, that is $\mathcal{F}(t_{i_{start}}|T,\mathcal{I}) = B - entity \quad and \quad \mathcal{F}(t_i|T,\mathcal{I}) = I - entity \ for \ i \in \{i_{start} + 1, \ldots, i_{end}\}$.

3   Approach

3.1   Motivation

Recent document analysis models such as LayoutLM models [9,35,36], Doc-Former [1], Lilt [32] and others used the datasets mentioned above to evaluate their models and benchmark them against other state-of-the-art works. However, these evaluation metrics are usually difficult to replicate on real-life business use-cases, particularly those obtained on SROIE and FUNSD. This discrepancy is, in part, due to the complexity of business use-cases and their lack of good-quality annotated data. However we also suspect that the current train and test splitting distribution of these datasets does not optimally evaluate the ability of models to generalize on unseen documents.

Even though it is a common practice in machine learning to keep similar distributions for both training and testing data, this practice is not optimal when benchmarking and comparing models that will later be finetuned on small datasets or inferred on out-of-domain data. By containing similar documents in both training and testing data, these datasets allow models to memorize predictions during training and simply infer them on test documents without evaluating their ability to understand and analyze new templates of documents. In real business use-cases, this is particularly harming long-term performances, as domain shift often occurs after a certain period of time, when new unseen templates are used.

3.2   Resampling Datasets

For each studied dataset, the current training and testing documents are thoroughly observed and analysed for homogeneous samples. We customize for every dataset a method to group similar documents and re-sample the training and testing splits to minimize template similarity and redundancy. We remind that the term template in this context refers to the disposition and layout of a document.

SROIE: Information extraction in this dataset is performed by extracting semantic entities from business receipts such as business' name, address, the order's date and total price. As shown in Fig. 2, receipts from the same business have a similar disposition, they contain the same business' name and address as well as the same template. The current official data split of SROIE does not account for this factor as same business receipts can be present in both train and test documents. We group same businesses and re-sample train-test splits while assuring that every group is present in only a single split. The sizes of groups of samples sharing the same template varies from 1 to 76 receipts and their distribution is described in Fig. 1 in a logarithmic scale.

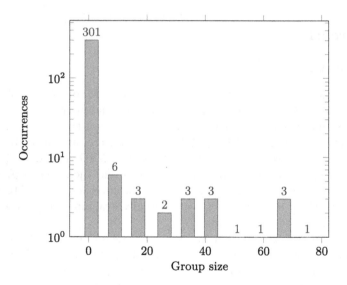

Fig. 1. SROIE: distribution of similarity groups, each group containing receipts of the same template

FUNSD: By investigating FUNSD, we observed that multiple forms of the same template were present in the dataset while being filled with different information, for instance a standard hospital application filled by different patients. Having

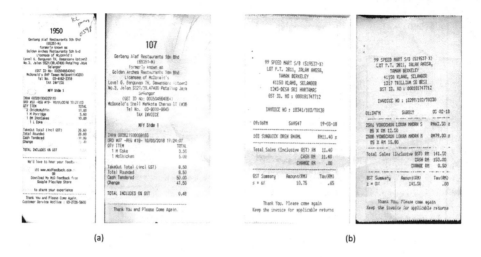

Fig. 2. SROIE: Receipts from two different business (a) and (b)

the exact same form template in both training and testing is a clear indicator of information redundancy as illustrated in the template comparison in Fig. 4. Based on the fact that same template forms share similar slot names (*questions*), we introduced a similarity metric on forms using the overlap of their question annotations. Based on this assumption, we propose the following overlap score for two forms:

$$Overlap(docA, docB) = \frac{Count(QuestionsA \cap QuestionsB)}{Max(len(QuestionsA), len(QuestionsB))}$$

where *QuestionsA* and *QuestionsB* are respectively the question annotations sets of document A and document B.

We have manually defined a set of template groups as ground truths and then evaluated different grouping similarity thresholds, eventually keeping a threshold of 0.7. This metric was next used to group forms of same templates. The sizes of groups in this case was far lower than that of SROIE groups as the biggest group of forms sharing the same template was limited to 4 forms. From 50 forms in the testing set, we found 8 (16%) that shared the same template with at least one training form. We resample the train and test splits accordingly, ensuring that no forms with the same template are present in both splits. The resampled splits can be found in https://github.com/Seif-Lat (Fig. 3).

3.3 Models

In the context of information extraction from documents seen as a token classification task, three approaches exist in the literature:

– Standard NLP models using textual information,
– Layout Aware models using both text and layout information,

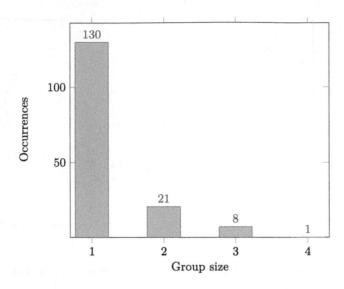

Fig. 3. FUNSD: distribution of similarity groups, each group containing forms of the same template

- Multi-modal approaches using text, layout and visual representations of tokens (Fig. 5).

In the context of our study, it is important to challenge the common evidence obtained using official datasets: multi-modal approaches are more effective than other approaches.

We studied the fine-tuning of the following pretrained models:

BERT. [4] is a bidirectional Transformer-based language model pretrained on a large corpus with Mask Language Modeling (MLM) and Next Sentence Prediction (NSP) tasks.

RoBERTa. [17] differs from BERT during pretraining with the use of a significantly larger dataset, a larger batch size and dynamic masking in the MLM task while dropping the NSP pretraining objective.

AlBERT. [14] is a scalable version of BERT that uses two methods to reduce the model memory footprint: sharing some of the model layers, and a factorized embedding parameterization.

Lilt. [33] is an approach decoupling layout and text representation in order to have the model using more layout information and being more language independent. It uses independent layout and text pretraining but also proposes a bidirection attention complementation mechanism in order to combine two flows: one for layout the other one for the text.

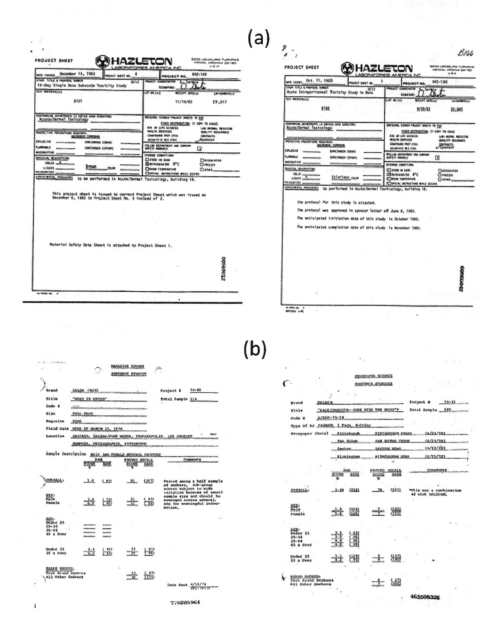

Fig. 4. FUNSD: Forms from two different templates (a) and (b)

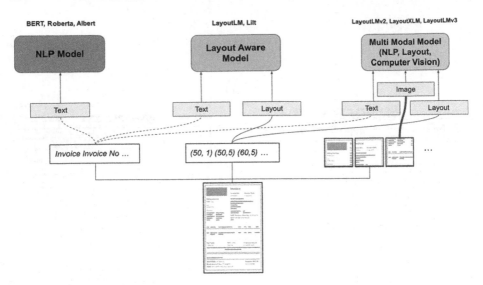

Fig. 5. Overview of document understanding models

LayoutLM. [36] introduced a multimodal pretraining approach combining text and layout features for document image understanding and information extraction tasks. It leverages both text and layout features and incorporates them into a single framework.

LayoutLMv2. [35] LayoutLMv2 is one of the first Transformers to use image feature during the pretraining process. They propose to use the output of a CNN architecture to create image token embeddings which gives useful information about the document layout.

LayoutXLM. [37] LayoutXLM has the same architecture as LayoutLMv2 but is pretrained on a multi-lingual dataset.

LayoutLMv3. [9] LayoutLMv3 uses a multi modal Transformer architecture and introduces an image reconstruction pretraining objective similar to text reconstruction in the masked language modeling objective.

4   Experiments

4.1   Experimental Setup

We use the same configuration when fine-tuning all the models. We use a batch size of 2 and an Adam optimizer with an initial learning rate of $2 * 10^{-5}$. We decrease the learning rate by half every 10 epochs without improvement in the validation F1 score. We stop the fine-tuning when the learning rate goes below 10^{-7}. We finally recover the model with the best validation F1.

For each experiment, the test set is defined initially, either the official testing set on the original datasets or our extracted testing set on the resampled datasets. We then generate four different splits from the remaining data as training and validation data with 80–20 ratio, we train and test each model on the four splits and present the average performance in the sections below. We perform this cross validation for a more robust model comparison as we have observed a shift of performance in consecutive trainings of the same model.

4.2 Results on Original Datasets

We train a group of models leveraging different modalities on receipt understanding and form understanding using the official splits of SROIE and FUNSD. Results are presented in Table 1. On form understanding (FUNSD), models using only the textual information of documents (BERT, AlBERT, RoBERTa) perform marginally worse than multi-modal models as their F1 scores are on average 20 points less. RoBERTa has the highest F1 (80.64) score among textual models and is 5 points behind the closest multi-modal model being LayoutXLM with 85.57 F1. Among multi-modal models, LayoutLMv2 and LayoutLMv3 have better scores than LayoutLM and LiLT since they also leverage visual attributes of forms. LayoutLMv3 has the highest F1 score of 88.81 thanks to its efficient multi-modal pre-training.

On receipt understanding (SROIE) multi-modal approaches also have higher metrics than textual models, however the discrepancy between them is much lower as the F1 increase between each multi-modal and the average score of all textual models varies from 2 to 4 and is considerably lower than what we observed on FUNSD. On this task LayoutLMv2 achieves the highest score amongst all models, reaching 96.14 F1. These results validate the importance of positional and visual attributes in information extraction tasks on visually rich documents.

4.3 Results on Resampled Datasets

After having resampled the splits of both FUNSD and SROIE, we train the same group of models and present results in Table 2. On form understanding, textual models are outperformed by multi-modal models and show a more important decrease in performance compared to results on the original split with an average drop of 3.5 F1 score compared to only 0.5 in multi-modal models.

On receipt understanding, F1 scores drastically drop compared to results on the original split in Table 1, BERT, AlBERT and RoBERTa drop on average 10.5 F1 points whereas multi-modal models drop only 7.5 F1 points on average. Scores on the adjusted splits show a higher discrepancy between the models and are more consistent with the modalities leveraged by each model and the efficiency of their pretraining. For instance LiLT marginally outperforms LayoutLM as its pretraining allows it to better leverage the positional information of documents. The adjusted split also makes the information extraction task more challenging

Table 1. Performance of state-of-the-art information extraction models on SROIE and FUNSD official testing sets.

Model	Params	SROIE			FUNSD		
		F1	Precision	Recall	F1	Precision	Recall
BERT$_{base}$	110M	92.47	92.36	92.68	61.03	60.29	61.90
AlBERT$_{base}$	12M	92.28	92.28	92.28	57.39	56.08	58.99
RoBERTa$_{base}$	125M	93.90	93.23	94.61	80.64	81.36	79.96
LayoutLM$_{base}$	112M	94.57	93.93	95.22	86.07	86.24	85.63
LiLT$_{base}$	131M	95.41	95.23	95.60	87.41	87.41	87.41
LayoutLMv2$_{base}$	200M	**96.14**	**96.39**	**95.94**	88.14	88.31	88.08
LayoutXLM$_{base}$	369M	94.75	94.57	94.94	85.57	85.87	85.46
LayoutLMv3$_{base}$	126M	95.11	94.87	95.70	**88.81**	**89.32**	**88.46**
BERT$_{large}$	351M	93.72	93.47	94.02	61.03	60.29	61.90
AlBERT$_{large}$	18M	88.96	87.86	90.20	58.81	57.46	60.39
RoBERTa$_{large}$	355M	94.99	94.91	95.09	82.43	82.99	81.78
LayoutLM$_{large}$	340M	94.70	94.31	95.10	84.29	84.60	84.10
LayoutLMv2$_{large}$	426M	**96.55**	**96.69**	**96.42**	88.79	89.06	88.75
LayoutLMv3$_{large}$	365M	95.87	95.71	96.03	**89.84**	**89.97**	**89.56**

Table 2. Performance of state-of-the-art information extraction models on SROIE and FUNSD **resampled** testing sets.

Model	Params	SROIE			FUNSD		
		F1	Precision	Recall	F1	Precision	Recall
BERT$_{base}$	110M	81.00	77.04	86.21	55.25	54.97	55.86
AlBERT$_{base}$	12M	79.86	77.48	82.97	53.66	52.63	54.99
RoBERTa$_{base}$	125M	86.05	84.15	88.20	78.71	79.16	78.59
LayoutLM$_{base}$	112M	84.80	82.18	88.02	85.88	85.85	86.11
LiLT$_{base}$	131M	**89.38**	86.94	**92.11**	84.76	85.12	84.91
LayoutLMv2$_{base}$	200M	87.92	86.67	89.34	88.61	88.65	88.94
LayoutXLM$_{base}$	369M	87.99	86.24	90.14	85.51	85.98	86.25
LayoutLMv3$_{base}$	126M	87.86	**87.29**	88.71	**89.07**	**89.26**	**89.16**
BERT$_{large}$	351M	81.15	77.53	85.84	55.90	55.46	56.59
AlBERT$_{large}$	18M	82.82	81.16	84.76	54.93	52.73	57.77
RoBERTa$_{large}$	355M	87.86	86.16	89.84	80.28	80.16	80.66
LayoutLM$_{large}$	340M	84.78	82.62	87.91	85.95	85.75	86.34
LayoutLMv2$_{large}$	426M	87.98	86.71	89.98	**90.14**	**90.19**	**90.31**
LayoutLMv3$_{large}$	356M	**88.47**	**87.00**	**90.13**	89.86	89.72	90.31

as the average F1 score drops from 94.34 to 85.60 and the highest reached F1 drops from 96.55 to 89.38.

These results show that the original splits of both SROIE and FUNSD contained data leaks that allowed models to infer on testing data without necessarily learning how to understand new templates. The adjusted splits evaluate more properly the generalization ability of models and their capacity to transfer knowledge to unseen templates.

5 Conclusion

In this paper, we showed that SROIE and FUNSD were featuring information redundancies between their training and testing sets. Having trained multiple state-of-the-art models on VrDU tasks, we showed that this information redundancy artificially increases the models performances. In particular, we observe that SROIE is still a challenging benchmark as the average F1 score drops from 96.38 on the official splits to 88.78 on the adjusted splits, proving that it remains a viable benchmark for upcoming works when it is sampled more carefully.

These findings demonstrate that generating Independent and Identically Distributed splits for evaluation datasets as is traditionally done is not an optimal approach, as it introduces a high memorization bias especially with large neural networks. The 0% overlap approach presented in this work is an example of an alternate strategy specific to evaluating models' generalization on unseen templates and is closer to real-world use-cases than traditional splits. Other criteria can also be explored for this same purpose such as a date based resampling.

References

1. Appalaraju, S., Jasani, B., Kota, B.U., Xie, Y., Manmatha, R.: Docformer: end-to-end transformer for document understanding. CoRR abs/2106.11539 (2021). https://arxiv.org/abs/2106.11539
2. Arpit, D., et al.: A closer look at memorization in deep networks. In: International Conference on Machine Learning, pp. 233–242. PMLR (2017)
3. Augenstein, I., Derczynski, L., Bontcheva, K.: Generalisation in named entity recognition: a quantitative analysis. Comput. Speech Lang. **44**, 61–83 (2017)
4. Devlin, J., Chang, M.W., Lee, K., Toutanova, K.: Bert: pre-training of deep bidirectional transformers for language understanding. In: Proceedings of the 2019 Conference of the North American Chapter of the Association for Computational Linguistics: Human Language Technologies, Volume 1 (Long and Short Papers) (2019). https://aclanthology.org/N19-1423.pdf
5. Dosovitskiy, A., et al.: An image is worth 16x16 words: transformers for image recognition at scale (2020). https://arxiv.org/abs/2010.11929
6. Gururangan, S., Swayamdipta, S., Levy, O., Schwartz, R., Bowman, S., Smith, N.A.: Annotation artifacts in natural language inference data. In: Proceedings of the 2018 Conference of the North American Chapter of the Association for Computational Linguistics: Human Language Technologies, Volume 2 (Short Papers), pp. 107–112. Association for Computational Linguistics, New Orleans, Louisiana, June 2018. https://doi.org/10.18653/v1/N18-2017, https://aclanthology.org/N18-2017

7. Hao, L., Gao, L., Yi, X., Tang, Z.: A table detection method for pdf documents based on convolutional neural networks. In: 2016 12th IAPR Workshop on Document Analysis Systems (DAS), pp. 287–292 (2016). https://doi.org/10.1109/DAS.2016.23

8. Harley, A.W., Ufkes, A., Derpanis, K.G.: Evaluation of deep convolutional nets for document image classification and retrieval. CoRR abs/1502.07058 (2015). http://arxiv.org/abs/1502.07058

9. Huang, Y., Lv, T., Cui, L., Lu, Y., Wei, F.: Layoutlmv3: pre-training for document AI with unified text and image masking. In: Proceedings of the 30th ACM International Conference on Multimedia, MM 2022, pp. 4083–4091. Association for Computing Machinery, New York, NY, USA (2022). https://doi.org/10.1145/3503161.3548112

10. Huang, Z., et al.: Icdar 2019 competition on scanned receipt OCR and information extraction, pp. 1516–1520 (2019). https://arxiv.org/pdf/2103.10213.pdf

11. Jaume, G., Kemal Ekenel, H., Thiran, J.P.: Funsd: a dataset for form understanding in noisy scanned documents. In: 2019 International Conference on Document Analysis and Recognition Workshops (ICDARW), vol. 2, pp. 1–6 (2019). https://doi.org/10.1109/ICDARW.2019.10029

12. Kim, G., et al.: Ocr-free document understanding transformer (2022)

13. Kim, W., Son, B., Kim, I.: Vilt: vision-and-language transformer without convolution or region supervision (2021). https://arxiv.org/abs/2102.03334

14. Lan, Z., Chen, M., Goodman, S., Gimpel, K., Sharma, P., Soricut, R.: Albert: a lite bert for self-supervised learning of language representations. arXiv preprint arXiv:1909.11942 (2019)

15. Lee, K., et al.: Pix2struct: screenshot parsing as pretraining for visual language understanding (2022)

16. Liu, X., Gao, F., Zhang, Q., Zhao, H.: Graph convolution for multimodal information extraction from visually rich documents. In: NAACL (2019)

17. Liu, Y., et al.: Roberta: a robustly optimized bert pretraining approach (2019). https://doi.org/10.48550/ARXIV.1907.11692, https://arxiv.org/abs/1907.11692

18. Mathew, M., Karatzas, D., Manmatha, R., Jawahar, C.V.: Docvqa: a dataset for VQA on document images. CoRR abs/2007.00398 (2020). https://arxiv.org/abs/2007.00398

19. Mghabbar, I., Ratnamogan, P.: Building a multi-domain neural machine translation model using knowledge distillation. In: Giacomo, G.D., et al. (eds.) ECAI 2020–24th European Conference on Artificial Intelligence, 29 August-8 September 2020, Santiago de Compostela, Spain, August 29 - September 8, 2020 - Including 10th Conference on Prestigious Applications of Artificial Intelligence (PAIS 2020), Frontiers in Artificial Intelligence and Applications, vol. 325, pp. 2116–2123. IOS Press (2020). https://doi.org/10.3233/FAIA200335

20. Moosavi, N.S., Strube, M.: Using linguistic features to improve the generalization capability of neural coreference resolvers. arXiv preprint arXiv:1708.00160 (2017)

21. Park, S., et al.: Cord: a consolidated receipt dataset for post-ocr parsing (2019)

22. Pradhan, S., Moschitti, A., Xue, N., Uryupina, O., Zhang, Y.: Conll-2012 shared task: modeling multilingual unrestricted coreference in ontonotes. In: Joint Conference on EMNLP and CoNLL-Shared Task, pp. 1–40 (2012)

23. Ramshaw, L., Marcus, M.: Text chunking using transformation-based learning. In: Third Workshop on Very Large Corpora (1995). https://aclanthology.org/W95-0107

24. Ren, S., He, K., Girshick, R., Sun, J.: Faster r-cnn: towards real-time object detection with region proposal networks. In: Cortes, C., Lawrence, N., Lee, D., Sugiyama, M., Garnett, R. (eds.) Advances in Neural Information Processing Systems, vol. 28. Curran Associates, Inc. (2015). https://proceedings.neurips.cc/paper/2015/file/14bfa6bb14875e45bba028a21ed38046-Paper.pdf
25. Sang, E.F., De Meulder, F.: Introduction to the conll-2003 shared task: language-independent named entity recognition. arXiv preprint cs/0306050 (2003)
26. Schreiber, S., Agne, S., Wolf, I., Dengel, A., Ahmed, S.: Deepdesrt: deep learning for detection and structure recognition of tables in document images. In: 2017 14th IAPR International Conference on Document Analysis and Recognition (ICDAR), vol. 01, pp. 1162–1167 (2017). https://doi.org/10.1109/ICDAR.2017.192
27. Soto, C., Yoo, S.: Visual detection with context for document layout analysis. In: Proceedings of the 2019 Conference on Empirical Methods in Natural Language Processing and the 9th International Joint Conference on Natural Language Processing (EMNLP-IJCNLP), pp. 3464–3470. Association for Computational Linguistics, Hong Kong, China, November 2019. https://doi.org/10.18653/v1/D19-1348, https://aclanthology.org/D19-1348
28. Stanislawek, T., et al.: Kleister: key information extraction datasets involving long documents with complex layouts. CoRR abs/2105.05796 (2021). https://arxiv.org/abs/2105.05796
29. Taillé, B., Guigue, V., Gallinari, P.: Contextualized embeddings in named-entity recognition: an empirical study on generalization. In: Jose, J.M., et al. (eds.) ECIR 2020. LNCS, vol. 12036, pp. 383–391. Springer, Cham (2020). https://doi.org/10.1007/978-3-030-45442-5_48
30. Toshniwal, S., Xia, P., Wiseman, S., Livescu, K., Gimpel, K.: On generalization in coreference resolution. In: Proceedings of the Fourth Workshop on Computational Models of Reference, Anaphora and Coreference, pp. 111–120. Association for Computational Linguistics, Punta Cana, Dominican Republic, November 2021. https://doi.org/10.18653/v1/2021.crac-1.12, https://aclanthology.org/2021.crac-1.12
31. Vu, H.M., Nguyen, D.T.: Revising FUNSD dataset for key-value detection in document images. CoRR abs/2010.05322 (2020), https://arxiv.org/abs/2010.05322
32. Wang, J., Jin, L., Ding, K.: Lilt: a simple yet effective language-independent layout transformer for structured document understanding (2022). https://doi.org/10.48550/ARXIV.2202.13669, https://arxiv.org/abs/2202.13669
33. Wang, J., Jin, L., Ding, K.: Lilt: a simple yet effective language-independent layout transformer for structured document understanding. arXiv preprint arXiv:2202.13669 (2022)
34. Weischedel, R., et al.: Ontonotes release 5.0 ldc2013t19. Linguistic Data Consortium, Philadelphia, PA 23 (2013)
35. Xu, Y., et al.: LayoutLMv2: multi-modal pre-training for visually-rich document understanding. In: Proceedings of the 59th Annual Meeting of the Association for Computational Linguistics and the 11th International Joint Conference on Natural Language Processing (Volume 1: Long Papers), pp. 2579–2591. Association for Computational Linguistics, August 2021. https://doi.org/10.18653/v1/2021.acl-long.201, https://aclanthology.org/2021.acl-long.201
36. Xu, Y., Li, M., Cui, L., Huang, S., Wei, F., Zhou, M.: Layoutlm: pre-training of text and layout for document image understanding. In: Proceedings of the 26th ACM SIGKDD International Conference on Knowledge Discovery & Data Mining, KDD 2020, pp. 1192–1200. Association for Computing Machinery, New York, NY, USA (2020). https://doi.org/10.1145/3394486.3403172

37. Xu, Y., et al.: Layoutxlm: multimodal pre-training for multilingual visually-rich document understanding. arXiv preprint arXiv:2104.08836 (2021)
38. Yu, W., Lu, N., Qi, X., Gong, P., Xiao, R.: Pick: processing key information extraction from documents using improved graph learning-convolutional networks (2020). https://arxiv.org/abs/2004.07464

Posters: Data and Synthesis

On Web-based Visual Corpus Construction for Visual Document Understanding

Donghyun Kim[1], Teakgyu Hong[2], Moonbin Yim[1], Yoonsik Kim[1], and Geewook Kim[1(✉)]

[1] NAVER CLOVA, Seongnam, South Korea
gwkim.rsrch@gmail.com
[2] Upstage AI, Yongin, South Korea

Abstract. In recent years, research on visual document understanding (VDU) has grown significantly, with a particular emphasis on the development of self-supervised learning methods. However, one of the significant challenges faced in this field is the limited availability of publicly accessible visual corpora or extensive collections of images with detailed text annotations, particularly for non-Latin or resource-scarce languages. To address this challenge, we propose Web-based Visual Corpus Builder (Webvicob), a dataset generator engine capable of constructing large-scale, multilingual visual corpora from raw Wikipedia HTML dumps. Our experiments demonstrate that the data generated by Webvicob can be used to train robust VDU models that perform well on various downstream tasks, such as DocVQA and post-OCR parsing. Furthermore, when using a dataset of 1 million images generated by Webvicob, we observed an improvement of over 13% on the DocVQA Task 3 compared to a dataset of 11 million images from the IIT-CDIP. The implementation of our engine is publicly available on https://github.com/clovaai/webvicob.

Keywords: Visual Document Understanding · Optical Character Recognition · Document Image Processing

1 Introduction

Language modeling has been a long-standing fundamental task in natural language processing (NLP). The trained language models (LMs) are utilized in a range of downstream NLP applications, such as information extraction (IE) [17] and question answering (QA) [9]. To build a powerful LM, recent methods utilize large-scale text corpus at the pretraining phase. The text corpus is generally constructed with a specialized engine or software. For example, WikiExtractor [1] extracts texts from Wikipedia HTML dumps and builds a clean text corpus.

Visual Document Understanding (VDU) [15,22,42] has been developed to conduct a wide range of real-world tasks on document images. For example,

---

T. Hong—This work is done at NAVER CLOVA.

© The Author(s), under exclusive license to Springer Nature Switzerland AG 2023
G. A. Fink et al. (Eds.): ICDAR 2023, LNCS 14189, pp. 297–313, 2023.
https://doi.org/10.1007/978-3-031-41682-8_19

Document Parsing [22,42] aims to extract some key texts from a document image [17,18]. Most recent VDU backbones can be regarded as an extension of LM.

Inspired by recent advances in LMs [9,40], recent VDU methods share a similar approach that (1) it first collects large-scale real document images, (2) conducts OCR on the images to extract texts, and (3) trains a BERT-like LM backbone on the extracted texts [15,16,41,42].

Although conventional VDU methods have shown promising results, several practical challenges exist, especially in the training dataset preparation phase. Most recent works [15,42] rely on a large-scale real image dataset. For example, IIT-CDIP dataset [27] consists of 11M industrial document images and is utilized in a range of VDU works. However, in most low-resourced languages, there is no public dataset like IIT-CDIP. In addition, off-the-shelf OCR engines (e.g., CLOVA OCR API,[1] MS Read API,[2] Amazon Textract[3]) are required to extract texts in the pre-processing. This often requires enormous costs.

Moreover, using OCR can have other negative consequences; the constructed data is strongly tied to the OCR engine, and OCR errors are propagated throughout the process. This problem becomes severe, especially in some non-Latin languages such as Korean and Japanese, where OCR is known to be complicated.

In this work, we propose an engine for building a web-based visual corpus. As shown in Fig. 2 and Fig. 3, the proposed Web-based Visual Corpus Builder (Webvicob) renders a web page into an image file and generates rich text annotations. With expressive Document Object Model (DOM) APIs, Webvicob produces accurate bounding boxes for all characters.

Moreover, Webvicob covers a wide range of word contexts. Wikipedia is a huge dataset consisting of over 270 languages and containing 60 million HTML documents. The results of PCA analysis on 3,626 samples are available in Sect. 5.3.

Compared to the traditional model trained on IIT-CDIP, the Webvicob-trained model shows a competitive result on DocVQA Task 1 and a higher score[4] on Task 3, showing the effectiveness of the de-biased Webvicob-based visual corpora.

The proposed method is simple yet effective. We show that the Webvicob-generated corpus is critical in building a powerful VDU backbone. Through extensive experiments and analyses, we show the effectiveness of Webvicob. The contributions of this work can be summarized as follows:

1. We propose Webvicob, which can be used for pretraining the visual document understanding models. Webvicob provides rich annotations, including character, word, line, and paragraph information.
2. Webvicob provides support for a wide range of fonts, allowing visual documents with identical content to appear differently. Additionally, we have taken

[1] https://clova.ai/OCR.
[2] https://docs.microsoft.com/en-us/azure/cognitive-services/computer-vision.
[3] https://learn.microsoft.com/ko-kr/azure/cognitive-services/computer-vision/overview-ocr.
[4] https://rrc.cvc.uab.es/?ch=17&com=evaluation&task=3.

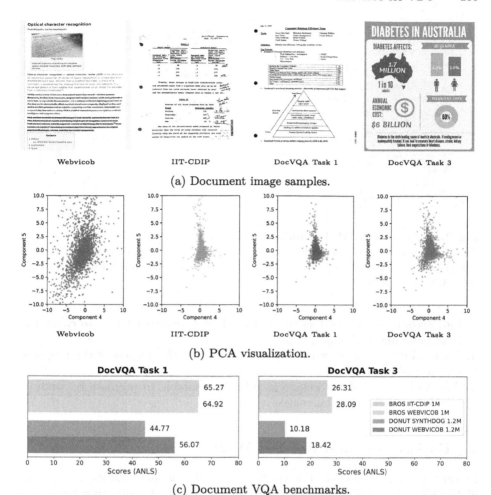

(a) Document image samples.

(b) PCA visualization.

(c) Document VQA benchmarks.

Fig. 1. Image Samples and Visualization of Main Results. (a) Samples of Webvicob, IIT-CDIP [27], DocVQA Task 1 [39], and Task 3 [32] are shown, respectively. (b) Visualization of principal component analysis (PCA) with bag of words (BoW) representations. The visualization was carried out using the 4th and 5th principal components. The analysis method and further visualization results can be found in Sect. 5.3. (c) Results on two benchmarks with pretrained BROS$_{BASE}$ [15] and Donut$_{Proto}$ [22] are shown. As can be seen in (a) and (b), DocVQA Task 1 has a similar distribution to IIT-CDIP, while Task 3 does not. This also affects the final score.

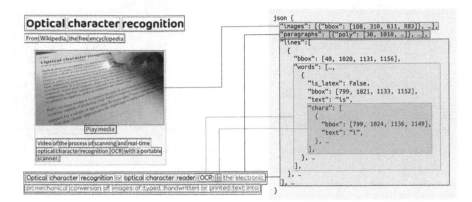

Fig. 2. A sample of Webvicob dataset. The dataset contains large-scale web document images with hierarchical text annotations (i.e., character, word, line, and paragraph-level annotations). If box contains LaTeX [34] (i.e., math formula), we set "is_latex" value as True.

into account the characteristics of each font to construct precise character-level bounding box annotations.

3. We conduct extensive experiments to verify the effectiveness of the proposed engine and dataset.
4. The source code of our engine is publicly available to promote research on VDUs in low-resource languages.

2   Background and Related Work

In this section, we introduce a traditional VDU pipeline and datasets. Most of the current methods share a similar approach of (1) collecting large-scale real images, (2) conducting OCR on the images, and (3) training a BERT-like backbone on the extracted texts [15, 42].

2.1   VDU Backbones

Inspired by BERT [9] and the recent advancements in language modeling, a range of BERT-like Transformer-based VDU backbones have been proposed [8,15,16,41,42]. For handling layout information of document images, spatial coordinates of OCR text boxes are fed to the VDU backbone [15,42,43]. Using visual encoders like ResNet [14], visual features of an input image are also being incorporated into the recent VDU backbones [16,41]. More recently, with the advances in Vision Transformer (ViT) [10], training a Transformer encoder-decoder VDU backbone without OCR has also been attempted [7,22,26]. Our engine can be used together in various VDU backbones. In this paper, we verified the performance of the Webvicob engine by pretraining BROS [15], LayoutXLM [43], and Donut [22].

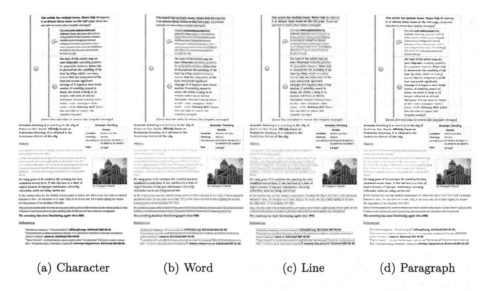

|(a) Character | (b) Word | (c) Line | (d) Paragraph |

Fig. 3. Visualization of Webvicob annotations. We use a colormap to show the order of box annotation.

2.2 Visual Corpus Construction for VDU

Most existing OCR annotated datasets have small sizes, leading to difficulties in training VDU backbones. To construct a rich corpus, in the traditional pipeline, large-scale real-world document images (e.g., IIT-CDIP) and an OCR engine (e.g., CLOVA OCR API (see footnote 1)) are used.

The quality of the OCR engine significantly affects the downstream processes [7, 22]. Hence, there have been difficulties in training and testing the VDU backbone. For example, since BROS [15] and LayoutLM [43] use different in-house OCRs, it has been difficult to make a fair comparison.

LayoutXLM [43] collects large-scale digital-born PDF data from the world wide web and extracts text annotations from the PDF via an open-source PDF renderer.[5] Although this showed another promising direction, it is not easy to follow the pipeline in practice. A practitioner has to collect the PDF data as there is no publicly available dataset ([43] did not open-source the dataset).

Moreover, since PDF files cannot easily be editable, augmentation is limited. Unlike the existing approach, Webvicob can efficiently modify and augment data (i.e., layout, background image) with javascript as Webvicob renders HTML directly. Moreover, Webvicob can easily be incorporated with the HTML dumps (e.g., Wikipedia dumps[6]), which are easily accessible and have been widely used in building powerful NLP backbones [9].

---

[5] https://github.com/pymupdf/PyMuPDF

[6] https://dumps.wikimedia.org.

3   Web-based Visual Corpus Builder

Webvicob uses HTML dumps and modifies Document Object Model (DOM) to generate data for pretraining VDU backbones with rich corpora. As seen in Fig. 2, Webvicob provides box annotations of characters, words, LaTeX [34], images, lines, and paragraphs, and also for images and LaTeXs.

In this section, we explain the generation procedures (Algorithm 1) in detail.

Algorithm 1. Get annotations from HTML
___
Input: `html`
Output: `image, annotations`

procedure `get_annotations_from_html(html)`
 1: `html ← add_spans(html)` {3.1}
    // From `<p>ab</p>` to `<p><span>a</span><span>b</span></p>`
 2: `driver ← get_selenium_driver(html)`
 3: `remove_unusable_elements(driver)` {3.1}
 4: `change_paragraph_fonts(driver)` {3.2}
 5: `image ← capture(driver)`
 6: `boxes ← get_glyph_box(driver)` {3.2}
 7: `annotations ← get_annots(boxes)` {3.1}
    return `image, annotations`

3.1   Annotation

Adding Spans. We can access each Element[7] using the DOM and find out the bounding box (bbox) of the Element through the `getBoundingClientRect()`[8] function. Webvicob modifies the HTML so that all characters can be bounded with a `<span>` tag to get the bbox of each character (i.e., from `<p>abc</p>` to `<p><span>a</span><span>b</span><span>c</span></p>`). With this procedure, the `getBoundingClientRect()` function can be applied to all spans, allowing us to obtain bounding box annotations for all characters.

Remove Unusable Elements. The essential step before creating data is to remove unusable Elements. For example, there are various ways to hide a specific Element in a web document. Even if the Element is invisible, the function `getBoundingClientRect()` returns results since Element occupies space. Specifically, the following Elements are removed:

- Pseudo Elements.[9]

---

[7] https://developer.mozilla.org/en-US/docs/Web/API/Element.
[8] https://developer.mozilla.org/en-US/docs/Web/API/Element/getBoundingClientRect.
[9] https://developer.mozilla.org/en-US/docs/Web/CSS/Pseudo-elements.

- Child Elements whose Element size is larger than the parent node.
- Elements that invisible style applied.
  (display: none / visibility: hidden / visibility: collapse / opacity: 0)
- Elements located outside the rendering screen.
- Element with placeholder attribute.

Construct Annotations. We construct word annotations and line annotations by calculating the spaces between the character boxes and the line boxes. We define LaTex Elements with "mwe-math-fallback-image-inline" className, and define image Elements with {image, canvas, SVG, video} tags. As Marku-pLM [21] did, paragraph annotations are extracted using a well-ordered tree structure. Elements with the same depth are grouped into one paragraph.

3.2 Rendering with Various Fonts

Random Paragraph Fonts. Visual diversity in pretraining datasets is generally associated with improved performance of VDU backbones [12,19,22,45].

For visual diversity, Webvicob renders HTML with various fonts for each paragraph (See Fig. 4a and 4b). We randomly select fonts from 3,567 Google-Fonts[10] in our experiments and analyses.

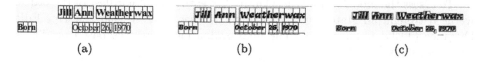

(a)   (b)   (c)

Fig. 4. Visualization of rendered images with various fonts. (a) Original font with `getBoundingClientRect()` results. (b) Random font with `getBoundingClientRect()` results. (c) Random font with actual glyph boxes.

Precise Bounding Boxes. The actual bounding box of the glyph and the result of `getBoundingClientRect()` are different. As shown in Fig. 4a and 4b, the result of `getBoundingClientRect()` has a large margin. We extract a ratio of the actual glyph box to the bounding box via rendering vector images in a font file with a Pygame FreeType handler.[11] Using the ratio, the final tight glyph box can be obtained (Fig. 4c).

4 Experiment and Analysis

4.1 Setup

Donut experiments are carried out using 8 NVIDIA V100 GPUs for a fair comparison with Donut$_{Proto}$, while other experiments are conducted using 4 NVIDIA A100 80G GPUs. We use mixed precision training technique [33].

---

[10] https://fonts.google.com.

[11] https://www.pygame.org/docs/ref/freetype.html.

304 Kim et al.

Employed Models. To show the effectiveness of Webvicob, We use three models with different properties.

BROS [15] was proposed with an effective pretraining method (i.e., area masking) and a relative positional encoding. To validate the effectiveness of Webvicob-generated data, we pretrain $BROS_{BASE}$ and measure performance on DocVQA Task 1 [39] and Task 3 [32].

LayoutXLM [43] is a widely-used multilingual VDU backbone. We reimplement and pretrain $LayoutXLM_{BASE}$ for eight languages to validate Webvicob in a multilingual scenario. FUNSD [20] and XFUND [43] datasets are used as benchmark datasets.

Donut [22] introduced a novel approach that utilizes images alone without relying on OCR results as input. We pretrain $Donut_{Proto}$ using data generated by Webvicob to demonstrate the versatility of the data for different architectures. This has been verified through experiments on DocVQA Task 1, DocVQA Task 3, FUNSD, and XFUND (Japanese) datasets.

Donut for Entity Recognition. Since Donut only takes an image input and cannot utilize the popular method of token classification using OCR information, we trained the model to decode sequences in a specific format (as depicted in Fig. 5) in order to extract all entity fields within the document.

To assess the model's ability to identify the entity set, regardless of the order, we used Hungarian Matching [25] to adjust the sequence so that the Tree Edit Distance (TED) is minimized. The final evaluation score was based on TED-based accuracy [18, 22, 47, 48].

Pretraining. In pretraining $BROS_{BASE}$, we used 6.4 million English samples generated by Webvicob and trained the model for 5 epochs. The training procedure involved randomly selecting 512 consecutive tokens from each document, similar to LayoutLMv2 [41]. The remaining training hyperparameters were kept the same as $BROS_{BASE}$.

As can be seen in Table 1 and Table 2, we pretrained $LayoutXLM_{BASE}$ using 18.6 million multilingual data generated by Webvicob, which included eight languages: English, Chinese, Japanese, Spanish, French, Portuguese, Italian, and German. The model was pretrained for 5 epochs with a batch size of 64, following the procedure described in LayoutXLM. In line with the multilingual

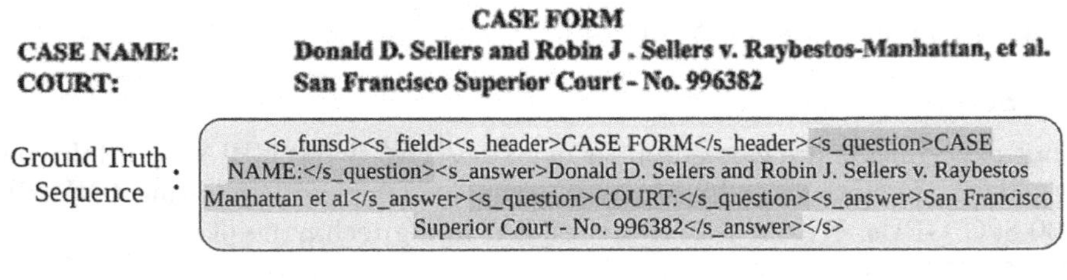

CASE FORM

CASE NAME: Donald D. Sellers and Robin J . Sellers v. Raybestos-Manhattan, et al.
COURT: San Francisco Superior Court - No. 996382

Ground Truth .
Sequence ·

<s_funsd><s_field><s_header>CASE FORM</s_header><s_question>CASE NAME:</s_question><s_answer>Donald D. Sellers and Robin J. Sellers v. Raybestos Manhattan et al</s_answer><s_question>COURT:</s_question><s_answer>San Francisco Superior Court - No. 996382</s_answer></s>

Fig. 5. A sample of ground truth for Donut Entity Recognition.

Table 1. Statistics of Webvicob and conventional datasets. ISO 639-1 language code is used for Language Support. Webvicob† is used in our multilingual experiments (Sect. 4.2). Webvicob‡ is a virtual dataset that can be constructed with Webvicob. It contains more than 60M samples with rich annotations and 270+ language supports. The total number of Wikipedia articles (60 M) can be found on the Wikipedia statistic webpage(https://en.wikipedia.org/wiki/List_of_Wikipedias).

Dataset	#Image			Annotation Level				Language Supports
	Train	Val	Test	Char	Word	Line	Para	
MSRA-TD500 [44]	300	0	200			✓		EN
IC15 [13]	1,000	0	500		✓			EN
CTW-1500 [46]	1,000	0	500			✓		EN
IC17 MLT [36]	7,200	1,800	9,000		✓			EN ZH JA KO BN AR
IC19 MLT [35]	10,000	0	10,000		✓			EN ZH JA KO BN AR HI
IC19 LSVT [38]	30,000	0	20,000			✓		EN ZH
IC19 ArT [4]	5,603	0	4,563		✓			EN ZH
Total-Text [5]	1,255	0	300		✓			EN ZH
TextOCR [37]	21,778	3,124	3,232		✓			EN
IntelOCR [24]	191,059	16,731	0		✓			EN
HierText [28]	8,281	1,724	1,634		✓	✓	✓	EN
SynthText [12]	858,750	0	0	✓	✓			EN
IIT-CDIP [27]	11,434,146	0	0					EN
OCR-IDL [2]	26,600,964	0	0		✓	✓		EN
Webvicob†	18,584,173	4000	4000	✓	✓	✓	✓	EN ZH JA ES FR PT IT DE
Webvicob‡	60,475,636			✓	✓	✓	✓	270+ languages

sampling strategy [3,6,43], each batch was sampled based on the language's frequency of occurrence in the data, with the probability $p_l \propto (\frac{N_l}{N})^\alpha$, where N is the total number of data and N_l is the number of data for language l. The parameter α was set to 0.7 to account for imbalanced data distribution.

The open-sourced Donut$_{\text{Proto}}$ was pretrained with 1.2 million samples generated by SynthDoG, including 400,000 samples each for Korean, Japanese, and English. The pretraining was performed for a duration of 5 d using 8 NVIDIA V100 GPUs, totaling 40 GPU days, with a batch size of 8. A similar approach was taken, where 400,000 samples each for Korean, Japanese, and English were generated by Webvicob and Donut$_{\text{Proto}}$ was trained for 3 d on 8 NVIDIA V100 GPUs (24 GPU days) with a batch size of 16. The images are cropped and then resized to a size of 2,048 in width and 1,536 in height for training purposes.

Finetuning. We finetune BROS with DocVQA datasets for 16K iterations. 64 batch size, Adam [23] optimizer, 5e-5 learning rate, and cosine annealing learning rate scheduler [29] are adapted.

Since the exact finetuning schedule has not been disclosed, we set up a "rough" finetuning schedule and compared results. For XFUND Semantic Entity Recognition (SER), we finetune LayoutXLM with 10K iterations, 64 batchsize, AdamW [30] optimizer, 1e-4 learning rate, linear decay learning rate, and

Table 2. Number of samples per language in our multilingual data. ISO 639-1 language code is denoted as "Code".

Language (Code)	#Train	#Val	#Test
English (EN)	6,403,095	500	500
Chinese (ZH)	1,248,720	500	500
Japanese (JA)	1,293,628	500	500
Spanish (ES)	1,712,046	500	500
French (FR)	2,409,552	500	500
Portuguese (PT)	1,087,824	500	500
Italian (IT)	1,747,412	500	500
German (DE)	2,681,896	500	500
Total	18,584,173	4,000	4,000

warmup [11] learning rate for 10% of total iterations. We use unilm library[12] for LayoutXLM finetuning experiments.

Donut$_{Proto}$ was finetuned for 300 epochs for the DocVQA tasks with an image size of 2,560 width and 2,048 height. For FUNSD and XFUND tasks, we finetuned the model for 2,000 epochs using images with a width of 2,048 and a height of 1,536. For all downstream tasks, 8 batch size, Adam [23] optimizer, 1.5e-5 learning rate, and cosine annealing learning rate scheduler are adapted.

4.2   Experimental Results

BROS. The corpus of DocVQA Task 1 and IIT-CDIP are very similar, while Webvicob-generated data has more various corpus (will be shown in Sect. 5.3).

The results also reflect this pattern. The results of Table 3 are categorized based on the quantity of data used in pretraining. Because of the significant amount of data specific to a particular domain within IIT-CDIP, it is possible to build a domain-specialized model. On the other hand, the extensive diversity of domains represented in Webvicob-generated data resulted in the development of a model that excels in the DocVQA Task 3.

The use of IIT-CDIP data for pretraining resulted in favorable performance on DocVQA Task 1. Conversely, when Webvicob data was used, DocVQA Task 3 performed much better. Surprisingly, the performance of Task 3 using Webvicob 1M has better performance than BROS using IIT-CDIP 11M. When a combination of both IIT-CDIP and Webvicob-generated data were used in equal portions for pretraining, there was a moderate enhancement in performance for both tasks.

---

[12] https://github.com/microsoft/unilm

Table 3. Quantitative comparison in DocVQA tasks with BROS$_{BASE}$ according to various settings of pretraining corpus. For an apples-to-apples comparison, we control the amount of pretraining data in experiments. The score is measured with Average Normalized Levenshtein Similarity (ANLS).

Model	IIT-CDIP	Webvicob	Task 1	Task 3
BROS	–	–	62.98	25.22
BROS	500K	–	65.01	25.12
BROS	–	500K	64.62	26.20
BROS	250K	250K	65.56	26.56
BROS	1M	–	65.27	26.31
BROS	–	1M	64.92	28.09
BROS	500K	500K	65.67	26.51
BROS	11M	–	68.07	24.76
BROS	–	6.4M	65.63	27.85

Table 4. Multitask finetuning Accuracy (F1) for Semantic Entity Recognition (SER) with FUNSD and XFUND datasets. We perform multitask finetuning LayoutXLM$_{BASE}$ under same setting with eight languages. Despite using only approximately 60% of the training data and iterations, the ultimate mean performance being competitive.

Model	FUNSD	ZH	JA	ES	FR	IT	DE	PT	Avg.
LayoutXLM (PDF 22M + IIT 8M)	0.785	0.897	0.787	0.744	0.791	0.811	0.832	0.824	0.785
LayoutXLM (Webvicob 18.6M)	0.727	0.893	0.794	0.686	0.749	0.765	0.760	0.764	0.767

LayoutXLM. We report the f1 scores of Semantic Entity Recognition (SER) for FUNSD and XFUND datasets. Table 4 shows f1 scores of multitask finetuning with eight languages. Under the same finetuning setup, Webvicob data shows comparable performance. Please note that the total number of pretraining iterations of LayoutXLM (Webvicob) is ~1.45M, much smaller than the iteration of LayoutXLM (PDF 22M + IIT 8M), which is ~2.34M.

Donut. We report finetuning results of Donut$_{Proto}$ on Table 5. Donut$_{Proto}$ trained using Webvicob data has achieved superior performance on all four tasks, despite 60% GPU training days compared to the Donut$_{Proto}$ trained with SynthDoG-generated data. While SynthDoG also made use of the Wikipedia corpus, the outstanding performance of Webvicob-generated data is noteworthy. We speculate that there are two reasons for this. First, the presence of text and images that are contextually relevant in Webvicob helps to accurately reflect the relationship between images and text. Second, the use of various types of real documents, such as tables, in the learning process can implicitly improve the learning of semantic information, such as key-value pairing.

Table 5. Finetuning results of Donut$_{\text{Proto}}$**.** The XFUND (JA) dataset is denoted as "JA", while DocVQA Task 1 and DocVQA Task 3 are denoted as "Task 1" and "Task 3" respectively. The evaluation of the Entity Recognition task in FUNSD and XFUND was conducted using TED-based accuracy. ANLS scores are used for the DocVQA tasks. Despite utilizing only 60% of GPU days, the Donut$_{\text{Proto}}$ model trained on Webvicob-generated data exhibited better performance than the Donut$_{\text{Proto}}$ model trained on SynthDoG-generated data.

Model	FUNSD	JA	Task 1	Task 3
Donut (Not Pretrained)	0.1390	0.1942	0.1391	0.0877
Donut (SynthDoG 1.2M)	0.6059	0.7817	0.4477	0.1018
Donut (Webvicob 1.2M)	0.7397	0.8455	0.5607	0.1842
Donut (SynthDoG 0.6M + Webvicob 0.6M)	0.7576	0.8488	0.5622	0.2039
Donut (SynthDoG 1.2M + Webvicob 1.2M)	0.7738	0.8611	0.5636	0.2165

5   Discussion

5.1   Scale-up Using CommonCrawl

Our initial goal was to handle tons of HTML data, such as CommonCrawl dataset[13] which consists of petabytes of data.

However, the massive variety in the HTML format made it difficult to design an integrated solution. Currently, Webvicob is only specialized in a Wikipedia format. Expanding the scope could be future work.

5.2   Webvicob with Augraphy

We expect Webvicob to easily be integrated with any augmentation project, such as Augraphy [31], to boost the performance further. As can be seen in Fig. 6, combining Webvicob and Augraphy can generate document images with rich visual effects.

5.3   PCA Analysis

We sampled 3,626 data points from each of the four datasets (Webvicob, IIT-CDIP, DocVQA Task 1, and DocVQA Task 3), creating a total of 14,504 vectors for PCA analysis. We constructed a vocabulary of 100,000 words using Wikipedia 1M data and represented the data using Bag of Words (BOW) representation, excluding stop words[14]. The Figs. 7, 8 in the ith row display the analysis between the ith and $(i+1)^{th}$ principal components. For example, the first row shows the plot of the 0th and 1st principal components.

---

[13] https://commoncrawl.github.io/cc-crawl-statistics/plots/crawlsize.
[14] https://www.nltk.org/index.html.

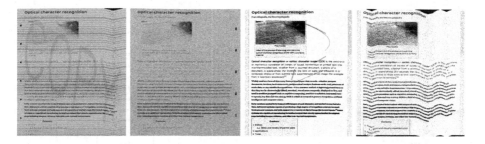

Fig. 6. Webvicob with Augraphy. Augraphy has the capability to produce a pencil sketch or folding appearance.

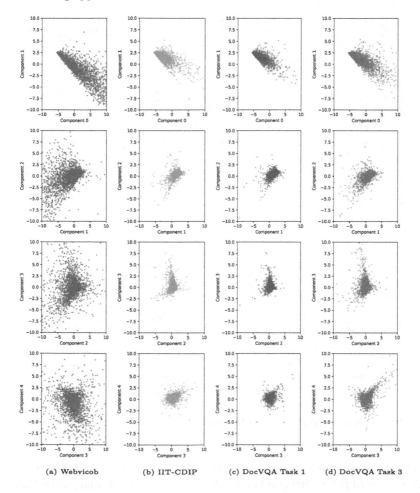

(a) Webvicob   (b) IIT-CDIP   (c) DocVQA Task 1   (d) DocVQA Task 3

Fig. 7. Visualization of PCA Results. Components 0, 1, 2, 3, 4

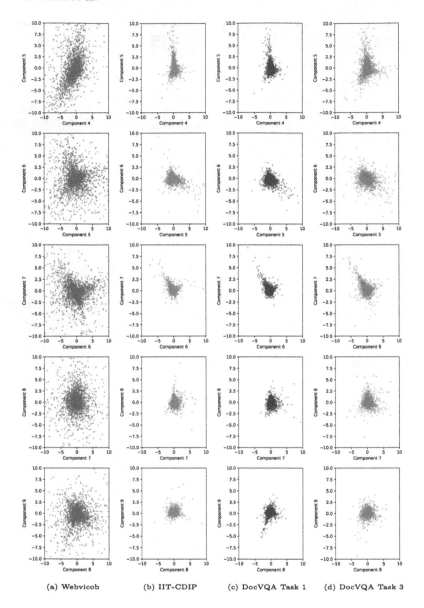

(a) Webvicob   (b) IIT-CDIP   (c) DocVQA Task 1   (d) DocVQA Task 3

Fig. 8. Visualization of PCA Results. Components 4, 5, 6, 7, 8, 9

We visualized the data using 10 principal components and found that the BOW vectors of IIT-CDIP and DocVQA Task 1 are very similar, while Webvicob contains a diverse range of BOW vectors.

Further visualization with Google vocabulary[15] can be found in https://github.com/clovaai/webvicob/blob/main/resources/PCA_GoogleVocab.png.

6 Conclusion

In this work, we propose an engine, Webvicob, that builds visual corpora from web resources. The constructed visual corpora can be utilized in building VDU backbones. In our experiments and analyses, we observe that the Webvicob-generated data helps the VDU backbone perform robustly across a variety of document formats and domains.

Acknowledgements. The authors thank NAVER CLOVA Text Vision Team and Information Extraction Team.

References

1. Attardi, G.: WikiExtractor. https://github.com/attardi/wikiextractor (2015)
2. Biten, A.F., Tito, R., Gomez, L., Valveny, E., Karatzas, D.: Ocr-idl: ocr annotations for industry document library dataset. In: European Conference on Computer Vision Workshop (ECCV Workshop) (2022)
3. Chi, Z., et al.: InfoXLM: an information-theoretic framework for cross-lingual language model pre-training. In: Annual Conference of the North American Chapter of the Association for Computational Linguistics (NAACL) (2021)
4. Chng, C.K., et al.: ICDAR2019 robust reading challenge on arbitrary-shaped text - RRC-art. In: International Conference on Document Analysis and Recognition (ICDAR) (2019)
5. Ch'ng, C.K., Chan, C.S., Liu, C.: Total-text: towards orientation robustness in scene text detection. International Journal on Document Analysis and Recognition (IJDAR) (2020)
6. Conneau, A., Lample, G.: Crosslingual language model pretraining. In: In Advances in Neural Information Processing Systems (NeurIPS) (2019)
7. Davis, B., Morse, B., Price, B., Tensmeyer, C., Wigington, C., Morariu, V.: End-to-end document recognition and understanding with dessurt. In: European Conference on Computer Vision (ECCV) (2022)
8. Deng, X., Shiralkar, P., Lockard, C., Huang, B., Sun, H.: Dom-lm: learning generalizable representations for html documents (2022). https://arxiv.org/abs/2201.10608
9. Devlin, J., Chang, M., Lee, K., Toutanova, K.: BERT: pre-training of deep bidirectional transformers for language understanding. In: Annual Conference of the North American Chapter of the Association for Computational Linguistics (NAACL) (2019)
10. Dosovitskiy, A., et al.: An image is worth 16x16 words: transformers for image recognition at scale. In: International Conference on Learning Representations (ICLR) (2021)

---

[15] https://github.com/first20hours/google-10000-english/blob/master/google-10000-english-no-swears.txt.

11. Goyal, P., et al.: Accurate, large minibatch SGD: training imagenet in 1 hour (2017). https://arxiv.org/abs/1706.02677
12. Gupta, A., Vedaldi, A., Zisserman, A.: Synthetic data for text localisation in natural images. In: IEEE Conference on Computer Vision and Pattern Recognition (CVPR) (2016)
13. Harley, A.W., Ufkes, A., Derpanis, K.G.: Evaluation of deep convolutional nets for document image classification and retrieval. In: International Conference on Document Analysis and Recognition (ICDAR) (2015)
14. He, K., Zhang, X., Ren, S., Sun, J.: Deep residual learning for image recognition. In: IEEE Conference on Computer Vision and Pattern Recognition (CVPR) (2016)
15. Hong, T., Kim, D., Ji, M., Hwang, W., Nam, D., Park, S.: BROS: a pre-trained language model for understanding texts in document. In: AAAI Conference on Artificial Intelligence (AAAI) (2022)
16. Huang, Y., Lv, T., Cui, L., Lu, Y., Wei, F.: LayoutLMv3: pre-training for document AI with unified text and image masking. In: ACM International Conference on Multimedia (ACM MM) (2022)
17. Hwang, W., et al.: Post-ocr parsing: building simple and robust parser via bio tagging. In: Workshop on Document Intelligence at NeurIPS (NeurIPS Workshop) (2019)
18. Hwang, W., Lee, H., Yim, J., Kim, G., Seo, M.: Cost-effective end-to-end information extraction for semi-structured document images. In: Empirical Methods in Natural Language Processing (EMNLP) (2021)
19. Jaderberg, M., Simonyan, K., Vedaldi, A., Zisserman, A.: Synthetic data and artificial neural networks for natural scene text recognition. In: International Conference on Neural Information Processing Systems Workshop (NIPS Workshop) (2014)
20. Jaume, G., Ekenel, H.K., Thiran, J.P.: Funsd: a dataset for form understanding in noisy scanned documents. In: ICDAR Workshop on Open Services and Tools for Document Analysis (ICDAR-OST) (2019)
21. Junlong, L., Xu, Y., Cui, L., Wei, F.: Markuplm: pre-training of text and markup language for visually rich document understanding. In: Annual Meeting of the Association for Computational Linguistics (ACL) (2022)
22. Kim, G., et al.: Donut: Document understanding transformer without OCR. In: European Conference on Computer Vision (ECCV) (2022)
23. Kingma, D.P., Ba, J.: Adam: A method for stochastic optimization. In: International Conference on Learning Representations (ICLR) (2015)
24. Krylov, I., Nosov, S., Sovrasov, V.: Open images v5 text annotation and yet another mask text spotter. In: Asian Conference on Machine Learning (ACML) (2021)
25. Kuhn, H.W.: The hungarian method for the assignment problem. Naval Research Logistics Quarterly (NRLQ) (1955)
26. Lee, K., et al.: Pix2struct: Screenshot parsing as pretraining for visual language understanding (2022). https://arxiv.org/abs/2210.03347
27. Lewis, D., Agam, G., Argamon, S., Frieder, O., Grossman, D., Heard, J.: Building a test collection for complex document information processing. In: International ACM SIGIR Conference on Research and Development in Information Retrieval (SIGIR) (2006)
28. Long, S., Qin, S., Panteleev, D., Bissacco, A., Fujii, Y., Raptis, M.: Towards end-to-end unified scene text detection and layout analysis. In: IEEE Conference on Computer Vision and Pattern Recognition (CVPR) (2022)
29. Loshchilov, I., Hutter, F.: SGDR: Stochastic gradient descent with warm restarts. In: International Conference on Learning Representations (ICLR) (2017)

30. Loshchilov, I., Hutter, F.: Decoupled weight decay regularization. In: International Conference on Learning Representations (ICLR) (2019)
31. Maini, S., Groleau, A., Chee, K.W., Larson, S., Boarman, J.: Augraphy: a data augmentation library for document images (2022). https://arxiv.org/abs/2208.14558
32. Mathew, M., Bagal, V., Tito, R., Karatzas, D., Valveny, E., Jawahar, C.: InfographicVQA. In: IEEE Winter Conference on Applications of Computer Vision (WACV) (2022)
33. Micikevicius, P., et al.: Mixed precision training. In: International Conference on Learning Representations (ICLR) (2018)
34. Mittelbach, F., Schöpf, R.: With latex into the nineties. In: TUGboat (1989)
35. Nayef, N., et al.: ICDAR2019 robust reading challenge on multi-lingual scene text detection and recognition - RRC-MLT-2019. In: International Conference on Document Analysis and Recognition (ICDAR) (2019)
36. Nayef, N., et al.: ICDAR2017 robust reading challenge on multi-lingual scene text detection and script identification - RRC-MLT. In: International Conference on Document Analysis and Recognition (ICDAR) (2017)
37. Singh, A., Pang, G., Toh, M., Huang, J., Galuba, W., Hassner, T.: TextOCR: towards large-scale end-to-end reasoning for arbitrary-shaped scene text. In: IEEE Conference on Computer Vision and Pattern Recognition (CVPR) (2021)
38. Sun, Y., et al.: ICDAR 2019 competition on large-scale street view text with partial labeling - RRC-LSVT. In: International Conference on Document Analysis and Recognition (ICDAR) (2019)
39. Tito, R., Mathew, M., Jawahar, C.V., Valveny, E., Karatzas, D.: ICDAR 2021 competition on document visual question answering. In: Lladós, J., Lopresti, D., Uchida, S. (eds.) ICDAR 2021. LNCS, vol. 12824, pp. 635–649. Springer, Cham (2021). https://doi.org/10.1007/978-3-030-86337-1_42
40. Vaswani, A., et al.: Attention is all you need. In: International Conference on Neural Information Processing Systems (NIPS) (2017)
41. Xu, Y., et al.: LayoutLMv2: multi-modal pre-training for visually-rich document understanding. In: Annual Meeting of the Association for Computational Linguistics (ACL) (2021)
42. Xu, Y., Li, M., Cui, L., Huang, S., Wei, F., Zhou, M.: LayoutLM: pre-training of text and layout for document image understanding. In: Knowledge Discovery and Data Mining (KDD) (2019)
43. Xu, Y., et al.: LayoutXLM: multimodal pre-training for multilingual visually-rich document understanding (2021). https://arxiv.org/abs/2104.08836
44. Yao, C., Bai, X., Liu, W., Ma, Y., Tu, Z.: Detecting texts of arbitrary orientations in natural images. In: IEEE Conference on Computer Vision and Pattern Recognition (CVPR) (2012)
45. Yim, M., Kim, Y., Cho, H.C., Park, S.: SynthTIGER: synthetic text image generator towards better text recognition models. In: International Conference on Document Analysis and Recognition (ICDAR) (2021)
46. Yuliang, L., Lianwen, J., Zhang, S., Sheng, Z.: Detecting curve text in the wild: new dataset and new solution (2017). https://arxiv.org/abs/1712.02170
47. Zhang, K., Shasha, D.: Simple fast algorithms for the editing distance between trees and related problems. SIAM J. Comput. (SICOMP) **18**(6), 1245–1262 (1989)
48. Zhong, X., ShafieiBavani, E., Jimeno Yepes, A.: Image-based table recognition: data, model, and evaluation. In: Vedaldi, A., Bischof, H., Brox, T., Frahm, J.M. (eds.) European Conference on Computer Vision (ECCV) (2020)

Ambigram Generation by a Diffusion Model

Takahiro Shirakawa$^{(\boxtimes)}$ and Seiichi Uchida$^{(\boxtimes)}$ (iD)

Kyushu University, Fukuoka, Japan
takahiro.shirakawa@human.ait.kyushu-u.ac.jp, uchida@ait.kyushu-u.ac.jp

Abstract. *Ambigrams* are graphical letter designs that can be read not only from the original direction but also from a rotated direction (especially with 180°). Designing ambigrams is difficult even for human experts because keeping their dual readability from both directions is often difficult. This paper proposes an ambigram generation model. As its generation module, we use a diffusion model, which has recently been used to generate high-quality photographic images. By specifying a pair of letter classes, such as 'A' and 'B,' the proposed model generates various ambigram images which can be read as 'A' from the original direction and 'B' from a direction rotated 180°. Quantitative and qualitative analyses of experimental results show that the proposed model can generate high-quality and diverse ambigrams. In addition, we define *ambigramability*, an objective measure of how easy it is to generate ambigrams for each letter pair. For example, the pair of 'A' and 'V' shows a high ambigramability (that is, it is easy to generate their ambigrams), and the pair of 'D' and 'K' shows a lower ambigramability. The ambigramability gives various hints of the ambigram generation not only for computers but also for human experts. The code can be found at https://github.com/univ-esuty/ambifusion.

Keywords: Ambigram generation · Diffusion model · Ambigramability

1 Introduction

Ambigrams are graphical letter designs that can be read not only from the original direction but also from a rotated direction (especially with 180°). Figure 1 shows several examples of ambigram letters designed by human experts. The leftmost example in Fig 1 is an ambigram for 'A' and 'Y.' Hereafter, we denote such an ambigram as A↕Y. The middle and rightmost samples are B↕E and C↕D, respectively.

Designing ambigrams is difficult even for human experts due to the following two reasons. First, we must deform letter shapes to realize the dual readability from two directions. Moreover, the deformations for ambigrams are very different from those for general font styles. This means there is no standard way to design

G. A. Fink et al. (Eds.): ICDAR 2023, LNCS 14189, pp. 314–330, 2023.
https://doi.org/10.1007/978-3-031-41682-8_20

ambigrams. Second, creating their ambigrams is inherently difficult for many letter class pairs. For example, it will be almost impossible to design an ambigram D↕K, although it is easy for A↕V.

Designing ambigrams is also difficult for computers. One might think that generative adversarial networks (GANs) can generate new ambigrams by using ambigrams by human experts as "real" image samples. However, there is no large "real" ambigram dataset with sufficient style variations. Therefore, conventional font generation methodologies by GANs [1,5,27] cannot be used for ambigrams. Furthermore, computers will suffer from the above two difficulties, like human experts.

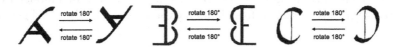

Fig. 1. Examples of ambigram letters designed by human experts. According to our notation, they are A↕Y, B↕E, and C↕D, respectively. These ambigrams are provided by the website *MakeAmbigrams* (https://makeambigrams.com/).

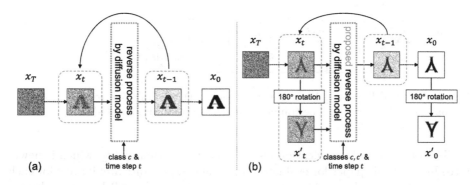

Fig. 2. Overview of (a) the standard diffusion model for image generation and (b) the proposed model for ambigram generation.

This paper proposes a diffusion model for this challenging task of generating ambigrams. Figure 2 shows an overview of (a) the standard diffusion model for image generation and (b) the proposed model for ambigram generation. The standard diffusion model generates a letter image x_0 of the class c by an iterative process called *reverse process*. In contrast, the proposed model generates a letter image x_0 for class c and its 180°-rotated version x_0' for class c'. For example, suppose two classes $c =$'A' and $c' =$'Y' are specified as conditions for the original and rotated directions, respectively. Then, the diffusion model is expected to produce an image x_0 of 'A' such that the 180°-rotated image x_0' is 'Y.'

An important property of the proposed model is that it does not need to have any real ambigram samples made by human experts. Instead, it needs just standard letter images of all classes. After learning from those images what shapes are readable as each letter class, the proposed model generates ambigrams with dual readability for c and c'. This property effectively generates a variety of unique and unexpected ambigrams, even in the current situation where reference ambigrams by human experts are not readily available enough.

In the experiments, we first observe that the proposed model can generate high-quality and various ambigrams. In addition to this qualitative observation, we conduct a quantitative evaluation by using the recognition accuracy of a letter classifier. If we achieve high accuracy, the generated ambigrams will have sufficient readability. The ambigrams obtained from the proposed model are then compared with those obtained from human experts and a GAN-based generator.

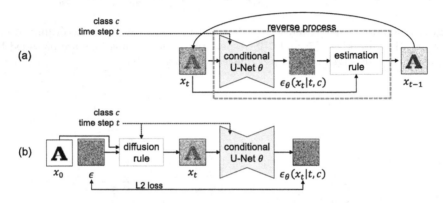

Fig. 3. Diffusion Denoising Probabilistic Model (DDPM). (a) The reverse process to estimate x_{t-1} from x_t. (b) The training process of the conditional U-Net.

The above letter classification results reveal *ambigramability*, which is newly introduced in this paper for evaluating how easy to generate ambigrams for each class pair. As noted above, generating ambigrams of A↕V will be easy, whereas D↕K is difficult. Therefore, the former has a higher ambigramability and the latter has a lower ambigramability. It is useful to know the ambigramability when we design a word or a phrase by ambigrams. When we find a word ambigram that needs a letter pair with low ambigramability, it will be better to think of another word or phrase (or use the capital letter instead of the lower case or vice versa). The ambigramability can be automatically evaluated by recognizing generated ambigrams by a letter classifier.

The main contributions of this paper are summarized as follows.

- To the authors' best knowledge, it is the first attempt to generate ambigrams automatically.
- We propose a new diffusion model for generating ambigrams in various styles without any real ambigram examples.

- We also propose "ambigramability" to measure the easiness of ambigram generation of each class pair objectively and automatically.
- Comparative studies reveal that ambigrams by the proposed model have more diversity while keeping their readability.

2 Related Work

2.1 Diffusion Denoising Probabilistic Model (DDPM)

Diffusion Denoising Probabilistic Model (DDPM) [7] is one of the most popular diffusion models [29]. DDPM assumes that a noise image x_T is derived by adding a small amount of Gaussian noise to the original (clean) image x_0 T times. (This iterative noise addition process is called *diffusion process*.) The highlight of DDPM is its *reverse process* to recover x_0 from x_T by iterating noise removal T times. Figure 3 (a) shows the reverse process. A conditional U-Net (with the weight parameter set denoted by θ) estimates the noise $\epsilon_\theta(x_t \mid t, c)$ of x_t under the condition that x_t is an image from class c after t times of noise addition. Then, x_{t-1} is estimated from x_t and $\epsilon_\theta(x_t \mid t, c)$ by an estimation rule[1]. By iterating this reverse process from $t = T$ to 1, we have x_0 as a generated image. Note that the generated image x_0 varies even from the same x_T because of the stochastic behavior of the estimation rule.

Training DDPM is equivalent to training the conditional U-Net θ. As shown in Fig. 3 (b), the U-Net is trained to output $\epsilon_\theta(x_t \mid t, c) \sim \epsilon$, where $\epsilon \sim \mathcal{N}(0, \mathbf{I})$ is the noise to prepare x_t from x_0 according to the diffusion rule defined in [7]. The conditions t and c are fed to the U-Net using an embedding technique like position embedding [26].

2.2 Letter Image Generation

For generating letter images (or font images), various image generation models, especially GANs [2,9,11,14,19], have been used. The GAN-based models can also generate letter images with conditions, such as class labels [5,10,12,16] or texts [18,28]. Recently, diffusion models have achieved high-quality photographic image generation [20,21,23]. All the above methods cannot be used for the ambigram generation directly; this is simply because i) they need real ambigrams as their training samples and ii) they have no mechanism to realize the dual readability as ambigrams. Consequently, to the authors' best knowledge, there has been no past attempt to generate ambigrams by computers. Note again that our model does not need any reference ambigram examples.

---

[1] Precisely speaking, the following equation [7] is used as the estimation rule in Fig. 3 (a):

$$x_{t-1} = \frac{1}{\sqrt{\alpha_t}} \left(x_t - \frac{1 - \alpha_t}{\sqrt{1 - \bar{\alpha}_t}} \epsilon_\theta(x_t \mid t, c) \right) + \sigma_t z,$$

where $\bar{\alpha}_t = \alpha_1 \cdots \alpha_t$, $\alpha_t = 1 - \beta_t$, and β_t is a constant defined by $(\beta_T - \beta_1)(t - 1)/(T-1) + \beta_1$ with hyperparameters β_1 and β_T. σ_t is also a time-dependent constant determined by β_t and z is a random vector from $\mathcal{N}(0, \mathbf{I})$. For more details, please refer to the original DDPM paper [7]..

3   Ambigram Generation by a Diffusion Model

3.1   Outline

Figure 4 shows the proposed reverse process to generate ambigrams. Inputs to the process are a noisy image x_t and its 180°-rotated version x_t'. If we aim to generate ambigrams of $c\updownarrow c'$, two class labels c and c' are given as conditions. Then, the reverse process will generate x_{t-1} with less noise than x_t. We expect that x_{t-1} becomes more readable as class c, whereas its 180°-rotated version (i.e., x_{t-1}') does c'. Preparing a noise image x_T and then repeating this reverse process until $t = 0$, we will have noise-free x_0 and x_0' (as shown in the rightmost part of Fig. 2(b)) that form an ambigram $c\updownarrow c'$.

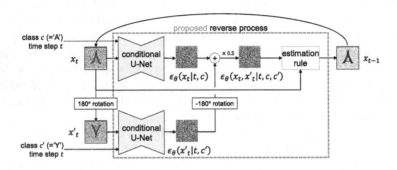

Fig. 4. The proposed reverse process for the ambigram generation.

The main difference between the proposed model of Fig. 4 and the standard DDPM of Fig. 3 (a) is that the former estimates two noise images $\epsilon_\theta(x_t \mid t, c)$ and $\epsilon_\theta(x_t' \mid t, c')$ by applying the same U-Net θ to x_t and x_t'. Then, these noise images are simply averaged to have a single noise image $\epsilon_\theta(x_t, x_t' \mid t, c, c')$:

$$\epsilon_\theta(x_t, x_t' \mid t, c, c') = \left(\epsilon_\theta(x_t \mid t, c) + \operatorname*{Rot}_{-180°} \left[\epsilon_\theta(x_t' \mid t, c') \right] \right) / 2. \tag{1}$$

Note that $\epsilon_\theta(x_t' \mid t, c')$ is rotated with $-180°$ (by the operator Rot[ ]) before taking average, for having the original direction as $\epsilon_\theta(x_t \mid t, c)$. The averaged noise image is useful to generate x_{t-1} with an appearance of c and (rotated) c', that is, useful to generate an ambigram $c\updownarrow c'$.

Training the proposed model of Fig. 4 is rather straightforward. We use the training process of the standard DDPM, shown in Fig. 3(b), for the proposed model. Only the difference from Fig. 3(b) is that the proposed model uses the averaged noise image $\epsilon_\theta(x_t, x_t' \mid t, c, c')$ instead of $\epsilon_\theta(x_t \mid t, c)$. The averaged noise image guides x_{t-1} to the intermediate direction between c and c'. Before this training, we also perform a pretraining step, as will explain below. As already emphasized, we use ordinary font images as x_0 and do not use any ambigram examples.

3.2 Details

Pretraining for Ordinary Letter Image Generation. Before generating ambigrams by the model of Fig. 4, its conditional U-Net θ is pretrained to generate ordinary (i.e., non-ambigram) letter images. In other words, we first train the U-Net θ in the standard DDPM scheme of Fig. 3 (b). We also use an ordinary font dataset for pretraining. The pretrained U-Net can produce various ordinary fonts with the reverse process of Fig. 3 (a).

Classifier-Free Guidance. In the reverse process, we employ *classifier-free guidance* [8]. In our ambigram generation task, this technique can control the trade-off between the readability and diversity of the generated ambigrams. By setting its *guidance strength* parameter s larger, the generated ambigrams become more readable with less diversity. In the later experiment, we will observe the effect of the parameter value s.

Re-Spacing Technique. *Re-spacing* [4,17] is a technique to accelerate the reverse process by skipping some time steps t. In general, diffusion models (including DDPM) need to set T at a large value (say, 1,000) for appropriate results. This means we must repeat the reverse process 1,000 times to generate just one image. The re-spacing technique allows t to decrease with Δ interval, and therefore the total repetition time is reduced from T to T/Δ. In [4], it is reported that this technique keeps the quality of generated images.

Data Augmentation by Horizontal Shift. A simple but effective technique to generate more readable and diverse ambigrams is data augmentation by horizontal shift. The augmented image of x_0 is just a horizontally-shifted version of x_0. Roughly speaking, without this technique, ambigrams E↕P will become like 'Ħ,' that is, 'E' and 'd' will completely overlap. In contrast, with the technique, we will have an ambigram like 'Æ.' Such designs are highly readable and common in ambigram designs by human experts. In the experiment, we will observe the actual effect of the data augmentation.

4 Evaluation of Ambigrams and Ambigramability

4.1 Evaluating Ambigrams by a Classifier

For evaluating the generated ambigram x_0 of $c↕c'$, we will check the readability of x_0 as c, as well as that of x_0' as c'. For example, we can treat an ambigram x_0 of C↕D as readable when x_0 and its 180°-rotated version x_0' are readable as 'C' and 'D,' respectively. This readability is objectively and automatically evaluated by a conditional binary classifier $\mathcal{C}(x, c) \in \{\texttt{True}, \texttt{False}\}$. The classifier returns $\texttt{True}$ when the input image x is acceptable as the input class label c and returns $\texttt{False}$ otherwise. Consequently, if $\mathcal{C}(x_0, c) = \texttt{True}$ and $\mathcal{C}(x_0', c') = \texttt{True}$, it is a *successful* ambigram of $c↕c'$ with dual readability.

The classifier is an MLP whose input is a concatenation of two vectors: one from a standard CNN [24] for the image input x_0 and the other from a fixed random vector pre-assigned to the class c. The CNN for x_0 is pre-trained with the standard letter image classification task, and the final layer is discarded to use the CNN as a feature extractor.

4.2 Ambigramability

One of the highlights of this paper is to introduce *ambigramability*, which evaluates how each alphabet letter pair is easy to generate ambigrams. As noted in Sect. 1, ambigramability heavily depends on class pairs; for example, A↕V will have a higher ambigramability, and D↕K have a lower ambigramability. Determining the pair-wise ambigramability automatically and objectively will help further ambigram designs.

Ambigramability for the pair of c and c' is calculated by a sufficient number N of ambigrams $c\updownarrow c'$ and the classifier in Sect. 4.1. Specifically, ambigramability is defined as the number of successful ambigrams among N ambigrams. For simplicity, we use $N = 100$ in the later experiment. Note that the ambigramability values for $c\updownarrow c'$ and $c'\updownarrow c$ will be similar but not the same due to the stochastic behavior of the diffusion model.

5   Experimental Results

5.1   Experiment Setup

Dataset. As described in Sect. 3, it is enough to prepare a standard (i.e., non-ambigram) font image dataset to train the proposed model. In the following experiment, we use 52 letter images of Latin alphabet classes ('A,'$\cdots$,'Z,''a,'$\cdots$,'z') from MyFonts dataset [3]. The MyFonts dataset contains about 20,000 fonts, and we remove illegible, collapsed, or excessively decorated fonts before training. Consequently, 11,991 fonts were used to train the model. Each image is binary and 64×64 pixels. We apply the data augmentation to all 52 classes of alphabets (A-z, a-z) while using 40% of training data. The amount of horizontal shifts is randomly determined within 25% of the image size.

Implementation Details. The model of the conditional U-Net is similar to a model used in [4]. This U-Net is more elaborated from the original U-Net [22] and consists of a stack of residual layers with an attention mechanism. We use an adaptive group normalization [4] for embedding time step t and class conditions c, c'. The model was trained end-to-end with 500K iterations and batch size 64. As an optimizer, Adam [13,15] is used with its learning rate 10^{-4}. As the parameters specific to DDPM, we use $T = 1,000$, $\beta_1 = 10^{-4}$, and $\beta_T = 0.02$ for the pretraining step and the ambigram generation step. We also set $\Delta = 20$ for re-spacing and $s = 5.0$ for classifier-free guidance. A preliminary experiment sets these parameters.

Evaluation Metrics. Using the trained classifier $\mathcal{C}(x, c)$ and $N = 100$ generated ambigrams for each class pair, the ambigramability score ($\uparrow$) is calculated for the quality evaluation. The binary classifier $\mathcal{C}(x, c)$ is trained by the MyFonts dataset. After training, its test accuracy was 98.70%. Similar to the past attempts to generate letter images by GANs, we also use Frechet Inception Distance (FID$\downarrow$) [6] for evaluating the quality and diversity of the generated ambigrams. The FID score of the generated ambigrams is calculated by comparing their feature distribution with that of the MyFonts dataset on the Inception-V3 [25].

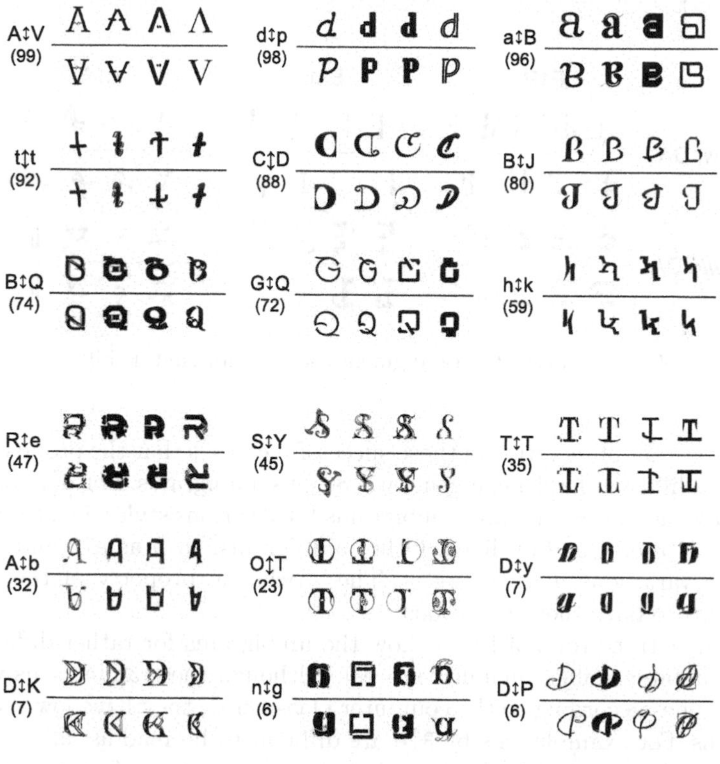

Fig. 5. Ambigrams generated by the proposed model. The parenthesized number is the ambigramability score ($\uparrow$) of the letter pair. The upper three rows are rather easy class pairs (with higher ambigramability scores), and the lower three are not. Note that all these ambigrams are successful ones.

5.2 Qualitative Evaluations

Examples of Generated Ambigrams. Figure 5 shows the ambigrams generated by the proposed model. Four ambigram examples are shown for each class pair $c \updownarrow c'$. The lower samples (x_0') are 180°-rotated versions of the upper samples (x_0). We expect that the upper samples are readable as c and the lower c'.

The parenthesized number is the ambigramability score $\in [0, N(= 100)]$ of the letter pair. Note that all ambigrams in this figure were successful ones by the classifier $\mathcal{C}$.

The upper three rows of Fig. 5 show the ambigrams for *easy* pairs with higher ambigramability scores. These ambigrams realize their dual readability from both directions. For the pair A↕V, the generated images of 'A' have a thin or no horizontal stroke; consequently, their 180°-rotated versions are readable as 'V.' Similarly, for the pair C↕D, the generated images of 'C' have a thin or short vertical stroke to make their 180°-rotated versions readable as 'D'. For t↕t, the horizontal stroke often comes in the middle of the ambigrams, even though such a style is not common in the non-ambigram fonts used for training.

Fig. 6. Effect of data augmentation by horizontal shift.

An important observation of those successful cases is that the proposed model based on a diffusion model can generate various ambigrams. For d↕p, one of the easiest class pairs, the generated ambigrams have various styles, i.e., calligraphic, pixel art, slab-serif, and outlined. Other ambigrams, such as a↕B and B↕Q, also have wide variations in their styles. The stochastic property of the diffusion model helps to have those variations.

The lower three rows of Fig. 5 show the ambigrams for rather difficult class pairs (with lower ambigramability scores). Although these ambigrams were still successful ones according to the computer classifier $\mathcal{C}$, they have low readability for humans. For example, 'e's by R↕e are difficult to be read as 'e.'

On the other hand, it is also true that the generated ambigrams give various inspirations to humans, even when they have low ambigramability scores. For example, the examples of S↕Y and D↕K show that there are still possibilities to design ambigrams for these pairs by using heavy but careful deformations. The ambigrams T↕T are also inspiring; our model discovers that having a short serifed horizontal stroke and a long straight horizontal stroke can realize the dual readability of T↕T.

Effect of Data Augmentation by Horizontal Shift. Figure 6 shows the ablation study for the data augmentation by horizontal shift. As expected, this data augmentation increases the degree of freedom in ambigram design. For example, the generated ambigrams of E↕P suggest that the data augmentation is effective in having more readable ambigrams. (Later in Fig 10, we will observe that the ambigrams of E↕P by human experts also look like 'Æ,' instead of 'Ⅎ.') Similar positive effects are found in E↕L and A↕A. Note that even with data augmentation, the model can generate ambigrams that would be generated without data augmentation. (The model can generate 'Ⅎ' if it has dual readability.) In other words, the model automatically chooses the better one. Consequently, the data augmentation helps to increase the readability and diversity of generated ambigrams.

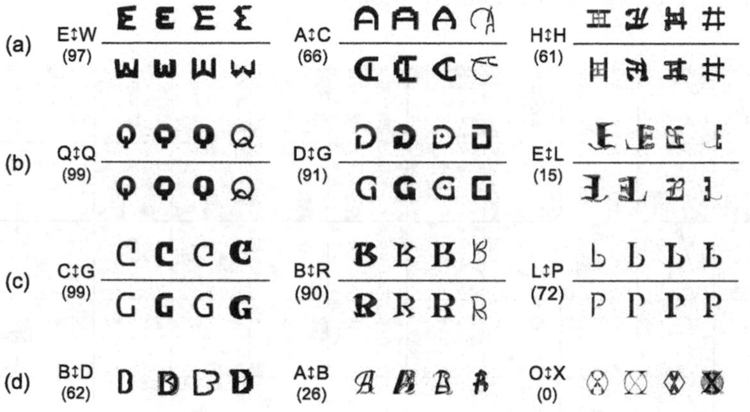

Fig. 7. Ambigrams for (a) 90°-rotation, (b) horizontal flip, (c) vertical flip, and (d) overlay.

Generating Different Types of Ambigrams. Our model can generate different types of ambigrams by replacing the ±180°-rotation module in Fig. 4 with another transformation module. Figure 7 shows the ambigram examples of four different types. Ambigrams of (a), (b), and (c) are generated by ±90°-rotation, horizontal flip, and vertical flip, respectively, instead of ±180°-rotation. Consequently, they become readable as another letter after the transformations. Samples of Fig. 7 (d) are generated without any transformation — therefore, they seem like overlaid letters (rather than ambigrams) and keep readability as two letters from their original direction.

5.3 Quantitative Evaluations

Ambigramability Scores. Figure 9 shows the ambigramability scores for all pairs of 52 Latin alphabet letters. Since $N = 100$, the score will range in $[0, 100]$. If the score at the (c, c')-th element is high, generation of ambigrams of $c \updownarrow c'$ is easy.

First, Fig. 9 proves that successful ambigrams are available for most class pairs. Almost all pairs have non-zero ambigramability, and most have more than

30. In fact, the average score is about 37. Several pairs, such as b↕q, f↕J, and H↕H, have the highest score of 100. Moreover, we often find unexpectedly-high cases. For example, although generating B↕Q and G↕Q seems difficult, they have rather high scores of 74 and 72, respectively. These positive results suggest that our model will be useful to give inspiration to human experts for creating ambigrams.

We also find that several pairs, such as n↕P, O↕X, and y↕y, have the lowest score of zero. Such negative information is also useful for designing word ambigrams. For example, the zero ambigramability of n↕P suggests that designing a word ambigram of Pen↕Pen is difficult and better to be avoided. Note that Fig. 7 (d) shows a possible remedy for O↕X; it is not an ambigram but has dual readability as 'O' and 'X' without rotation.

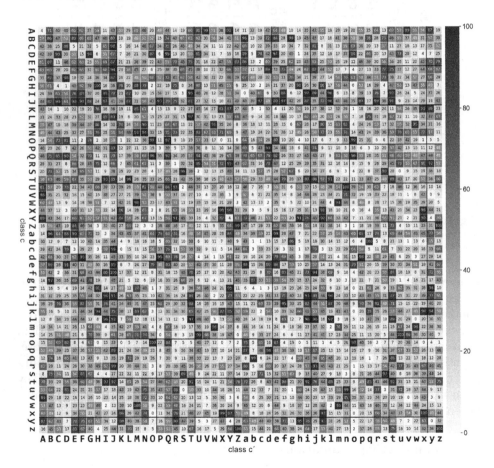

Fig. 8. The trade-off between the readability and diversity of generated ambigrams by controlling the guidance strength parameter s.

In ambigrams designed by human experts, capital and small letters are used interchangeably. For example[2] a word ambigram caNDy↕caNDy is used instead

---

[2] https://www.basilemorin.com/ambigrams.html,.

of candy↕candy. This is because ambigrams a↕D (88) and N↕N (94) are easier than a↕d (21) and n↕n (44), respectively. (The parenthesized number is the corresponding ambigramability score.) Fig 5 shows other prominent examples, such as d↕p (98) and D↕P (6). By referring to these differences, we can design various word-wise ambigrams with appropriately chosen capital and small letters.

Trade-Off Between Readability and Diversity. As described in Sect. 3.2, we introduce the classifier-free guidance to control the trade-off between read-

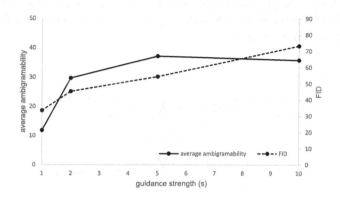

Fig. 9. Ambigramability score of each letter pair $c↕c'$. A pair with a darker cell has a higher ambigramability score; that is, generating their ambigrams is rather easy.

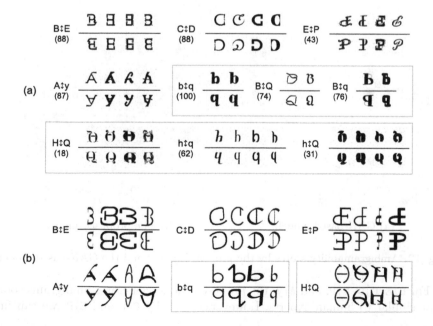

Fig. 10. Qualitative comparison of ambigrams by (a) the proposed model and (b) human experts. Since we could not find a sufficient number of ambigrams by human experts, we do not show the ambigramability scores for (b).

ability and diversity. Figure 8 shows that the guidance strength s can control the trade-off. As s increases, both the average ambigramability and FID increase; this means that the readability of the generated ambigrams increases, but their diversity decreases. We, therefore, use $s = 5$ in the above experiments as the best compromise for the trade-off.

5.4   Comparative Studies

Comparison with Ambigrams by Human Experts. Figure 10 compares ambigrams by (a) the proposed model and (b) human experts. The latter samples were collected from the website `MakeAmbigrams`[3] where eight types of man-made ambigram fonts are provided.

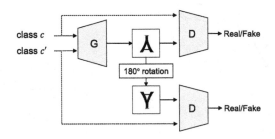

Fig. 11. A GAN-based ambigram generator.

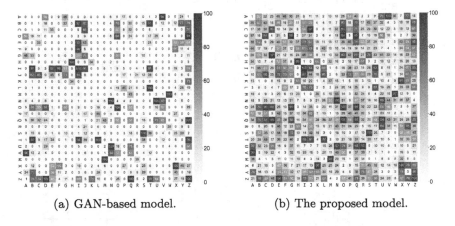

(a) GAN-based model.   (b) The proposed model.

Fig. 12. Ambigramability scores by the proposed model and the GAN-based model.

For B↕E, C↕D, E↕P, and A↕y, the proposed model generates ambigrams somewhat similar to human experts. However, especially C↕D and E↕P, we can find

---

[3] https://makeambigrams.com/,.

different types of ambigrams between our model and humans. These facts imply that human experts and our models can work complementary to design readable and diverse ambigrams.

It is interesting to note that the class pairs are often limited in the ambigrams by human experts. For example, B↕Q is not prepared, and therefore, we need to use b↕q instead. (Recall the discussion on interchangeability between capitals and small letters in Sect. 5.3.) This might be because the design of B↕Q was not straightforward for human experts. In contrast, for our model, the ambigrams of B↕Q and B↕q are rather easy because these ambigramability scores are 74 and 76, respectively. Figure 5, as well as Fig. 10, shows the examples of B↕Q generated by the proposed model. A similar observation can be done for h↕q, which is difficult for human experts (and therefore only H↕Q is designed) but not for our model.

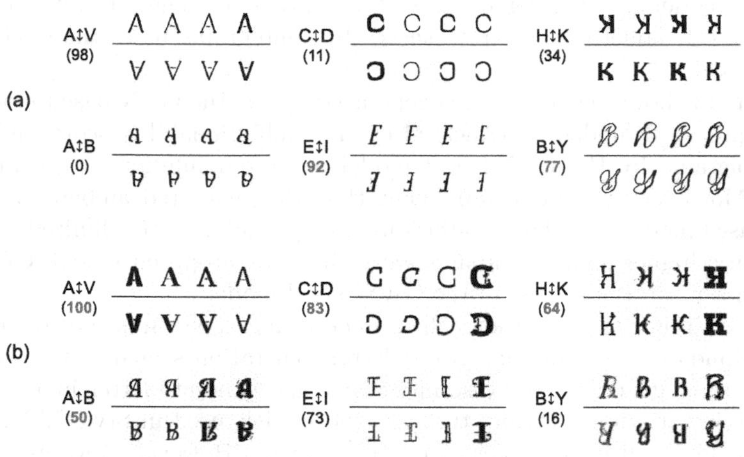

Fig. 13. Comparison of ambigrams by (a) the GAN-based model and (b) the proposed model. Ambigramability scores in red indicate whether (a) or (b) is the winner for a certain $c \uparrow c'$.

Comparison with Ambigrams by GAN. To the authors' best knowledge, there has been no past attempt to generate ambigrams by computers. Therefore, there are neither baselines nor state-of-the-art methods which are comparable with our model. Considering this fact, we prepare our own comparative model, using GAN instead of DDPM, to validate the proposed model.

Figure 11 shows the GAN-based model. This model consists of one image generator and two discriminators. The generator is trained (i.e., updated) especially when the discriminator does not determine the generated image as a real sample of class c and/or the 180°-rotated version as c'. The discriminator is trained when it cannot discriminate a generated image and its rotated image from a real image of a class c and c'. The model structures of the generator and discriminator are the same as StyleGAN2 [12]. This GAN-based model also can be trained end-to-end.

In this comparative study, we use Google Fonts[4] instead of the MyFonts dataset. This is because the GAN-based model trained by the MyFonts dataset tends to generate hardly-readable noisy images. The MyFonts dataset contains various fonts, including heavily decorative fonts. These variations are beneficial to train the model for the ambigrams generation. However, GAN (especially its generator) cannot deal with the scattered distribution due to the variations. In contrast, Google Fonts contains about 3,000 fonts, most of which have standard shapes and higher legibility. The GAN-based model could perform much better with Google Fonts datasets than the MyFont. Another treatment is to limit the number of classes to 26 (i.e., 'A'-'Z') instead of 52. This limitation also helped the GAN-based model.

Figure 12 shows the ambigramability scores evaluated by the same classifier. The GAN-based model has very low scores for many pairs. In contrast, the proposed model achieves higher scores for many pairs. These results show that the proposed model can generate successful ambigrams more than the GAN-based model.

Figure 13 shows the ambigrams generated by (a) the GAN-based model and (b) the proposed model. As indicated by the ambigramability scores of Fig. 12, the ambigrams by the GAN-based model are often inferior to the proposed model. Moreover, it is more prominent that the generated ambigrams by the GAN-based model show fewer variations. In E↕I and B↕Y, this limited variation results in a higher ambigramability score than the proposed model. Of course, it does not mean the success of the GAN-based model.

This is similar to a situation called *mode collapse*, which is specific to GAN. In the standard (i.e., non-ambigram) letter generation scenario, we confirmed that the same generator and discriminator could generate letter images in sufficient style variations. (Again, they are the models used in StyleGAN2, which is famous for good image generation performance.) However, if we use them in the framework of Fig. 11 for ambigram generation, variations in the generated images decrease drastically. This might be because it is difficult for the generator to find the narrow distributions of ambigrams, which must satisfy dual readability. In other words, DDPM could avoid the mode collapse problem and thus generate not only readable but also diverse images.

6 Conclusion

In this paper, we tackled a novel image generation task, i.e., ambigram generation, and proposed a diffusion model. The proposed model generates high-quality and diverse ambigrams. For example, if we specify two letter classes, 'B' and 'a', as conditions, the proposed model generates ambigrams that can be read as 'B' in their original direction and as 'a' from a direction rotated 180° degrees. An important property of the proposed model is that it does not require any reference ambigrams; it generates ambigrams only using the readability of standard letter images.

---

[4] https://fonts.google.com/.

Qualitative and quantitative evaluations show that the proposed model can generate ambigrams with sufficient readabilities and variations for many class pairs. At the same time, we also reveal that the ambigrams for specific class pairs, such as 'D' and 'K', are difficult. We proposed *ambigramability* as a metric to automatically and objectively evaluate how easy it is to generate readable ambigrams for individual class pairs.

A limitation of the current model is that it does not generate word-wise ambigrams. Although they can be realized by arranging the letter-wise ambigrams generated by the current model, it would be more efficient to generate word-wise ambigrams at once. Applying the proposed model to non-Latin alphabets will be another future work.

References

1. Azadi, S., Fisher, M., Kim, V.G., Wang, Z., Shechtman, E., Darrell, T.: Multi-content GAN for few-shot font style transfer. In: CVPR (2018)
2. Brock, A., Donahue, J., Simonyan, K.: Large scale GAN training for high fidelity natural image synthesis. arXiv preprint arXiv:1809.11096 (2018)
3. Chen, T., Wang, Z., Xu, N., Jin, H., Luo, J.: Large-scale tag-based font retrieval with generative feature learning. In: ICCV (2019)
4. Dhariwal, P., Nichol, A.: Diffusion models beat GANs on image synthesis. In: NeurIPS (2021)
5. Hayashi, H., Abe, K., Uchida, S.: GlyphGAN: style-consistent font generation based on generative adversarial networks. Knowl.-Based Syst. **186**, 104927 (2019)
6. Heusel, M., Ramsauer, H., Unterthiner, T., Nessler, B., Hochreiter, S.: GANs trained by a two time-scale update rule converge to a local nash equilibrium. In: NeurIPS (2017)
7. Ho, J., Jain, A., Abbeel, P.: Denoising diffusion probabilistic models. In: NeurIPS (2020)
8. Ho, J., Salimans, T.: Classifier-free diffusion guidance. arXiv preprint arXiv:2207.12598 (2022)
9. Karras, T., Aila, T., Laine, S., Lehtinen, J.: Progressive growing of GANs for improved quality, stability, and variation. In: ICLR (2018)
10. Karras, T., et al.: Alias-free generative adversarial networks. In: NeurIPS (2021)
11. Karras, T., Laine, S., Aila, T.: A style-based generator architecture for generative adversarial networks. In: CVPR (2019)
12. Karras, T., Laine, S., Aittala, M., Hellsten, J., Lehtinen, J., Aila, T.: Analyzing and improving the image quality of styleGAN. In: CVPR (2020)
13. Kingma, D.P., Ba, J.: Adam: a method for stochastic optimization. arXiv preprint arXiv:1412.6980 (2014)
14. Lee, K., Chang, H., Jiang, L., Zhang, H., Tu, Z., Liu, C.: ViTGAN: training GANs with vision transformers. arXiv preprint arXiv:2107.04589 (2021)
15. Loshchilov, I., Hutter, F.: Decoupled weight decay regularization. arXiv preprint arXiv:1711.05101 (2017)
16. Mirza, M., Osindero, S.: Conditional generative adversarial nets. arXiv preprint arXiv:1411.1784 (2014)
17. Nichol, A.Q., Dhariwal, P.: Improved denoising diffusion probabilistic models. In: ICML (2021)

18. Patashnik, O., Wu, Z., Shechtman, E., Cohen-Or, D., Lischinski, D.: StyleCLIP: text-driven manipulation of styleGAN imagery. In: ICCV (2021)
19. Radford, A., Metz, L., Chintala, S.: Unsupervised representation learning with deep convolutional generative adversarial networks. arXiv preprint arXiv:1511.06434 (2015)
20. Ramesh, A., Dhariwal, P., Nichol, A., Chu, C., Chen, M.: Hierarchical text-conditional image generation with clip latents. arXiv preprint arXiv:2204.06125 (2022)
21. Rombach, R., Blattmann, A., Lorenz, D., Esser, P., Ommer, B.: High-resolution image synthesis with latent diffusion models. In: CVPR (2022)
22. Ronneberger, O., Fischer, P., Brox, T.: U-Net: convolutional networks for biomedical image segmentation. In: MICCAI (2015)
23. Saharia, C., et al.: Photorealistic text-to-image diffusion models with deep language understanding. arXiv preprint arXiv:2205.11487 (2022)
24. Simonyan, K., Zisserman, A.: Very deep convolutional networks for large-scale image recognition. arXiv preprint arXiv:1409.1556 (2014)
25. Szegedy, C., Vanhoucke, V., Ioffe, S., Shlens, J., Wojna, Z.: Rethinking the inception architecture for computer vision. In: CVPR (2016)
26. Vaswani, A., et al.: Attention is all you need. In: NeurIPS (2017)
27. Xie, Y., Chen, X., Sun, L., Lu, Y.: DG-Font: deformable generative networks for unsupervised font generation. In: CVPR (2021)
28. Xu, T., et al.: AttnGAN: fine-grained text to image generation with attentional generative adversarial networks. In: CVPR (2018)
29. Yang, L., et al.: Diffusion models: a comprehensive survey of methods and applications. arXiv preprint arXiv:2209.00796 (2022)

Analyzing Font Style Usage and Contextual Factors in Real Images

Naoya Yasukochi[1]([✉]), Hideaki Hayashi[2], Daichi Haraguchi[1],
and Seiichi Uchida[1]

[1] Kyushu University, Fukuoka, Japan
`naoya.yasukochi@human.ait.kyushu-u.ac.jp`
[2] Osaka University, Suita, Japan

Abstract. There are various font styles in the world. Different styles give different impressions and readability. This paper analyzes the relationship between font styles and contextual factors that might affect font style selection with large-scale datasets. For example, we will analyze the relationship between font style and its surrounding object (such as "bus") by using about 800,000 words in the Open Images dataset. We also use a book cover dataset to analyze the relationship between font styles with book genres. Moreover, the meaning of the word is assumed as another contextual factor. For these numeric analyses, we utilize our own font-style feature extraction model and word2vec. As a result of co-occurrence-based relationship analysis, we found several instances of specific font styles being used for specific contextual factors.

Keywords: Font style · Contextual factors · Scene text · Typographic design

1 Introduction

We have various font styles worldwide, giving different impressions and readability. Graphic designers and typographers carefully select appropriate fonts to visualize textual information. For example, when they select a font in a context where some "stable" impression is required, they might select a sans-serif font with thick strokes and might not choose a serif font with thin strokes. In this case, some contextual factors determine the fonts selected.

The purpose of this paper is to understand the *relationship between the style of the selected font and several contextual factors*. Figure 1 shows two scenarios for this purpose. The first scenario uses scene images. In (a), the word "TOWN" is shown on a bus. In this case, two possible contextual factors that determine the font style of "TOWN" are i) the object "bus" where this word is printed and ii) the meaning of the word "town." The second scenario uses book covers. The style of the word "HAT" in (b) might be determined by considering i) the genre of this book ("Children's Book") and/or ii) the meaning of the word

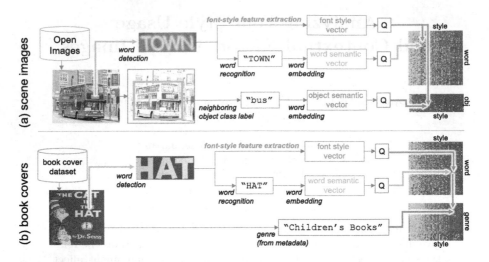

Fig. 1. Two scenarios of font style usage analysis. (a) Scene images. (b) Book covers. 'Q' represents quantization by clustering. The rightmost heat maps are the main subjects to be analyzed.

"hat." By collecting actual font usage like above and summarizing them as histograms (like heat maps in the rightmost of Fig. 1), we will be able to capture the relationship between the font style and those contextual factors. The captured relationship will be useful as empirical knowledge for future typographic designs and, simultaneously, as important statistics to understand human behavior on font selection.

However, there are difficulties in the relationship analysis. First, many exceptional cases will hide the weak relationship. For example, the advertisement texts on a bus will not have direct relevance with their contextual factor, the object "bus." Similarly, the word "Dr." in the author's name of the book will have no relationship with the book genre. The font styles of "Gorgeous hotel" and "Cheap hotel" will be different, although the meaning of the word "Hotel" is the same—therefore, in this case, the word meaning is not a strong factor in determining the style. Second, we need to use some numeric ways to handle font styles and word meanings, especially since there is no standard numeric description of font styles.

In this paper, we tackle the first difficulty by using large-scale datasets. For example, in the scene image scenario, we use the Open Images dataset [9], which contains more than one million images. By using large-scale datasets, we can cancel the perturbations by exceptional cases and will find even a weak trend in the relationship. We also use a simple quantization scheme based on clustering for accumulating weak relationships into stronger ones. For the second difficulty, we implement a font-style feature extractor based on a convolutional neural network (CNN) and prepare a numeric style vector for each word. Similarly, we use word2vec to have a semantic representation of each word.

The main contributions of this paper are summarized as follows.

- To the best of the authors' knowledge, this is the first attempt to analyze the word-wise relationship between font style and contextual factors (object around the word (e.g., "bus"), genre, and meaning of the word) with large-scale datasets.
- We set two very different scenarios with scene images and book cover images because they provide different contextual factors.
- The co-occurrence-based analysis with scene images revealed that font styles in scene images showed co-occurrences with objects surrounding the word. For example, handwriting styles and whiteboards tend to co-occur.
- Our analysis of the book cover dataset revealed numerous co-occurrences between font styles and words/genres, with consistent tendencies between word-style and genre-style relationships. For example, the "Mystery, Thriller, & Suspense" genre tends to use condensed sans-serif fonts that co-occur with suspenseful words.

2   Related Work

2.1   Font Style Embedding and Application

Regarding font-related tasks, many font style embedding and extraction approaches have been proposed. Here, we survey various font-related tasks and their font style embedding and extraction approaches.

Visual font recognition (VFR) is a task that aims at automatic recognition of the fonts in real images. Chen *et al.* [2] are the pioneer in this task and proposed a handcrafted font embedding approach based on the local image descriptors using max pooling. Following this, DeepFonts [22], a deep learning-based approach using a convolutional neural network (CNN) to recognize fonts, was proposed. Wang *et al.* also tackled the VFR task for Chinese fonts using CNNs.

Font retrieval and recommendation are also hot topics in font-related tasks [3,12,13,16]. To retrieve or recommend fonts, query font images or tags associated with fonts are mainly used. Several researchers employed deep CNNs such as ResNet-18 and ResNet-50 to embed font styles into a feature space [3,12,13]. In particular, [12,13] used the font style and word embeddings by word2vec [14,15] and achieved font retrieval and recommendation from query font images and tags.

In addition, we introduce three font-related tasks in real images. For web designs, a task aiming at estimating a font that matches a target web design has been proposed [24]. As a task using movie posters, a novel task that estimates time periods of the release date of a movie by using fonts on a query movie poster has been tackled [21]. Jiang *et al.* [7] proposed a method for automatically identifying header and body font pairs from PDF pages. In all the above three tasks, CNNs were used to extract and embed font styles.

As we reviewed above, deep learning-based approaches were mainly used to embed font styles regardless of the tasks. Following the existing trials, we also used a deep learning model to extract font style features.

2.2 Multimodal Analysis of Fonts

Kulahcioglu *et al.* conducted two interesting multimodal analyses related to fonts [10,11]. One is the analysis of the relationship between fonts and semantic attributes, *e.g.,* "angular," "artistic," and so on [10]. The other is the analysis of the effectiveness of fonts in word clouds [11].

Shinahara *et al.* [17] analyzed the correlation between fonts in book covers, their colors, and the genres of books. Their analysis is similar to ours; however, there are clear differences. First, they classified font images into predefined six font styles by simply matching query font images to reference font images, whereas we obtained font style embeddings by using a CNN, and then we conducted clustering of the style embeddings to 64 clusters. From this difference, we can analyze the relationship between fonts and contextual factors by paying attention to more detailed styles. Second, we do not address font colors yet analyze the correlation between fonts and word semantics in book covers. Our study aims to understand the trend of font style usage; therefore, we use semantics that might correlate with font styles.

2.3 Word Embedding and Its Multimodal Analysis

Word embedding, a technique that transforms the semantics of a word into a numerical vector, constitutes one of the fundamental methodologies in recent natural language processing. Among the variety of word embedding methods, word2vec [14,15] stands out as the most prominent one. The underlying principles of word2vec include two core techniques: continuous bag-of-words and Skip-gram. Both methods are based on the distributional hypothesis that words appearing in the same sentence are semantically related. In the learning process of Skip-gram, the surrounding words of the target word are inferred from the target word, while the target word is assumed to be inferred from the input of the surrounding words in the learning process of continuous bag-of-words.

In addition, a pre-trained natural language processing model known as bidirectional encoder representations from transformers (BERT) [4] has exhibited superior performance over word2vec and thus has been increasingly used in recent studies. The advantage of BERT is its capability to address ambiguity. While word2vec provides a unique word embedding for a given input, BERT generates diverse word embeddings for the same word depending on the context.

Several types of research use word embedding for multimodal analysis between words and other modalities [6,19]. In this paper, we employ word2vec to obtain static word embeddings for analyzing the co-occurrence between words and font styles. This allows us to quantitatively analyze the relationship between words and font styles.

3 Analysis of Font Style Usage in Real Images

3.1 Experimental Strategy

This study aims to reveal how font styles are used in real-world images. In particular, it is considered useful to investigate the co-occurrence of font styles with written words and contextual factors in the images, such as neighboring objects in scene images and metadata in document images. A straightforward way to analyze this co-occurrence involves detecting the text in the image and identifying the font using a classifier. However, a perfect font classifier is difficult to attain, while relatively accurate font classifiers such as DeepFont [22] have been proposed. In addition, a font classifier needs to be fine-tuned to the dataset, but image datasets that contain both font labels and rich meta-data are unavailable.

Therefore, we employed a semi-manual strategy based on feature extraction and clustering. After detecting texts in the real images and cropping them as image patches, we first extract features from the image patches using a CNN so that fonts can be well separated in the feature space. We then perform clustering in the feature space. We manually analyze the font family of each cluster and examine the co-occurrence of the font styles and contextual factors for each cluster. This process inevitably produces clusters containing random fonts, which are manually discarded.

The experiment consists of the following three parts: (a) verification of the trained style feature extractor, (b) analysis of font style usage in scene images, and (c) analysis of font style usage in book cover images. In part (a), we trained a style feature extractor with a synthetic font image dataset and evaluated it using another font image dataset with font labels. Part (b) involves analyzing how font styles are used in scene images according to words and neighboring objects. Part (c) investigates font usage in book cover images based on words and book genre.

3.2 Image Dataset

AdobeVFR Real [22]. AdobeVFR real is a large-scale real-world text image dataset. This dataset contains 4,384 text images collected from various typography forums, which covers 617 font classes. Each image has a hand-annotated font label inspected by independent experts. AdobeVFR real is provided as a subset of the AdobeVFR dataset, which includes an unlabeled font image dataset and synthetic font image dataset in addition to AdobeVFR real. Considering the purpose of our experiment, we only employed AdobeVFR real, which is a real-world and labeled image dataset.

In the verification of the style feature extractor using AdobeVFR real, we preprocessed the images according to the following procedure: First, we resized each image to be with a height of 64 pixels. If the width was smaller than 128 pixels, the same images were concatenated horizontally until the width reached 128 pixels or greater. When inputting the images into the style feature extractor, we cropped images with a size of 64 pixels × 128 pixels from the concatenated images at a random location.

Open Images v4 [9]. Open Images v4 (hereinafter referred to as Open Images) is a large-scale scene image dataset annotated with image-level labels, object bounding boxes, and visual relationships. This dataset contains 14,610,229 object bounding boxes for 600 object classes. We filtered out images without bounding boxes from the original dataset, which consisted of approximately 9M images, resulting in a final set of 1,743,042 images.

We used this dataset to analyze how fonts are used in the scene images according to the words and the objects neighboring the words. In this analysis, we detected texts in the image and investigated the relationship between used font styles and words/objects.

Book Cover Dataset [17]. This dataset contains 207,572 book cover images collected from Amazon.com along with their corresponding metadata, such as titles and genres. This dataset covers 32 genres, and each image belongs to a single genre.

We used texts, including book titles, to analyze the use of font styles and their relationship to words and book genres. After detecting and cropping the word on book covers as a patch image, images with a height of smaller than 15 pixels were discarded. We compared the text recognition results with the title information in the meta-data of the book cover and selected words that match the titles. We then excluded stop words, numbers, and compound words. We standardized the spelling of words, such as tenses, using a lemmatizer. To eliminate special proper nouns and coined words, we employ words included in the British national corpus (BNC)[1]. Eventually, the number of words used in this experiment resulted in 4,092.

3.3 Text Detection and Recognition

For text detection and recognition for the Open Images and book cover datasets, we followed the procedures adopted in [19]. We employed CRAFT [1] to detect texts in the images. CRAFT provides word bounding boxes, which are rotatable, thereby allowing the detection of rotated texts. For character recognition, we used a combination of models called "TPS-ResNet-BiLSTM-Attn," which is proposed in [1]. This model employs thin-plate spline (TPS) removing spatial distortion, ResNet, BiLSTM, and attention-based sequence prediction (Atten) for feature extraction, sequence modeling, and word recognition.

3.4 Font Style Feature Extraction

The CNN model for extracting font style features (hereinafter, referred to as a style feature extractor) is constructed by training a CNN as a font classifier. A CNN trained as a classifier is supposed to acquire an embedding from an input image space into a feature space in which features are separable for each class. This nature can be used to obtain a feature embedding suitable for clustering. This idea of using a font classifier as a feature extractor for clustering is inspired by the method in [7].

---

[1] http://www.kilgarriff.co.uk/bnc-readme.html.

Fig. 2. Examples from the image dataset for training the style feature extractor. Images are generated by SynthTIGER [23].

We employed ResNet-18 [5] as the backbone model for the style feature extractor. We modified the network structure by inserting a fully-connected layer with a 200-dimensional output and a ReLU activation function between the average pooling layer and the final layer of the original model to obtain 200-dimensional style features whose dimension is the same as that of the text feature. We also applied dropout [18] to the added fully-connected layer to improve the generalization capability. After the network is trained as a classifier, we removed the final layer and treated the output from the added fully-connected layer as the style feature.

We synthesized a text image dataset using SynthTIGER [23] and used it as a training dataset for the style feature extractor. Image examples from our dataset are shown in Fig. 2. We selected 200 fonts in Google Fonts[2] and generated 1,000 images for each font; namely, 200,000 images were obtained in total. Words were randomly selected from the corpus of MJsynth. To ensure the diversity of fonts, 40, 40, 40, and 80 fonts were randomly selected from serif, sans-serif, handwriting, and display fonts, respectively. We generated images with a height of 64 pixels while maintaining the aspect ratio; if the width was smaller than 128 pixels, the same images were concatenated horizontally until the width reached 128 pixels or greater[3]. When inputting the images into the style feature extractor, we cropped images with a size of 64 pixels × 128 pixels from the original images at a random location to unify image sizes and improve the generalization capability. To avoid non-italic fonts becoming italicized, we removed the process of tilting texts from the original implementation.

In the training of the style feature extractor, the adam optimizer [8] with a learning rate of 0.001 was employed. The batch size was set to 64, and the maximum number of training epochs was set to 100 with early stopping.

3.5 Word Embedding

We employed word2vec [15] to vectorize the object class names for the Open Images dataset and detected words for both datasets. Word2vec is a well-known word embedding method that can represent each word as a 200-dimensional semantic vector used in [19].

---

[2] https://fonts.google.com.

[3] We tried padding to unify the width of the images, but this concatenation-based method was better for style feature extraction.

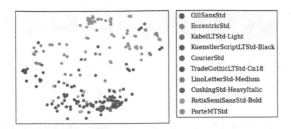

Fig. 3. Visualization of the style features of ten randomly selected fonts from the Adobe VFR dataset based on t-SNE.

4   Experimental Results

4.1   Verification of the Trained Style Feature Extractor

The purpose of this experiment is to demonstrate the validity of the style feature extractor explained in Sect. 3.4 using a dataset with font names as ground truth. A style feature extractor trained with SynthTIGER is expected to produce features that separate styles well. However, it is unknown whether the extractor will perform similarly on datasets other than SynthTIGER.

Therefore, we evaluate the feature extractor using the following procedure: First, we extract style features from the AdobeVFR real dataset using the style feature extractor trained with SynthTIGER. We then perform clustering on the obtained style features. The clustering result is qualitatively and quantitatively evaluated by visualizing the extracted features via t-SNE and calculating the adjusted mutual information (AMI) between the clusters and the font labels. The AMI is a variation of mutual information that measures the strength of the relationship between two indices while correcting the effect of the random agreement. If this index is sufficiently higher than zero, the clustering result is at least not random with respect to the ground truth; thus, feature vectors that reflect font style information are supposed to be extracted. The numbers of clusters were set to 64 and 614, which are the number of clusters to be used in the following experiments and the ground-truth number of classes, respectively.

Figure 3 visualizes the style features of the AdobeVFR real dataset extracted by the style feature extractor. To enhance clarity, ten fonts were randomly selected from the original dataset of 614 fonts, as the complete visualization became excessively cluttered. This figure shows that the style features are separated for each font to some extent, thereby suggesting that the features are clusterable.

The AMI scores are 0.2645 and 0.1590 when the numbers of clusters are 64 and 614, respectively. The scores are sufficiently higher than zero, and thus there is a non-random relationship between the clustering results and ground truth of font labels, suggesting that the extracted features reflect the font style information.

Table 1. Top 30 frequent words in Open Images Dataset. Words are displayed only if they are included in the British national corpus (BNC).

1-10th: 'new,' 'photography,' 'one,' 'state,' 'park,' 'make,' 'go,' 'world,' 'get,' 'day'
10-20th: 'stop,' 'open,' 'city,' 'free,' 'time,' 'police,' 'life,' 'may,' 'service,' 'good'
20-30th: 'first,' 'house,' 'work,' 'school,' 'love,' 'photo,' 'big,' 'way,' 'book,' 'use'

Table 2. Top 30 frequent objects in Open Images Dataset.

1-10th: 'poster,' 'building,' 'clothing,' 'book,' 'man,' 'person,' 'tree,' 'car,' 'woman,' 'bottle'
10-20th: 'house,' 'bus,' 'plant,' 'truck,' 'drink,' 'girl,' 'boat,' 'van,' 'beer,' 'train'
20-30th: 'table,' 'boy,' 'wine,' 'camera,' 'television,' 'window,' 'suit,' 'box,' 'food,' 'toy'

4.2 Analysis of Font Style Usage in Scene Images

To reveal how font styles are used in real images according to words and surrounding objects, we analyzed the Open images dataset. As the foundational information of this dataset, the histogram of the top 30 frequent words is shown in Table 1. It should be noted that words that are not included in BNC are excluded because the original list of the detected words includes many meaningless such as a single letter of the alphabet. Table 2 shows the top 30 frequent objects in the Open Image dataset. The total number of detected object types was 578.

The procedure of this experiment is as follows. First, we detect scene texts in individual images as described in Sect. 3.3 and cropped them as patches. We then extract style features using the style feature extractor, which is explained in Sect. 3.4. We also extracted word features based on word2vec as explained in Sect. 3.5. The analysis based on the extracted features is two-fold: word–style analysis and object–style analysis. The details of each analysis are described below. We used k-means clustering, and the number of clusters in each clustering trial was determined based on the gap statictics [20].

Word–Style Analysis. In this analysis, we examine the relationship between words and font styles detected in the images. First, we performed clustering using the word features to examine what kind of word groups are created. Second, we also clustered the style features to confirm what style features are contained in the dataset. Finally, we analyzed the co-occurrence between words and styles by visualizing the heat map of two feature vectors.

Figure 4 shows a heat map of font style clusters and word clusters for Open Images. To avoid meaningless co-occurrences caused by multiple occurrences of images with the same word in the same font, such as repeated logos, we considered any word that was significantly larger in scale (specifically, more than 10 times the median) than the number of occurrences of other words, as a single instance in each cluster. The heat map was normalized for each row by dividing the value in each cell by the total row sum. The procedures of avoiding meaningless co-occurrences and normalization are consistently applied to all subsequent heat maps.

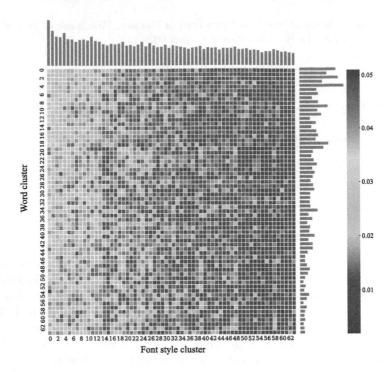

Fig. 4. Heat map of word clusters and font style clusters for Open Images.

Table 3. Characteristic co-occurrences and their details in Fig. 4.

Word cluster ID	Style cluster ID	Members	Estimated meaning	Font styles
12	4, 11, 13	'much,' 'ordinary,' 'little,' 'many,' 'big'	Adjective & Adverb	Serif
19	5, 18, 19	'service,' 'software,' 'digital,' 'retail,' 'technology'	IT & Service	Sans-serif
23	21, 22, 23	'station,' 'seat,' 'vehicle,' 'bus,' 'tunnel'	Vehicles	Sans-serif

Table 3 shows the significant combinations identified in Fig. 4. There were not so many co-occurrences between words and font styles, but we found some characteristic co-occurrences that suggest the relationship between words and font style usage.

Our findings in this analysis are summarized as follows:

- Adjectives and adverbs tend to be written in serif fonts.
- IT and service-related terms tend to be written in sans-serif fonts.
- Vehicle-related terms tend to be written in sans-serif fonts.

Object–Style Analysis. We investigated how font styles are used in scene images according to the objects surrounding the words, we analyzed the co-occurrence of objects and font styles. We conducted this analysis using two approaches: object-wise and object cluster-wise. In the object-wise approach, we examined the co-occurrence of font style clusters with individual names of the top 40 most frequently occurring objects. In the object cluster-wise approach, we analyzed the co-occurrence of font style clusters with clusters of objects that

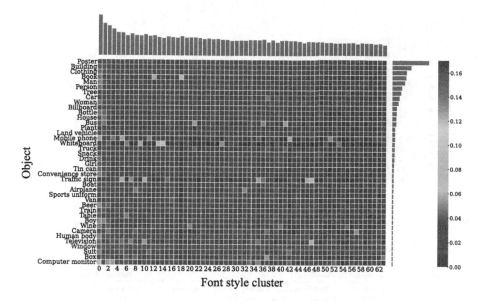

Fig. 5. Heat map of font style clusters and object names for the Open Images dataset.

Table 4. Characteristic combinations and their members in Fig. 5.

Object label	Style cluster ID	Font styles
Book	12, 18	Serif
Bus	21, 36, 41	Sans-serif condensed
		Sans-serif with narrow kerning
Whiteboard	14, 27, 45	Handwriting
Mobile phone	7, 42, 51	Sans-serif
Traffic sign	7, 10, 47	Sans-serif
Television	10, 47	Sans-serif
Computer monitor	33, 35, 42	Sans-serif

share similar meanings. To obtain object clusters, we performed clustering on the object features, which are the word2vec embedding of object names, and obtained 64 clusters.

However, due to a significant imbalance in the number of members and total instances for each cluster, we excluded clusters with the number of instances below the median and used the remaining 20 clusters for the analysis.

Figure 5 shows a heat map of the object names and font style clusters. There were more co-occurrences than in the word–style relationship. Some characteristic co-occurrences are summarized in Table 4.

Based on these results, our findings are summarized as follows:

- Texts on books tend to be written in serif fonts.
- Texts on buses involve sans-serif type fonts such as sans-serif condensed and sans-serif with narrow kerning.

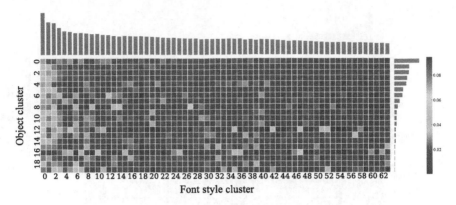

Fig. 6. Heat map of font style clusters and object clusters for the Open Images dataset.

Table 5. Characteristic co-occurrences and their details in Fig. 6.

Object cluster ID	Style cluster ID	Members	Estimated meaning	Font styles
12	5, 42, 51	'Mobile phone,' 'Corded phone,' 'Telephone'	Phone	Sans-serif
16	10, 46, 47	'Traffic sign,' 'Stop sign,' 'Tick'	Sign	Sans-serif
19	11, 35, 52	'Computer monitor,' 'Home appliance,' 'Calculator'	House appliances	Sans-serif

- Characteristic fonts that co-occur with mobile phones are sans-serif, sans-serif regular, and sans-serif thin; text written on mobile phones tends to be in sans-serif type fonts.
- Texts written on whiteboards tend to be in cursive style.
- Traffic signs tend to co-occur with sans-serif fonts.
- Words nearby televisions are written in sans-serif fonts.
- Computer monitors often involve sans-serif fonts.

Figure 6 shows a heat map of the object clusters and font style clusters, and Table 5 summarizes noteworthy co-occurrences. While there were only a few explicit instances of co-occurrence, some remarkable patterns emerged. For example, clusters containing words related to house appliances tended to use sans-serif fonts.

4.3   Analysis of Font Style Usage in Book Cover Images

We investigated how font styles are used in book cover images according to words and genres. We employed a similar procedure to that described in Sect. 4.2. We detected texts in the title region in each book cover image and cropped them as patches. We then extract style features using the style feature extractor and word features based on word2vec. Besides word–style analysis the same as in Sect. 4.2, we also conducted genre–style analysis using the pre-defined genres of the book cover dataset.

Word–Style Analysis. In this analysis, we examined the relationship between words and font styles detected in the images. We performed clustering on the

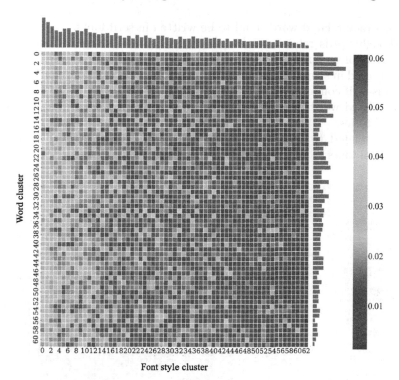

Fig. 7. Heat map of the word clusters and font style clusters in the book cover dataset

Table 6. Characteristic co-occurrences and their details in Fig. 7.

Word cluster ID	Style cluster ID	Members	Estimated meaning	Font styles
8	5, 9	'book,' 'poetry,' 'novel,' 'diary'	Press cover	Serif
11	9, 10, 17	'empire,' 'king,' 'queen,' 'castle'	Aristocracy	Serif
33	10, 17, 32	'prayer,' 'sacred,' 'god,' 'saint'	Religion	Serif
34	25, 26, 34	'computer,' 'application, 'software,' 'data'	IT	Sans-serif
40	27, 35, 43	'kill,' 'crime,' 'murder,' 'spy'	Suspense	Sans-serif condensed
46	11, 27, 39	'battle,' 'war,' 'violence,' 'army'	War	Sans-serif condensed
58	10, 16, 25	'gene,' 'molecular,' 'serum,' 'chemical'	Chemistry	Serif, Sans-serif

word features and style features and then analyzed the co-occurrence between words and styles by visualizing the heat map of two feature vectors. We performed clustering on the word features and style features and obtained 62 and 63 clusters respectively.

Figure 7 shows a heat map of font style clusters and word clusters for the book cover dataset.

Table 6 summarizes characteristic co-occurrences in Fig. 7. Our findings in this analysis are summarized as follows:

- Religion-related words tend to be written in serif fonts.
- IT-related words tend to be written in sans-serif fonts.

- Aristocracy-related words tend to be written in serif fonts.
- Although words related to the press cover a wide range of fonts, they are relatively often written in serif.
- Words used in suspense stories such as kill, suicide, and murder are all written in sans-serif condensed.
- War-related words are often written in sans-serif condensed, as well as suspense stories.
- Words related to chemistry and medicine tend to be written in serif and sans-serif fonts.

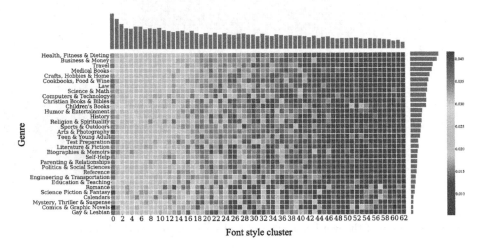

Fig. 8. Heat map of the genres and font style clusters in the book cover dataset

Table 7. Characteristic co-occurrences and their details in Fig. 8.

Genre	Style cluster ID	Font styles
Law	6, 10	Serif
Computers & Technology	4, 8, 13	Sans-serif
Christian Books & Bibles	9, 15, 17	Serif
Children's Books	51, 53	Handwriting, Fancy
History	5, 6, 9	Serif
Religion & Spirituality	5, 10, 17	Serif
Sports & Outdoors	11, 23, 27	Sans-serif
Literature & Fiction	10, 17, 24	Serif
Engineering & Transportation	4, 8, 23	Sans-serif
Mystery, Thriller & Suspense	11, 27, 35	Sans-serif condensed

Genre–Style Analysis. We investigated how font styles are used in book cover images according to the book genre. We examined the co-occurrence of the book genres and font styles by visualizing the heat map of them. Thirty-two genres provided in the book cover dataset are used. We used the clustering results of the style features in the word–style analysis above.

Figure 8 shows a heat map of genres and style clusters for the book cover dataset.

Our findings in this analysis are summarized as follows:

- "Law," "Christian Books & Bibles," "History," "Religion & Spirituality," and "Literature & Fiction" include many serif fonts.
- "Business & Money," "Computers & Technology," "Sports & Outdoors," "Test Preparation," "Engineering & Transportation," "Education & Teaching," "Mystery, Thriller & Suspense" often involve sans-serif fonts.
- "Children's Books" and "Comics & Graphic Novels" use handwriting and fancy fonts more than other genres.

It is noteworthy that some findings in Table 7 are consistent with those observed in the word-style analysis in Table 6. Serif fonts are frequently used in "Christian books & Bibles," which suggests a strong correlation with the finding that religious word clusters co-occur with serif fonts. Additionally, the "Mystery, Thriller, & Suspense" genre tends to use condensed sans-serif fonts that co-occur with suspenseful words.

5   Conclusion

In this paper, we aimed at understanding the relationship between the style of the selected font and several contextual factors. For this purpose, we conducted a co-occurrence-based analysis of font styles and contextual factors such as used words, surrounding objects, and book genres for scene images and book cover images.

In the analysis of font styles in scene texts using the Open Images, there were not so many co-occurrences between words and font styles; however, we found some interesting co-occurrences such as adjectives and adverbs which are often written in serif fonts. For objects and font styles, we found many co-occurrences. For example, text on traffic signs tends to be written in sans-serifs, texts on book pages tend to be written in serifs, and mobile phones, television, and computer monitor tend to be written in sans-serifs

As for the book cover, we found many co-occurrences between words and font styles more than in the Open Images. The clusters of religious words tended to use serif fonts, while the clusters of IT words tended to use sans serif fonts. In addition, the clusters of suspense and war-related words tended to use a condensed sans-serif font. In terms of genres and font styles, we found that fonts such as handwriting and fancy tend to be used more in children's books than in other genres. We also found some similarities between word–style analysis and genre–style analysis. For example, the "Christian books & Bibles" genre often uses serif fonts that co-occur with the religious word cluster, and the "Mystery, Thriller & Suspense" genre frequently involves condensed sans-serif fonts that co-occur the suspenseful word cluster.

In future work, we will conduct a composite analysis using multiple contextual factors. For example, we can calculate the conditional probabilities $p(\text{style} \mid \text{contextual factor})$ and investigate which factor strongly affects the usage of font styles. We will also conduct experiments on other types of datasets such as document images.

Acknowledgment. This work was supported in part by JSPS KAKENHI Grant Numbers JP22H00540 and JP21H03511.

References

1. Baek, Y., Lee, B., Han, D., Yun, S., Lee, H.: Character region awareness for text detection. In: Proceedings of the Conference on Computer Vision and Pattern Recognition (CVPR), pp. 9365–9374 (2019)
2. Chen, G., Yang, J., Jin, H., Brandt, J., Shechtman, E., Agarwala, A., Han, T.X.: Large-scale visual font recognition. In: Proceedings of the Conference on Computer Vision and Pattern Recognition (CVPR), pp. 3598–3605 (2014)
3. Choi, S., Matsumura, S., Aizawa, K.: Assist users' interactions in font search with unexpected but useful concepts generated by multimodal learning. In: Proceedings of the International Conference on Multimedia Retrieval, pp. 235–243 (2019)
4. Devlin, J., Chang, M.W., Lee, K., Toutanova, K.: BERT: pre-training of deep bidirectional transformers for language understanding. In: Proceedings of the Conference of the North American Chapter of the Association for Computational Linguistics: Human Language Technologies, pp. 4171–4186 (2019)
5. He, K., Zhang, X., Ren, S., Sun, J.: Deep residual learning for image recognition. In: Proceedings of the Conference on Computer Vision and Pattern Recognition (CVPR), pp. 770–778 (2016)
6. Ikoma, M., Iwana, B.K., Uchida, S.: Effect of text color on word embeddings. In: Bai, X., Karatzas, D., Lopresti, D. (eds.) DAS 2020. LNCS, vol. 12116, pp. 341–355. Springer, Cham (2020). https://doi.org/10.1007/978-3-030-57058-3_24
7. Jiang, S., Wang, Z., Hertzmann, A., Jin, H., Fu, Y.: Visual font pairing. IEEE Trans. Multimedia **22**(8), 2086–2097 (2019)
8. Kingma, D.P., Ba, J.: Adam: a method for stochastic optimization. arXiv preprint arXiv:1412.6980 (2014)
9. Krasin, I., et al.: OpenImages: a public dataset for large-scale multi-label and multi-class image classification. Dataset available from https://storage.googleapis.com/openimages/web/index.html (2017)
10. Kulahcioglu, T., De Melo, G.: Predicting semantic signatures of fonts. In: Proceedings of the International Conference on Semantic Computing (ICSC), pp. 115–122. IEEE (2018)
11. Kulahcioglu, T., De Melo, G.: Paralinguistic recommendations for affective word clouds. In: Proceedings of the International Conference on Intelligent User Interfaces, pp. 132–143 (2019)
12. Kulahcioglu, T., De Melo, G.: Fonts like this but happier: a new way to discover fonts. In: Proceedings of the International Conference on Multimedia, pp. 2973–2981 (2020)

13. Matsumura, S., Choi, S., Aizawa, K.: Font search across various languages based on multimodal learning. In: Proceedings of the Conference on Multimedia Information Processing and Retrieval (MIPR), pp. 173–176. IEEE (2020)
14. Mikolov, T., Chen, K., Corrado, G.S., Dean, J.: Efficient estimation of word representations in vector space. In: International Conference on Learning Representations (2013)
15. Mikolov, T., Sutskever, I., Chen, K., Corrado, G.S., Dean, J.: Distributed representations of words and phrases and their compositionality. In: Proceedings of the Advances in Neural Information Processing Systems (NIPS) 26 (2013)
16. O'Donovan, P., Lībeks, J., Agarwala, A., Hertzmann, A.: Exploratory font selection using crowdsourced attributes. ACM Trans. Graph. **33**(4), 1–9 (2014)
17. Shinahara, Y., Karamatsu, T., Harada, D., Yamaguchi, K., Uchida, S.: Serif or sans: visual font analytics on book covers and online advertisements. In: Proceedings of the International Conference on Document Analysis and Recognition (ICDAR), pp. 1041–1046. IEEE (2019)
18. Srivastava, N., Hinton, G., Krizhevsky, A., Sutskever, I., Salakhutdinov, R.: Dropout: a simple way to prevent neural networks from overfitting. J. Mach. Learn. Res. **15**(1), 1929–1958 (2014)
19. Takeshita, K., Shioyama, J., Uchida, S.: Label or message: a large-scale experimental survey of texts and objects co-occurrence. In: Proceedings of the International Conference on Pattern Recognition (ICPR), pp. 6227–6234. IEEE (2021)
20. Tibshirani, R., Walther, G., Hastie, T.: Estimating the number of clusters in a data set via the gap statistic. J. Roy. Statist. Soc. Ser. B (Statist. Methodol.) **63**(2), 411–423 (2001)
21. Tsuji, K., Uchida, S., Iwana, B.K.: Using robust regression to find font usage trends. In: Barney Smith, E.H., Pal, U. (eds.) ICDAR 2021. LNCS, vol. 12917, pp. 126–141. Springer, Cham (2021). https://doi.org/10.1007/978-3-030-86159-9_9
22. Wang, Z., et al.: DeepFont: identify your font from an image. In: Proceedings of the International Conference on Multimedia, pp. 451–459 (2015)
23. Yim, M., Kim, Y., Cho, H.-C., Park, S.: SynthTIGER: synthetic text image GEneratoR towards better text recognition models. In: Lladós, J., Lopresti, D., Uchida, S. (eds.) ICDAR 2021. LNCS, vol. 12824, pp. 109–124. Springer, Cham (2021). https://doi.org/10.1007/978-3-030-86337-1_8
24. Zhao, N., Cao, Y., Lau, R.W.: Modeling fonts in context: Font prediction on web designs. In: Computer Graphics Forum, pp. 385–395. Wiley Online Library (2018)

CCpdf: Building a High Quality Corpus for Visually Rich Documents from Web Crawl Data

Michał Turski[1,2]([✉]), Tomasz Stanisławek[1], Karol Kaczmarek[1,2], Paweł Dyda[1,2], and Filip Graliński[1,2]

[1] Snowflake, Warsaw, Poland
[2] Adam Mickiewicz University, Poznań, Poland
{michal.turski,tomasz.stanislawek,karol.kaczmarek,
pawel.dyda,filip.gralinski}@snowflake.com

Abstract. In recent years, the field of document understanding has progressed a lot. A significant part of this progress has been possible thanks to the use of language models pretrained on large amounts of documents. However, pretraining corpora used in the domain of document understanding are single domain, monolingual, or nonpublic. Our goal in this paper is to propose an efficient pipeline for creating a big-scale, diverse, multilingual corpus of PDF files from all over the Internet using Common Crawl, as PDF files are the most canonical types of documents as considered in document understanding. We analyzed extensively all of the steps of the pipeline and proposed a solution which is a trade-off between data quality and processing time. We also share a CCpdf corpus in a form or an index of PDF files along with a script for downloading them, which produces a collection useful for language model pretraining. The dataset and tools published with this paper offer researchers the opportunity to develop even better multilingual language models.

Keywords: Natural Language Processing · language models · dataset construction · document understanding

1 Introduction

Natural Language Processing (NLP) in recent years has made significant progress thanks to using language models such as GPT-3 [6] or T5 [23]. Usually these models are trained in a two-step process. The first part is pretraining, which utilizes a large corpus of text, and the second step is finetuning on a final task. Recent works demonstrate a considerable impact of pretraining on the final performance of a model [10,17,27]. For instance, GPT-3 was pretrained on a combination of texts from Common Crawl, WebText2, two book corpora, and the English Wikipedia (499 billion tokens in total) [6] while T5 was pretrained on the C4 corpus, which is 750 GB of data [23].

Work done while at Applica.ai, later acquired by Snowflake.

G. A. Fink et al. (Eds.): ICDAR 2023, LNCS 14189, pp. 348–365, 2023.
https://doi.org/10.1007/978-3-031-41682-8_22

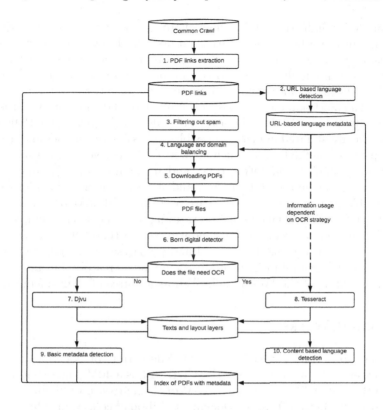

Fig. 1. The full flow of the process. Cylinders represent data, rectangles represent processing steps, and arrows represent data flow. A solid line indicates that the information is always used, and a dashed line represents data usage dependent on the processing strategy.

The recent progress in document understanding (defined as "capacity to convert a document into meaningful information" [5]) has been possible thanks to 2D language models such as LayoutLM [29,30], LAMBERT [9], or TILT [21]. Similarly to the models mentioned above, they also need large amounts of data for pretraining. The input to these models is a multi-modal representation of a document, e.g. tokens with their positions and images of pages.

The World Wide Web abounds in multi-modal documents, which contain enormous amounts of information. This information can be used in multiple domains: NLP, law, knowledge extraction, history, and many more. Yet, this aspect of the Internet remains relatively unexplored. So far, attempts of document dataset creation have been focused on either single domain (e.g. medical[1],

---

[1] https://www.ncbi.nlm.nih.gov/pmc/tools/openftlist/.

academic [3], or industrial [14]), while there has been no all-over-the-Internet approach. On the other hand, existing all-over-the Internet corpora (e.g. Web 1T 5-gram[2]) were focused on text only, not on multi-modal documents.

A document is a multi-modal form of communication: to interpret documents properly, we have to understand not only text, but also the layout and graphical elements. The most popular and portable multi-modal document format is PDF. In this study, we aim to describe a carefully designed pipeline for PDF corpus creation. We investigated numerous possible processing techniques and described their impact on the final data, which allowed us to achieve satisfactory trade-off between data quality, computing time, and monetary cost. The dataset itself (in the form of an index of PDF files and a script for downloading them) is also available at our website[3]. We share a corpus of 14.5M pages. It is useful as a dataset for 2D language model pretraining, but may also be employed as a source for derived datasets, in the same way as the IIT-CDIP dataset [14] was used to create many diverse challenges. Finally, analysis of the collected PDF files themselves yields helpful insight for language model creators, but also enhances our understanding of the World Wide Web as a source of PDF documents.

2   Related Works

The general problem of creating a large-scale corpus of documents has been studied extensively in recent years. IIT-CDIP [14] is a 40M pages (but according to the authors of OCR-IDL [4] only 35.5M of them are still reachable) dataset of reports from the Legacy Tobacco Documents Library[4] collection, which was later reused to prepare a 400k page document classification dataset [12]. Also, OCR-IDL [4] reused IIT-CDIP to publish a 26M page dataset with high-quality OCR output. DoRe [19] is a French dataset of 2350 annual reports from 336 companies, unfortunately the data weren't shared publicly. There are also two layout analysis datasets based on scientific articles: Docbank [15] and Publaynet [33]. Their volumes are 500k and 360k pages, respectively. In addition, Ammar *et al.* [3] provided corpora of scientific documents together with a literature graph (defined as "a directed property graph which summarizes key information in the literature and can be used to answer the queries mentioned earlier as well as more complex queries" [3]). The National Library of Medicine has shared a PMC Open Access Subset[5] which is a corpus of open-access, open-licensed medical publications. Allison *et al.* [2] proposed a pipeline for creating a corpus of PDFs sourced from the Internet. The goal of this work is to "identify key edge cases or common deviations from the format's specification". They also provide analyses of files in their corpus. All of these datasets are single-domain or single-language collections (usually both), while our aim is to create a diverse, multilingual dataset. There exists only one publication presenting such a dataset [31], but the authors limited

---

[2] https://catalog.ldc.upenn.edu/LDC2006T13.
[3] https://github.com/applicaai/CCpdf.
[4] https://industrydocuments.ucsf.edu/tobacco/.
[5] https://ncbi.nlm.nih.gov/pmc/tools/openftlist/.

themselves to describing the data processing pipeline without analyzing their decisions. Also, their dataset was not shared.

Attempts have also been undertaken to create diversified corpora of texts sourced from the Internet. For instance, in CCNet [28], Common Crawl was used to create curated monolingual corpora in more than 100 languages. Also Schwenk *et al.* used Common Crawl in CCMatrix [24], but their purpose was to extract parallel sentences in different languages. The result was 10.8 billion parallel sentences in 90 languages. Another study in this vein is Smith *et al.* [25], whose method allowed to extract a 278 million token corpus of parallel English-French, English-Spanish, and English-German texts. In CCQA [13], a method for composing multilingual question-answering task using Common Crawl was proposed. The authors shared 130 million question-answer pairs. Liu and Curran [16] used Open Directory Project[6] to extract a topic-diverse English corpus of 10 billion words. To pretrain the T5 language model [23], the authors extracted a 750 GB English text corpus, called C4, employing Common Crawl. Dodge *et al.* [7] explored this dataset further and analyzed the effects of the applied filtering. A similar pipeline to that used for C4 was applied to create the mT5 [32] training corpus, which is a multilingual version of T5. The proposed corpus has 6.3 trillion tokens. Qi *et al.* [22] crawled 10 million images with captions from the Internet and used it to pretrain the multi-modal ImageBERT model. C4Corpus [11] (not to be confused with C4 proposed by Raffel *et al.*, described above) utilized Common Crawl resources to provide multilingual (more than 50 languages) over 10 billion token corpus to the community. The Pile [8] is a 885 GB text corpus composed of 22 different datasets, and one of its subparts are texts from Common Crawl. Abadji *et al.* [1] proposed a document-oriented multilingual 12 GB corpus of texts from Common Crawl with quality annotations. It must be noted that the authors define the term "document" as a long, coherent piece of text, not as a PDF file, as we do in this study. Luccioni and Viviano [18] analyzed Common Crawl in terms of undesirable content, including hate speech and sexually explicit content, and investigated different filtering methods.

Table 1. Comparison of existing publicly available corpora. *Numbers of valid documents/pages according to the authors of OCR-IDL [4].

Dataset	Documents	Pages	Avg pages per doc	Languages	Domains	Years
IIT-CDIP	6.5M*	35.5M*	5.5	1	Industry documents	1990s
OCR-IDL	4.6M	26M	5.7	1	Industry documents	1990s
CCpdf	1.1M	14.5M	12.9	11	Multi-domain	Mostly 2010–2022

---

[6] http://odp.org.

3   Collecting and Processing PDFs

In this section we describe how we addressed the challenge of finding, down-loading, and processing a great volume of PDF documents. The full process is presented in Fig. 1.

3.1   Common Crawl

As our input we used web indexes created by Common Crawl[7]. Common Crawl is a project of The Internet Archive[8] – an organization dedicated to providing a copy of the Internet to the community. They crawl webpages and save them into crawls dumps. A crawl dump contains billions of webpages (hundreds of terabytes of uncompressed data) and a new dump has been published nearly every month since March 2014. Some earlier, more irregular dumps starting from 2008 are also available.[9] Each dump also contains an index of the crawled pages.

We decided to simply use the latest (and the largest) dump available at the time of writing this paper - the May 2022 dump.[10] It contains 3.45 billion web pages, which amounts to 462 TB of uncompressed content. It would obviously be possible to apply the extraction procedure described in this paper to all crawls to obtain an even larger collection of PDFs, which would also allow for a diachronic analysis, but we wanted to focus on the most recent documents.

Note that dumps contain only files considered as text files by the Common Crawl web robot. Mostly these are web pages in the HTML format, but, for-tunately, PDFs are also treated as text files, being derivative of the PostScript page description language. This is not the case with, for instance, images, Excel files, DOCX files. Consequently, such files cannot be amassed using the methods described in the aforementioned papers.

3.2   PDF Links Extraction

We experimented with two methods for extracting links to PDF files (step 1 in Fig. 1):

1. using CDX files, i.e., index server files provided by Common Crawl;
2. looking for links to PDF files in WARC, i.e., raw crawl data files.

The first method is simpler, as CDX files are easy to download and take up only 225 GB in total. The second method might yield more links to PDF files, but:

– it is impossible for us to download all WARCs. Only a limited number of them can be processed, though still a significant number of PDF links can be added even if a small percentage of all WARC files are processed,

---

[7] https://commoncrawl.org.
[8] https://archive.org/.
[9] https://commoncrawl.org/the-data/get-started/.
[10] https://commoncrawl.org/2022/06/may-2022-crawl-archive-now-available/.

– there is lower probability that the file linked is available at all, be it in the crawl dump or simply at the original address.

In CDX files, the MIME type of a captured file is specified, and we limited ourselves to the `application/pdf` type.

Hence, in this paper, we focus on the first method, which allows to speed up the whole processing pipeline.

3.3 URL-Based Language Detection

We decided to limit our investigation to the following set of 11 languages: Arabic, Dutch, English, French, German, Italian, Japanese, Polish, Portuguese, Russian, and Spanish.

When deciding whether to process a given URL, we applied a number of simple heuristics to determine the language. For example, we assumed that PDFs from `.pl` domains are Polish unless there is `lang=en` inside the URL etc. Note that this is a preliminary filter; later, when the contents have been downloaded, we do a proper language detection (see Sect. 3.9).

In August 2018, Common Crawl added language metadata to CDX files.[11] Unfortunately, the Compact Language Detector 2 employed there is applicable only for plain texts or HTMLs, and only a small percentage of PDF links contained the language metadata; therefore, it was unusable for our purposes.

This step of the pipeline is presented as block 2 in Fig. 1.

3.4 Filtering Out Spam

One of the challenges to be tackled in Web information retrieval or when creating a massive text corpus sourced from the Web is the problem of (web) spam and, more generally, low quality pages (step 3 in Fig. 1). Web spam is usually related to black-hat search engine optimization, i.e., creating link farms of web pages with automatically or semi-automatically generated content. It turns out that PDF files found on the Internet have the advantage of a relatively low percentage of spam, especially when compared to HTML web pages. More generally, we believe PDF files usually contain more formal content as most of them are business, legal, or scientific documents.

Still, some spam PDFs were found in Common Crawl dumps. Fortunately, the way in which spammers operate is rather homogeneous. A typical telltale of a spammy PDF was a long name composed of lower-case letters interspersed with hyphens. A regular expression was written to detect suspicious URLs, and if a domain happened to contain a large percentage of such URLs, it was assumed to be spammy as a whole and totally discarded. Thanks to this simple heuristic, in a sample of 1k documents we manually annotated (see Sect. 3.9) we found no spam PDFs.

---

[11] https://commoncrawl.org/2018/08/august-2018-crawl-archive-now-available/.

3.5   PDF Data Download Methods

In order to ensure diversity, we downloaded at most three PDF files from each domain for a language in a random but reproducible manner. For English and German this number was lowered to, respectively, one and two, as PDFs in these two languages are much more numerous compared to others. This limitation also serves as a filter against anomalies such as millions of PDFs coming from a single domain; especially a spammy one, if not detected with the procedure described in Sect. 3.4. Balancing is represented as step 4 in Fig. 1.

The files were downloaded from the original URLs (step 5 in Fig. 1). Optionally, one could extract the file from a Common Crawl dump, especially if the file is not available at the original site. We provide a script to extract PDF files directly from the dump; fortunately, one does not need to download the whole dump to extract a file.

There is, however, one serious issue with extracting PDFs from crawl dumps: all files are truncated by the crawler to 1 MB. This limit is quite high for HTML pages, but unfortunately rather low for PDF files. This means that only small-sized PDF files can be extracted from Common Crawl dumps; larger ones have to be downloaded from the original sites.

The final and intermediary statistics for the files downloaded are presented in Table 2.

Table 2. Number of documents per processing step and language. Percentage values show success rates of downloading (in the downloaded column) or downloading and processing together (in the processed column). The success rate for processing a downloaded document equals 94.94%.

	URLs found	Anti-spam filtered	Domain balanced	Language balanced	Successfully downloaded	Successfully processed
ar	65395	65374	13142	13142	11 710 (89.10%)	10 826 (82.38%)
de	1661317	1659713	320978	200000	182 607 (91.30%)	172 668 (86.33%)
en	11515766	11501781	952776	200000	182 071 (91.04%)	175 440 (87.72%)
es	871843	871478	106143	106143	93 163 (87.77%)	88 952 (83.80%)
fr	654250	653120	143020	143020	129 927 (90.85%)	121 905 (85.24%)
it	831344	831026	129610	129610	119 731 (92.38%)	114 265 (88.16%)
ja	1160543	1160410	151686	151686	139 990 (92.29%)	134 310 (88.54%)
nl	339519	338946	92372	92372	84 848 (91.85%)	79 720 (86.30%)
pl	438770	438531	85635	85635	79 668 (93.03%)	75 374 (88.02%)
pt	697535	697285	73130	73130	64 725 (88.51%)	61 405 (83.97%)
ru	628473	628061	105535	105535	91 708 (86.90%)	85 552 (81.07%)
all	18864755	18845725	2174027	1300273	1 180 148 (90.76%)	1 120 417 (86.17%)

3.6   Born Digital Scanner

To process correctly all kinds of documents in the document understanding domain we need to extract tokens from PDF files with their bounding boxes sorted properly, i.e., according to the reading order. The most common approach

[9, 21, 29, 30] is to process each PDF file with the use of some OCR engine, e.g. Tesseract [26], Amazon Textract[12], Microsoft Azure Computer Vision API,[13] or Google Vision API[14]. This method simplifies the processing pipeline and removes the need to understand the complicated PDF file format.

The biggest challenge in direct text and layout extraction lies in processing image content since there is no easy way to detect whether an image contains text. On the other hand, some documents lack pictures altogether; instead they contain textual information in the PDF file structure. We call them *documents that do not require OCR*. From such documents text can be extracted along with bounding boxes using dedicated Python libraries, such as pdfminer.six[15], pdfplumber,[16] or a DjVu-based tool[17]. Direct text extraction using these tools leads to the reduction of the processing time and improvement of the quality of the extracted data by preventing OCR errors. Therefore, we decided to introduce a mechanism, called the Born Digital detector, for finding these kinds of documents (step 6 in Fig. 1).

3.7 Born Digital Detection Heuristics

In order to detect documents that do not need to be processed with an OCR pipeline, we created a fast, simple heuristic-based classifier:

- *Visible Text Length > 100* - Visible text in the document contains more than 100 characters
- *Hidden Text Length = 0* - There is no hidden text in the document
- *Image Count = 0* - There are no images in the document

Used statistics (*Visible Text Length, Hidden Text Length, Image Count*) were extracted using *Digital-born PDF Scanner*[18] tool written by us.

Our simple method was able to classify 219 documents out of 967 as born-digital files that do not require OCR. (In other words, we can skip the time-consuming OCR process for more than 1 out of 5 PDF files). To check quality of our heuristic we manually annotated the same sample of documents. The precision of the proposed method was 93.15%. All errors (15) were caused by adding a background with logo text to the file. In the future, we can also improve that kind of cases by extracting metadata information about PDF file background as well.

---

[12] https://aws.amazon.com/textract/.
[13] https://docs.microsoft.com/en-us/azure/cognitive-services/computer-vision/overview-ocr.
[14] https://cloud.google.com/vision/docs/pdf.
[15] https://github.com/pdfminer/pdfminer.six.
[16] https://github.com/jsvine/pdfplumber.
[17] http://jwilk.net/software/pdf2djvu, https://github.com/jwilk/ocrodjvu.
[18] https://github.com/applicaai/digital-born-pdf-scanner.

Table 3. Results for the Born Digital detector mechanism.

	Gold standard #	Born Digital detector			
		Precision	Recall	F1-score	TP + FP #
born digital, OCR not required	471 (48.71%)	93.15	43.31	59.13	219 (22.65%)
born digital, OCR required	321 (33.20%)	–	–	–	–
scan	175 (18.10%)	–	–	–	–
all	967	–	–	–	–

3.8 OCR Processing

One of the initial steps of the PDF processing pipeline is the URL based language detection method (see Sect. 3.3). Information about the language of the document is needed for filtering documents for specific languages and also by the OCR tool. In the next step (see Sect. 3.7), we select PDF files for processing either by the DjVu-based tool (if the file is born digital then it does not require OCR) or by Tesseract OCR [26]. The result is hocr files containing extracted text with its bounding boxes. This form of data serves as the input to the subsequent processing and analyzing steps.

Table 4. Comparison of resource utilization for different strategies of the text extraction from PDF files. *for Azure OCR we used 4 CPU (which is minimal recommendation for container in version 3.2) and multiplied the number by 4.

Strategy name	Processing time (using 1 CPU)		Additional cost	
	1k files	in relation to fastest	Single page	1k files
DjVu-based tool + Born-digital detector	5.6 h	1x	-	-
Tesseract + URL based LD	23.7 h	4x	-	-
Tesseract + Build-in LD mechanism	75.9 h	14x	-	-
MSOCR + Build-in LD mechanism	16.7 h*	3x	0.001$	13$

Possible Alternatives. In a typical scenario of extracting text with bounding boxes from a PDF file, researchers use a custom OCR engine [9,21,29,30], e.g. Tesseract, Microsoft Azure Computer Vision API, Amazon Textract, or Google Vision API. However, when we want to process millions of PDF files, we need to think about the utilization of resources in the context of time and money. Additionally, contrary to previous work, the language of a PDF file that we want to process is unknown. Therefore, to choose the most economical option, we tested the following strategies:

1. *DjVu-based tool with a born-digital detector* – for details, please see Sect. 3.7
2. *Tesseract with URL based Language Detection (LD)* – described at the beginning of this section
3. *Tesseract with a built-in LD mechanism* – in this strategy, we use the Tesseract OCR [26] engine with a built-in language detection mechanism

4. *Azure CV API with a built-in LD mechanism* – in this strategy we use Microsoft Azure Computer Vision API[19] with a built-in language detection mechanism

We achieved the shortest processing time (see Table 4) with the DjVu-based tool and a born-digital detector (see Sect. 3.7), which followed from the fact that we did not need to run any ML models. Also quality of output from the DjVu-based tool is better than from any OCR engine, because it extracts real content of a file and does not introduce any processing noise. *Azure CV API* and *Tesseract with URL based language detection* are the slowest OCR engines with 3-4 longer processing time. It's turn out that the slowest processing time has strategy is *Tesseract with build-in LD mechanism* and, therefore, we will not apply it in our final pipeline.

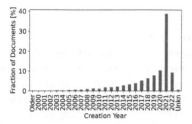

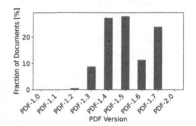

Fig. 2. Distribution of the analyzed sample in terms of creation year.

Fig. 3. Distribution of the analyzed sample according to PDF version.

3.9 Language Identification

In our final processing pipeline we used two language detection mechanisms:

1. URL-based method. Described in Sect. 3.3.
2. Content based method. We used the *langdetect*[20] library to detect language based on its text content extracted in the previous step.

We tested the quality of our language detection methods on ~1k manually annotated documents (Table 5). Both of our mechanisms can detect only a single language but, in reality, we found out that 27 documents had multiple languages (in 23 cases one of them was English). Fortunately, detecting a single language allowed us to predict the language correctly for almost all documents (97.3%).

With the use of the URL based method, we achieved a 90.51% F1-score on average, which seems reasonably good when we take into account the simplicity of the method. The content based method works better in general with an F1-score of 94.21% on average. The single exception here is the Japanese language. Both mechanisms produced the least satisfactory results for two languages: Arabic and English. It turned out that many documents from the *.ar* domain were actually in English. Therefore, for the content based mechanism we wrongly processed the PDF files with the Arabic Tesseract model.

---

[19] https://docs.microsoft.com/en-us/azure/cognitive-services/computer-vision/overview-ocr.
[20] https://pypi.org/project/langdetect/.

Additionally, we found out that when we used the proper Tesseract model our results increased drastically to an F1-score of 98.05% on average. The main reason why this happened was the fact that the language identification mechanism was working on the proper alphabet.

Possible Alternatives. In Table 6 we present the results for different language identification tools. All of them achieved similar F1-scores, of which *spacy* (94.33%) and *langdetect* (94.21%) performed best. When we also take into consideration the processing time, it turns out that *gcld3* was the best one with a huge advantage over the second tool, which was the *langdetect* library. Therefore, we decided to balance quality and resource utilization and use *langdetect* as our main tool for language identification.

Table 5. Quality of the language identification methods verified on 996 manually annotated documents.

	Gold standard #	URL based method			Content based method					
					All documents			Proper Tesseract lang		
		Precision	Recall	F1	Precision	Recall	F1	Precision	Recall	F1-score
ar	20	46.51	95.24	62.50	44.19	95.00	60.32	100.0	100.0	100.0
de	94	94.68	92.71	93.68	98.94	98.94	98.94	100.0	100.0	100.0
en	119	80.46	58.82	67.96	94.34	84.03	88.89	98.55	98.55	98.55
es	75	94.52	93.24	93.88	98.65	97.33	97.99	98.57	98.57	98.57
fr	108	93.94	86.92	90.29	100.0	91.67	95.65	100.0	100.0	100.0
it	101	93.20	95.05	94.12	98.97	95.05	96.97	94.79	94.79	94.79
jp	108	100.0	98.10	99.04	100.0	89.81	94.63	92.38	92.38	92.38
nl	90	84.91	100.0	91.84	98.86	96.67	97.75	96.67	96.67	96.67
pl	88	95.56	100.0	97.73	98.86	98.86	98.86	98.86	98.86	98.86
pt	83	94.38	98.82	96.55	97.62	98.80	98.21	98.78	98.78	98.78
ru	78	96.34	98.75	97.53	97.47	98.72	98.09	100.0	100.0	100.0
other	2	0	0	0	0	0	0	0	0	0
no text	3	0	0	0	0	0	0	0	0	0
multi	27	0	0	0	0	0	0	0	0	0
all	996	88.59	92.51	90.51	93.45	94.99	94.21	98.05	98.05	98.05

Table 6. Comparison of the quality and processing time of different language identification tools.

Tool name	F1-scores for content based method												Processing time	
	ar	de	en	es	fr	it	jp	nl	pl	pt	ru	all	1k files	1M files
langdetect[a]	60.32	98.94	88.89	97.99	95.65	96.97	94.63	97.75	98.86	98.21	98.09	**94.21**	**0.28 min**	**4.67 h**
lingua-py[b]	60.32	97.90	88.79	96.69	94.74	95.92	98.59	96.77	99.44	98.18	97.47	94.05	2.57 min	42.8 h
spacy[c]	60.32	98.94	87.33	97.99	94.79	97.49	98.59	97.18	100.0	98.18	97.47	**94.33**	3.62 min	60.3 h
gcld3[d]	59.01	98.94	89.91	97.96	96.15	97.49	92.16	97.73	99.44	98.78	98.07	94.08	**0.03 min**	**0.33 h**

[a] https://pypi.org/project/langdetect/
[b] https://github.com/pemistahl/lingua-py
[c] https://spacy.io/universe/project/spacy_fastlang
[d] https://pypi.org/project/gcld3/

3.10   Produced Index

As a result of our pipeline, we created an index of successfully downloaded and processed files. We decided to download up to 200k documents per language to share a reasonably sized corpus, with a good diversity of languages. It gives an acceptably good trade-off between the balance of languages and the size of the dataset. Statistics about the index are presented in Table 2.

A comparison of our dataset to existing corpora is presented in Table 1. The corpus we provided is smaller than the previous ones considering the total number of documents and pages. Still, language models will benefit in many aspects, (1) understanding long-distance relationships as the dataset has, on average, the longest documents compared to previous works, (2) multi-language training as we selected 11 different languages, (3) multi-domain training as we sourced documents from different websites all over the Internet, (4) document understanding of recently created documents (which may differ from the old ones in terms of language, layout, and graphical style) as the majority of files in our corpus were produced after 2010 (in IIT-CDIP, the most popular corpus so far, all the documents were created in the 90s).

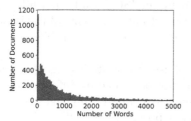

Fig. 4. Distribution of word count per document.

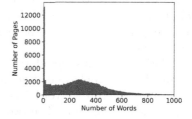

Fig. 5. Distribution of word count per page.

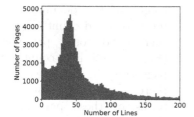

Fig. 6. Distribution of the number of lines per page.

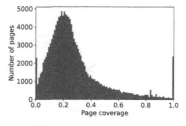

Fig. 7. Distribution of text coverage of page.

4   Exploration of PDFs

Since we provide a large scale, highly diversified collection of PDFs downloaded from all over the Internet, we want to provide some insight into the properties of PDF files which are accessible on the Internet. To do so we randomly picked 1k documents in each of the languages in our corpus (11k documents at total) and analyzed them in terms of various properties.

Firstly, we analyzed them in terms of their creation date, the outcome of which is presented in Fig. 2. For this analysis we used the `CreationDate` field of metadata. Since most documents come with this field filled in, we were able to read the creation date for more than 99.4% of our sample. However, sometimes unreasonable values such as 1442 occurred as the creation year. As we can see, our corpus contains relatively new documents. It is an important point, because language evolves constantly, and three years ago terms such as "lockdown" or "post-pandemic" were absent from documents. Since we want our language models to represent current language and document types correctly, hence a distribution like that in Fig. 2 is desired. The spike for the year 2021 was probably caused by the use of the Common Crawl dump from May 2022. We assume that crawlers from Common Crawl usually tend to find files that are a few months old, which often means that they are from the previous year.

We also analyzed the documents in terms of the exact version of PDF standard used. The data is presented in Fig. 3. As we can see, the majority of our sample (above 76%) are PDFs prior to the 1.7 version. It is an important property, because versions 1.7 and 2.0 are defined by an ISO standard, while the older ones were defined only by Adobe proprietary standards. Some of the issues that we experienced during processing may have been caused by problems with older standards.

A PDF file contains metadata about the tool used to create it (Fig. 8). There are many different tools, and often the same tool was described by different values (for instance Microsoft Word has different names in many languages despite being the same program). The two most popular providers of PDF tools found in our corpus are Microsoft (29.8% of the sample) and Adobe (21.3% of the sample).

Fig. 8. Tools used to create PDFs.

Other properties that we were interested in were the length of the documents in terms of word count (Fig. 4), their word count per page (Fig. 5), and line count per page (Fig. 6). As we can see, there is great variability in terms of these parameters. For instance, there are many documents and pages with almost no text. Up to our manual check, most of the documents with little text are graphically rich, for instance, technical drawings, or infographics. The typical value of words per page is between 0 and 500, and the typical value of lines per page is between 0 and 55.

To provide some insight into the layout of the documents, we checked to what extent each page was covered by bounding boxes of tokens. We may look at it as part of the text coverage parameter. Distribution of this value is presented in Fig. 7. 76.2% of our sample fell into the range of 5% to 40% with respect to that parameter. Similarly to the previously described properties, once again we see a peak for empty pages. There is also a peak for pages fully or almost fully covered by text.

We were also interested in the ratio of page dimensions. 99.7% of x/y ratios were in the range of 0.4 to 2; the smallest value being 0.09, and the largest – 4.79. In our sample, 65.0% were pages with the dimension ratio close to $\sqrt{2}$, which is a standard ratio for the A, B and C paper series. 86.9% of them were vertical pages, and 13.1% – horizontal ones. Also, the LETTER format was popular; it comprised 10.6% of the sample: 92.9% documents were vertical, and 7.1% horizontal. In total, the A, B, C, and LETTER series comprised 75.5% of the sample.

To gather more information about the layout of the documents, we created heatmaps of token bounding boxes for vertical and horizontal pages (Figs. 9 and 10, respectively). As we can see, layouts with two columns of text are fairly popular, especially for horizontal pages. Also, text occurs more frequently on the right side of a page.

Fig. 9. Heatmap for vertical pages (brighter means more tokens, darker – fewer tokens).

Fig. 10. Heatmap for horizontal pages (brighter means more tokens, darker – fewer tokens).

5   Discussion

In this study we analyzed a pipeline for creating a corpus of documents. According to our experiments, the most effective way of OCR processing of PDF files is a two-step procedure. The first step consists in the classification of the files according to whether they need an OCR engine or simple text extraction is sufficient. In the second step, we process the file with either an OCR engine (in our case Tesseract) or an extraction tool (in our case the DjVu-based tool). In the former scenario, we also discovered that predefining the OCR language speed up the process substantially; unfortunately, this comes at some cost in terms of data quality. However, this cost may be mitigated by a simple heuristic which filters out documents where the predefined OCR language did not match the one discovered by the language detector. We also analyzed different language detection tools in terms of output quality and processing time, and discovered that the langdetect tool offered the best trade-off between these values.

One of the limitations of this research study was that we focused only on the processing pipeline without analyzing the impact of each project decision on the final language model. However, this kind of study would be very expensive, as it would require multiple pretrainings of a language model. Language model pretraining is a costly process in terms of money, time, and environmental impact [20].

Also, conclusions drawn from the analysis of our sample can hardly be generalized to the whole content of the Internet and only provide some insight, rather undisputed knowledge. This follows from the filtering procedure: we decided to down-sample document-rich domains and languages, therefore, statistics calculated on the whole content of the Internet may differ from the ones presented in this work.

The approach which we used to create the dataset may be reused to all of the previous Common Crawl dumps in the WARC format, of which there are 84 in total. We decided to limit ourselves to one dump only due to computational and storage limitations. One with enough computing resources may easily reproduce our pipeline and create a corpus up to 84 times larger.

6 Conclusions

Large corpora of documents are crucial for 2D language model pretraining. Recent approaches to their creation have had limitations in terms of diversity and multilinguality. Diversity of the dataset is a crucial property, as data used in the training phase impact the biases of the model. Efficient design of a pipeline for creating such a corpus has not been studied before. In this work we addressed those limitations by designing a process of downloading diversified samples of PDFs and their efficient processing. To obtain documents we used Common Crawl, which is a popular source of data for language model pretraining, but has rarely been used in the context of 2D language models. The PDF files used for this project were balanced across languages and domains, which guarantees diversity with respect to layouts and topics. To make the processing pipeline efficient in terms of computing time and data quality, we tested different strategies of OCR processing, i.e. usage of the embedded textual layer for documents not requiring OCR, and predefining the OCR language. The language detection step was also carefully analyzed.

The result of this work is an index of PDF files with their URL addresses and metadata, and the script for downloading it is available at our repository[21]. The supplied data were analyzed in terms of not only document length and layout, but also metadata connected to the PDF format (i.e., the PDF version and the creator tool), which can help understand better the dataset itself, but also give an insight into the content of the Internet.

Acknowledgments. The Smart Growth Operational Programme partially supported this research under projects no. POIR.01.01.01-00-0877/19-00 (*A universal platform for robotic automation of processes requiring text comprehension, with a unique level of implementation and service automation*) and POIR.01.01.01-00-1624/20 (*Hiper-OCR - an innovative solution for information extraction from scanned documents*).

References

1. Abadji, J., Suarez, P.O., Romary, L., Sagot, B.: Towards a cleaner document-oriented multilingual crawled corpus. ArXiv abs/2201.06642 (2022)
2. Allison, T., et al.: Research report: Building a wide reach corpus for secure parser development. In: 2020 IEEE Security and Privacy Workshops (SPW), pp. 318–326 (2020). https://doi.org/10.1109/SPW50608.2020.00066
3. Ammar, W., et al.: Construction of the literature graph in semantic scholar. In: Proceedings of the 2018 Conference of the North American Chapter of the Association for Computational Linguistics: Human Language Technologies, Volume 3 (Industry Papers), pp. 84–91. Association for Computational Linguistics, New Orleans - Louisiana (2018). https://doi.org/10.18653/v1/N18-3011. https://aclanthology.org/N18-3011
4. Biten, A.F., Tito, R., Gomez, L., Valveny, E., Karatzas, D.: OCR-IDR: OCR annotations for industry document library dataset. arXiv preprint arXiv:2202.12985 (2022)
5. Borchmann, Ł., et al.: DUE: end-to-end document understanding benchmark. In: NeurIPS Datasets and Benchmarks (2021)
6. Brown, T., et al.: Language models are few-shot learners. In: Larochelle, H., Ranzato, M., Hadsell, R., Balcan, M., Lin, H. (eds.) Advances in Neural Information Processing Systems. vol. 33, pp. 1877–1901. Curran Associates, Inc. (2020). https://proceedings.neurips.cc/paper/2020/file/1457c0d6bfcb4967418bfb8ac142f64a-Paper.pdf
7. Dodge, J., et al.: Documenting large webtext corpora: a case study on the colossal clean crawled corpus. In: Proceedings of the 2021 Conference on Empirical Methods in Natural Language Processing, pp. 1286–1305. Association for Computational Linguistics, Online and Punta Cana, Dominican Republic (2021). https://doi.org/10.18653/v1/2021.emnlp-main.98. https://aclanthology.org/2021.emnlp-main.98
8. Gao, L., et al.: The Pile: an 800gb dataset of diverse text for language modeling. arXiv preprint arXiv:2101.00027 (2020)
9. Garncarek, Ł, et al.: LAMBERT: layout-aware language modeling for information extraction. In: Lladós, J., Lopresti, D., Uchida, S. (eds.) ICDAR 2021. LNCS, vol. 12821, pp. 532–547. Springer, Cham (2021). https://doi.org/10.1007/978-3-030-86549-8_34

---

[21] https://github.com/applicaai/CCpdf.

10. Gururangan, S., et al.: Don't stop pretraining: adapt language models to domains and tasks. In: Proceedings of the 58th Annual Meeting of the Association for Computational Linguistics (2020). https://doi.org/10.18653/v1/2020.acl-main.740. http://dx.doi.org/10.18653/v1/2020.acl-main.740
11. Habernal, I., Zayed, O., Gurevych, I.: C4Corpus: Multilingual web-size corpus with free license. In: Proceedings of the Tenth International Conference on Language Resources and Evaluation (LREC2016), pp. 914–922. European Language Resources Association (ELRA), Portorož, Slovenia (2016). https://aclanthology.org/L16-1146
12. Harley, A.W., Ufkes, A., Derpanis, K.G.: Evaluation of deep convolutional nets for document image classification and retrieval. In: ICDAR (2015)
13. Huber, P., et al.: CCQA: a new web-scale question answering dataset for model pre-training (2021)
14. Lewis, D., Agam, G., Argamon, S., Frieder, O., Grossman, D., Heard, J.: Building a test collection for complex document information processing. In: Proceedings of the 29th Annual International ACM SIGIR Conference on Research and Development in Information Retrieval, pp. 665–666. SIGIR 2006, Association for Computing Machinery, New York, NY, USA (2006). https://doi.org/10.1145/1148170.1148307
15. Li, M., et al.: DocBank: a benchmark dataset for document layout analysis (2020)
16. Liu, V., Curran, J.R.: Web text corpus for natural language processing. In: 11th Conference of the European Chapter of the Association for Computational Linguistics. Association for Computational Linguistics, Trento, Italy (2006). https://www.aclweb.org/anthology/E06-1030
17. Liu, Y., et al.: RoBERTa: a robustly optimized BERT pretraining approach (2019)
18. Luccioni, A.S., Viviano, J.D.: What's in the box? An analysis of undesirable content in the Common Crawl corpus. In: ACL (2021)
19. Masson, C., Paroubek, P.: NLP analytics in finance with DoRe: a French 250M tokens corpus of corporate annual reports. In: Proceedings of The 12th Language Resources and Evaluation Conference, pp. 2261–2267. European Language Resources Association, Marseille, France (2020). https://www.aclweb.org/anthology/2020.lrec-1.275
20. Patterson, D.A., et al.: Carbon emissions and large neural network training. ArXiv abs/2104.10350 (2021)
21. Powalski, R., Borchmann, Ł., Jurkiewicz, D., Dwojak, T., Pietruszka, M., Pałka, G.: Going full-TILT boogie on document understanding with text-image-layout transformer. In: Lladós, J., Lopresti, D., Uchida, S. (eds.) ICDAR 2021. LNCS, vol. 12822, pp. 732–747. Springer, Cham (2021). https://doi.org/10.1007/978-3-030-86331-9_47
22. Qi, D., Su, L., Song, J., Cui, E., Bharti, T., Sacheti, A.: ImageBERT: cross-modal pre-training with large-scale weak-supervised image-text data (2020)
23. Raffel, C., et al.: Exploring the limits of transfer learning with a unified text-to-text transformer (2019)
24. Schwenk, H., Wenzek, G., Edunov, S., Grave, E., Joulin, A.: CCMatrix: mining billions of high-quality parallel sentences on the web. In: ACL (2021)
25. Smith, J.R., Saint-Amand, H., Plamada, M., Koehn, P., Callison-Burch, C., Lopez, A.: Dirt cheap web-scale parallel text from the common Crawl. In: Proceedings of the 51st Annual Meeting of the Association for Computational Linguistics (Volume 1: Long Papers), pp. 1374–1383. Association for Computational Linguistics, Sofia, Bulgaria (2013). https://www.aclweb.org/anthology/P13-1135
26. Smith, R.: Tesseract open source OCR engine (2022). https://github.com/tesseract-ocr/tesseract

27. Turc, I., Chang, M.W., Lee, K., Toutanova, K.: Well-read students learn better: On the importance of pre-training compact models. arXiv preprint arXiv:1908.08962v2 (2019)
28. Wenzek, G., et al.: CCNet: extracting high quality monolingual datasets from web crawl data (2019)
29. Xu, Y., et al.: LayoutLMv2: multi-modal pre-training for visually-rich document understanding. In: ACL-IJCNLP 2021 (2021)
30. Xu, Y., Li, M., Cui, L., Huang, S., Wei, F., Zhou, M.: LayoutLM: pre-training of text and layout for document image understanding. In: Proceedings of the 26th ACM SIGKDD International Conference on Knowledge Discovery & Data Mining (2020)
31. Xu, Y., et al.: LayoutXLM: multimodal pre-training for multilingual visually-rich document understanding. arXiv preprint arXiv:2004.21040 (2021)
32. Xue, L., et al.: mT5: a massively multilingual pre-trained text-to-text transformer. In: NAACL (2021)
33. Zhong, X., Tang, J., Yepes, A.J.: PubLayNet: largest dataset ever for document layout analysis. In: 2019 International Conference on Document Analysis and Recognition (ICDAR), pp. 1015–1022. IEEE (2019). https://doi.org/10.1109/ICDAR.2019.00166

ESTER-Pt: An Evaluation Suite for TExt Recognition in Portuguese

Moniele Kunrath Santos[✉] ⓘ, Guilherme Bazzo ⓘ, Lucas Lima de Oliveira ⓘ, and Viviane Pereira Moreira ⓘ

Institute of Informatics, Federal University of Rio Grande Do Sul, Porto Alegre, Brazil
{mksantos,gtbazzo,llolveira,viviane}@inf.ufrgs.br
http://www.inf.ufrgs.br

Abstract. Optical Character Recognition (OCR) is a technology that enables machines to read and interpret printed or handwritten texts from scanned images or photographs. However, the accuracy of OCR systems can vary depending on several factors, such as the quality of the input image, the font used, and the language of the document. As a general tendency, OCR algorithms perform better in resource-rich languages as they have more annotated data to train the recognition process. In this work, we propose ESTER-Pt, an **E**valuation **S**uite for **TE**xt **R**ecognition in Portuguese. Despite being one of the largest languages in terms of speakers, OCR in Portuguese remains largely unexplored. Our evaluation suite comprises four types of resources: synthetic text-based documents, synthetic image-based documents, real scanned documents, and a hybrid set with real image-based documents that were synthetically degraded. Additionally, we provide results of OCR engines and post-OCR correction tools on ESTER-Pt, which can serve as a baseline for future work.

Keywords: OCR evaluation · Digitization errors · Post-OCR correction

1 Introduction

Optical Character Recognition (OCR) is commonly used to extract the textual contents of scanned texts. The output of OCR can be noisy, especially when the quality of the input is poor and this can impact downstream tasks in Natural Language Processing (NLP). Most research on the evaluation of OCR quality was done over datasets in English. This represents a limitation since OCR is typically language-dependent.

A recent survey on post-OCR processing [37] suggests that upcoming work should focus on other languages. The main limitation for increasing language diversity is that creating datasets demands a significant human annotation effort. Nevertheless, there are some important efforts in this direction. Among those is the ICDAR competition on post-OCR text correction [41]. The organizers focused on increasing the representativity of other languages by assembling

G. A. Fink et al. (Eds.): ICDAR 2023, LNCS 14189, pp. 366–383, 2023.
https://doi.org/10.1007/978-3-031-41682-8_23

datasets for post-OCR correction in Bulgarian, Czech, Dutch, English, Finish, French, German, Polish, Spanish, and Slovak. Other multilingual datasets have also been released [2, 9].

In this work, we create resources for Portuguese, which is the 6th largest world language in the number of native speakers. Yet, it is underrepresented in terms of linguistic resources. There has been some recent work on assessing the impact of OCR errors in Portuguese [3, 38, 48] but these works were limited to a single task (*i.e.,* information retrieval). Furthermore, the only dataset for OCR in Portuguese we found was recently released and consists of only 170 annotated sentences [38].

Our main contribution is ESTER-Pt an Evaluation Suite for TExt Recognition in Portuguese. ESTER-Pt consists of four datasets of typeset documents, which are freely available[1].:

- Synthetic text-based documents: text files with artificially inserted errors and their clean version (*i.e.,* ground truth);
- Synthetic image-based documents: image-based files generated from clean texts;
- Real image-based PDFs that were synthetically degraded with their human-checked ground truths;
- Real image-based PDFs with their human-checked ground truths.

In order to serve as baselines for future work, we provide results of OCR engines and post-OCR correction on ESTER-Pt. The error rates for the different OCR engines varied significantly. In addition, the correction tools were not able to improve performance.

The remainder of this paper is organized as follows. In Sect. 2, we survey the existing literature on datasets for OCR evaluation. Section 3 details our synthetic datasets. Section 4 explains how we created the hybrid image-based dataset. The dataset with real image-based documents is described in Sect. 5. Experimental results for digitization and post-OCR correction are presented in Sect. 6. Finally, Sect. 7 concludes this paper.

2   Related Work

OCR errors can impact several NLP tasks. A number of recent works were devoted to assessing this impact in a variety of downstream tasks such as named entity recognition [13, 15, 19, 20, 22, 23], text classification [22, 51], and information retrieval [3, 27, 38, 47]. These works observed that the quality of digitization has an effect on the outputs of the downstream tasks and that strategies to reduce errors are effective. However, in order to improve OCR quality, intrinsic experiments that compare the obtained output against the ground truth are crucial. These experiments can show the causes of errors to the designers and help them refine the OCR engines. This type of experiment requires datasets

---

[1] https://zenodo.org/record/7872951#.ZEueOXbMJhE.

Table 1. OCR datasets currently available grouped by language. The type of the dataset informs whether it is Synthetic Text-Based (STB), Synthetic Image-Based (SIB), or Real Image-Based (RIB)

Language	Dataset	Type	Size
Arabic	Doush et al. [12]	SIB	8,994 pages
	Hegghammer [22]	SIB+RIB	4400 pages
Bulgarian	ICDAR2019 [41]	RIB	200 sent.
Chinese	CAMIO [2]	RIB	2,500 pages
Czech	ICDAR2019 [41]	RIB	200 sent.
Dutch	ENP [9]	RIB	19 pages
	DBNL [10]	RIB	220 sent.
	ICDAR2019 [41]	RIB	200 sent.
English	CAMIO [2]	RIB	2,500 pages
	OCR-IDL [4]	RIB	70M docs
	ICDAR2017 [8]	RIB	6M chars
	ENP [9]	RIB	50 pages
	RDD&TCP [11]	STB	10,800,000 lines
	Overproof-2 [16]	RIB	67,000 words
	Overproof-3 [16]	RIB	18,000 words
	Pontes et al. [21,28]	SIB	370 docs
	Hegghammer [22]	SIB+RIB	14,168 pages
	noisy MUC3&4 [24]	SIB	1,297 docs
	TREC-5 [26]	RIB	55,600 docs
	MiBio [31]	RIB	84,492 words
	ALTA 2017 [32]	RIB	8,000 docs
	ACL collection [36]	RIB	18,849 docs
	ICDAR2019 [41]	RIB	200 sent.
	Text+Berg Corpus [49]	RIB	1,683 sent.
	RETAS[50]	RIB	100 books
Estonian	ENP [9]	RIB	50 pages
Farsi	CAMIO [2]	RIB	2,500 pages
Finish	ENP [9]	RIB	31 pages
	ICDAR2019 [41]	RIB	393 sent.
French	ICDAR2017 [8]	RIB	6M chars
	ENP [9]	RIB	50 pages
	OCR17 [17]	RIB	30,000 lines
	ICDAR2019 [41]	RIB	3340 sent.
	Text+Berg Corpus [49]	RIB	682,452 sent.
	RETAS [50]	RIB	20 books

Language	Dataset	Type	Size
German	ENP [9]	RIB	169 pages
	ICDAR2019 [41]	RIB	10,088 sent.
	GT4HistOCR [46]	RIB	650 pages
	Text+Berg Corpus [49]	RIB	1,178,906 sent.
	RETAS [50]	RIB	20 books
Greek	GRPOLY-DB [18]	RIB	353 pages
	Polyton-DB [43]	SIB+RIB	353 pages
Hindi	CAMIO [2]	RIB	2,500 pages
Italian	Text+Berg Corpus [49]	RIB	17,723 sent.
Japanese	CAMIO [2]	RIB	2,500 pages
Kannada	CAMIO [2]	RIB	2,500 pages
Korean	CAMIO [2]	RIB	2,500 pages
Latin	Frankfurter OCR-Korpus [14]	STB	5,213 words
	GT4HistOCR [46]	RIB	300 pages
Latvian	ENP [9]	RIB	41 pages
Polish	ENP [9]	RIB	37 pages
	ICDAR2019 [41]	RIB	200 sent.
Portuguese	Oliveira et al. [38]	RIB	170 sent.
Russian	CAMIO [2]	RIB	2,500 pages
	ENP [9]	RIB	5 pages
Romansh	Text+Berg Corpus [49]	RIB	3,607 sent.
Spanish	ICDAR2019 [41]	RIB	200 sent.
	RETAS [50]	RIB	20 books
Serbian	ENP [9]	RIB	50 pages
Slovak	ICDAR2019 [41]	RIB	200 sent.
Sanskrit	Maheshwari et al. [29]	RIB	218,000 sent.
Swedish	ENP [9]	RIB	19 pages
Tamil	CAMIO [2]	RIB	2,500 pages
Thai	CAMIO [2]	RIB	2,500 pages
Ukrainian	ENP [9]	RIB	4 pages
Urdu	CAMIO [2]	RIB	2,500 pages
Vietnamese	CAMIO [2]	RIB	2,500 pages
Yiddish	ENP [9]	RIB	3 pages

with the input documents and their ground truth digitizations. Several such datasets were put together throughout the years. A survey by Nguyen et al. [37] on post-OCR correction identified 17 available datasets covering 12 languages. We took that survey as a starting point and identified nine additional available datasets [2, 4, 9, 12, 18, 22, 28, 29, 43]. The complete list of datasets organized by language is given in Table 1. We categorized each dataset according to its type, i.e., whether it is Synthetic Text-Based (STB), Synthetic Image-Based (SIB), or Real Image-Based (RIB). The sizes of the datasets are also given. There are also other works that mention OCR datasets but they do not seem to have made the datasets available [7, 30, 39, 42]

Overall, the 26 existing datasets cover 34 languages. As expected, English is the language with the largest number of datasets (17). Regarding their types,

Table 2. Statistics of the Synthetic Datasets ESTER-Pt$_{SIB}$ and ESTER-Pt$_{STB}$

Pages	5,000
Sentences	224,810
Words	5,723,565
Unique Words	469,509
Characters	27,308,006

17 datasets are RIB, five are SIB, two are STB, and two are hybrid (SIB+RIB). The datasets have a wide variation in terms of size, ranging from a few sentences to several thousand pages.

We can see that the distribution of the languages in the datasets does not reflect the distribution of speakers of the language. Widely spoken languages such as Chinese and Spanish, are underrepresented with only one and two datasets, respectively. The same can be said for Portuguese, which is the most widely spoken language in the Southern Hemisphere and the sixth worldwide in terms of the number of native speakers (over 265 million)[2] – there is a single small dataset for OCR in Portuguese with only 170 annotated sentences [38]. In this work, we address this issue by creating ESTER-Pt, an evaluation suite with real and synthetic OCR datasets for Portuguese.

3   Synthetic Datasets

Synthetic datasets have been widely used for OCR evaluation [11,12,14,22,24, 28] as they allow the generation of large datasets without the need for laborious human annotation for creating the ground truth. This alternative can be especially appealing for training data generation.

ESTER-Pt synthetic datasets have 5,000 pages with texts collected from the Portuguese Wikipedia[3]. The selected texts were taken by randomly sampling articles with at least 500 words. These texts were used as the basis for our STB and SIB datasets, which are described in the next subsections. The complete statistics on ESTER-Pt synthetic datasets are given in Table 2.

3.1   ESTER-Pt Synthetic Text-based

In order to generate our ESTER-Pt$_{STB}$ dataset, we took the clean raw texts from Wikipedia-Pt and inserted errors that simulate the pattern of OCR errors. The types of errors include character exchanges, splitting/joining words, and the addition of symbols. Character exchanges are by far the most frequent error (90%), followed by splitting words (5%), joining words (4.9%), and the addition of symbols (0.1%).

---

[2] https://en.unesco.org/sites/default/files/accord_unesco_langue_portuguaise_confe rence_generale_eng.pdf.

[3] https://dumps.wikimedia.org/ptwiki/.

Pearl Harbor ou o Porto das Pérolas é uma base naval dos Estados Unidos
e o quartel-general da frota norte-americana do Pacífico, na ilha de O'ahu,
no Havaí, perto de Honolulu. Em livros mais antigos em português,
usa-se por vezes a designação Porto das Pérolas. Antes de ser utilizado
como base militar pela marinha norte-americana [...]

(a) Clean Text (ground truth)

Pearl Harbar ou o Porto das Pérolas é uma bose naval des Estados Unidos
e o quartel-general u frota norte-americana ro Pacífico, na ilha te O'ahu,
no Hevaí, perto de Honolulu . Em livros mois antigos em pertuguês,
usa-se por vozes a designação Porto das Pérolas. Antos de ser utilizado
como base militar pela marinha norte-americana [...]

(b) Text with synthetic errors

Fig. 1. A sample from ESTER-Pt Synthetic text-based dataset. The inserted errors are highlighted in red (replaced content) and blue (lost content) (Color figure online)

We relied on a list of common character exchanges that was assembled based on the pattern of exchanges that occur in OCR engines applied to Portuguese texts. For example, "morno" → "momo" and "realmente" → "tealmente". The observed frequency of each exchange is also recorded.

In addition to the list with character exchanges, we also input the desired error rate. In the ESTER-Pt$_{STB}$, the error rate was 15%, which means that each word in the text has a 15% chance of having an error inserted. Using the error rate, we iterate through every candidate word in the document. The word is selected to be modified with a probability equivalent to the given error rate. If the word is selected, then the type of error to be inserted is determined according to the observed frequencies of errors found in real OCR tools.

This is achieved by drawing a random float between 0 and 1 and matching it against the corresponding error frequency. Values under 0.1% correspond to an erroneous symbol being added to the text (*i.e.,* commas, apostrophes, *etc.*). These symbols usually represent dirt or printing errors in scanned documents. To simulate such errors, we placed a symbol in an arbitrary position within the word. Values between 0.1% and 5% mean two sequential words would be joined. The connection between the words would happen on either side, chosen at random, unless it was a border word (at the beginning or end of the text). Additionally, words could also be split into two. This happens between 5% and 10% values, and the word is split in a random position along its length. If the randomly drawn float was above 10% then we exchanged characters based on the list of character exchanges. The exchange was made using tournament selection in ten rounds according to the frequency of the exchange. An excerpt of a document in ESTER-Pt is shown in Fig. 1.

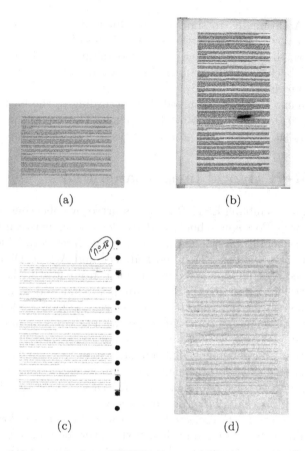

(a)   (b)

(c)   (d)

Fig. 2. Sample of documents from ESTER-Pt$_{SIB}$. (*a*) 'century school book' font on 'JulesVerne-01' background, (*b*) 'dejavu sans' font on 'manuscript-with-borders-03' background, (*c*) 'Dyuthi' font on 'bg-1', and (*d*) 'Latin Modern mono Caps' font on 'vesale-01' background.

3.2 ESTER-Pt Synthetic Image-based

ESTER-Pt$_{SIB}$ uses the same Wikipedia articles from the text-based dataset. We used the open-source software DocCreator [25] for creating synthetic document images. This tool has been used in recent OCR dataset designs [21,28,39]. Using the original Wikipedia text (which served as the ground truth) as input, we generated images that simulate real scanned pages. Examples of synthetic image-based documents are given in Fig. 2. In order to ensure a higher diversity, we selected ten different types of fonts, namely Century school book, Dejavu sans, Dyuthi, Latin Modern mono Caps, Latin Modern Roman Dunhill, LM Roman Unslanted, LM Sans Quot, TeX Gyre Chorus, URW Bookman, and URW Gothic. The tool also allows using different styles such as bold, oblique, italic, and regular. Additionally, to simulate the different types of backgrounds that can occur in real scanned pages, we used five types of image backgrounds (bg-1, bg-2, JulesVerne-01, manuscript-with-borders-03 and versale-01). The different types of fonts, backgrounds, and degradation were uniformly distributed among the pages in our dataset.

The document images have mixed sizes according to the background image, varying from 1920×1370 to 1160×1790 pixels. The size of the margins varied between 50 and 100 pixels. Font size varied from 9 to 18 points.

4   ESTER-Pt Hybrid Image-Based PDFs

The process for assembling ESTER-Pt$_{HIB}$ started by collecting real PDF documents. Those were Portuguese books and articles written between 1949 and 1999. The documents consist of scanned pages that are freely available on Brazilian government websites, namely Domínio Público[4] and Acervo Digital[5]. The documents selected to compose our dataset come from five books [1,6,40,44,45] and three letters written by the Brazilian politician and journalist Joaquim Nabuco [33–35]. From the collected corpus, we randomly selected 224 pages with mainly textual contents, omitting pages with tables, summaries, and appendices. Although all books were scanned, in some cases the PDF was searchable. This means that the documents had already gone through OCR. In order to get image-only PDFs, we transformed them into images and then back to PDFs. The resulting dataset consists of image-based PDFs without searchable contents. The statistics on ESTER-Pt$_{HIB}$ are in Table 3.

The ground truth for the selected pages was obtained as follows. For the letters by Joaquim Nabuco, we found their textual version on the Domínio Público website. For the books, we extracted the texts from the original searchable PDF documents and then manually revised their contents so that they match the image-based documents. Nine human volunteers carried out the revision.

To add further noise to the documents and increase digitization difficulty, we applied different types of degradation using DocCreator [25]. First, we randomly divided the dataset into eight subsets containing 28 pages each. Then, a single degradation was applied to each of the subsets. Thus, we have a homogeneous distribution of image deterioration across the dataset. The degradation models used in this work are described in more detail below.

- *Ink degradation*: simulates character deterioration found in old documents by creating small ink spots near random characters or erasing some ink area around the characters. The variability parameters are a minimum number and a maximum number of degradation occurrences, varying from one to ten (one meaning very low occurrences and ten very high occurrences). Within this range, a random number is chosen to define how many times this deformation will appear in the document. We used a degradation range of four to ten.

---

[4] http://www.dominiopublico.gov.br/.
[5] http://bndigital.bn.gov.br/acervodigital/.

(a) without degradation

(b) after applying degradation to simulate holes

Fig. 3. Sample of a document in ESTER-Pt$_{HIB}$

- *Phantom Character*: is a typical defect that often appears in documents that were manually printed several times. The phantom character model provides an algorithm that reproduces such ink apparitions around the characters. This model has an occurrence frequency parameter, which ranges from 0 to 1 (0 meaning very low frequency and 1 very high frequency). We set the phantom character frequency parameter to 0.85.
- *Bleed-through*: this degradation mimics the verso ink that seeps through the current side. To give this effect, the template randomly selects a document from a pre-selected set as a front image that will be overlaid with the target image. The parameter for this distortion is randomly chosen among opacity levels ranging from 1 to 100 (where 1 represents almost no visibility of the overlaid document and 100 is the maximum visibility of the degradation). We used values between 27 and 50.
- *Gaussian noise*: is a common digital image damage, caused mainly by poor illumination and high temperatures. It generally refers to a random variation in the brightness or color of an image. Its parameters are a variance threshold ranging from 0 to infinity, where larger values indicate more noise. We set the variance between [0.2,0.5].
- *Salt and pepper noise*: is a random disturbance in a digital image, caused when only some of the pixels are noisy. The percentage of pixels to be salted (whitened) varied between 0.1 and 0.2.
- *Rotation*: Images were rotated from −7.5 to 7.5°C.
- *Adaptive blur*: this is a very common defect in scanned documents. We set the blur radius between one and four, where a larger radius means lower contrast.

Table 3. Statistics of the hibrid image-based dataset ESTER-Pt$_{HIB}$

Pages	224
Sentences	4,450
Words	114,283
Unique Words	24,873
Characters	528,580

– *Paper holes*: this image perturbation simulates holes found in real old documents. Holes are represented by black regions, as illustrated in Fig. 3. In addition, it is possible to set the areas where the holes will appear in the document. We applied holes in the center, border, and corner zones, with a minimum parameter of two and a maximum parameter of ten in a range that goes from one to ten (one meaning very low occurrences and ten very high occurrences).

5   ESTER-Pt Real Image-Based PDFs

Our final dataset, ESTER-Pt$_{RIB}$, consists of real image-based PDFs corresponding to scanned books written between 1800 and 1910. More specifically, the books primarily consist of novel and poetry. The nine books we collected were "A Cidade do Vício" (1882), "Annos de Prosa" (1863), "Carlota Ângela" (1858), "Histórias Brasileiras" (1874), "Histórias Sem Data" (1884), "Nossa Gente (1900)", "O Guarany" (1857), "Poesias Herculano" (1860), and "Relíquias de Casa Velha" (1906). The printed digitized version of these books was obtained from the collection of the Acervo Digital[6].

For the ground truths, we resorted to the repository of Project Gutenberg[7], an initiative dedicated to digitizing and archiving books and cultural works. we collected the ground truth data for the nine books written. The ground truths consist of ebooks that correspond to the same edition, year, and volume as the scanned books. The data collected from the Gutenberg repository includes page markings that enabled us to create a dataset comprising 2,028 pages. The complete statistics for ESTER-Pt$_{RIB}$ are given in Table 3.

Ground truth generation consisted in matching the page markings from the ebooks with the printed pages. The following steps were performed:

– We first converted the ebook file from HTML to plain text.
– Then, we separated the pages using the page numbering indicated in the text by curly brackets and created a text document for each page.
– Next, we cleaned up the text to remove extra spaces and the specific special symbols '', '#', '_' and '*', which originate from HTML.

---

[6] http://bndigital.bn.gov.br/acervodigital/.
[7] https://www.gutenberg.org/.

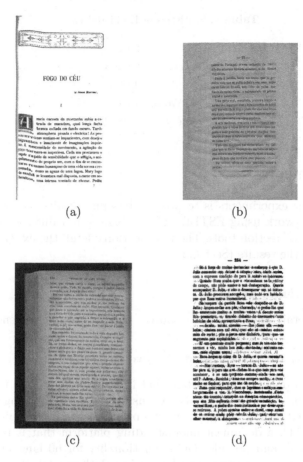

(a)   (b)

(c)   (d)

Fig. 4. Sample of documents from ESTER-Pt$_{RIB}$. (a) page 82 from 'Nossa Gente', (b) page 7 from 'O Guarany', (c) page 108 from 'Reliquias de Casa Velha', and (d) page 230 from 'Annos de Prosa'.

– In the digitized PDFs, we divided the book into pages and then converted each page into an image using a resolution of 130DPI.

Each book contains approximately 200 pages. However, "O Guarany" could only be entirely matched because the ground truth does not include page numbers for the complete book. As a result, only 50 pages of this book were included in the dataset (Table 4).

Figure 4 shows samples of pages from different books in ESTER-Pt$_{RIB}$. It shows documents with different styles that may improve the generalization capability of machine learning algorithms.

Table 4. Statistics of ESTER-Pt$_{RIB}$

Pages	2,028
Sentences	24,003
Words	488,448
Unique Words	73,198
Characters	2,485,988

6 Experiments

The goal of our experiments is to provide baseline results to serve as a reference for future work using ESTER-Pt. We have tested different OCR engines and post-OCR correction tools. The next sections detail the experimental setup (Sect. 6.1) and the results (Sect. 6.2).

6.1 Experimental Setup

This section details the tools and resources used in our experiments.

OCR Engines: We experimented with two OCR engines:

- Tesseract[8], a popular OCR system that is extensively used in the related literature [2,20,22]. Processing was done using version 5.2.0 using the extractor model trained on Portuguese texts. No preprocessing was applied.
- Document AI[9], a document understanding platform that is part of Google Cloud Services and provides OCR functionality for 60 languages, including Portuguese. The extraction was done using Python v3.9.12 and the Python library google-cloud-vision (v3.3.1) to access the Document AI REST API.

Post-OCR Correction Tools: The following tools were used to try and correct the OCR errors:

- SymSpell[10], a language-independent spelling corrector. This tool is based on the Symmetric Delete algorithm, which provides fast processing. It performs dictionary lookups of unigrams and bigrams using predefined lists and Levenshtein's edit distance.
- sOCRates [48], a post-OCR text corrector for Portuguese texts. sOCRates has an error detection phase that is based on a BERT classifier trained to identify sentences with errors. Then, for the sentences identified as containing errors, a second classifier using format, semantic, and syntactic similarity features suggests the corrections.

---

[8] https://github.com/tesseract-ocr/tesseract.
[9] https://cloud.google.com/document-ai.
[10] https://github.com/wolfgarbe/SymSpell.

Evaluation Metrics: In order to evaluate the results of the OCR and correction tools, we calculated the standard metrics *Character Error Rate* (CER) and *Word Error Rate* (WER) [5]. The OCREvaluation script[11] was used to perform the calculations. These metrics consider the number of differences at character and word level between the output of the tools and the ground truth.

6.2   Results

The results are shown in Table 5. OCR is only applicable to the ESTER-Pt$_{SIB}$, ESTER-Pt$_{HIB}$, and ESTER-Pt$_{RIB}$ since the documents in ESTER-Pt$_{STB}$ are already in plain text. The correction tools were applied to all datasets. Looking at the error rates in the different datasets, we can see that ESTER-Pt$_{SIB}$ is the noisiest of our datasets, with a WER of 40 in Tesseract and 28 in Document AI. As expected, the WER in ESTER-Pt$_{STB}$ is very close to the desired error rate that was used to create the dataset (15%).

Table 5. Error rates for OCR digitization and Post-OCR correction on ESTER-Pt. Best scores in bold.

Dataset	OCR Engine	CER	WER
ESTER-Pt$_{STB}$	Text with errors	**2.85**	14.86
	+sOCRates	5.12	**11.97**
	+Symspell	5.61	17.07
ESTER-Pt$_{SIB}$	Tesseract	**27.91**	40.44
	+sOCRates	29.18	**39.90**
	+SymSpell	31.20	45.87
	Document AI	**18.05**	**28.03**
	+sOCRates	19.44	28.71
	+SymSpell	20.78	31.69
ESTER-Pt$_{HIB}$	Tesseract	**23.11**	**27.93**
	+sOCRates	24.78	28.70
	+SymSpell	26.02	31.49
	Document AI	**5.25**	**8.58**
	+sOCRates	7.16	10.36
	+SymSpell	8.93	14.36
ESTER-Pt$_{RIB}$	Tesseract	**6.36**	**15.72**
	+sOCRates	9.88	22.34
	+SymSpell	10.10	24.58
	Document AI	**3.71**	**9.49**
	+sOCRates	7.50	17.43
	+SymSpell	8.08	20.04

---

[11] https://github.com/impactcentre/ocrevalUAtion.

	(a)	(b)
Image	 Ge~~túlio Vargas,~~	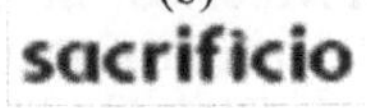 **sacrifício**
Tesseract	Ge?	sacrificio
+sOCRates	Ge?	sacrificio
+SymSpell	Ge.	sacrificio
Doc AI	Getulle-Vergas	sacrificio
+sOCRates	Getúlio-varguense	sacrificio
+SymSpell	Getu le Vergas	sacrificio
Ground Truth	Getúlio Vargas	sacrifício
	(c)	(d)
Image	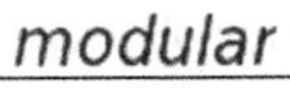 *modular*	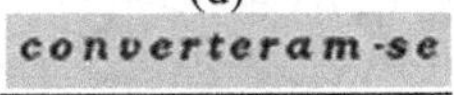 *converteram-se*
Tesseract	modular	converteram-se
+sOCRates	modelar	envaretaram-se
+SymSpell	modular	converteram se
Doc AI	modular	converteram-se
+sOCRates	modelar	envaretaram-se
+SymSpell	modular	converteram se
Ground Truth	modular	converteram-se

Fig. 5. Excerpts of words from pages in ESTER-Pt$_{SIB}$ and their extracted versions. In a (a), none of the tools was able to match the ground truth. In (b), Document AI made a mistake in the accented character and it was fixed by both correctors. In (c) and (d) sOCRates inserted an error in a word that had been correctly recognized by the OCR engines.

Comparing the OCR engines, we can see that Document AI consistently out-performed Tesseract by a wide margin, especially in ESTER-Pt$_{HIB}$. Still, it is worth mentioning that we did not explore the full potential of Tesseract. Applying preprocessing operations such as rescaling, binarization, noise removal, and deskewing would certainly yield better scores. In addition, Tesseract an open-source engine while Document AI is paid. As a general tendency, we observed that many recognition errors in both engines involved accented characters (*e.g.,* à, é, ó) that are very frequent in Portuguese. In Tesseract, common confusions occur between 't' and 'l', and also between 'f' and 't'. The errors in Document AI are more frequent in named entities. This is expected as it is known that named entities represent a challenge for OCR. Regarding processing time, Tesseract was approximately five times faster than Document AI. While Document AI took 15 min to process a sample of 100 pages from ESTER-Pt$_{HIB}$, Tesseract took only three minutes.

The post-OCR correction tools were not able to improve results. The CER scores were always worse with the correction tools. In terms of WER, sOCRates was able to slightly improve results in ESTER-Pt$_{STB}$ and in ESTER-Pt$_{SIB}$ with Tesseract. An important challenge this dataset poses to correction systems is that the orthography of several words in older books changed over time. Thus, a correction system may interpret the different orthographies as typos and try

Table 6. Error Rates on ESTER-Pt$_{HIB}$ for the different types of degradation

Degradation	Tesseract		Document AI	
	CER	WER	CER	WER
Ink degradation	8.02	26.63	2.00	5.66
Phantom character	1.37	4.0	0.91	1.99
Bleed-through	1.43	3.36	12.68	17.17
Gaussian noise	0.92	2.51	1.31	2.23
Salt and pepper noise	99.27	99.40	4.34	9.87
Rotation	54.32	55.99	3.34	4.71
Adaptive blur	2.14	5.58	1.09	2.08
Paper Hole	18.08	27.67	15.87	23.45

to correct them. Nevertheless, they managed to fix some OCR errors as can be seen in Fig. 5. Comparing sOCRates and SymSpell, the former was better, especially in terms of WER. On the other hand, SymSpell has a significantly lower processing time.

Table 6 shows error rates for both OCR engines in ESTER-Pt$_{HIB}$ grouped by degradation type. We can see that the engines were affected differently by the types of degradation. For Tesseract, Salt and Pepper noise was the most problematic, while, for Document AI, Paper Holes impacted the most. Tesseract was more robust to Bleed-trough than Document AI, and the situation reverses for Rotations.

Experiments carried out by Bazzo et al. [3] found that the quality of the ranking produced by information retrieval systems is significantly affected by WER scores above 5%. Looking at the results in Table 5, we see that the best WER score was 8.58. Thus, if the documents in ESTER-Pt were used for information retrieval, it is likely that some relevant documents would not be retrieved. This shows that there are still opportunities for improvement in OCR tools.

7 Conclusion

OCR quality depends on the availability of corpora and annotated datasets. While some languages have more resources, some are clearly underrepresented – that is the case for Portuguese. In order to close that gap, in this paper, we describe the creation of ESTER-Pt, an evaluation suite for text recognition in Portuguese. ESTER-Pt is composed of four types of datasets, synthetic text-based, synthetic image-based, real image-based, and a hybrid version in which real images were synthetically degraded. The synthetic datasets have 5K pages, the real dataset has 2K, and the hybrid dataset has 224 pages.

We performed experiments with two OCR engines and two post-OCR tools on ESTER-Pt. The results showed that even the best error rates are still within ranges that would significantly impact downstream tasks. Thus, we conclude that there is still room for improving the results of digitization and correction tools. In particular, we found that accented characters and named entities tend to pose challenges for these tools.

In future work, we plan to use ESTER-Pt as a source of training data for machine learning models for text recognition.

Acknowledgment. This work has been financed in part by CAPES Finance Code 001 and CNPq/Brazil.

References

1. Almeida, H.D.: Augusto dos Anjos - Um Tema para Debates. Apex (1970)
2. Arrigo, M., Strassel, S., King, N., Tran, T., Mason, L.: CAMIO: a corpus for OCR in multiple languages. In: Proceedings of the Thirteenth Language Resources and Evaluation Conference, pp. 1209–1216 (2022)
3. Bazzo, G.T., Lorentz, G.A., Vargas, D.S., Moreira, V.P.: Assessing the impact of OCR errors in information retrieval. In: European Conference on Information Retrieval, pp. 102–109 (2020)
4. Biten, A.F., Tito, R., Gomez, L., Valveny, E., Karatzas, D.: OCR-IDL: OCR annotations for industry document library dataset. arXiv preprint arXiv:2202.12985 (2022)
5. Carrasco, R.C.: An open-source OCR evaluation tool. In: Proceedings of the First International Conference on Digital Access to Textual Cultural Heritage, pp. 179–184 (2014)
6. de Carvalho, G.V.: Biografia da Biblioteca Nacional, 1807–1990. Editora Irradiação Cultural (1994)
7. Chen, J., et al.: Benchmarking Chinese text recognition: datasets, baselines, and an empirical study. arXiv preprint arXiv:2112.15093 (2021)
8. Chiron, G., Doucet, A., Coustaty, M., Moreux, J.P.: ICDAR2017 competition on post-OCR text correction. In: 2017 14th IAPR International Conference on Document Analysis and Recognition (ICDAR), vol. 1, pp. 1423–1428 (2017)
9. Clausner, C., Papadopoulos, C., Pletschacher, S., Antonacopoulos, A.: The ENP image and ground truth dataset of historical newspapers. In: 2015 13th International Conference on Document Analysis and Recognition (ICDAR), pp. 931–935. IEEE (2015)
10. DBNL: DBNL OCR data set, June 2019. https://doi.org/10.5281/zenodo.3239290
11. Dong, R., Smith, D.A.: Multi-input attention for unsupervised OCR correction. In: Proceedings of the 56th Annual Meeting of the Association for Computational Linguistics (Volume 1: Long Papers), pp. 2363–2372 (2018)
12. Doush, I.A., AIKhateeb, F., Gharibeh, A.H.: Yarmouk arabic OCR dataset. In: 2018 8th International Conference on Computer Science and Information Technology (CSIT), pp. 150–154 (2018)
13. Dutta, H., Gupta, A.: PNRank: unsupervised ranking of person name entities from noisy OCR text. Decis. Support Syst. **152**, 113662 (2022)
14. Eger, S., vor der Brück, T., Mehler, A.: A comparison of four character-level string-to-string translation models for (OCR) spelling error correction. Prague Bull. Math. Linguist. **105**(1), 77 (2016)

15. Ehrmann, M., Hamdi, A., Pontes, E.L., Romanello, M., Doucet, A.: Named entity recognition and classification on historical documents: a survey. arXiv preprint arXiv:2109.11406 (2021)
16. Evershed, J., Fitch, K.: Correcting noisy OCR: context beats confusion. In: Proceedings of the First International Conference on Digital Access to Textual Cultural Heritage, pp. 45–51 (2014)
17. Gabay, S., Clérice, T., Reul, C.: OCR17: ground truth and models for 17th c. French Prints (and hopefully more), May 2020. https://hal.science/hal-02577236
18. Gatos, B., et al.: Grpoly-db: an old Greek Polytonic document image database. In: 2015 13th International Conference on Document Analysis and Recognition (ICDAR), pp. 646–650. IEEE (2015)
19. Gupte, A., et al.: Lights, camera, action! a framework to improve NLP accuracy over OCR documents (2021)
20. Hamdi, A., Jean-Caurant, A., Sidère, N., Coustaty, M., Doucet, A.: Assessing and minimizing the impact of OCR quality on named entity recognition. In: Hall, M., Merčun, T., Risse, T., Duchateau, F. (eds.) TPDL 2020. LNCS, vol. 12246, pp. 87–101. Springer, Cham (2020). https://doi.org/10.1007/978-3-030-54956-5_7
21. Hamdi, A., Pontes, E.L., Sidere, N., Coustaty, M., Doucet, A.: In-depth analysis of the impact of OCR errors on named entity recognition and linking. Nat. Lang. Eng. 1–24 (2022)
22. Hegghammer, T.: OCR with tesseract, amazon textract, and google document AI: a benchmarking experiment. J. Comput. Soc. Sci. , 1–22 (2021). https://doi.org/10.1007/s42001-021-00149-1
23. Huynh, V.-N., Hamdi, A., Doucet, A.: When to use OCR post-correction for named entity recognition? In: Ishita, E., Pang, N.L.S., Zhou, L. (eds.) ICADL 2020. LNCS, vol. 12504, pp. 33–42. Springer, Cham (2020). https://doi.org/10.1007/978-3-030-64452-9_3
24. Jean-Caurant, A., Tamani, N., Courboulay, V., Burie, J.C.: Lexicographical-based order for post-OCR correction of named entities. In: 2017 14th IAPR International Conference on Document Analysis and Recognition (ICDAR), vol. 1, pp. 1192–1197. IEEE (2017)
25. Journet, N., Visani, M., Mansencal, B., Van-Cuong, K., Billy, A.: DocCreator: a new software for creating synthetic ground-truthed document images. J. Imaging 3(4), 62 (2017)
26. Kantor, P.B., Voorhees, E.M.: The TREC-5 confusion track: comparing retrieval methods for scanned text. Inf. Retrieval 2, 165–176 (2000)
27. Kettunen, K., Keskustalo, H., Kumpulainen, S., Pääkkönen, T., Rautiainen, J.: OCR quality affects perceived usefulness of historical newspaper clippings-a user study. arXiv preprint arXiv:2203.03557 (2022)
28. Linhares Pontes, E., Hamdi, A., Sidere, N., Doucet, A.: Impact of OCR quality on named entity linking. In: Jatowt, A., Maeda, A., Syn, S.Y. (eds.) ICADL 2019. LNCS, vol. 11853, pp. 102–115. Springer, Cham (2019). https://doi.org/10.1007/978-3-030-34058-2_11
29. Maheshwari, A., Singh, N., Krishna, A., Ramakrishnan, G.: A benchmark and dataset for post-OCR text correction in sanskrit. arXiv preprint arXiv:2211.07980 (2022)
30. Martínek, J., Lenc, L., Král, P.: Training strategies for OCR systems for historical documents. In: MacIntyre, J., Maglogiannis, I., Iliadis, L., Pimenidis, E. (eds.) AIAI 2019. IAICT, vol. 559, pp. 362–373. Springer, Cham (2019). https://doi.org/10.1007/978-3-030-19823-7_30

31. Mei, J., Islam, A., Moh'd, A., Wu, Y., Milios, E.: Post-processing OCR text using web-scale corpora. In: Proceedings of the 2017 ACM Symposium on Document Engineering, pp. 117–120 (2017)
32. Molla, D., Cassidy, S.: Overview of the 2017 ALTa shared task: correcting OCR errors. In: Proceedings of the Australasian Language Technology Association Workshop 2017, pp. 115–118 (2017)
33. Nabuco, J.: Um estadista do Império: Nabuco de Araujo: sua vida, suas opiniões, sua época, por seu filho Joaquim Nabuco (Tomo 3). H. Garnier, Rio de Janeiro (1897)
34. Nabuco, J.: Cartas aos abolicionistas ingleses. Joaquim Nabuco, Massangana (1985)
35. Nabuco, J.: O abolicionismo. Centro Edelstein (2011)
36. Nastase, V., Hitschler, J.: Correction of OCR word segmentation errors in articles from the ACL collection through neural machine translation methods. In: Proceedings of the Eleventh International Conference on Language Resources and Evaluation (LREC 2018) (2018)
37. Nguyen, T.T.H., Jatowt, A., Coustaty, M., Doucet, A.: Survey of post-OCR processing approaches. ACM Comput. Surv. (CSUR) **54**(6), 1–37 (2021)
38. de Oliveira, L.L., et al.: Evaluating and mitigating the impact of OCR errors on information retrieval. Int. J. Digit. Libr. **24**, 45–62 (2023). https://doi.org/10.1007/s00799-023-00345-6
39. Pack, C., Liu, Y., Soh, L.K., Lorang, E.: Augmentation-based pseudo-ground truth generation for deep learning in historical document segmentation for greater levels of archival description and access. J. Comput. Cult. Heritage (JOCCH) **15**(3), 1–21 (2022)
40. Ribeiro, N.: Albrecht Dürer: o apogeu do Renascimento alemão (1999)
41. Rigaud, C., Doucet, A., Coustaty, M., Moreux, J.P.: ICDAR 2019 competition on post-OCR text correction. In: 2019 International Conference on Document Analysis and Recognition (ICDAR), pp. 1588–1593 (2019)
42. Saini, N., et al.: OCR synthetic benchmark dataset for Indic languages. arXiv preprint arXiv:2205.02543 (2022)
43. Simistira, F., Ul-Hassan, A., Papavassiliou, V., Gatos, B., Katsouros, V., Liwicki, M.: Recognition of historical Greek Polytonic scripts using LSTM networks. In: 2015 13th International Conference on Document Analysis and Recognition (ICDAR), pp. 766–770. IEEE (2015)
44. Sodré, N.W.: Brasil: radiografia de um modelo. Vozes (1975)
45. Sodré, N.W.: História da imprensa no Brasil. Mauad Editora Ltda (1998)
46. Springmann, U., Reul, C., Dipper, S., Baiter, J.: Ground truth for training OCR engines on historical documents in German fraktur and early modern Latin. J. Lang. Technol. Comput. Linguis. **33**(1), 97–114 (2018)
47. van Strien, D., Beelen, K., Ardanuy, M.C., Hosseini, K., McGillivray, B., Colavizza, G.: Assessing the impact of OCR quality on downstream NLP tasks. In: Proceedings of the 12th International Conference on Agents and Artificial Intelligence, ICAART, pp. 484–496 (2020)
48. Vargas, D.S., de Oliveira, L.L., Moreira, V.P., Bazzo, G.T., Lorentz, G.A.: sOCRates-a post-OCR text correction method. In: Anais do XXXVI Simpósio Brasileiro de Bancos de Dados, pp. 61–72 (2021)
49. Volk, M.: The text+berg corpus: an alpine French-German parallel resource (2011)

50. Yalniz, I.Z., Manmatha, R.: A fast alignment scheme for automatic OCR evaluation of books. In: 2011 International Conference on Document Analysis and Recognition, pp. 754–758. IEEE (2011)
51. Zosa, E., Mutuvi, S., Granroth-Wilding, M., Doucet, A.: Evaluating the robustness of embedding-based topic models to OCR noise. In: Ke, H.-R., Lee, C.S., Sugiyama, K. (eds.) ICADL 2021. LNCS, vol. 13133, pp. 392–400. Springer, Cham (2021). https://doi.org/10.1007/978-3-030-91669-5_30

Augraphy: A Data Augmentation Library
for Document Images

Alexander Groleau[1], Kok Wei Chee[1], Stefan Larson[2(✉)], Samay Maini[1],
and Jonathan Boarman[1]

[1] Sparkfish LLC, Addison, USA
augraphy@sparkfish.com
[2] Vanderbilt University, Nashville, USA
stefan.larson@vanderbilt.edu

Abstract. This paper introduces *Augraphy*, a Python library for constructing
data augmentation pipelines which produce distortions commonly seen in real-
world document image datasets. *Augraphy* stands apart from other data augmentation tools by providing many different strategies to produce augmented versions of clean document images that appear as if they have been altered by standard office operations, such as printing, scanning, and faxing through old or dirty
machines, degradation of ink over time, and handwritten markings. This paper
discusses the *Augraphy* tool, and shows how it can be used both as a data augmentation tool for producing diverse training data for tasks such as document
denoising, and also for generating challenging test data to evaluate model robustness on document image modeling tasks.

1 Introduction and Motivation

The *Augraphy* library is designed to introduce realistic degradations to images of documents, producing training datasets for supervised learning by pairing synthetic dirty
documents with their clean ground truth document image sources. The need for such a
data augmentation tool in document image analysis is twofold: existing document analytics software must be robust against various forms of noise, and noise-free ground
truth images are unavailable for most of the documents recorded within human history.
Automated recovery of knowledge stored in such documents is thwarted by the presence
of distortion artifacts, which typically increase with time. As such, there is an ongoing
and growing need to improve both document restoration techniques and software processes, unlocking contents for use in search or other tasks and providing challenging
data for tooling development. *Augraphy* answers this demand by supporting document-
modality tasks such as reconstruction (Fig. 1), denoising (Fig. 5), and robustness testing
(Fig. 6).

An ongoing digital transformation has reduced everyday reliance on paper. However, major industries such as healthcare [36], insurance [15], legal [46], financial services [10], government [1], and education [2] continue to produce physical documents
that are later digitally archived. Paper documents accumulate physical damage and alterations through printing, scanning, photocopying and regular use. Real-world processes

G. A. Fink et al. (Eds.): ICDAR 2023, LNCS 14189, pp. 384–401, 2023.
https://doi.org/10.1007/978-3-031-41682-8_24

introduce many types of distortions that result in color changes, shadows in scanned documents, and degraded text where low ink, annotations, highlighters, or other markings add noise that hinder semantic access by digital means.

Noisy Original   Clean Reproduction   *Augraphy* Reproduction

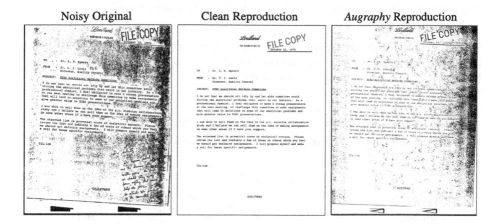

Fig. 1. *Augraphy* can be used to introduce noisy perturbations to document images like the original noise seen above, a real-life sample from RVL-CDIP. We demonstrate this by reproducing a clean image, then introducing noise with *Augraphy*.

Document image analysis tasks are impacted by the presence of such noise; high-level tasks like document classification and information extraction are frequently expected to perform well even on noisily-scanned document images. For instance, the RVL-CDIP document classification corpus [17] consists of scanned document images, many of which have substantial amounts of scanner-induced noise, as does the FUNSD form understanding benchmark [24]. Other intermediate-level tasks like optical character recognition (OCR) and page layout analysis may perform optimally if noise in a document image is minimized [9,39,42].

Attempts have been made to denoise document images in order to improve document tasks [7,14,30,37,38]. However, they have lacked the benefit of large datasets with ground truth images to train those denoisers. A tool such as *Augraphy* that helps to produce copious amounts of training data with document noise artifacts is essential in this space to developing advanced document binarization and denoising processes.

For this reason we introduce *Augraphy*, an open-source Python-based data augmentation library for generating versions of document images that contain realistic noise artifacts commonly introduced via scanning, photocopying, and other office procedures. *Augraphy* differs from most image data augmentation tools by specifically targeting the types of alterations and degradations seen in document images.

Augraphy offers *30* individual augmentation methods out-of-the-box across three "phases" of augmentations, and these individual phase augmentations can be composed together along with a "paper factory" step where different paper backgrounds can be added to the augmented image. The resulting images are realistic, noisy versions of clean documents, as evidenced in Fig. 1, where we apply *Augraphy* augmentations to a clean document image in order to mimic the types of noise seen in a real-world noisy document image from RVL-CDIP.

Several research efforts have used *Augraphy*: Larson et al. (2022) [31] used *Augraphy* to mimic scanner-like noise for evaluating document classifiers trained on RVL-CDIP; Jadhav et al. (2022) [22] used *Augraphy* to generate noisy document images for training a document denoising GAN; and Kim et al. (2022) [28] used *Augraphy* as part of a document generation pipeline for document understanding tasks.

Table 1. Comparison of various image-based data augmentation libraries. Number of augmentations is a rough count, and many augmentations in other tools are what *Augraphy* calls Utilities. Further, many single augmentations in *Augraphy* — geometric transforms, for example — are represented by multiple classes in other libraries.

Library	Number of Augmentations	Document Centric	Pipeline Based	Python	License
Augmentor [4]	27	✗	✓	✓	MIT
Albumentations [5]	216	✗	✓	✓	MIT
imgaug [26]	168	✗	✓	✓	MIT
Augly [40]	34	✗	✓	✓	MIT
DocCreator [25]	12	✓	✗	✗	LGPL-3.0
***Augraphy* (ours)**	*30*	✓	✓	✓	**MIT**

2   Related Work

This section reviews prior work on data augmentation and available tooling for robustness testing, particularly in the context of document understanding, processing, and analysis tasks.

Table 1 compares *Augraphy* with other image augmentation libraries and tools, highlighting that most existing data augmentation libraries do not specifically cater to the corruptions typically encountered in document image analysis datasets. Instead of general image manipulation, *Augraphy* focuses on document-specific distortions and alterations that are better-suited to challenging and evaluating OCR, object-detection, and other models with images which still retain sufficient quality to remain human-readable. Crucially, *Augraphy* is also designed to be easily integrated into machine learning pipelines alongside other popular Python packages used in machine learning and document analysis.

2.1   Data Augmentation

A wide variety of data augmentation tools and pipelines exist for machine learning tasks, including natural language processing (e.g., [12, 13, 48]), audio and speech processing (e.g., [29, 34, 35]), and computer vision and image processing.

As it pertains to computer vision, popular tools such as *Augly* [40], *Augmentor* [4], *Albumentations* [5], and *imgaug* [26] provide image-centric augmentation strategies including general-purpose transformations like rotations, warps, masking, and color modifications. In contrast, as seen in Table 1, only *Augraphy* offers a data augmentation library that supports imitating the corruptions commonly seen in document analysis corpora and is also easily integrated in modern machine learning training processes which rely on Python, including use of specialized paper, ink and post-processing pipelines not seen elsewhere. Our own bst-effort attempts to use *Albumentations* to recreate the document in Fig. 1 yielded results so poor and unlike the target that we could not in good faith include them in this paper.

DocCreator [25] is the next best option, providing an image-synthesizing tool for creating synthetic images that mimic common corruptions seen in document collections. However, *DocCreator* differs from *Augraphy* in several crucial ways. The first difference is that *DocCreator*'s augmentations are meant to imitate those seen in historical (e.g., ancient or medieval) documents, while *Augraphy* is intended to replicate a wider variety of both historical and modern scenarios. *DocCreator* was also written in the C++ programming language as a monolithic *what-you-see-is-what-you-get (WYSI-WYG)* tool, and does not have a scripting or API interface to enable use in a broader machine learning pipeline. *Augraphy*, on the other hand, is written in Python and can be easily integrated into machine learning model development and evaluation pipelines, alongside other Python packages like PyTorch [41].

2.2 Robustness Testing

Testing a model's robustness with challenging datasets is a vital step to evaluate a model's resilience against real-world scenarios, helping developers identify and address weaknesses for enhanced effectiveness in document processing tasks.

Prior work for evaluating image classification and object detection models includes image blur; contrast, brightness and color alterations; masking; geometric transforms; pixel-level noise; and compression artifacts (e.g., [11, 19–21, 27, 44, 47]). More specific to the document understanding field, recent prior work has used basic noise-like corruptions to evaluate the robustness of document classifiers trained on RVL-CDIP [43]. Our paper also uses robustness testing of OCR and face-detection models, as a way to showcase *Augraphy's* efficacy in producing document-centric distortions which challenge these models while remaining human-readable.

3   Document Distortion, Theory and Technique

As referenced in the *Related Work* section of this paper, many general-purpose image distortion techniques exist for data augmentation and robustness testing. While these techniques are useful in general image analysis and understanding, they bear little resemblance to the features commonly found in real-world documents.

Augraphy's suite of augmentations was designed to faithfully reproduce the level of complexity in the document lifecycle as seen in the real world. Such real-world conditions and features have direct implementations either within the library or on the development roadmap.

Some techniques exist for introducing these features into images of documents, including but not limited to the following:

1. Text can be generated independently of the paper texture, overlaid onto the "paper" by a number of blending functions, allowing use of a variety of paper textures.
2. Similarly, any markup features may be generated and overlaid by the same methods.
3. Documents can be digitized with a commercial scanner, or converted to a continuous analog signal and back with a fax machine.
4. The finished document image can be used as a texture and attached to a 3D mesh, then projected back to 2 dimensions to simulate physical deformation. DocCreator's 3D mesh support here is excellent and has inspired *Augraphy* to add similar functionality in its roadmap. In the meantime, *Augraphy* currently offers support for paper folds and book bindings.

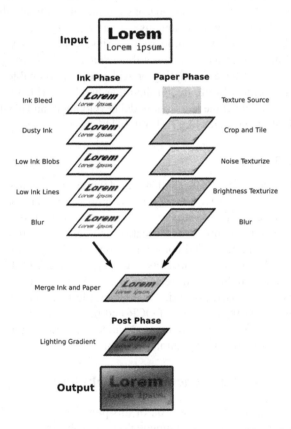

Fig. 2. Visualization of an *Augraphy* pipeline, showing the composition of several image augmentations together with a specific paper background

Augraphy attempts to decompose the lifetime of features accumulating in a document by separating the pipeline into *three phases*: **ink**, **paper**, and **post**. The ink phase exists to sequence effects which specifically alter the printed ink — like bleedthrough from too much ink on page, extraneous lines or regions from a mechanically-faulty printer, or fading over time — and transform them prior to "printing".

The paper phase applies transformations to the underlying paper on which the ink gets printed; here, a `PaperFactory` generator creates a random texture from a set of given texture images, as well as effects like random noise, shadowing, watermarking, and staining. After the ink and paper textures are separately computed, they are merged in the manner of Technique 1 from above, simulating the printing of the document.

After "printing", the document enters the post phase, wherein it may undergo modifications that might alter an already-printed document out in the world. Augmentations are available here which simulate the printed page being folded along multiple axes, marked by handwriting or color highlighter, faxed, photocopied, scanned, photographed, burned, stained, and so on. Figure 2 shows the individual phases of an example pipeline combining to produce a noised document image.

4 Augraphy

Augraphy is a lightweight Python package for applying realistic perturbations to document images. It is registered on the Python Package Index (PyPI) and can be installed simply using **pip install augraphy** .

Augraphy requires only a few commonly-used Python scientific computing or image handling packages, such as NumPy [18], and has been tested on Windows, Linux, and Mac computing environments and supports recent major versions of Python 3. Below is a basic out-of-the-box *Augraphy* pipeline demonstrating its use:

```
import augraphy; import cv2
pipeline = augraphy.default_augraphy_pipeline()
img = cv2.imread("image.png")
data = pipeline.augment(img)
augmented_img = data["output"]
```

Listing 1.1. Augmenting a document image with *Augraphy*.

Augraphy is designed to be immediately useful with little effort, especially as part of a preprocessing step for training machine learning models, so care was taken to establish good defaults. *Augraphy* provides *30* unique augmentations, which may be sequenced into pipeline objects which carry out the image manipulation. The default *Augraphy* pipeline (used in the code snippet above) makes use of all of these augmentations, with starting parameters selected after manual visual inspection of thousands of images. Users of the library can define directed acyclic graphs of image transformations via the `AugraphyPipeline` API, representing the passage of a document through real-world alterations.

4.1 *Augraphy* Augmentations

Augraphy provides *30* out-of-the-box augmentations, listed in Table 2 (with some examples shown in Fig. 3). As mentioned before, Ink Phase augmentations include those that imitate noisy processes that occur in a document's life cycle when ink is printed on paper. These augmentations include `BleedThrough`, which imitates what happens when ink bleeds through from the opposite side of the page. Another, `LowInkLines`, produces a streaking behavior common to printers running out of ink.

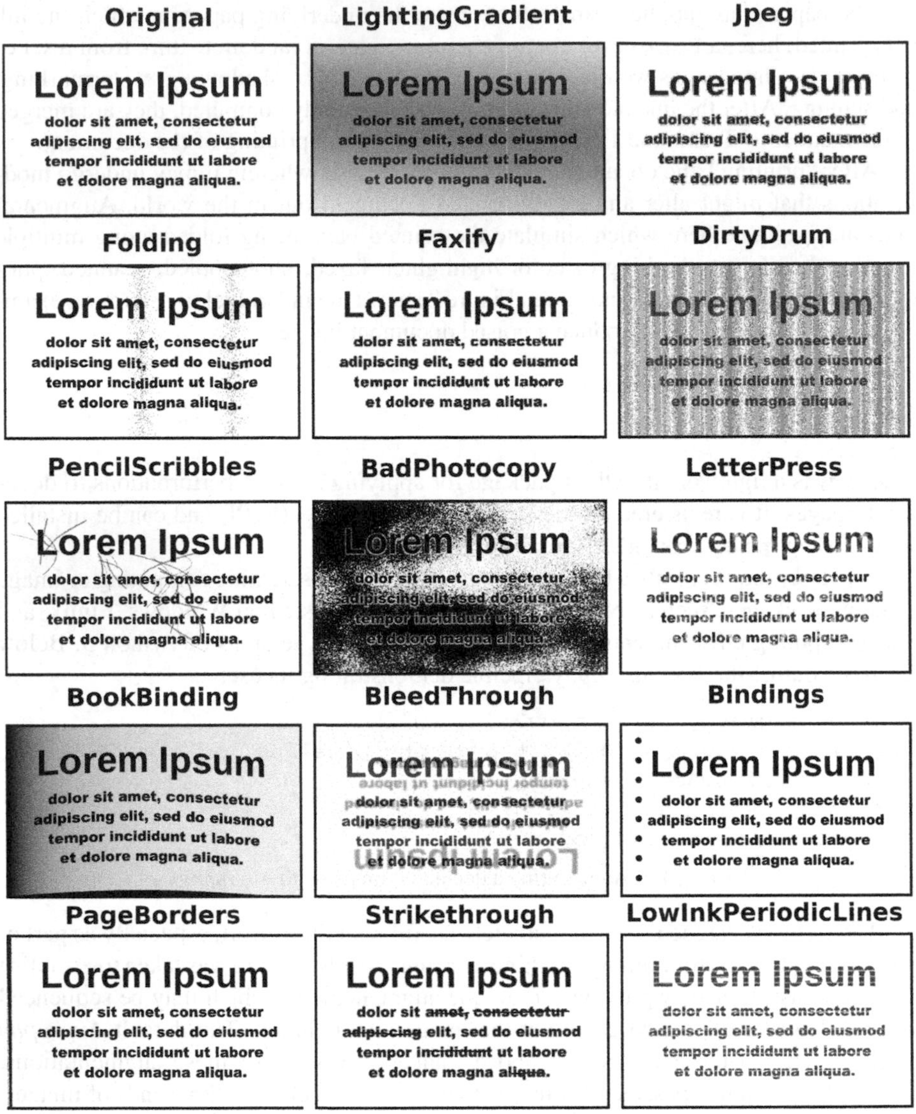

Fig. 3. Examples of some *Augraphy* augmentations.

Augmentations provided by the Paper Phase include `BrightnessTexturize`, which introduces random noise in the brightness channel to emulate paper textures, and `Watermark`, which imitates watermarks in a piece of paper. Finally, the Post Phase includes augmentations that imitate noisy-processes that occur after a document has been created. Here we find `BadPhotoCopy`, which uses added noise to generate an effect of a dirty copier, and `BookBinding`, which creates an effect with shadow and curved lines to imitate how a page from a book might appear after capture by a flatbed scanner.

Table 2. Individual *Augraphy* augmentations for each augmentation phase, in suggested position within a pipeline. Augmentations that work well in more than one phase are listed in the last column.

Ink Phase	Paper Phase	Post Phase	Multiple
BleedThrough	ColorPaper	BadPhotoCopy	BrightnessTexturize
LowInkPeriodicLines	Watermark	BindingsAndFasteners	DirtyDrum
LowInkRandomLines	Gamma	BookBinding	DirtyRollers
InkBleed	LightingGradient	Folding	Dithering
Letterpress	SubtleNoise	JPEG	Geometric
	DelaunayTessellation	Markup	NoiseTexturize
	VoronoiTesselation	Faxify	Scribbles
	PatternGenerator	PageBorder	LinesDegradation
			Brightness

In addition to document-specific transformations, *Augraphy* also includes transformations like blur, scaling, and rotation, reducing the need for additional general-purpose augmentation libraries. Descriptions of all *Augraphy* augmentations are available online, along with the motivation for their development and usage examples.[1]

4.2 The Library

There are four "main sequence" classes in the *Augraphy* codebase, which together provide the bulk of the library's functionality, and when composed generate realistic synthetically-augmented document images:

Augmentation. The Augmentation class is the most basic class in the project, and essentially exists as a thin wrapper over a probability value in the interval [0,1]. Every augmentation contained in a pipeline has its own chance of being applied during that pipeline's execution.

AugmentationResult. After an augmentation is applied, the output of its execution is stored in an AugmentationResult object and passed forward through the pipeline. These also record a full copy of the augmentation runtime data, as well as any metadata that might be relevant for debugging or other advanced use.

AugmentationSequence. A list of Augmentations — intended to be applied in sequence — determines an AugmentationSequence, which is itself both an Augmentation and callable (behaves like a function). In practice, these model the pipeline phases discussed previously with lists of Augmentation constructor calls which produce callable Augmentation objects of the various flavors explored in AugmentationSequences applied in each AugmentationPipeline phase, and in each case yield the image, transformed by some of the Augmentations in the sequence.

AugmentationPipeline. The bulk of the heavy lifting in *Augraphy* resides in the Augmentation pipeline, which is our abstraction over one or more events in a physical document's life.

---
[1] https://augraphy.readthedocs.io/en/latest/.

Consider the following sequence:

1. Ink is adhered to paper material during the initial printing of a document.
2. The document is attached to a public bulletin board in a high-traffic area.
3. The pages are annotated, defaced, and eventually torn away from their securing staples, flying away in the wind.
4. Fifty years later, the tattered pages are discovered and turned over to library archivists.
5. These conservationists use delicate tools to carefully position and record images of the document, storing these in a digital repository.

An `AugmentationPipeline` represents such document lifetimes by composing augmentations modeling each of these individual events, while collecting runtime metadata about augmentations and their parameters and storing copies of intermediate images, allowing for inspection and fine-tuning to achieve outputs with desired features, facilitating (re)production of documents as in Fig. 1.

Realistically reproducing effects in document images requires careful attention to how those effects are produced in the real world. Many issues, like the various forms of misprint, only affect text and images on the page. Others, like a coffee spill, change properties of the paper itself. Further still, there are transformations like physical deformations which alter the geometry and topology of both the page material and the graphical artifacts on it. Effectively capturing processes like these in reproducible augmentations means separating our model of a document pipeline into ink, paper, and post-processing layers, each containing some augmentations that modify the document image as it passes through. With *Augraphy*, producing realistically-noisy document images can be reduced to the definition and application of one or more *Augraphy* pipelines to some clean, born-digital document images.

There are also two classes that provide additional critical functionality in order to round out the *Augraphy* base library:

OneOf. To model the possibility that a document image has undergone one and only one of a collection of augmentations, we use `OneOf`, which simply selects one of those augmentations from the given list, and uses this to modify the image.

PaperFactory. We often print on multiple sizes and kinds of paper, and out in the world we certainly *encounter* such diverse documents. We introduce this variation into the `AugmentationPipeline` by including `PaperFactory` in the `paper` phase of the pipeline. This augmentation checks a local directory for images of paper to crop, scale, and use as a background for the document image. The pipeline contains edge detection logic for lifting only text and other foreground objects from a clean image, greatly simplifying the "printing" onto another "sheet", and capturing in a reproducible way the construction method used to generate the NoisyOffice database [49]. Taken together, `PaperFactory` makes it trivial to re-print a document onto other surfaces, like hemp paper, cardboard, or wood.

Interoperability and flexibility are core requirements of any data augmentation library, so *Augraphy* includes several utility classes designed to improve developer experience:

ComposePipelines. This class provides a means of composing two pipelines into one, allowing for the construction of complex multi-pipeline operations.

Foreign. This class can be used to wrap augmentations from external projects like *Albumentations* and *imgaug*.

OverlayBuilder. This class uses various blending algorithms to fuse foreground and background images together, which is useful for "printing" or applying other artifacts like staples or stamps.

NoiseGenerator. This class uses make blobs algorithm to generate mask of noises in different shape and location.

5 Document Denoising with *Augraphy*

Augraphy aims to facilitate rapid dataset creation, advancing the state of the art for document image analysis tasks. This section describes a brief experiment in which *Augraphy* was used to augment a new collection of clean documents, producing a new corpus which fills a feature gap in existing public datasets. This collection was used to train a denoising convolutional neural network capable of high-accuracy predictions, demonstrating *Augraphy*'s utility for robust data augmentation.

All code used in these experiments is available in the augraphy-paper GitHub repository.[2]

5.1 Model Architecture

To evaluate *Augraphy*, we used an off-the-shelf Nonlinear Activation-Free Network (NAFNet) [8], making only minor alterations to the model's training hyperparameters, changing the batch size and learning epoch count to fit our training data.

5.2 Data Generation

Despite recent techniques ([6,33]) for reducing the volume of input data required to train models, large datasets remain king; feeding a model more data during training can help ensure better latent representations of more features, improving robustness of the model to noise and increasing its capacity for generalization to new data.

To train our model, we developed the *ShabbyPages* dataset [16], a real-world document image corpus with *Augraphy*-generated synthetic noise. Gathering the 600 ground-truth documents was the most challenging part, requiring the efforts of several crowdworkers across multiple days. Adding the noise to these images was trivial however; *Augraphy* makes it easy to produce large training sets. For additional realistic variation, we also gathered 300 paper textures and used the PaperFactory augmentation to "print" our source documents onto these.

The initial 600 PDF documents were split into their component pages, totaling 6202 clean document images exported at 150 pixels per inch. We iteratively developed an *Augraphy* pipeline which produced satisfactory output with few overly-noised images, then ran each of the clean images through this to produce the base training set. Patches were taken from each of these images and used for the final training collection.

---

[2] https://github.com/sparkfish/augraphy-paper.

Another new dataset — *ShabbyReal* — was produced as part of the larger *Shabby-Pages* corpus [16]; this data was manually produced by applying sequences of physical operations to real paper. *ShabbyReal* is fully out-of-distribution for this experiment, and also has a very high degree of diversity, providing good conditions for evaluating *Augraphy*'s effect.

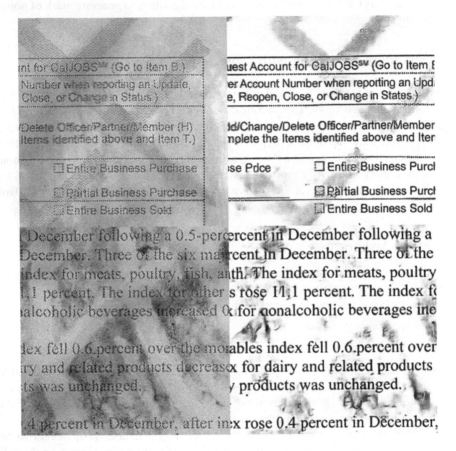

Fig. 4. Sample *ShabbyReal* noisy images (left) and their denoised counterparts (right).

5.3   Training Regime

The NAFNet was given a large sample to learn from: for each image in the *ShabbyPages* set, we sampled ten patches of 400×400 pixels, using these to train the instance across just 16 epochs.

The model was trained with mean squared error as the loss function, using the Adam optimizer and cosine annealing learning rate scheduler, then evaluated with the mean average error metric.

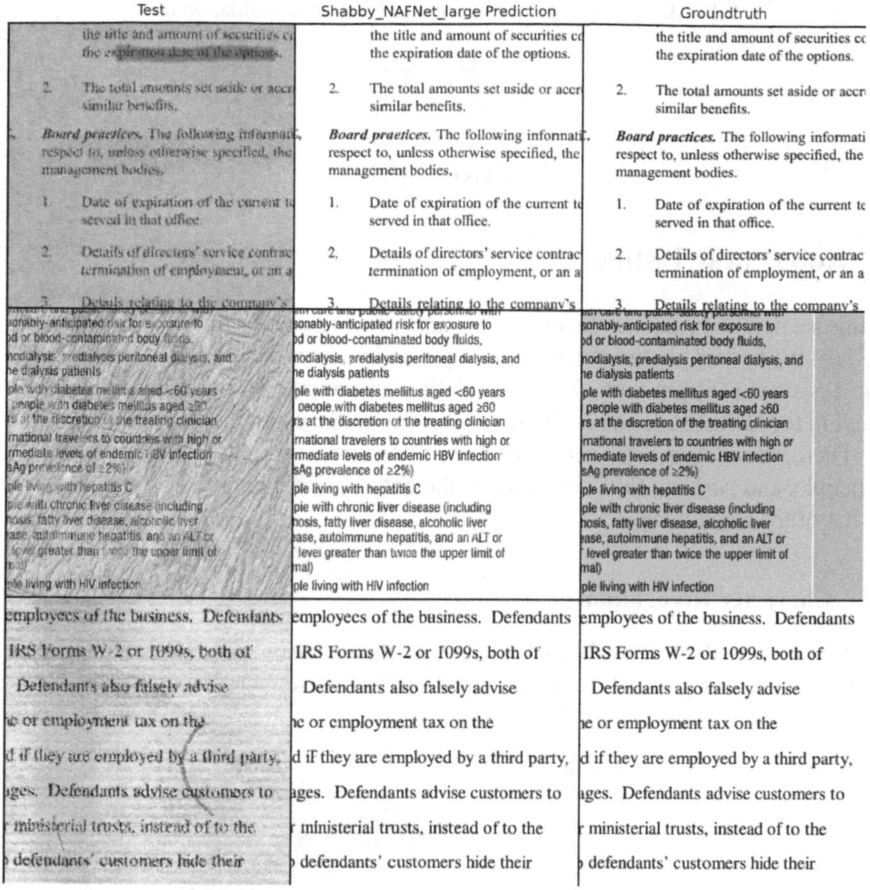

Test	Shabby_NAFNet_large Prediction	Groundtruth

Fig. 5. Sample *ShabbyPages* noisy images (left), denoised by Shabby_NAFNet (center), and groundtruth (right).

5.4 Results

Sample predictions from the model on the validation task are presented in Figs. 4 and 5. The SSIM score indicates that these models generalize very well, though they do over-compensate for shadowing features around text, by increasing the line thickness in the prediction, resulting in a bold font.

To compare the model's performance on the validation task, we considered the root mean square error (RMSE), structural similarity index (SSIM), and peak signal-to-noise ratio (PSNR) metrics.

Quantitative results using these metrics are displayed in Table 3. As can be seen in Fig. 4, the model was able to remove a significant amount of blur, line noise, shadowing from tire tracks, and watercoloring.

396   A. Groleau et al.

Table 3. Denoising performance on *ShabbyReal* validation task

Metric	Score
SSIM	0.71
PSNR	32.52
RMSE	6.52

6 Robustness Testing

In this section, we examine the use of *Augraphy* to add noise which challenges existing models. We first use *Augraphy* to add noise to an image of text with known groundtruth. The Tesseract [45] pre-trained OCR model's performance on the clean image is compared to the OCR result on the *Augraphy*-noised image using the Levenshtein distance.

Then, we add *Augraphy*-generated noise to document images which contain pictures of people, and perform face-detection on the output, comparing the detection accuracy to that of the clean images.

6.1 Character Recognition

We first compiled 15 ground-truth, noise-free document images from a new corpus of born-digital documents, whose ground-truth strings are known. We then used Tesseract

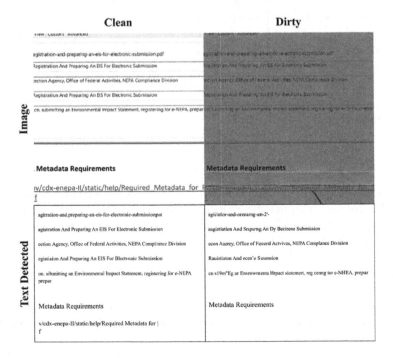

Fig. 6. Sample *ShabbyReal* clean/noisy image pair, and the OCR output for each. The computed Levenshtein distance between the OCR result strings is 203.

to generate OCR predictions on these noise-free documents, as a baseline for comparison. We considered these OCR predictions as the ground-truth labels for each document. Next, we generated noisy versions of the 15 documents by running them through an *Augraphy* pipeline, and again used Tesseract to generate OCR predictions on these noisy documents. We compared the word accuracy rate on the noisy OCR results versus the ground-truth noise-free OCR results, and found that the noisy OCR results were on average 52% less accurate, with a range of up to 84%. This example use-case demonstrates the effectiveness of using *Augraphy* to create challenging test data that remains human-readable, suitable for evaluating OCR systems.

6.2 Face Detection Robustness Testing

In this section, we move beyond text-related tasks to robustness analysis of image classifiers and object detectors. In particular, we examine the robustness of face detection models on *Augraphy*-altered images containing faces, such as newspapers [32] and legal identification documents [3], which are often captured with noisy scanners.

Table 4. Face detection performance.

Model	Accuracy
Azure	57.1%
Google	60.1%
Amazon	49.1%
UltraFace	4.5%

We begin by sampling 75 images from the FDDB face detection benchmark [23], using *Augraphy* to generate 10 altered versions of each image, manually removing any augmented image where the face(s) is not reasonably visible. We then test four face detection models on the noisy and noise-free images. These models are: proprietary face detection models from Google,[3] Amazon,[4] and Microsoft,[5] and lastly the freely-available UltraFace[6] model.

Table 4 shows face detection accuracy of both models on the noisy test set, and example images where no faces were detected by the Microsoft model are shown in Fig. 7. We see that all four models struggle on the *Augraphy*-augmented data, with the proprietary models seeing detection performances drop to between roughly 50-60%, and UltraFace detecting only 4.5% of faces that it found in the noise-free data.

---

[3] https://cloud.google.com/vision/docs/detecting-faces.

[4] https://aws.amazon.com/rekognition/.

[5] https://azure.microsoft.com/en-us/services/cognitive-services/face.

[6] https://github.com/onnx/models/tree/main/vision/body_analysis/ultraface.

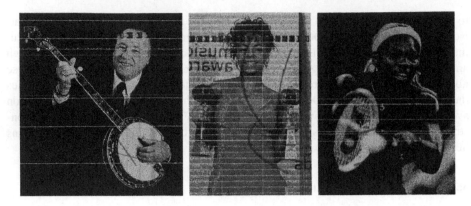

Fig. 7. Example augmented images that yielded false-positive predictions with the Azure face detector.

7   Conclusion and Future Work

This paper presents *Augraphy*, a framework licensed under the MIT open source license for generating realistic synthetically-augmented datasets of document images. Other available tools were examined and found to lack needed features, motivating the creation of this library specifically supporting alterations and degradations seen in document images.

We described how to use *Augraphy* to create a new document image dataset containing synthetic real-world noise, then compared results obtained by training a convolutional NAFNet instance on this corpus. Finally, we examined *Augraphy*'s efficacy in generating confounding data for testing the robustness of document vision models.

Future work on the *Augraphy* library will focus on adding new types of augmentations, increasing performance to enable faster creation of larger datasets, providing more scale-invariant support for all document image resolutions, and responding to community-initiated feature requests.

References

1. Albano, G.L., Sparro, M.: The role of digitalization in public service provision: evidence and implications for the efficiency of local government services. Local Gov. Stud. **44**(5), 613–636 (2018)
2. Alkhezzi, F., Alsabawy, A.Y.: Factors influencing the implementation of learning management systems in higher education: a case study. Educ. Inf. Technol. **25**(4), 2827–2845 (2020)
3. Arlazarov, V., Bulatov, K., Chernov, T., Arlazarov, V.: MIDV-500: a dataset for indentity document analysis and recognition on mobile devices in video stream. Comput. Optics **43**(5), 818–824 (2019)
4. Bloice, M.D., Roth, P.M., Holzinger, A.: Biomedical image augmentation using Augmentor. Bioinformatics **35**(21), 4522–4524 (2019)
5. Buslaev, A., Iglovikov, V.I., Khvedchenya, E., Parinov, A., Druzhinin, M., Kalinin, A.A.: Albumentations: fast and flexible image augmentations. Information **11**(2), 125 (2020)

6. Cao, Y., Yu, H., Wu, J.: Training vision transformers with only 2040 images. CoRR abs/2201.10728 (2022)

7. Castro-Bleda, M.J., España-Boquera, S., Pastor-Pellicer, J., Zamora-Martínez, F.: The Noisy-Office database: a corpus to train supervised machine learning filters for image processing. Comput. J. **63**(11), 1658–1667 (2019)

8. Chen, L., Chu, X., Zhang, X., Sun, J.: Simple baselines for image restoration. arXiv preprint arXiv:2204.04676 (2022)

9. Cheriet, M., Kharma, N., Liu, C.L., Suen, C.Y.: Character Recognition Systems: a Guide for Students and Practitioners. Wiley (2007)

10. Chuen, D.L.K., Deng, R.H.: Handbook of Blockchain, Digital Finance, and Inclusion: Cryptocurrency, FinTech, InsurTech, Regulation. Mobile Security, and Distributed Ledger. Academic Press, ChinaTech (2017)

11. Dodge, S., Karam, L.: Understanding how image quality affects deep neural networks. In: Proceedings of the 2016 Eighth International Conference on Quality of Multimedia Experience (QoMEX) (2016)

12. Fadaee, M., Bisazza, A., Monz, C.: Data augmentation for low-resource neural machine translation. In: Proceedings of the 55th Annual Meeting of the Association for Computational Linguistics (Volume 2: Short Papers) (2017)

13. Feng, S.Y., et al.: A survey of data augmentation approaches for NLP. In: Findings of the Association for Computational Linguistics: ACL-IJCNLP 2021. Association for Computational Linguistics (2021)

14. Gangeh, M.J., Plata, M., Motahari Nezhad, H.R., Duffy, N.P.: End-to-end unsupervised document image blind denoising. In: Proceedings of the 2021 IEEE/CVF International Conference on Computer Vision (ICCV) (2021)

15. Gatzert, N., Schmeiser, H.: The impact of various digitalization aspects on insurance value: an enterprise risk management approach. Geneva Papers Risk Insur.-Issues Pract. **41**(3), 385–405 (2016)

16. Groleau, A., Chee, K.W., Larson, S., Maini, S., Boarman, J.: ShabbyPages: a reproducible document denoising and binarization dataset. arXiv preprint arXiv:2303.09339 (2023)

17. Harley, A.W., Ufkes, A., Derpanis, K.G.: Evaluation of deep convolutional nets for document image classification and retrieval. In: Proceedings of the International Conference on Document Analysis and Recognition (ICDAR) (2015)

18. Harris, C.R., et al.: Array programming with NumPy. Nature **585**(7825), 357–362 (2020)

19. Hendrycks, D., Dietterich, T.: Benchmarking neural network robustness to common corruptions and perturbations. In: Proceedings of the International Conference on Learning Representations (ICLR) (2019)

20. Homeyer, A., et al.: Recommendations on test datasets for evaluating ai solutions in pathology. arXiv preprint arXiv:2204.14226 (2022)

21. Hosseini, H., Xiao, B., Poovendran, R.: Google's cloud vision API is not robust to noise. arXiv preprint arXiv:1704:05051 (2017)

22. Jadhav, P., Sawal, M., Zagade, A., Kamble, P., Deshpande, P.: Pix2Pix generative adversarial network with ResNet for document image denoising. In: Proceedings of the 4th International Conference on Inventive Research in Computing Applications (ICIRCA) (2022)

23. Jain, V., Learned-Miller, E.: FDDB: a benchmark for face detection in unconstrained settings. Tech. Rep. UM-CS-2010-009, University of Massachusetts, Amherst (2010)

24. Jaume, G., Ekenel, H.K., Thiran, J.P.: FUNSD: a dataset for form understanding in noisy scanned documents. In: Accepted to ICDAR-OST (2019)

25. Journet, N., Visani, M., Mansencal, B., Van-Cuong, K., Billy, A.: DocCreator: a new software for creating synthetic ground-truthed document images. J. Imag. **3**(4), 62 (2017)

26. Jung, A.B., et al.: imgaug. https://github.com/aleju/imgaug (2020). Accessed 01 Feb 2020

27. Karahan, S., Kilinc Yildirum, M., Kirtac, K., Rende, F.S., Butun, G., Ekenel, H.K.: How image degradations affect deep CNN-based face recognition? In: Proceedings of the 2016 International Conference of the Biometrics Special Interest Group (BIOSIG) (2016)
28. Kim, D., Hong, T., Yim, M., Kim, Y., Kim, G.: Technical report on web-based visual corpus construction for visual document understanding. arXiv preprint arXiv:2211.03256 (2022)
29. Ko, T., Peddinti, V., Povey, D., Khudanpur, S.: Audio augmentation for speech recognition. In: Proceedings of Interspeech 2015 (2015)
30. Kulkarni, M., Kakad, S., Mehra, R., Mehta, B.: Denoising documents using image processing for digital restoration. In: Swain, D., Pattnaik, P.K., Gupta, P.K. (eds.) Machine Learning and Information Processing. AISC, vol. 1101, pp. 287–295. Springer, Singapore (2020). https://doi.org/10.1007/978-981-15-1884-3_27
31. Larson, S., Lim, G., Ai, Y., Kuang, D., Leach, K.: Evaluating out-of-distribution performance on document image classifiers. In: Proceedings of the Thirty-sixth Conference on Neural Information Processing Systems Datasets and Benchmarks Track (2022)
32. Lee, B.C.G., et al.: The newspaper navigator dataset: extracting headlines and visual content from 16 million historic newspaper pages in chronicling America. In: Proceedings of the 29th ACM International Conference on Information & Knowledge Management (CIKM) (2020)
33. Lee, S.H., Lee, S., Song, B.C.: Vision transformer for small-size datasets. arXiv preprint arXiv:2112.13492 (2021)
34. Maguolo, G., Paci, M., Nanni, L., Bonan, L.: Audiogmenter: a MATLAB toolbox for audio data augmentation. Applied Computing and Informatics (2021)
35. McFee, B., Humphrey, E., Bello, J.: A software framework for musical data augmentation. In: Muller, M., Wiering, F. (eds.) Proceedings of the 16th International Society for Music Information Retrieval Conference, ISMIR 2015 (2015)
36. Menachemi, N., Collum, T.H.: Benefits and drawbacks of electronic health record systems. Risk Manage. Healthcare Policy 4, 47–55 (2011)
37. Mohamed, S.S.A., Rashwan, M.A.A., Abdou, S.M., Al-Barhamtoshy, H.M.: Patch-based document denoising. In: Proceedings of the 2018 International Japan-Africa Conference on Electronics, Communications and Computations (JAC-ECC) (2018)
38. Mustafa, W.A., Kader, M.M.M.A.: Binarization of document image using optimum threshold modification. J. Phys. Confer. Ser. 1019, 012022 (2018)
39. O'Gorman, L., Kasturi, R.: Document image analysis. IEEE Computer Society (1997)
40. Papakipos, Z., Bitton, J.: AugLy: data augmentations for robustness. arXiv preprint arXiv:2201:06494 (2022)
41. Paszke, A., et al.: PyTorch: an imperative style, high-performance deep learning library. In: Advances in Neural Information Processing Systems 32, pp. 8024–8035 (2019)
42. Rotman, D., Azulai, O., Shapira, I., Burshtein, Y., Barzelay, U.: Detection masking for improved OCR on noisy documents. arXiv preprint arXiv:2205.08257 (2022)
43. Saifullah, Siddiqui, S.A., Agne, S., Dengel, A., Ahmed, S.: Are deep models robust against real distortions? A case study on document image classification. In: Proceedings of the 26th International Conference on Pattern Recognition (ICPR) (2022)
44. Schömig-Markiefka, B., et al.: Quality control stress test for deep learning-based diagnostic model in digital pathology. Modern pathol.: an official journal of the United States and Canadian Academy of Pathology, Inc 34(12), 2098–2108 (2021)
45. Smith, R.: An overview of the tesseract OCR engine. In: Proceedings of the Ninth International Conference on Document Analysis and Recognition (ICDAR 2007), vol. 2, pp. 629–633. IEEE (2007)
46. Staudt, R.W., Medeiros, A.P.: Access to justice and technology clinics: a 4% solution. Chicago-Kent Law Rev. 88(3), 695–728 (2015)
47. Vasiljevic, I., Chakrabarti, A., Shakhnarovich, G.: Examining the impact of blur on recognition by convolutional networks. arXiv preprint arXiv:1611.05760 (2016)

48. Wei, J., Zou, K.: EDA: easy data augmentation techniques for boosting performance on text classification tasks. In: Proceedings of the 2019 Conference on Empirical Methods in Natural Language Processing and the 9th International Joint Conference on Natural Language Processing (EMNLP-IJCNLP). Association for Computational Linguistics (2019)
49. Zamora-Martínez, F., España Boquera, S., Castro-Bleda, M.: Behaviour-based clustering of neural networks applied to document enhancement. In: Computational and Ambient Intelligence (2007)

TextREC: A Dataset for Referring Expression Comprehension with Reading Comprehension

Chenyang Gao, Biao Yang, Hao Wang, Mingkun Yang, Wenwen Yu, Yuliang Liu(✉), and Xiang Bai

Huazhong University of Science and Technology, Wuhan, China
{m202172425,hust_byang,wanghao4659,yangmingkun,
wenwenyu,ylliu,xbai}@hust.edu.cn

Abstract. Referring expression comprehension (REC) aims at locating a specific object within a scene given a natural language expression. Although referring expression comprehension has achieved tremendous progress, most of today's REC models ignore the scene texts in images. Scene text is ubiquitous in our society, and frequently critical to understand the visual scene. To study how to comprehend scene text in the referring expression comprehension task, we collect a novel dataset, termed TextREC, in which most of the referring expressions are related to scene text. Our TextREC dataset challenges a model to recognize scene text, relate it to the referring expressions, and select the most relevant visual object. We also propose a text-guided adaptive modular network (TAMN) to comprehend scene text associated with objects in images. Experimental results reveal that current state-of-the-art REC methods fall short on the TextREC dataset, while our TAMN gets inspiring results by integrating scene text.

Keywords: Referring expression comprehension · Scene text representation · Multi-modal understanding

1 Introduction

Referring expression comprehension (REC) [17] aims at locating a specific object within a scene given a natural language expression. It is a fundamental issue in the field of human-computer interaction and also a bridge between computer vision and natural language processing. Although referring expression comprehension has achieved tremendous progress, most of today's REC models ignore the scene texts in images. However, scene text is indispensable and more natural for distinguishing different objects. Considering the situation in Fig. 1, it is difficult to detect the target man using basic visual attributes, since the players wear the same uniform and their position is constantly changing during a football match. But the target man can be easily and naturally detected with the guidance of scene text.

Scene text is ubiquitous in our society, which conveys rich information for understanding the visual scene [27]. As the COCO-Text dataset [36] suggests,

G. A. Fink et al. (Eds.): ICDAR 2023, LNCS 14189, pp. 402–420, 2023.
https://doi.org/10.1007/978-3-031-41682-8_25

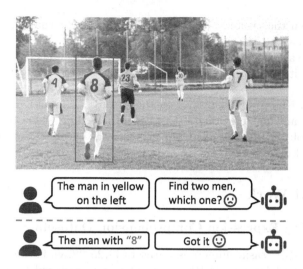

Fig. 1. This paper introduces a novel dataset to study integrating scene text in the referring expression comprehension task. For the above example, the scene text "8" provides crucial information that naturally distinguishes different players.

about 50% of the images contain scene text in large-scale datasets such as MS COCO [24] and the percentage increases sharply in urban environments. To move towards human-oriented referring expression comprehension, it is necessary to integrate scene text in existing REC pipelines. Scene text can provide more discriminative information so that the target object can be more easily specified. For example, "get a bottle of Coca-Cola from the fridge" is more precise for a robot to find the target object and more user-friendly. In literature, there are many studies successfully using scene text for vision-language tasks, *e.g.*, visual question answering [34], image captioning [33], cross-modal retrieval [27,37], and fine-grained image classification [16]. Therefore, explicitly utilizing scene text should be a natural step toward a more reasonable REC model.

To study how to comprehend scene text associated with objects in images, we collect a new dataset named TextREC. It contains 24,352 referring expressions and 36,083 scene text instances on 8,690 images, and most of the referring expressions are related to scene text. Our TextREC dataset challenges a model to recognize scene text, relate it to the referring expressions, and choose the most relevant visual object, requiring semantic and visual reasoning between multiple scene text tokens and visual entities. Besides, we also evaluate the performance of different state-of-the-art REC models, from which we observe the limited performance due to ignoring the scene texts contained in images. To this end, we propose a **T**ext-guided **A**daptive **M**odule **N**etwork (**TAMN**) to address this issue. The contributions of this paper are threefold:

– We introduce a novel dataset (TextREC) in which most of the referring expressions are related to scene text. Our TextREC dataset requires a model to leverage the additional modality provided by scene text so that the relationship between the visual objects in images and the textual semantic referring expression can be identified properly.

- We propose a text-guided adaptive modular network (TAMN) to utilize scene text, relate it to the referring expressions, and select the most relevant visual object.
- Substantial experimental results on the TextREC dataset demonstrate that it is important and meaningful to take into account scene text for locating the target object, meanwhile demonstrating the excellent performance of our TAMN in this task.

2   Related Work

2.1   Referring Expression Comprehension Datasets.

To tackle the REC task, numerous datasets [3,25,28,39,42,45] have been constructed. The first large-scale REC dataset was introduced by Kazemzadeh et al. [17], which is collected by applying a two-player game named ReferIt Game on the ImageCLEF IAPR [8] dataset. Unlike ReferIt Game, RefCOCOg [28] is collected in a non-interactive setting based on the MSCOCO [24] images. Ref-COCO [45] and RefCOCO+ [45] are also collected using ReferIt Game on the MSCOCO images. Due to the non-interactive setting, the referring expressions in RefCOCOg are longer and more complex than those in RefCOCO and Ref-COCO+. The above datasets are collected in real-world images. While Liu et al. [25] consider using synthesized images and carefully design templates to generate referring expressions, resulting in a synthetic dataset named CLEVR-Ref+. Wang et al. [39] point out that commonsense knowledge is important to identify the objects in the images in our daily life. They also collect a dataset based on Visual Genome [18], named KB-Ref. To answer each referring expression, at least one piece of commonsense knowledge should be included. Chen et al. [3] and Yang et al. [42] adopt the expression template and scene graphs provided in [11,18] to generate referring expressions in the real-world images. Recently, Bu et al. [2] collect a dataset based on various image sources to highlight the importance of scene text.

2.2   Vision-Language Tasks with Text Reading Ability

With the maturity of reading scene text (OCR) [6,19–23,26,31,32,41,46], vision-language tasks with text reading ability become an active research field. Several existing datasets [1,29,30,34,35,40] study the task of Visual Question Answering with Text Reading Ability. These datasets require understanding the scene text in the image when answering the questions. Similarly, to enhance scene text comprehension in an image, a new task named image captioning with reading comprehension and a corresponding dataset called TextCaps [33] is proposed.

Existing works [7,10,15,33,34,38,47] propose various network architectures to utilize scene text information. LoRRA [34] adds an OCR attention branch

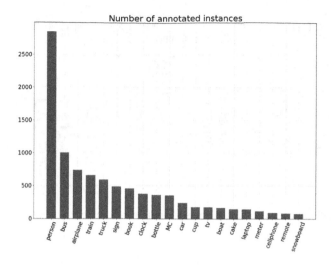

Fig. 2. Number of annotated instances per category.

to the VQA model [13], to select an answer either from a fixed vocabulary or detected OCR tokens. M4C [10] utilizes a multi-modal Transformer encoder to encode the question, image and scene text jointly, then generates answers through a dynamic pointer network. M4C-Captioner [33] directly remove question input in the aforementioned M4C model to solve the text-based image captioning task. SA-M4C [15] proposes a spatially aware self-attention layer that ensures each input focuses on a local context rather than dispersing attention amongst all other entities in a standard self-attention layer. MM-GNN [7] utilizes graph neural networks to build three separate graphs for different modalities. Then the designed aggregators use multi-modal contexts to obtain a better representation for the downstream VQA. SSBaseline [47] designs three simple attention blocks to suppress irrelevant features. LSTM-R [38] constructs the geometrical relationship between OCR tokens through the relation-aware pointer network.

3   TextREC Dataset

Our dataset enables referring expression comprehension models to conduct spatial, semantic, and visual reasoning between multiple scene text tokens and visual objects. In this section, we describe the process of constructing our TextREC dataset. We start by describing how to select the images used in TextREC. We then explain the pipeline for collecting the referring expressions related to scene texts. Finally, we provide statistics and an analysis of our TextREC.

3.1   Images

In order to make full use of the annotations of existing datasets, we rely on the MSCOCO 2014 train images (Creative Commons Attribution 4.0 License). Since

Fig. 3. Wordcloud visualization of most frequent scene text tokens contained in the referring expressions.

the goal of our dataset is to integrate scene text in existing REC pipelines, we are more interested in the images that contain scene texts. To select images containing scene texts, we use COCO-Text [36], which is a scene text detection and recognition dataset based on the MSCOCO dataset. We select images containing at least one non-empty legible scene text instance. Through the visualization of the result images, we notice that some scene text instances are too small and difficult to recognize. So we further add a constraint to the images to filter out the scene text instances with an area smaller than 100 pixels. Filtering these out results in 10,752 images, which form the basis of our TextREC dataset.

3.2 Referring Expressions

In the second stage, we collect referring expressions for objects in the above images. Different from the traditional referring expression comprehension task, in most cases, the target object can be uniquely specified with scene text. For example, if we want to ground NO.13 player in a football match, only using the number 13 on the player's clothes is sufficient. So in the referring expressions, we want to include scene text as much as possible, ignoring appearance information and location information. As a result, we choose some simple templates to generate referring expressions. We can get the bounding box of each object based on MSCOCO annotations. According to the bounding box, we can find the scene text instances contained in this bounding box. For each selected scene text, we generate referring expressions using two templates: *"The object with <OCR string> on it"* and *"The <category name> with <OCR string> on it"*. Among these templates, *<OCR string>* will be replaced by the scene text instance in the images, and *<category name>* will be replaced by the category name of the object. However, the referring expressions generated through the two templates may not refer to the corresponding objects. The reason is that the scene text instance is contained in the object's bounding box but irrelevant to the object. As shown in Fig. 7, the scene text instances are contained in the

Table 1. Comparison between standard benchmarks and the proposed TextREC.

Dataset	Total Images	Annotations	
		Scene Text Related Expressions	Scene Text
ReferItGame [17]	20,000	×	×
RefCOCOg [28]	26,711	×	×
RefCOCO [45]	19,994	×	×
RefCOCO+ [45]	19,992	×	×
Clevr-ref+ [25]	85,000	×	×
KB-Ref [39]	24,453	×	×
Ref-Reasoning [42]	113,000	×	×
Cops-Ref [3]	113,000	×	×
TextREC	**8,690**	✓	✓

red bounding boxes, but irrelevant to the corresponding objects. To address this issue, we develop an annotation tool using Tkinter to check the plausibility of each referring expression. Finally, we manually filter out 48,704 valid referring expressions from 61,000 expressions.

3.3 Statistics and Analysis

Our TextREC dataset contains 8,690 images, 36,083 scene text instances, 10,450 annotated bounding boxes belonging to 50 categories and 48,704 referring expressions (each template has 24,352 referring expressions). We also compare our TextREC dataset with standard benchmarks in the referring expression comprehension task. As shown in Table 1, our dataset is the only benchmark containing both scene text related expressions and scene text annotations.

We also analyze the number of annotated instances per category to see which categories are most likely to contain scene texts. The top-20 categories and their corresponding instance numbers are shown in Fig. 2. It can be observed that the category of person is most likely to contain scene texts. This is not surprising since people usually wear clothing with various logos such as "nike" or "adidas". The category of the vehicle also tends to contain scene texts. The bus often indicates its route using some characters, and the airplane also indicates which airline it belongs to.

Moreover, we visualize word clouds for the scene text tokens contained in the referring expressions. As shown in Fig. 3, most scene text tokens are meaningful. The most frequent word is "stop" since one category of MSCOCO is stop sign. The second most frequent word is "police" because police vehicles appear frequently in our dataset.

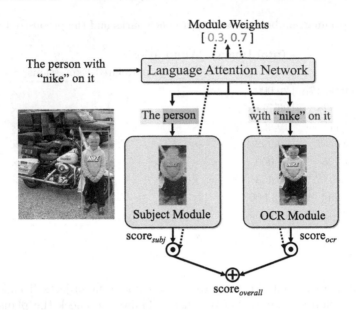

Fig. 4. Our model learns to parse an expression into the subject module and the text-guided matching module using the language attention network. Then computes an individual matching score for each module. For simplicity, we refer to the **text-guided matching module** as the **OCR module** for short.

4   Method

In this section, we introduce our Text-Guided Adaptive Modular Network (TA-MN) to align the referring expressions with the scene texts. The overall framework is shown in Fig. 4. Given a referring expression r and a candidate object o_i as input, where i represents the i-th object in the image, we start with the language attention network to parse the expressions into the subject module and the text-guided matching module. Then we use the text-guided matching module to calculate a matching score for o_i with respect to the weighted referring expression r. Finally, we take this matching score along with the score from the subject module proposed in MAttNet [44]. The overall matching score between o_i and r is the weighted combination of these two scores.

4.1   Language Attention Network

Similar to CMN [9] and MAttNet [44], we utilize the soft attention mechanism over the word sequence to attend to the relevant words automatically. As shown in Fig. 5, given a expression of T words $r = \{m_t\}_{t=1}^{T}$, we first embed each word m_t to a vector e_t using an one-hot word embedding. Then a bi-directional LSTM is applied to encode the context for each word. To get the final representation

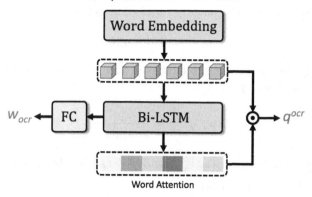

Fig. 5. The illustration of the language attention network.

for each word, we concatenate the hidden state in both directions:

$$e_t = \text{embedding}(m_t)$$
$$\vec{h}_t = \text{LS}\vec{\text{T}}\text{M}(e_t, \vec{h}_{t-1})$$
$$\reflectbox{$\vec{\reflectbox{$h$}}$}_t = \text{LS}\overleftarrow{\text{T}}\text{M}(e_t, \overleftarrow{h}_{t+1})$$
$$h_t = [\vec{h}_t, \overleftarrow{h}_t].$$

The attention weight over each word m_t for the text-guided matching module is obtained through a learned linear prediction over h_t followed by a softmax function:

$$a_t = \frac{\exp\left(\text{FC}(h_t)\right)}{\sum_{k=1}^{T}\exp\left(\text{FC}(h_k)\right)}$$

The language representation of the text-guided matching module is obtained by the weighted sum of word embeddings:

$$q^{ocr} = \sum_{t=1}^{T} a_t e_t$$

Finally, we utilize another two fully-connected layers to get the weights w_{ocr} and w_{subj} for our text-guided matching module and subject module:

$$[w_{ocr}, w_{subj}] = \text{softmax}(\text{FC}([h_0, h_T]))$$

4.2   Text-Guided Matching Module

Our text-guided matching module is illustrated in Fig. 6. Given a candidate o_i and all the ground truth scene text instances $\{p_n\}_{n=1}^{N}$ contained in the bounding box of o_i, we first encode each scene text instance p_n to a vector using the same word embedding layer of the language attention network.

$$u_n = \text{embedding}(p_n)$$

Then we compute the cosine similarity between each word embedding of the scene text instance and q^{ocr}:

$$S(u_n, q^{ocr}) = \frac{u_n^T q^{ocr}}{||u_n||||q^{ocr}||}$$

The similarity score between $\{u_n\}_{n=1}^N$ and q^{ocr} can be obtained by choosing the largest score in $\{S(u_n, q^{ocr})\}_{n=1}^N$:

$$S(u, q^{ocr}) = \max_{1 \leq n \leq N} S(u_n, q^{ocr})$$

This score is not sufficient as the matching score between o_i and r. We will illustrate the reasons with a few specific examples. As shown in Fig. 7, a scene text instance may exist both in the bounding boxes of two different objects. But it only relates to one object (green box). If we use $S(u, q^{ocr})$ as the matching score, another unrelated object (red box) may mismatch with the expression. To address this problem, the algorithm should find the association between the scene text and object. For example, "NIKE" is unlikely to appear on a motorcycle, but can appear on a person. So we further add a confidence score to $S(u, q^{ocr})$:

$$S(f_{obj}, q^{ocr}) = \frac{f_{obj}^T q^{ocr}}{||f_{obj}||||q^{ocr}||} \tag{1}$$

where f_{obj} is the visual representation of the candidate object extracted in the subject module. This confidence score can drive the model to learn the association between the scene text and object.

The final matching score of our text-guided matching module can be obtained by multiplying $S(u, q^{ocr})$ with its confidence score:

$$S(o_i|q^{ocr}) = S(f_{obj}, q^{ocr})S(u, q^{ocr})$$

4.3   Learning Objective

Assume we get $S(o_i|q^{ocr})$ and $S(o_i|q^{subj})$ from our proposed text-guided matching module and the subject module proposed in MAttNet [44]. We also get the module weights w_{ocr} and w_{subj} for the text-guided matching module and the subject module in the language attention network. The overall matching score for candidate object o_i and referring expression r is:

$$S(o_i|r) = w_{ocr}S(o_i|q^{ocr}) + w_{subj}S(o_i|q^{subj})$$

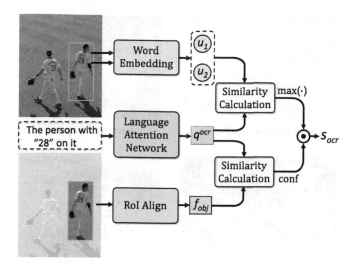

Fig. 6. The illustration of proposed text-guided matching module, "conf" refers to the confidence score calculated in Eq. 1.

Inspired by the triplet loss for the image retrieval task, for each positive pair (o_i, r_i), we randomly sample two negative pairs (o_i, r_j) and (o_k, r_i). r_j is the expression matched with other object in the same image of o_i, and o_k is other object in the same image of r_i. The combined hinge loss is calculated as follows:

$$L_{rank}^{overall} = \sum_i \lambda_1 [\delta + S(o_i|r_j) - S(o_i|r_i)]_+$$
$$+ \sum_i \lambda_2 [\delta + S(o_k|r_i) - S(o_i|r_i)]_+$$

where δ is a margin hyper-parameter and $[\cdot]_+ = \max(\cdot, 0)$. To stabilize the training procedure, we further add a hinge loss to the text-guided matching module:

$$L_{rank}^{ocr} = \sum_i \lambda_3 [\delta + S(o_i|q_j^{ocr}) - S(o_i|q_i^{ocr})]_+$$
$$+ \sum_i \lambda_4 [\delta + S(o_k|q_i^{ocr}) - S(o_i|q_j^{ocr})]_+$$

The final loss function is summarized as follows:

$$L = L_{rank}^{ocr} + L_{rank}^{overall}$$

5 Experiment

In this section, we first introduce the experiment setting. Then we evaluate the TAMN and several state-of-the-art REC methods on our TextREC dataset.

The object with "rapid" on it   The object with "cingular" on it

The teddy bear with "love" on it   The object with "nike" on it

Fig. 7. The motivation of adding the confidence score in our OCR module.

Furthermore, we conduct ablation studies to demonstrate the effectiveness of each component in our TAMN. We also explore more templates and a new test setting. Finally, the attention weights for each word in the referring expressions are visualized to demonstrate the effectiveness of the language attention network.

5.1   Dataset and Evaluation Protocol

We evaluate our text-guided adaptive modular network on the TextREC dataset. From Fig. 2, it can be observed that the categories of the dataset follow a long-tailed distribution. To ensure that the test set contains rare categories, we divide our dataset according to the ratio of instances of each category to the total, resulting in train and test splits with image numbers 7,422 and 1,268.

Following the standard evaluation setting [28], we compute the Intersection over Union (IoU) ratio between the ground truth and predicted bounding box. We regard the detection as a true positive If IoU is greater than 0.5, otherwise it is a false positive. For each image, we then compute the precision@1 measure according to the confidence score. The final performance is obtained by averaging these scores over all images.

5.2   Implementation Details

The detection model we adopt is Mask R-CNN. We follow the same implementation as MattNet [44]. The detection model is trained on a union of MSCOCO's 80k train and 35k subset of val (trainval35k) images excluding the test images in our TextREC dataset. We use the ground truth bounding boxes during training. In the test stage, we utilize the Mask R-CNN mentioned above to generate boxes. Our model is optimized with Adam optimizer and the batch size is set to 15. The initial learning rate is 0.0004. Moreover, the model is trained for 50 epochs with a learning rate decay by a factor of 2 every 16 epochs. The size of the word embedding and the hidden state of the bi-LSTM is set to 512. The size of the word embedding for the scene text is also set to 512. We set the output of all fully-connected layers within our model to be 512-dimensional. For the hyper-parameters in the loss functions, we set $\lambda_1 = 1$ and $\lambda_2 = 1$ in $L_{rank}^{overall}$. In addition, we set $\lambda_3 = 1$ and $\lambda_4 = 1$ in L_{rank}^{ocr}.

Table 2. Performance of the baselines on our TextREC dataset. TAMN significantly benefits from scene text input and achieves the highest precision@1 (%) score, suggesting that it is important to integrate scene text for the referring expression comprehension task.

Model	Template1	Template2
TransVG [4]	50.1	54.0
MAttNet [44]	52.3	60.5
QRNet [43]	52.7	59.1
Mdetr [14]	54.4	63.3
TAMN (ours)	**77.8**	**80.8**

5.3 Performance of the Baselines on TextREC Dataset

To illustrate the gap between the traditional REC datasets and our TextREC dataset, we conduct experiments with different state-of-the-art REC methods. As shown in Table 2, current state-of-the-art methods [4,14,43,44] fall short on our TextREC dataset. The results indicate that these methods ignore scene text in images, while our TAMN gets inspiring results by integrating scene text. This clearly verifies that it is important and meaningful to take into account scene text for the referring expression comprehension task.

5.4 Ablation Studies

The Subject Module and OCR Module. As shown in Fig. 4, our TAMN consists of two modules: the subject module and OCR module. We test the performance only with each module and the results are shown in Table 3. Compared with the only subject module, adding our OCR module gives 26.4 and 20.2 performance improvement in template1 and template2, respectively. Cooperating with our OCR module, the subject module gives 1.5 and 2.7 performance improvement in template1 and template2, respectively. These verify the effectiveness of the subject module and OCR module. Moreover, for our TAMN,

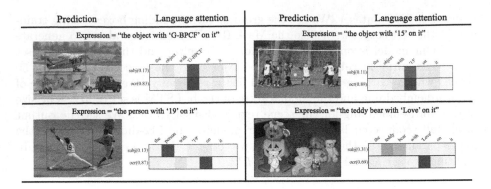

Fig. 8. The visualization results of the word attention in the language attention network.

Table 3. Ablation studies on different modules in our framework. The precision@1 (%) is reported.

Subject	OCR	Template1	Template2
✓	×	51.4	60.6
×	✓	76.3	78.1
✓	✓	**77.8**	**80.8**

Table 4. Ablation studies on different OCR systems. "GT" denotes using grounding truth scene text annotations.

Model	Template1	Template2
PaddleOCR [5]	63.2	69.8
EasyOCR [12]	67.1	72.7
GT	77.8	80.8

we compute the similarity score of each module to the overall score over the whole test set. The experimental results are summarized as follows: in template1, our OCR module makes the dominating contribution (**97.1%**) to the overall score. The contribution (**2.90%**) of the subject module can be ignored. When the expression form transfers to template2, the contribution of our OCR module decreases from **97.1%** to **70.0%**. While the contribution of the subject module increases from **2.90%** to **30.0%**. Our OCR module still accounts for the majority. The reason is that scene texts provide more information than the object categories in most cases. These clearly demonstrate the effectiveness of our proposed OCR module.

The Confidence Score in Our OCR Module. As shown in Fig. 6, we add a confidence score by calculating the similarity between the RoI feature of the candidate object and the scene text embedding. To verify the effectiveness of this confidence score, we conduct ablation experiments which are shown in Table 5. It can be observed that adding the confidence score gains 1.8% and 3.3% performance improvement in template1 and template2 only using the OCR module. We also test the effectiveness of adding the confidence score in our whole framework. It can be observed that adding the confidence score gains 1.9 and 1.3 performance improvement in template1 and template2. These results clearly verify the effectiveness of adding this confidence score.

Table 5. Ablation studies on the confidence score in our OCR module. The precision@1 (%) is reported.

Model	Confidence	Template1	Template2
OCR	×	74.5	74.8
OCR	✓	76.3	78.1
TAMN	×	75.9	79.5
TAMN	✓	77.8	80.8

Different OCR Systems. We conduct ablation studies to see the performance using different OCR systems. Results in Table 4 show that the performance of scene text detection and recognition methods has a great impact on the final results. The reason why EasyOCR has better performance is that the text spotting precision of EasyOCR is 6.8% higher than that of PaddleOCR.

Templates in Different Forms. We conduct ablation studies to see the performance using different templates. As shown in Table 6, it can be observed that the performance is very close with different templates as long as they contain the same amount of information (<category name> or <OCR string>). For example, in row 1, 3, and 5, the performance differences are within 0.3 in terms of precision@1 measure. Similarly, in row 2 and 4, the performance differences are also within 0.3.

Table 6. Ablation studies on the templates in different forms. The precision@1 (%) is reported.

Templates	Pre@1
The object with <OCR string> on it	77.8
The <category name> with <OCR string> on it	**80.8**
Object with <OCR string>	77.6
<category name> with <OCR string>	80.5
<OCR string>	77.5
The object	51.2

New Test Setting. In the traditional referring expression comprehension datasets, one referring expression only has one corresponding bounding box in an image. However, in our TextREC dataset, one referring expression can have multiple corresponding bounding boxes. For example, we may ask "The object with 'police' on it", there can be more than one police car in the image. It is necessary to find all the objects that match the description. Therefore, we propose a new test setting that calculates the precision, recall, and F1 score. This can be done by setting a threshold on the confidence of all detected bounding boxes. We set 0.75 for template1 and 0.35 for template2 due to their different score distributions. Then we take the selected boxes to match the ground truth bounding boxes to get the true positives, false positives, and false negatives. We test our TAMN on this new setting and the results are shown in Table 7. We believe this new setting can offer more comprehensive evaluations on the models.

Table 7. The performance of our TAMN in the new test setting. The precision, recall and F1-Score (%) are reported.

Template	Threshold	Precision	Recall	F1-Score
Template1	0.70	76.8	75.2	76.0
	0.75	78.6	73.8	76.1
	0.80	80.1	72.2	75.9
Template2	0.30	73.0	83.8	78.1
	0.35	81.8	76.2	78.9
	0.40	86.7	66.2	75.1

5.5 Visualization Analysis

To verify the effectiveness of the language attention network. We visualize the attention weight for each word in the referring expressions. As shown in Fig. 8, both the subject module and the OCR module focus on the scene texts in template1. When the expression form transfers to template2, the OCR module still focuses on the scene texts. However, the subject module changes to focus on the category name. For example, in the sentence "the object with '15' on it", the subject module focuses on the "15". While, it focuses on the "person" in the sentence "the person with '19' on it". It is reasonable since the only discriminate information is the scene text in template1.

6 Conclusion

In this paper, we point out that most of the existing REC models ignore scene text which is naturally and frequently employed to refer to objects. To address this issue, we construct a new dataset termed TextREC, which studies how to comprehend the scene text associated with objects in an image. We also propose a text-guided adaptive modular network (TAMN) that explicitly utilizes scene text, relates it to the referring expressions, and chooses the most relevant visual object. Experimental results on the TextREC dataset show that the current state-of-the-art REC methods fail to achieve the expected results, but our TAMN achieves excellent results. The ablation studies also show that it is important to take into account scene text for the referring expression comprehension task.

Acknowledgements. This work was supported by the National Science Fund for Distinguished Young Scholars of China (Grant No.62225603), the Young Scientists Fund of the National Natural Science Foundation of China (Grant No.62206103), and the National Natural Science Foundation of China (No.622061 04).

References

1. Biten, A.F., et al.: Scene text visual question answering. In: Proceedings of the IEEE/CVF International Conference on Computer Vision, pp. 4291–4301 (2019)
2. Bu, Y., et al.: Scene-text oriented referring expression comprehension. IEEE Transactions on Multimedia (2022)
3. Chen, Z., Wang, P., Ma, L., Wong, K.Y.K., Wu, Q.: Cops-Ref: a new dataset and task on compositional referring expression comprehension. In: Proceedings of the IEEE/CVF Conference on Computer Vision and Pattern Recognition, pp. 10086–10095 (2020)
4. Deng, J., Yang, Z., Chen, T., Zhou, W., Li, H.: TransVG: end-to-end visual grounding with transformers. In: Proceedings of the IEEE/CVF International Conference on Computer Vision, pp. 1769–1779 (2021)
5. Du, Y., et al.: PP-OCR: a practical ultra lightweight OCR system. arXiv preprint arXiv:2009.09941 (2020)
6. Fang, S., Xie, H., Wang, Y., Mao, Z., Zhang, Y.: Read like humans: autonomous, bidirectional and iterative language modeling for scene text recognition. In: Proceedings of the IEEE/CVF Conference on Computer Vision and Pattern Recognition, pp. 7098–7107 (2021)
7. Gao, D., Li, K., Wang, R., Shan, S., Chen, X.: Multi-modal graph neural network for joint reasoning on vision and scene text. In: Proceedings of the IEEE/CVF Conference on Computer Vision and Pattern Recognition, pp. 12746–12756 (2020)
8. Grubinger, M., Clough, P., Müller, H., Deselaers, T.: The IAPR TC-12 benchmark: a new evaluation resource for visual information systems. In: International Workshop ontoImage, vol. 2 (2006)
9. Hu, R., Rohrbach, M., Andreas, J., Darrell, T., Saenko, K.: Modeling relationships in referential expressions with compositional modular networks. In: Proceedings of the IEEE/CVF Conference on Computer Vision and Pattern Recognition, pp. 1115–1124 (2017)
10. Hu, R., Singh, A., Darrell, T., Rohrbach, M.: Iterative answer prediction with pointer-augmented multimodal transformers for textVQA. In: Proceedings of the IEEE/CVF Conference on Computer Vision and Pattern Recognition, pp. 9992–10002 (2020)
11. Hudson, D.A., Manning, C.D.: GQA: a new dataset for compositional question answering over real-world images **3**(8). arXiv preprint arXiv:1902.09506 (2019)
12. JaidedAI: EasyOCR (2022). https://github.com/JaidedAI/EasyOCR
13. Jiang, Y., Natarajan, V., Chen, X., Rohrbach, M., Batra, D., Parikh, D.: Pythia v0. 1: the winning entry to the VQA challenge 2018. arXiv preprint arXiv:1807.09956 (2018)

14. Kamath, A., Singh, M., LeCun, Y., Synnaeve, G., Misra, I., Carion, N.: MDETR-modulated detection for end-to-end multi-modal understanding. In: Proceedings of the IEEE/CVF International Conference on Computer Vision, pp. 1780–1790 (2021)

15. Kant, Y., et al.: Spatially aware multimodal transformers for TextVQA. In: Vedaldi, A., Bischof, H., Brox, T., Frahm, J.-M. (eds.) ECCV 2020. LNCS, vol. 12354, pp. 715–732. Springer, Cham (2020). https://doi.org/10.1007/978-3-030-58545-7_41

16. Karaoglu, S., Tao, R., Gemert, J.C.V., Gevers, T.: Con-text: text detection for fine-grained object classification. IEEE Trans. Image Process. **26**, 3965–3980 (2017)

17. Kazemzadeh, S., Ordonez, V., Matten, M., Berg, T.: ReferitGame: referring to objects in photographs of natural scenes. In: Proceedings of the 2014 Conference on Empirical Methods in Natural Language Processing, pp. 787–798 (2014)

18. Krishna, R., et al.: Visual genome: connecting language and vision using crowd-sourced dense image annotations. Int. J. Comput. Vision **123**(1), 32–73 (2017)

19. Liao, M., Pang, G., Huang, J., Hassner, T., Bai, X.: Mask TextSpotter v3: segmentation proposal network for robust scene text spotting. In: Vedaldi, A., Bischof, H., Brox, T., Frahm, J.-M. (eds.) ECCV 2020. LNCS, vol. 12356, pp. 706–722. Springer, Cham (2020). https://doi.org/10.1007/978-3-030-58621-8_41

20. Liao, M., Shi, B., Bai, X.: Textboxes++: a single-shot oriented scene text detector. IEEE Trans. Image Process. **27**(8), 3676–3690 (2018)

21. Liao, M., Shi, B., Bai, X., Wang, X., Liu, W.: Textboxes: a fast text detector with a single deep neural network. In: Proceedings of the AAAI Conference on Artificial Intelligence, vol. 31 (2017)

22. Liao, M., Wan, Z., Yao, C., Chen, K., Bai, X.: Real-time scene text detection with differentiable binarization. In: Proceedings of the AAAI Conference on Artificial Intelligence, vol. 34, pp. 11474–11481 (2020)

23. Liao, M., Zou, Z., Wan, Z., Yao, C., Bai, X.: Real-time scene text detection with differentiable binarization and adaptive scale fusion. IEEE Trans. Pattern Anal. Mach. Intell. **45**(1), 919–931 (2022)

24. Lin, T.-Y., et al.: Microsoft COCO: common objects in context. In: Fleet, D., Pajdla, T., Schiele, B., Tuytelaars, T. (eds.) ECCV 2014. LNCS, vol. 8693, pp. 740–755. Springer, Cham (2014). https://doi.org/10.1007/978-3-319-10602-1_48

25. Liu, R., Liu, C., Bai, Y., Yuille, A.L.: CLEVR-Ref+: diagnosing visual reasoning with referring expressions. In: Proceedings of the IEEE/CVF Conference on Computer Vision and Pattern Recognition, pp. 4185–4194 (2019)

26. Lyu, P., Liao, M., Yao, C., Wu, W., Bai, X.: Mask TextSpotter: an end-to-end trainable neural network for spotting text with arbitrary shapes. In: Ferrari, V., Hebert, M., Sminchisescu, C., Weiss, Y. (eds.) Computer Vision – ECCV 2018. LNCS, vol. 11218, pp. 71–88. Springer, Cham (2018). https://doi.org/10.1007/978-3-030-01264-9_5

27. Mafla, A., de Rezende, R.S., G'omez, L., Larlus, D., Karatzas, D.: StacMR: scene-text aware cross-modal retrieval. In: 2021 IEEE Winter Conference on Applications of Computer Vision, pp. 2219–2229 (2021)

28. Mao, J., Huang, J., Toshev, A., Camburu, O., Yuille, A.L., Murphy, K.: Generation and comprehension of unambiguous object descriptions. In: Proceedings of the IEEE/CVF Conference on Computer Vision and Pattern Recognition, pp. 11–20 (2016)

29. Mathew, M., Karatzas, D., Jawahar, C.: DocVQA: A dataset for VQA on document images. In: Proceedings of the IEEE/CVF Winter Conference on Applications of Computer Vision, pp. 2200–2209 (2021)

30. Mishra, A., Shekhar, S., Singh, A.K., Chakraborty, A.: OCR-VQA: visual question answering by reading text in images. In: 2019 International Conference on Document Analysis and Recognition, pp. 947–952. IEEE (2019)
31. Shi, B., Bai, X., Yao, C.: An end-to-end trainable neural network for image-based sequence recognition and its application to scene text recognition. IEEE Trans. Pattern Anal. Mach. Intell. **39**(11), 2298–2304 (2016)
32. Shi, B., Yang, M., Wang, X., Lyu, P., Yao, C., Bai, X.: ASTER: an attentional scene text recognizer with flexible rectification. IEEE Trans. Pattern Anal. Mach. Intell. **41**(9), 2035–2048 (2018)
33. Sidorov, O., Hu, R., Rohrbach, M., Singh, A.: TextCaps: a dataset for image captioning with reading comprehension. In: Vedaldi, A., Bischof, H., Brox, T., Frahm, J.-M. (eds.) ECCV 2020. LNCS, vol. 12347, pp. 742–758. Springer, Cham (2020). https://doi.org/10.1007/978-3-030-58536-5_44
34. Singh, A., et al.: Towards VQA models that can read. In: Proceedings of the IEEE/CVF Conference on Computer Vision and Pattern Recognition, pp. 8317–8326 (2019)
35. Tanaka, R., Nishida, K., Yoshida, S.: VisualMRC: machine reading comprehension on document images. In: Proceedings of the AAAI Conference on Artificial Intelligence, vol. 35, pp. 13878–13888 (2021)
36. Veit, A., Matera, T., Neumann, L., Matas, J., Belongie, S.: Coco-text: Dataset and benchmark for text detection and recognition in natural images. arXiv preprint arXiv:1601.07140 (2016)
37. Wang, H., Bai, X., Yang, M., Zhu, S., Wang, J., Liu, W.: Scene text retrieval via joint text detection and similarity learning. In: Proceedings of the IEEE/CVF Conference on Computer Vision and Pattern Recognition, pp. 4556–4565 (2021)
38. Wang, J., Tang, J., Yang, M., Bai, X., Luo, J.: Improving OCR-based image captioning by incorporating geometrical relationship. In: Proceedings of the IEEE/CVF Conference on Computer Vision and Pattern Recognition, pp. 1306–1315 (2021)
39. Wang, P., Liu, D., Li, H., Wu, Q.: Give me something to eat: referring expression comprehension with commonsense knowledge. In: Proceedings of the 28th ACM International Conference on Multimedia, pp. 28–36 (2020)
40. Wang, X., et al.: On the general value of evidence, and bilingual scene-text visual question answering. In: Proceedings of the IEEE/CVF Conference on Computer Vision and Pattern Recognition, pp. 10126–10135 (2020)
41. Yang, M., et al.: Reading and writing: Discriminative and generative modeling for self-supervised text recognition. In: Proceedings of the 30th ACM International Conference on Multimedia, pp. 4214–4223 (2022)
42. Yang, S., Li, G., Yu, Y.: Graph-structured referring expression reasoning in the wild. In: Proceedings of the IEEE/CVF Conference on Computer Vision and Pattern Recognition, pp. 9952–9961 (2020)
43. Ye, J., et al.: Shifting more attention to visual backbone: query-modulated refinement networks for end-to-end visual grounding. In: Proceedings of the IEEE/CVF Conference on Computer Vision and Pattern Recognition, pp. 15502–15512 (2022)
44. Yu, L., et al.: MattNet: modular attention network for referring expression comprehension. In: Proceedings of the IEEE/CVF Conference on Computer Vision and Pattern Recognition, pp. 1307–1315 (2018)

45. Yu, L., Poirson, P., Yang, S., Berg, A.C., Berg, T.L.: Modeling context in referring expressions. In: Leibe, B., Matas, J., Sebe, N., Welling, M. (eds.) ECCV 2016. LNCS, vol. 9906, pp. 69–85. Springer, Cham (2016). https://doi.org/10.1007/978-3-319-46475-6_5
46. Yu, W., Liu, Y., Hua, W., Jiang, D., Ren, B., Bai, X.: Turning a clip model into a scene text detector. arXiv preprint arXiv:2302.14338 (2023)
47. Zhu, Q., Gao, C., Wang, P., Wu, Q.: Simple is not easy: a simple strong baseline for textvqa and textcaps. In: Proceedings of the AAAI Conference on Artificial Intelligence, vol. 35, pp. 3608–3615 (2021)

SIMARA: A Database for Key-Value Information Extraction from Full-Page Handwritten Documents

Solène Tarride[1]([✉]) [iD], Mélodie Boillet[1,2] [iD], Jean-François Moufflet[3],
and Christopher Kermorvant[1,2] [iD]

[1] TEKLIA, Paris, France
starride@teklia.com
[2] LITIS, Normandy University, Rouen, France
[3] Archives Nationales, Paris, France

Abstract. We propose a new database for information extraction from historical handwritten documents. The corpus includes 5,393 finding aids from six different series, dating from the 18th-20th centuries. Finding aids are handwritten documents that contain metadata describing older archives. They are stored in the National Archives of France and are used by archivists to identify and find archival documents. Each document is annotated at page-level, and contains seven fields to retrieve. The localization of each field is not available in such a way that this dataset encourages research on segmentation-free systems for information extraction. We propose a model based on the Transformer architecture trained for end-to-end information extraction and provide three sets for training, validation and testing, to ensure fair comparison with future works. The database is freely accessible (https://zenodo.org/record/7868059).

Keywords: Open dataset · Key-value extraction · Historical Document · Segmentation-free Approach

1 Introduction

After transforming information retrieval on the web, machine learning and deep learning techniques are now commonly used in archives [3]. The major difference with the natively digital data of the web or traditional information systems, is that the majority of documents kept in archives are physical documents. These documents have been massively digitised in recent years, but only a few percent of the total collections have been digitised and as the volumes are colossal, complete digitisation of the archives is not envisaged. It is therefore necessary to select the documents to be digitised, according to their usefulness. Among the documents kept by the archives, finding aids play a special role. These documents have been produced by the archive services for decades to facilitate access to the information contained in the documents. They are the main gateway to the mass of documents held and are therefore consulted as a priority for any research.

G. A. Fink et al. (Eds.): ICDAR 2023, LNCS 14189, pp. 421–437, 2023.
https://doi.org/10.1007/978-3-031-41682-8_26

Their digitisation and conversion to digital format of the indexing information they contain are therefore essential to improve the service provided to users.

While the automatic processing of original historical documents usually seeks to produce a textual transcription of their content [11], the same cannot be said of finding aids: for the latter, the automatic processing must make it possible to extract the indexing information they contain in a typed and standardised form, so that it can be inserted into a database. The processing is therefore more about extracting textual information than about pure transcription. The automatic processing chains for these documents must therefore include, in addition to the traditional stages of document analysis and handwriting recognition, information or entity (named entity) extraction [7] and normalisation stages.

Paradoxically, while finding aids are central to accessing archival information, they have hardly been studied in the document analysis community: no such database is publicly available and document recognition systems are not evaluated on this type of document. It is to fill this gap that we make available to the community the SIMARA database described in this article. In addition, we also wish to introduce to the community a new type of task for automatic processing models: a task for extracting key-value information from historical documents by processing the whole page. This task differs from traditional transcription or named entity extraction tasks in the following aspects:

- the entire document must be considered to perform the information extraction (full page)
- among all the text contained in the document, only a part has to be extracted
- each extracted information must be typed, it is a key (type)-value association
- the position of the information on the document is not necessary
- the extracted information (value) must be normalized to satisfy the constraints of the database it will enrich.

In this paper, we propose two main contributions. First, we present SIMARA, a new dataset of finding aids images, annotated with the text transcriptions and their corresponding fields at page-level. We also present an end-to-end baseline model trained on SIMARA, which performs, in a single forward step, the extraction of the target key-value information.

This paper is organized as follows. In Sect. 2, we review the different datasets related to SIMARA as well as the recently proposed models for extracting text and entities from full-page images. Section 3 presents the SIMARA dataset and the different series in detail. Finally, in Sect. 4, we present first experiments and results of training automatic models for key-value information extraction from SIMARA images.

2   Related Work

The SIMARA dataset addresses a rather different task compared to what can be found in the current literature. The challenge is to recognize handwritten text from full pages, but also the fields (key) corresponding to each piece of text (value). These fields are at the frontier between named entities and layout-based tokens. Many public datasets have been introduced for historical document image recognition. All these datasets have been listed and grouped by subtask through the tremendous work carried out by Nikolaidou et al. [12] in their systematic review.

In this section, we present some of the datasets that are the most similar to SIMARA. We also detail the various methods that have been proposed for end-to-end handwriting recognition from full pages, with or without entity extraction.

2.1 Datasets

Many datasets have been proposed for automatic handwriting recognition but, to our knowledge, none contain digitised finding aids. Generally, datasets can be classified into two main types according to their level of annotation:

Datasets Designed for Text-Line Recognition: several datasets were designed for handwriting recognition from line images. As a result, the main objective is to recognize text from pre-segmented text lines. For example, the Norhand database [2] contains documents from the 19th-20th centuries written in Norwegian, but only include text line images and their corresponding transcriptions. Similarly, the IAM [9] database have been designed for text recognition from line images. Most algorithms evaluated on this database are trained on text-lines, although full pages are available and reading order can be retrieved easily.

Datasets Designed for Full Page Recognition: other datasets include full page images and annotations, including transcriptions and detailed layout information. This information is generally encoded in an XML PAGE ground truth file. For example, the following datasets provide complete ground truth annotation at page-level: READ 2016 (German, 15-19th century) [16], Digital Peter (Russian, 18th century) [13], ESPOSALLES (Catalan, 18th century) [8], HOME-ALCAR (Latin, 12-14th centuries) [15], and POPP (French, 19-20th centuries) [4]. Some of them also include named entity annotations. In ESPOSALLES [8], two labels are associated to each word: a *category* and a *person*. As a result, this database allows training end-to-end models on full pages for handwriting recognition and named entity recognition. In POPP [4], there are no explicit named entities, but they can be retrieved using the column separator token included in transcriptions. Finally, in HOME-ALCAR [15], two named entities (place and person) are provided.

Although full page-level annotations are available for these datasets, most researchers focus on text line detection or text line recognition.

2.2 Models for End-to-end Page Recognition

In this section, we present automatic methods for end-to-end handwritten page recognition, with or without entity extraction. We focus only on systems that can handle full pages for documents.

Two main approaches have been proposed so far. First, *segmentation-based* methods combine multiple models, each designed to tackle a specific task: text line detection, text recognition, and optionally named entity recognition. This type of approach is very popular as it is simpler to implement. The other approach consists in performing *segmentation-free* handwriting recognition, and optionally named entity recognition, directly from full pages.

Segmentation-Based Models. Constum et al. [4] perform table detection from double page scans, page classification, text line detection and handwriting recognition on French census images from the POPP dataset. Similarly, Tarride et al. [17] propose a complete workflow for end-to-end recognition of parish registers from Quebec. Their workflow includes page classification, text line detection, text recognition, and named entity recognition.

Segmentation-Free Models. Start Attend Read (SAR) was the first HTR model proposed for paragraph recognition. The model is an attention-based CRNN-CTC network that attends over characters and manages to read multiple lines. Start Follow Read (SFR) [18] combines a Region Proposal Network (RPN) combined with a CRNN-CTC network. As a result, it jointly learns line detection and recognition. OrigamiNet [19] is a CRNN-CTC network that unfolds multi-line features into a single line feature. Vertical Attention Network (VAN) [6] is an attention-based model composed of an FCN encoder, a hybrid attention module, and a recurrent decoder. This model works at line or paragraph level and takes advantage of line breaks in the transcription to attend over line features. All these models are able to recognize text from single pages or paragraphs, but cannot deal with multi-column text-blocks. The Document Attention Network (DAN) [5] is the first model that can effectively process full documents, including documents featuring multiple columns and complex reading order. It is based on the Transformer architecture and jointly learns characters and special tokens that represent layout information. This model reaches state-of-the-art results on RIMES [10] and READ [16] Finally, Rouhou et al. [14] introduce a Transformer model for combined HTR and NER on records on the ESPOSALLES database, where named entities are also represented by special tokens. They show that extracting information directly from records allows the model to benefit from more contextual information compared to a model trained on text lines. This model still requires record segmentation, but could potentially be applied to full pages.

3   The SIMARA Dataset

In this section, we introduce the SIMARA dataset, illustrated in Fig. 1. We describe the source of the documents, their usage, their structure and their content. The database containing both the images and the annotations is freely available at https://zenodo.org/record/7868059.

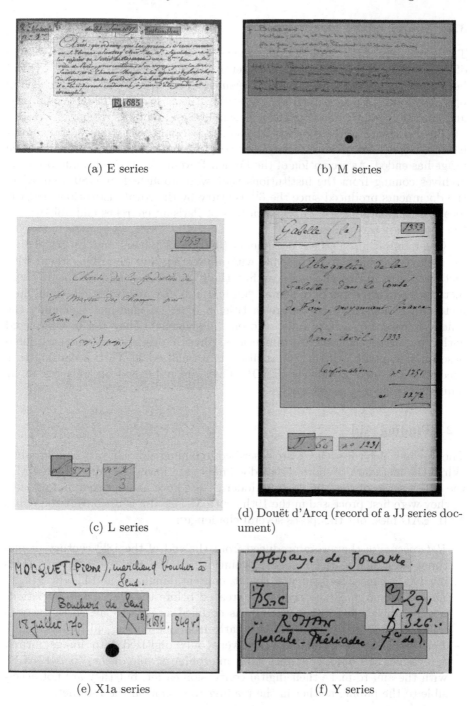

(a) E series

(b) M series

(c) L series

(d) Douët d'Arcq (record of a JJ series document)

(e) X1a series

(f) Y series

Fig. 1. Illustration of the different series of the SIMARA dataset and the information they contain. **Legend:** ▮ date; ▯ title; ▮ analysis; ▮ arrangement; ▮ serie; ▮ article_number; ▮ volume_number.

3.1 Source of the Documents

The documents of the SIMARA dataset were produced by the National Archives of France. The National Archives are a service in charge of preserving the archival heritage of the nation[1]. Their main missions are collecting the documents having both legal and historical interest, preserving them, whatever their medium is, and communicate them to the citizens, scholars and administrations.

The National Archives collect the documents from the current central administrations (presidency, ministries, public corporations) once their administrative usage has ended. As a creation of the French Revolution, they also inherited the archives coming from the institutions that were abolished in 1789: mainly all the documents produced since the 6th century by the royal institutions and the local Parisian administration, the charters of Parisian churches and abbeys and the Parisian notaries files.

Communicating all these documents (380 km of papers and 18 To of digital archives) would not be possible without sorting and describing them with metadata sets, so that the users can find the files that may interest them. These metadata sets are commonly named "finding aids" by archivists. The minimum metadata to identify archives are: a title, a date, an identifier (generally the unique reference of a box, a file, or even a document). They can be completed with other information such as dimensions, physical descriptions, and additional descriptive content. All this information is formalized in XML according to a specific DTD used worldwide: EAD[2]. They are published on websites to be requested through a search engine.

3.2 Finding Aids

The way in which the archives are described in finding aids can vary considerably. While the metadata of more recent documents are formalized directly in XML, the oldest series of the Middle and Modern Ages have handwritten finding aids, as they were first processed in the 19th century. The aim is to convert these into XML EAD files, but this poses several challenges:

- *Heterogeneity*: they were produced from the end of the 18th century (sometimes even before the French Revolution) to the mid 20th century. This involves a huge discrepancy in layout, content, language, and handwriting.
- *Lack of standardization*: they were created before the digital age, in periods when archival description was not as codified as it is today, so the way in which content is presented varies greatly from one finding aid to another.
- *Inaccessibility*: these finding aids were only digitized into image format (grayscale, 300 dpi, sometimes from microfilm or microfiche) in 2017-2019, with the aim to make their digital conversion easier, but they are not accessible to the public, neither in the reading room nor on the Internet.

---

[1] https://www.archives-nationales.culture.gouv.fr/en_GB/web/guest/qui-sommes-nous.

[2] Encoded Archival Description. Official site: https://www.loc.gov/ead/.

– *Quantity*: the index files are estimated to consist of around 800,000 paper cards. Using traditional methods, the digital conversion of these documents would take several years: manually typing the content and then manually encoding it into XML is a lengthy process that can be daunting.

The result of these considerations was the need for the use of automatic processes for the conversion of this material into digital data.

Table 1. Description of the information contained in the different series of the SIMARA dataset.

Serie	Information to extract
E	- Date (`date`)
	- Analysis of the decision (`title`)
	- 1st part of the unique identifier (`serie`, `article_number`)
	- 2nd part of the unique identifier (`volume_number`)
	- Original reference (`arrangement`)
	- Number of the card (`analysis`)
L	- Year of the charter (`date`)
	- Main analysis of the charter (`title`)
	- Further details (`analysis`)
	- Reference of the charter (`serie`, `article_number`, `volume_number`)
M	- Name and information on the person (`title`)
	- Description and proof of nobility (`analysis`)
X1a	- Name of the first party (`title`)
	- Name of the second party (`analysis`)
	- Date of the trial (`date`)
	- Reference of the judgement (`serie`, `article_number`, `volume_number`)
Y	- Name of the person (`title`)
	- Information about the person (`analysis`)
	- Date (`date`)
	- Reference (`serie`, `article_number`, `volume_number`)
Douet d'Arcq	- Heading of the record (`title`)
	- Complements and information (`analysis`)
	- Date (`date`)
	- Reference (`serie`, `article_number`, `volume_number`)

3.3 Description of the Handwritten Finding Aids

In this section, we describe the different series of finding aids included in the SIMARA datasetand the difficulties they present. Table 1 summarizes the fields to be extracted in each series. We have also highlighted these fields for each series in Fig. 1.

E series ("King's councils")

Type of archives: registers in which the decisions of the King's councils were recorded in the 17th and 18th centuries.
Finding aid: 40 480 index cards from the 18th century, containing an analysis of each decision of the King's councils. This series is difficult to read due to the old French script and language.

L series ("Spiritual life of churches and abbeys")

Type of archives: medieval charters of the Paris church of Saint-Martin-des-Champs.
Finding aid: 623 index cards containing an analysis of each charter. They were written in the mid 19th century. The main difficulties lie in the heterogeneous layout of the information and recent corrections of the identifiers (the old 19th identifier was crossed out with a pencil and corrected by a contemporary hand).

M series ("Proofs of nobility")

Type of archives: documents relating to the Knights of Malta and containing evidence of their nobility.
Finding aid: 4 847 cards arranged in alphabetical order. They were written around the 1950s.s. These documents do not present any particular difficulties.

X1a series ("Parliament")

Type of archives: registers in which were recorded the judgements from the Parliament, which was the highest court of justice.
Finding aid: 101 036 cards containing an analysis of 18th century trials involving two parties. These cards are sorted in alphabetic order by the name of the parties. They were written in the 20th century, and do not present any particular difficulties.

Y series ("Châtelet")

Type of archives: notarial deeds of the 18th century recorded in the Châtelet, which was a Parisian institution where the notaries had to report all the transactions they had recorded.
Finding aid: 61 878 cards containing an analysis of the notarial deeds. The cards are sorted in alphabetical order by the name of the person who obtained the deed. They were written in the 1950s. The main difficulties are the poor digitization quality, the handwriting style, the heterogeneous layout as well as the density of the text.

Douët d'Arcq file

Type of archives: unlike the previous archives, several series are indexed by this file, mainly the "historical series" (J, K, L, M) which mainly correspond to the oldest and most prestigious documents (medieval royal archives, charters from Parisian abbeys and churches, University of Paris, Templars and Hospitallers...). Different kind of documents (charters, registers, accounts) are covered.

Finding aid: 118 093 cards created in the mid 19th century, and divided into three subsets: a file for personal names, another one for geographical names and a last one ordered by subjects. We only focus on the latter (6 284 cards).

3.4   Finding Aids and Original Documents

To give the reader a clearer idea of what finding aids and original records are, Fig. 2 shows examples of finding aids and reproductions of the original documents they describe.

We have highlighted the main categories of information in the records, as shown in Fig. 1, and also where some of this information was found in the original documents.

3.5   Production of the Annotated Data

The approach chosen for the generation of the annotated data of the SIMARA dataset was different from the traditionally used approaches. Usually, as processing chains are composed of different models for line detection, handwriting recognition and named entity extraction, it is necessary to generate specific annotations for each model: line positions on the image, transcribed text lines and entities positioned on the text. Each of these types of annotations requires a specific interface, and this interface is also different from the interface used to validate the model predictions in a production phase. The consequence of these multiple interfaces and multiple types of training data is that the data validated in production cannot be used to improve the models. To re-train the models, it is necessary to repeat annotation campaigns, in parallel with the validation tasks, which leads to a duplication of tasks. The approach chosen in this work is radically different: the annotation interface used to generate the first data needed to train the models is the same as the one used to validate the data in production. In this way, all production data can be used to iteratively improve the models.

The originality of the SIMARA dataset is that the ground-truth was easily created by filling a form designed by the archivist. Each annotator was in charge of typing the ground-truth in the form by searching the information on the image of the finding aid. The annotation interface is shown on Fig. 3.

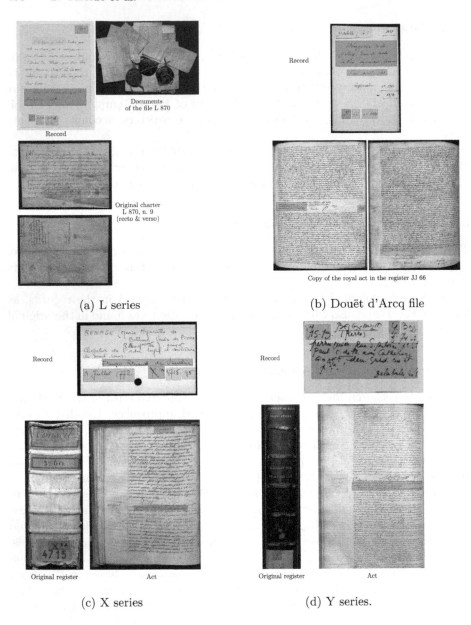

(a) L series

(b) Douët d'Arcq file

(c) X series

(d) Y series.

Fig. 2. Sample of records with the original document for four series.

Once the models are trained, they are used in the production phase to pre-fill the form with suggestions. In this phase, the task of the annotators is different since they have to read, confirm or correct the information suggested in each field. This step is certainly the most time-consuming in the process, but it is crucial since the corrections made by the annotator are used to improve the

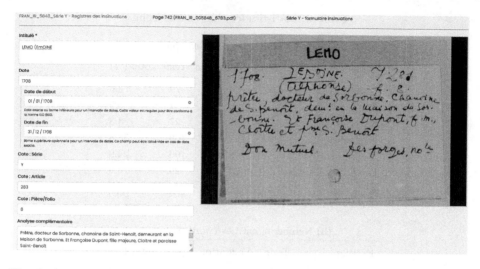

Fig. 3. Annotator interface: the same interface is used to produce the ground-truth data and to validate the suggestions of the model in the production phase.

information extraction model. The validation interface is the same as for the ground-truth generation.

The quality of the transcription was ensured by a process of double validation in case of doubt: when the annotators were uncertain about a piece of information, they could indicate it to an archivist for verification and validation.

4   Experimental Results

In this section, we report the results of the experiments conducted on the SIMARA dataset with our proposed system for full page key-value information extraction.

4.1   Datasets

Our training dataset evolved during the course of the project, as more and more documents were annotated and became available for training. As we trained models at different stages of the annotation procedure, we trained three models on different splits:

- *Split-v1*: 780 images for training, 50 for validation and 50 for testing;
- *Split-v2*: 3659 images for training, 783 for validation and 784 for testing;
- *Split-v3*: 3778 images for training, 811 for validation and 804 for testing (see Table 2).

Table 2. Statistics for the *Split-v3* split

(a)Statistics for the *Split-v3* split

Series	Train	Validation	Test	Total (%)
E series	322	64	79	8.6
L series	38	8	4	0.9
M series	128	21	27	3.3
X1a series	2209	491	469	58.8
Y series	940	205	196	24.9
Dout s'Arcq series	141	22	29	3.5
Total	3778	811	804	100.0

(b) Number of entities (word count)

Entities	Train	Validation	Test	Total (%)
date	8406	1814	1799	10.4
title	35531	7495	8173	44.4
serie	3168	664	676	3.9
analysis	25988	5130	5602	31.8
volume_number	3913	808	813	4.8
article_number	3181	665	678	3.9
arrangement	644	122	153	0.8
Total	80831	16698	17894	100.0

4.2 Key-Value Information Extraction Model

We train a model based on the DAN architecture [5] for key-value information extraction from full pages on SIMARA. DAN[3] is an open source attention-based Transformer model for handwritten text recognition that can work directly on pages. It is trained with the cross-entropy loss function, and its last layer is a linear layer with a softmax activation to compute probabilities for each character of the vocabulary. We address the task of key-value information extraction by encoding each field with a special token located at the beginning of each section, as illustrated in Table 3. A complete description and evaluation of this model is presented in [1].

4.3 Metrics

HTR Metrics: The quality of handwriting recognition is evaluated using the standard Character Error Rate (CER) and Word Error Rate (WER). The full text is evaluated, but named entities are ignored for this kind of the evaluation.

---

[3] https://github.com/FactoDeepLearning/DAN.

Table 3. Example of transcription used to train our baseline model. Special tokens are localized at the beginning of each text section to characterize the corresponding field. Note that the fields always appear in the same order in transcriptions, even if it does not correspond to the reading order on the images.

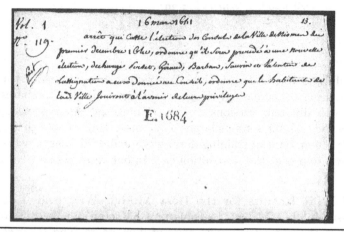

Transcription

`<reference_number>n° 119<date>16 mars 1641 <analysis>13`
`<title>Arrêt qui casse l'élection des consuls de la`
`ville de Nîmes du premier décembre 1640, ordonne qu'il`
`sera procédé à une nouvelle élection, décharge Perset,`
`Giraud, Barban, Saurin et Valentin de l'assignation à eux`
`donnée au Conseil, ordonne que les habitants de ladite`
`ville jouiront à l'avenir de leurs privilèges. <serie>E`
`<article_number>1684 <arrangement>Volume 1`

NER Metrics: We use the Nerval[4] evaluation toolkit to evaluate Named Entity Recognition results. Nerval is able to deal with noisy text by aligning the automatic transcription with the ground truth at character level. Predicted and ground-truth entities are considered a match if their label is similar and their edit distance below a threshold, set to 30% in our experiments. From this alignment, precision, recall and F1-score are computed to evaluate Named Entity Recognition.

4.4   Results

In this section, we present the performance reach by our baseline model for key-value information extraction on SIMARA. First, we study the impact of the training dataset size, then we detail the results obtained with the best model.

---

[4] https://gitlab.com/teklia/ner/nerval.

Table 4. Results of the three models on the test set.

Model name	Split	CER (%)	WER (%)	N images
Model-v1	*Split-v1*	12.08	22.41	50
Model-v2	*Split-v2*	7.39	15.07	784
Model-v3	*Split-v3*	**6.46**	**14.79**	804

Impact of Training Dataset Size. We trained three models on three different data splits, using an increasingly bigger training set. Table 4 compares the results obtained with different versions of the training set. As expected, the results improve as the training set gets larger. It is interesting to note that even a small increase in the amount of training data, e.g. a hundred images between *Split-v2* and *Split-v3*, improves the recognition rate in our case by one CER point.

Detailed HTR Results for the Best Model. We provide detailed results for the best model (*Model-v3*) in Table 5a. Character and Word Error Rates are computed on the test set for each series of documents. The *L series* is the best recognized, although it is not well represented in the training set. It should be noted that this series only has 4 documents in the test set, which limits the statistical interpretation of this observation. The most difficult series is Douët d'Arcq, with a WER close to 25%. This can be explained by the fact that this series has a unique layout and few examples in the training set. The other most difficult series are the Y series (high CER) and the E series (high WER) which, although highly represented in the training set, present some difficulties in terms of layout and language respectively. Overall, text recognition results are very acceptable, with a WER below 15%.

Detailed NER Results for the Best Model. These documents are difficult to evaluate from a transcription point of view because the reading order of the machine does not necessarily coincide with that of the annotator. Therefore, a high error rate at character or word level does not necessarily indicate poor transcription quality, but rather differences in the reading order.

An evaluation of the detection of key-value information (named entity), allows to overcome the problems of reading order. The Table 5b shows the results of automatic extraction with the *Model-v3*, detailed by type of information. We note very good performances on numerical information such as date, serial number and item number. Textual information such as the title and the analysis are also extracted with very good rates. Only information that is not very present in the training set, such as the arrangement field, performs less well. In general, more than 95% of the information is extracted correctly.

Table 5. Detailed evaluation results for *Model-v3*.

(a) Character and Word Error Rates are computed for each serie on the test set. Special tokens encoding each field are removed at this stage of the evaluation.

Series	CER (%)	WER (%)	N images
E series	6.89	18.73	79
L series	**3.33**	**9.35**	4
M series	7.70	15.87	27
X1A series	4.97	11.16	469
Y series	8.23	16.19	196
Douët d'Arcq	10.24	24.97	29
Total	6.46	14.79	804

(b) NER precision, recall and F1-score are computed for each field on the test set.

Tag	Predicted	Matched	Precision (%)	Recall (%)	F1 (%)	N entities
date	1808	1784	98.67	**99.17**	**98.92**	1799
title	8059	7650	94.92	93.60	94.26	8173
serie	677	664	98.08	98.22	98.15	676
analysis	5449	5248	96.31	93.68	94.98	5602
article_number	672	664	**98.81**	97.94	98.37	678
volume_number	805	776	96.40	95.45	95.92	813
arrangement	206	119	57.77	77.78	66.30	153
Total	17676	16905	95.64	94.47	95.05	17894

5   Conclusion

In this article, we have described SIMARA, a new public database consisting of images of digitised finding aids with the transcription of the information they contain. This database has allowed us to define a new type of task for automatic historical document processing systems. This task consists in the extraction of key-value information from document pages. We proposed a baseline system based on a Transformers neural network, allowing the extraction of this information in a single pass. The performances of this model constitute reference results, allowing to evaluate the progress of models which will be proposed to solve this task on the SIMARA database.

References

1. Tarride, S., Boillet, M., Kermorvant, C.: Key-value information extraction from full handwritten pages. In: International Conference on Document Analysis and Recognition (2023)
2. Beyer, Y., Kåsen, A.: NorHand/Dataset for handwritten text recognition in Norwegian (2022). https://doi.org/10.5281/zenodo.6542056

3. Colavizza, G., Blanke, T., Jeurgens, C., Noordegraaf, J.: Archives and AI: an overview of current debates and future perspectives. J. Comput. Cult. Heritage **15**(1), 1–15 (2021)
4. Constum, T., et al.: Recognition and information extraction in historical handwritten tables: toward understanding early 20th century Paris census. In: 15th International Workshop on Document Analysis Systems (DAS), pp. 143–157 (2022). https://doi.org/10.1007/978-3-031-06555-2_10
5. Coquenet, D., Chatelain, C., Paquet, T.: DAN: a segmentation-free document attention network for handwritten document recognition. IEEE Trans. Pattern Anal. Mach. Intell. (2023). https://doi.org/10.1109/TPAMI.2023.3235826
6. Coquenet, D., Chatelain, C., Paquet, T.: End-to-end handwritten paragraph text recognition using a vertical attention network. In: IEEE Transactions on Pattern Analysis and Machine Intelligence, vol. 45, pp. 508–524 (2023). https://doi.org/10.1109/TPAMI.2022.3144899
7. Cunha1, L.F., Ramalho, J.C.: Fine-tuning BERT models to extract named entities from archival finding aids. In: 26th International Conference on Theory and Practice of Digital Libraries (2022)
8. Fornés, A., et al.: ICDAR 2017 competition on information extraction in historical handwritten records. In: International Conference on Document Analysis and Recognition, pp. 1389–1394 (2017). https://doi.org/10.1109/ICDAR.2017.227
9. Marti, U.V., Bunke, H.: The IAM-database: an English sentence database for offline handwriting recognition. In: International Journal on Document Analysis and Recognition, vol. 5, pp. 39–46 (2002). https://doi.org/10.1007/s100320200071
10. Menasri, F., Louradour, J., Bianne-Bernard, A.L., Kermorvant, C.: The A2iA French handwriting recognition system at the Rimes-ICDAR2011 competition. In: Proceedings of SPIE - The International Society for Optical Engineering, vol. 8297, p. 51 (2012). https://doi.org/10.1117/12.911981
11. Muehlberger, G., et al.: Transforming scholarship in the archives through handwritten text recognition: Transkribus as a case study. J. Document. **75**, 954–976 (2019). https://doi.org/10.1108/JD-07-2018-0114
12. Nikolaidou, K., Seuret, M., Mokayed, H., Liwicki, M.: A survey of historical document image datasets. Int. J. Doc. Anal. Recogn. (IJDAR) **25**(4), 305–338 (2022). https://doi.org/10.1007/s10032-022-00405-8
13. Potanin, M., et al.: Digital peter: new dataset, competition and handwriting recognition methods. In: The 6th International Workshop on Historical Document Imaging and Processing, pp. 43–48. HIP 2021, Association for Computing Machinery, New York, NY, USA (2021). https://doi.org/10.1145/3476887.3476892
14. Rouhou, A.C., Dhiaf, M., Kessentini, Y., Salem, S.B.: Transformer-based approach for joint handwriting and named entity recognition in historical document. Patt. Recogn. Lett. **155**, 128–134 (2022). https://doi.org/10.1016/j.patrec.2021.11.010. https://www.sciencedirect.com/science/article/pii/S0167865521004013
15. Stutzmann, D., Torres Aguilar, S., Chaffenet, P.: HOME-Alcar: aligned and annotated cartularies (2021). https://doi.org/10.5281/zenodo.5600884
16. Sánchez, J.A., Romero, V., Toselli, A.H., Vidal, E.: ICFHR2016 competition on handwritten text recognition on the read dataset. In: 2016 15th International Conference on Frontiers in Handwriting Recognition (ICFHR), pp. 630–635 (2016). https://doi.org/10.1109/ICFHR.2016.0120
17. Tarride, S., Maarand, M., Boillet, M., et al.: Large-scale genealogical information extraction from handwritten Quebec parish records. IJDAR (2023). https://doi.org/10.1007/s10032-023-00427-w

18. Wigington, C., Tensmeyer, C., Davis, B., Barrett, W., Price, B., Cohen, S.: Start, follow, read: end-to-end full-page handwriting recognition. In: Ferrari, V., Hebert, M., Sminchisescu, C., Weiss, Y. (eds.) ECCV 2018. LNCS, vol. 11210, pp. 372–388. Springer, Cham (2018). https://doi.org/10.1007/978-3-030-01231-1_23
19. Yousef, M., Bishop, T.: OrigamiNet: weakly-supervised, segmentation-free, one-step, full page text recognition by learning to unfold. In: IEEE Conference on Computer Vision and Pattern Recognition (CVPR), pp. 14698–14707 (2020)

Diffusion Models for Document Image Generation

Noman Tanveer[1(✉)], Adnan Ul-Hasan[1], and Faisal Shafait[1,2(✉)]

[1] National Center of Artificial Intelligence (NCAI), National University of Sciences and Technology (NUST), Islamabad, Pakistan
nomantanveer021@gmail.com, {adnan.ulhassan, faisal.shafait}@seecs.edu.pk
[2] School of Electrical Engineering and Computer Science (SEECS), National University of Sciences and Technology (NUST), Islamabad, Pakistan

Abstract. Image generation has got wide attention in recent times; however, despite advances in image generation techniques, document image generation having wide industry application has remained largely neglected. The previous research on structured document image generation uses adversarial training, which is prone to mode collapse and overfitting, resulting in lower sample diversity. Since then, diffusion models have surpassed previous models on conditional and unconditional image generation. In this work, we propose diffusion models for unconditional and layout-controlled document image generation. The unconditional model achieves state-of-the-art FID 14.82 in document image generation on DocLayNet. Furthermore, our layout-controlled document image generation models beat previous state-of-the-art in image fidelity and diversity. On the PubLayNet dataset, we get an FID score of 15.02. On the complicated DocLayNet dataset, we obtained an FID score of 20.58 with 256×256 resolution for conditional image generation.

Keywords: Document Image Generation · Diffusion Models · Latent Diffusion Models

1 Introduction

Document image understanding is one of the most important objectives of the Document AI community. Often there is a need to augment real data for training and fine-tuning with generated data. So far generative models are not able to achieve document image generation with readable text. However, end-to-end document generation is useful for training document region detection models. Building and maintaining large datasets is expensive and often prohibitive for most businesses. We propose a document image generation model which acts as augmentation for training document region detection models and serves as a basis for end-to-end readable document image generation. The proposed generation model can be used to augment small datasets for improved performance on region detection. So far not much work has been done to address

G. A. Fink et al. (Eds.): ICDAR 2023, LNCS 14189, pp. 438–453, 2023.
https://doi.org/10.1007/978-3-031-41682-8_27

this need. According to Office Document Architecture (ODA), a document can be expressed with two formalisms: an image for displaying and its textual representation [31] both of these can be met by a digitized image. Organizations in the US spend 120 Billion dollars on printed forms alone, most of which becomes obsolete within 3 months [26]. Furthermore, the environmental impact of deforestation caused by the paper industry has forced organizations to consider going paperless [27]. This has further increased the importance of digital documents. Current template-based approaches, however, are unable to cater to the huge variety of documents that need to be digitized and provide an otherwise restrictive use of the document space in terms of layout and content as specified by the ODA [28,31]. A controllable document image generation system can address both of these problems.

Generative models have achieved ground-breaking results in unconditioned and text-conditioned image generation over the past few years. Notably Dall-E [1], Dall-E-2 [2], StyleGan3 [41], Imagen [3], VQ-VAE2 [4] have achieved significant results. In parallel, however, document image generation has not received much attention apart from a few works [5,29].

Controlled document image generation has traditionally been restricted by two factors: difficulties in text generation and the difficulty in the presentation of the information. the first problem has gotten a lot of attention [30,45,46], lately and has largely been resolved. To address the second problem i.e. the visual presentation of document layout information in the document form, we propose a layout-controlled document image generation diffusion model. Although previous generative models i.e. StyleGANs [41,47], VAEs [4] performed well in generating high-resolution images on everyday objects i.e. COCO and ImageNet datasets [23,24], these models have been out-performed by diffusion models on fidelity and diversity metrics [3,13,14]. As far as we know, no work has been done in the document domain using diffusion models [5,29].

There are multiple approaches to image generation. The most popular ones include Generative Adversarial Networks (GANs), Variational Auto-encoders (VAEs), Auto-regressive models, and flow-based models. Each of these has specific trade-offs associated with it. GANs although good at high-resolution image generation and fast generation, require adversarial training and suffer from Mode-collapsed, adversely affecting the sample diversity. VAEs offer fast image generation and higher sample diversity but the quality of the generated samples is not as good as the GANs. Auto-regressive models although performing well at generating high-quality samples are very slow at inference times. Similarly, flow-based models offer explicit distribution modeling of the data but impose strong restrictions on the architecture of the models. There also exist several mixed approaches i.e. VQ-GANs [32], Energy-based models [33] and Auto-regressive Generative models [34] that try to combine these approaches and try to improve on the individual approaches.

One of the latest generative approaches, the diffusion models, has recently gotten much attention from the research community due to their improvement over the previous approaches in terms of sample quality i.e. Fréchet Inception Distance (FID) and Diversity Score, and achieving state-of-the-art results on a

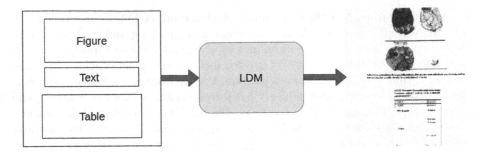

Fig. 1. Layout guided document image synthesis. Given a layout, the model generates a sample from the training data distribution $p(x)$ with the given layout.

number of datasets [18–20, 23, 48]. At training time diffusion models progressively add noise to the image and create a Markov Chain of increasingly noisy images according to a noising schedule that controls the amount of the noise to the point when the images are indistinguishable from the random noise.

A stateful model then learns to reverse the noising process at each step and is trained to predict the noise added at the corresponding step. At inference time, the models can then generate highly diverse samples from random noise. A trade-off associated with diffusion models was that they took huge computing resources to train. This was addressed by the Latent Diffusion Models (LDMs) [21] that propose to apply diffusion models on the latent space of an auto-encoder instead of the full-resolution images. LDMs offer similar sample quality for a fraction of the computing cost by ignoring the pixel-level details whose contribution to the semantic understanding of the image is minuscule. LDMs can further be used for conditional image generation by using embeddings from domain-specific encoders and applying cross attention at the latent space of the diffusion model (Fig. 1).

DocSynth [5] proposed controlled document image generation similar to 1 using GANs for the first time. They conditioned image generation with the desired layout of the image. In addition to using complicated architecture and a number of auxiliary losses, they trained their models on 64×64 and 128×128 resolutions. Using diffusion models for this task can result in higher fidelity and diversity.

The following are the major contributions of this work:

1. We propose a diffusion model for controlled document image generation, guided by the desired document layout.
2. We demonstrate that the samples generated by our model can improve the performance of a visual document region detector.
3. We demonstrate the ability of the diffusion model to generate diverse document images, by unconditional training on DocLayNet.

The rest of the paper is divided into the following sections. Section 2 provides the necessary background in generative diffusion models. Section 3 discusses the components of the proposed architecture. Sections 4 and 5 provide experimental details and results respectively. Lastly, Sect. 6 gives the conclusion.

2   Background

In the limited work done in document image generation, models use adversarial training e.g. Generative Adversarial Models (GANs). Adversarial training is prone to over-fitting and mode collapse and requires a lot of hyperparameter tuning to get right. Although some improvements have been made in this regard [36,37], this still leads to lower sample diversity in GANs. Diffusion Models (DMs) have recently addressed these issues and achieved state-of-the-art results in Image Generation [14]. We propose a diffusion model for layout-controlled document image generation building on the following previous work in generative and multimodal-distribution modeling.

Deep image generation models were popularized by the introduction of Generative Adversarial Models [38]. Since then, GANs have undergone several modifications and fine-tuned many times (BigGANs [39], StyleGANs [41] and Hierarchial GANs [40]) to improve the results notably. In GANs, two models are trained in an adversarial manner to generate high-fidelity images; as a result, they are difficult to train and suffer from mode collapse issues. Recently, diffusion models that are a subclass of the likelihood-based models have achieved better results than GANs. In addition, since these models are not trained adversarially, their training is stable and scales favorably with compute.

Diffusion Models (DMs) are a type of latent variable model in which the latent variables are of the same dimension as the original data distribution $p(x)$ [11]. DMs define a Markov Chain $x_1, x_2, ..., x_n$ of these latents. The forward process adds Gaussian noise to the image sample according to a schedule $\beta_1, \beta_1, ..., \beta_n$ as (Fig. 2):

$$q(x_{1:T}|x_0) = \prod_i^T q(x_t|x_{t-1}) \tag{1}$$

where:

$$q(x_t|x_{t-1}) = \mathcal{N}(x_t; \sqrt{1-\beta_t} * x_{t-1}, \beta_t I) \tag{2}$$

Here β represents the schedule variable from a noise scheduler that controls the amount of noise to be added at each step [11] and q represents the latent distribution of the noised data. For a long Markov Chain, the noised sample is indistinguishable from random noise. In the reverse (generation/regeneration) process a model (usually, U-Net for images) learns to reverse these Gaussian transitions. The final distribution is given as (Fig. 3):

$$p_\theta(x_{0:T}) = p(x_T) \prod_{t=1}^T p_\theta(x_{t-1}|x_t) \tag{3}$$

where:

$$p_\theta(x_{t-1}|x_t) = \mathcal{N}(x_{t-1}; \mu_\theta(x_t, t), \sum_\theta(x_t, t)) \tag{4}$$

Here p is the data distribution and $\sum$ represents the variance matrix. Where all the transitions have a shared set of parameters, i.e., $t = [1-T]$. Additionally,

Fig. 2. The forward noising process of a diffusion model for $t = [1 - T]$. At each step Gaussian noise is added according to a schedule, such that at step T the sample $z[t]$ is indistinguishable from random noise.

since the model is stateful, it can be trained in a stochastic manner. The model is trained by minimizing the variational lower bound on negative log-likelihood using the reparameterization trick.

Diffusion models require a huge amount of computing resources since they were applied on the pixel space. Sampling an image from a diffusion model requires [25-1,000] passes through the U-Net. High computing requirements made it prohibitively expensive to train high-resolution DMs. Hence [21] introduced Latent Diffusion Models (LDMs) that applied multistage diffusion on the latent space of an auto-encoder model instead of applying it on the full image resolution. Hence the diffusion model only models semantic information instead of spending significant compute on high-frequency details while resulting in faster training and sampling. LDMs provide the best trade-offs in image fidelity and compute requirements at the moment and have been widely used in many applications. Diffusion models can be trained conditionally through classifier-guidance [14] and classifier-free guidance [15]. Classifier guidance provides a method analogous to truncation in GANs by stepping in the direction of classifier gradients to improve image fidelity. Classifier-free guidance instead simultaneously trains a conditional and unconditional model and sweeps over their weight to generate high-fidelity images. LDMs [21] introduced a generalized conditioning mechanism where embeddings from a multi-modal-conditioning model (e.g., CLIP) are used in a cross-attention mechanism to produce the desired conditioning results. Attention conditioning is compatible with classifier-free guidance and produces even better results when used together.

Diffusion models produce high fidelity but GANs still have faster inference sample rates because they produce samples in a single step as opposed to diffusion models which take [25-1,000] steps for sampling. Although, DMs work well for short conditioning embeddings. As of now, diffusion model conditioning does not work well for long sequences i.e. text to image generation. A better conditioning mechanism for diffusion models is an active area of research.

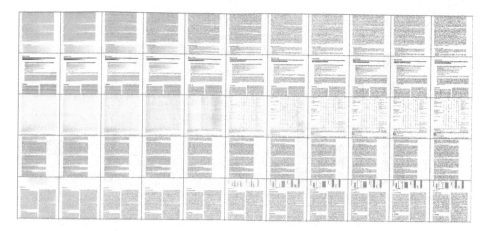

Fig. 3. A diffusion model (U-Net) learns to reverse the noising process from random noise in T steps $[T:0]$ such that t=0 corresponds to a sample from $p(x)$.

Controlled document image generation proposed by   [5] uses GANs. Later, [29] proposed a Graph Neural Network (GNN) for image synthesis. Our approach in addition to being much cleaner than these two is trained on higher 256×256 resolution. In addition to that, we can also control the fidelity-compute trade-off with a simple compression parameter f which controls the dimension at which the diffusion model is applied.

3   Methodology

The proposed latent diffusion model consists of three components: a diffusion model that generates samples in the latent space, a transformer that generates conditional embedding, and an auto-encoder to downsample and upsample document images to and from the latent space. First, an auto-encoder model is trained on document images separately. Secondly, the diffusion model and a BERT encoder are trained end-to-end with the regeneration loss (Fig. 4).

3.1   Diffusion Model

The diffusion model consists of a step-conditioned U-Net model. At any time step (t), the model takes a noised image $x(t)$ and predicts image $x(t-1)$ by passing the image and the corresponding time-step through a U-Net [11].

Diffusion is applied in the latent space of the auto-encoder instead of the full resolution of the image. This trick significantly reduces the compute requirements compared to applying the diffusion on the full resolution.

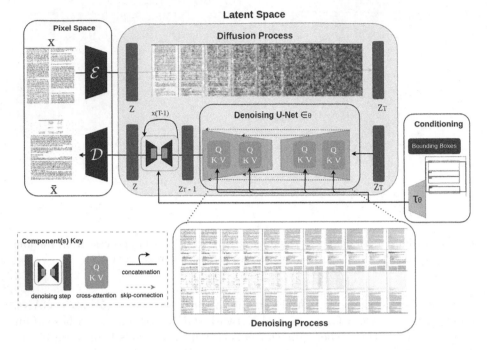

Fig. 4. The architecture of a latent diffusion model. Auto-encoder encodes pixel space to latent space and vice-versa: $z = \mathcal{E}(x)$, $x = \mathcal{D}(z)$, where $\mathcal{E}$ and $\mathcal{D}$ are the encoder and the decoder of the auto-encoder. $\in_\theta$ is the stateful denoising U-Net. The diffusion model generates models in the latent space, conditioned by the bounding-box encoder τ_θ using cross-attention.

3.2   Condition Encoder

We use a transformer [42] architecture (BERT-encoder with 16 layers) to encode bounding boxes and labels of a document. The encoder is fed with tokenized bounding boxes and labels of the region in a document. The encoder generates embedding of shape batch_size $\times$ 92 $\times$ 512. The encoder is trained end-to-end with a diffusion objective $e = \tau_\theta$(Bounding-Boxes), where e is encoder embedding for condition encoder τ_θ.

3.3   Auto-Encoder

A vector quantized auto-encoder [43] is trained separately on the images from a similar domain with reconstruction loss. We train it on the samples from the PubLayNet dataset [22]. At training time, each image is encoded before it is passed to the diffusion model. Similarly, at inference time, each image generated by the diffusion is decoded to a higher resolution by the decoder of the auto-encoder. The depth and scaling factor f of the encoder provides a way to trade off *compute* vs *fidelity* in the architecture. Regularization is applied to the latent space to avoid large variances in the latent representation.

$$z = \mathcal{E}(x) \tag{5}$$

$$x = \mathcal{D}(z) \tag{6}$$

The encoder $\mathcal{E}$ converts the image x to its latent space representation, whereas the decoder $\mathcal{D}$ converts a latent space image z to the pixel space. Layout to image generation is achieved by end-to-end training of the diffusion model and the BERT embeddings encoder. Tokenized bounding boxes and corresponding labels are passed to a 16-layer BERT transformer. Then cross-attention is performed between the encoded image and the embeddings of the bounding boxes to train the diffusion model and the BERT encoder simultaneously. Using this approach we generate 256 × 256 resolution images, which follow the layout passed at inference time.

4   Experimental Setup

The model is trained on an Nvidia-A6000 GPU with a batch size of 32, for 1.2M iterations. The learning rate of all the components is set to $10e^{-6}$ with a linear warmup for 10,000 steps. The auto-encoder is fine-tuned with L1-loss, where cross-attention is performed at resolutions of 8,4, and 2 respectively with 32 attention heads. The encoder scales a 256 × 256 image down to 64 × 64 resolution, which is then used to train the diffusion model. At inference time, an image is generated from random noise while cross-attention is performed at each decoding timestep (t) of the diffusion generation process. The generated image in the latent space is then upscaled through the decoder $\mathcal{D}$ to the 256 × 256 resolution. For the unconditional case, we train only the diffusion model on 64 × 64 in the pixel space, i.e., no downscaling or attention conditioning is applied. The unconditional diffusion is trained on DocLayNet for 200K steps.

4.1   Evaluation Metric

We have evaluated our model on the standard evaluation metric used in image generation: Fréchet Inception Distance (FID) which measures the distance between the Inception net embedding of the two sample distributions. As per the standard practice, the FID score is calculated on the training distribution. FID is used to measure the quality of the generated images using Inception-Net embedding and measuring their distance from the original images. Although it is not a perfect metric, it accounts for both sample diversity and fidelity. Since FID also accounts for sample diversity, we believe most of the improvement in results comes from the increased diversity of our samples.

4.2 Datasets

We train our model on standard large-scale document region detection datasets. We use DocLayNet to train the unconditional diffusion model. While the conditional model is trained on both PubLayNet and DocLayNet.

PubLayNet. [22] is an open dataset of 350,000 Layout annotated scanned documents. The annotated regions include title, text, table, figure, and list. The dataset mostly consists of medical publications.

DocLayNet. [25] is a human-annotated dataset collected from a variety of sources. It consists of 80,863 documents with diverse layouts. The dataset comes from financial reports, scientific articles, laws and regulations, government tenders, manuals, and patents.

4.3 Implementation Details and Hyperparameters

Latent space mapping is only performed for the conditional case. Unconditional generation is done on 64×64 images with pixel space diffusion. Each image is center-cropped to maintain the aspect ratio of the image while training. The diffusion model uses L1 loss for optimization using Weighted Adam optimizer [44]. In auto-encoder, the downsampling factor f provides a way to tradeoff compute and generation fidelity [21]. We use the downsampling factor (f) of 4 because it preserves the document details in a computationally efficient way. Our model has 300M trainable parameters; however, it requires orders of magnitude less compute than training a diffusion model on full 256×256 resolution. We use a linear scheduler for the diffusion model same as [21]. Attention in the diffusion model is applied at 2, 4, and 8 resolutions. Most other parameters are the same as [21].

5 Results and Analysis

Using this approach, for the conditional case, we have achieved a state-of-the-art FID score of 15.02 on the PubLayNet dataset, which is significantly better than DocSynth. In addition, we also train our model on the DocLayNet dataset that includes diverse and challenging layouts and achieve an FID score of 20.58. For unconditional training on DocLayNet, we get an FID score of 14.82. It can be seen from the Fig. 5, that the diffusion model captures the variety of samples and can model variance in the data distribution. For layout-conditioned image generation, we trained the auto-encoder with both KL-regularization and VQ-regularization. We found that VQ-regularization performs better than KL-regularization in agreement to the previous results in [21].

Fig. 5. Unsorted 64 × 64 images generated by the unconditional diffusion model on the DocLayNet dataset. The generated samples show that the diffusion model is able to capture a variety of document layouts, including graphics and color.

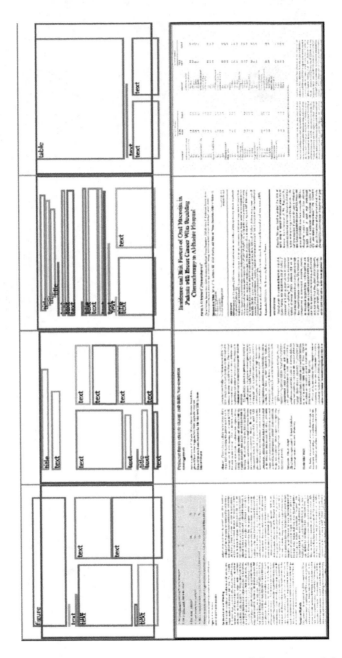

Fig. 6. Layout conditioned 256×256 images generated by our model for different layouts from the PubLayNet validation set. The generated samples are able to capture diverse layouts and generate high-fidelity samples, where the black bounding box represents the area considered for training after the center crop.

5.1   Quantitative Results

We have evaluated our model on FID score and compared it against DocSynth [5] a Generative Adversarial Network for layout-controlled image generation. Table 1 shows our results:

Table 1. The comparison of the FID score of Doc-Synth and our approach on Pub-LayNet and DocLayNet datasets

Dataset	Proposed		DocSynth [5]	
	128×128	256×256	128×128	256×256
PubLayNet	**12.34**	**15.02**	33.75	-
DocLayNet	**17.86**	**20.58**	-	-

Table 2. The results of training Faster-RCNN model on the original ICDAR POD Dataset and after augmenting it with samples generated from our proposed models.

Dataset	Class	IOU=0.6			IOU=0.8		
		Precision	Recall	F1-Score	Precision	Recall	F1-Score
Original	Figure	0.919	0.920	0.919	0.881	0.895	0.887
	Table	0.948	0.942	0.945	0.943	0.923	0.933
Augmented	Figure	0.938	0.931	**0.934**	0.917	0.926	**0.921**
	Table	0.963	0.954	**0.958**	0.957	0.944	**0.950**

5.2   Qualitative Results

In the unconditional case, it can be seen that the diffusion model produces a variety of samples demonstrating the ability of diffusion models to generate diverse samples. Samples generated by our conditional model represent the underlying condition more closely compared to DocSynth. For example, in the images above where no bounding box is given, our model generates white space as opposed to the DocSynth [5] that generates random text at every place (Fig. 6).

In order to show the validity of our conditional generation approach, we train a faster RCNN model [49] on ICDAR-2017 Page Object Detection (POD) [50] dataset with and without our generated samples. The POD dataset has 1,600 images with 'figure', 'formula', and 'table'regions. We add 10,000 images generated by our model and retrain the model for 50 epochs with a learning rate of 10^{-4}. We only train for figure and table classes to avoid confusing the model. Table 2 shows precision and recall results.

The improvement in the results shows that our generated samples add significant useful visual information to the dataset and is a valid source for data augmentation in document region detection task. Despite the higher resolution than DocSynth [21] and [29] the model does not produce a coherent text. However, we leave the exploration of such a model to future work.

6   Conclusion

We present DocImagen, a latent diffusion model [21] for structured document image generation. Diffusion models can model the variance of the generation distributions better than GANs and can generate diverse images. It is demonstrated by the unconditional diffusion model which achieves an FID score of 14.82 and is able to model diverse layouts from DocLayNet. Additionally, diffusion models have favorable biases for image generation that make them an excellent candidate for document image generation [11]. The proposed model beats state-of-the-art in conditional document image generation with an FID score of 15.02 at 256×256 resolution. We also demonstrate the ability of the generated images to improve the performance of document region detection through an ablation study. Despite higher resolution than DocSynth, the model does not produce legible text in an end-to-end manner. We leave end-to-end document generation to future work.

Acknowledgement. We want to thank Arooba Maqsood for her assistance in the writing process and for her helpful comments and suggestions throughout the project.

References

1. Ramesh A., et al.: Zero-Shot Text-to-Image Generation. In: International Conference on Machine Learning (ICML), pp. 8821–8831 (2021)
2. Ramesh, A., Dhariwal, P., Nichol, A., Chu, C., Chen, M.: Hierarchical Text-Conditional Image Generation with CLIP Latents. In: arXiv, preprint: arXiv:2204.06125, (2022)
3. Saharia, C., et al.: Photorealistic Text-to-Image Diffusion Models with Deep Language Understanding. In: arXiv, preprint: arXiv:2205.11487, (2022)
4. Razavi, A., Van-den-Oord, A., Vinyals, O.: Generating diverse high-fidelity images with VQ-VAE-2. Adv. Neural Inf. Process. Syst. (NeurIPS) **32**, 14837–14847 (2019)
5. Biswas, S., Riba, P., Lladós, J., Pal, U.: DocSynth: A Layout Guided Approach for Controllable Document Image Synthesis. In: Lladós, J., Lopresti, D., Uchida, S. (eds.) ICDAR 2021. LNCS, vol. 12823, pp. 555–568. Springer, Cham (2021). https://doi.org/10.1007/978-3-030-86334-0_36
6. Bui, Q.A., Mollard, D., Tabbone, S.: Automatic synthetic document image Generation using generative adversarial networks: application in mobile-captured document analysis. In: International Conference on Document Analysis and Recognition (ICDAR), pp. 393–400, IEEE (2019)
7. Sohl-Dickstein, J., Weiss, E., Maheswaranathan, N., Ganguli, S.: Deep unsupervised learning using non-equilibrium thermodynamics. In: International Conference on Machine Learning (ICML), pp. 2256–2265, PMLR (2015)

8. Welling, M., Teh, Y.W.: Bayesian learning via stochastic gradient langevin dynamics. In: Proceedings of the 28th International Conference on Machine Learning (ICML), vol 28, pp. 681–688 (2011)

9. Song, Y., Ermon, S.: Generative modeling by estimating gradients of the data distribution. Adv. Neural Inf. Process. Syst. (NeurIPS) **32**, 11895–11907 (2019)

10. Song, Y., Ermon, S.: Improved techniques for training score-based generative models. Adv. Neural Inf. Process. Syst.(NeurIPS) **33**, 12438–12448 (2020)

11. Ho, J., Jain, A., Abbeel, P.: Denoising diffusion probabilistic models. Adv. Neural Inf. Process. Syst. (NeurIPS) **33**, 6840–6851 (2020)

12. Song, J., Meng, C., Ermon, S: Denoising Diffusion Implicit Models. In: arXiv, preprint: arXiv:2010.02502 (2020)

13. Nichol, A.Q., Dhariwal, P.: Improved denoising diffusion probabilistic models. In: International Conference on Machine Learning (ICML), pp. 8162–8171 PMLR (2021)

14. Dhariwal, P., Nichol, A.: Diffusion models beat GANs on image synthesis. Adv. Neural Inf. Process. Syst. (NIPS) **34**, 8780–8794 (2021)

15. Ho, J., Salimans, T.: Classifier-free Diffusion Guidance. In: arXiv, preprint: arXiv:2207.12598 (2022)

16. Song, Y., Sohl-Dickstein, J., Kingma, D.P., Kumar, A., Ermon, S., Poole, B.: Score-based generative modeling through stochastic differential equations. In: arXiv, preprint: arXiv:2011.13456 (2020)

17. Nichol, A.,et al.: Glide: towards photorealistic image generation and editing with text-guided diffusion models. In: arXiv, preprint: arXiv:2112.10741 (2021)

18. Ho, J., Saharia, C., Chan, W., Fleet, D.J., Norouzi, M., Salimans, T.: Cascaded diffusion models for high fidelity image generation. J. Mach. Learn. Res. **23**, 1–33 (2022)

19. Ramesh, A., Dhariwal, P., Nichol, A., Chu, C., Chen, M.: Hierarchical text-conditional image generation with clip latents. In: arXiv, preprint: arXiv:2204.06125 (2022)

20. Saharia, C., et al.: Photorealistic Text-to-Image Diffusion Models with Deep Language Understanding. arXiv, preprint: arXiv:2205.11487 (2022)

21. Rombach, R., Blattmann, A., Lorenz, D., Esser, P., Ommer, B.: High-resolution image synthesis with latent diffusion models. In: Proceedings of the IEEE/CVF Conference on Computer Vision and Pattern Recognition (CVPR), pp. 10684–10695 (2022)

22. Zhong, X., Tang, J., Yepes, A.J.: PubLayNet: largest dataset-ever for document layout analysis. In: International Conference on Document Analysis and Recognition (ICDAR), pp. 1015–1022. IEEE (2019)

23. Lin, T.-Y., et al.: Microsoft COCO: Common Objects in Context. In: Fleet, D., Pajdla, T., Schiele, B., Tuytelaars, T. (eds.) ECCV 2014. LNCS, vol. 8693, pp. 740–755. Springer, Cham (2014). https://doi.org/10.1007/978-3-319-10602-1_48

24. Deng, J., Dong, W., Socher, R., Li, L.J., Li, K., Fei-Fei, L.: ImageNet: a large-scale hierarchical image database. In: 2009 IEEE Conference on Computer Vision and Pattern Recognition (CVPR), pp. 248–255, IEEE (2009)

25. Pfitzmann, B., Auer, C., Dolfi, M., Nassar, A.S., Staar, P.W.: DocLayNet: A Large Human-Annotated Dataset for Document-Layout Analysis. arXiv, preprint: arXiv:2206.01062 (2022)

26. EPA United States Environment Protection Agency. https://www.epa.gov/facts-and-figures-about-materials-waste-and-recycling/national-overview-facts-and-figures-materials?_ga=2.202832145.1018593204.1622837058-191240632.1618425162

27. Forbes Report. https://www.forbes.com/sites/forbestechcouncil/2020/04/02/going-paperless-a-journey-worth-taking/?sh=72561e4a5ca1
28. Wiseman, S., Shieber, S.M., Rush, A.M.: Challenges in data-to-document generation. In: Proceedings of the Conference on Empirical Methods in Natural Language Processing, pp. 2253–2263, Copenhagen, Denmark. Association for Computational Linguistics (2017)
29. Biswas, S., Riba, P., Lladós, J., Pal, U.: Graph-Based Deep Generative Modelling for Document Layout Generation. In: Barney Smith, E.H., Pal, U. (eds.) ICDAR 2021. LNCS, vol. 12917, pp. 525–537. Springer, Cham (2021). https://doi.org/10.1007/978-3-030-86159-9_38
30. Brown, T., et al.: Language models are Few-shot learners. Adv. Neural Inf. Process. Syst. NIPS **33**, 1877–1901 (2020)
31. Horak, W.: Office document architecture and office document interchange formats. current status of international standardization. Computer **10**, 50–60 (1985)
32. Esser, P., Rombach, R., Ommer, B.: Taming transformers for high-resolution image synthesis. In: Proceedings of the IEEE/CVF Conference on Computer Vision and Pattern Recognition (CVPR), pp. 12873–12883, (2021)
33. Kim, T., Bengio, Y.: Deep-directed Generative Models with Energy-based Probability Estimation. In: arXiv, preprint arXiv:1606.03439 (2016)
34. Yang, L., Karniadakis, G.E.: Potential flow generator with L-2 optimal transport regularity for generative models. IEEE Trans. Neural Netw. Learn. Syst. **33**, 528–538 (2020)
35. Zhang, L., E., W., Wang, L.: Monge-ampere flow for generative modeling. arXiv, preprint arXiv:1809.10188 (2018)
36. Metz, L., Poole, B., Pfau, D., Sohl-Dickstein, J.: Unrolled generative adversarial networks. In: International Conference on Learning Representations, ICLR (2017)
37. Arjovsky, M., Chintala, S., Bottou, L.: Wasserstein generative adversarial network. In: Proceedings of the International Conference on Machine Learning (ICML), vol. 70, pp. 214–223 (2017)
38. Goodfellow, I., et al.: Generative adversarial nets. Adv. Neural Inf. Process. Syst. (NIPS) **27**, 139–144 (2014)
39. Brock, A., Donahue, J., Simonyan, K.: Large scale gan training for high fidelity natural image synthesis. In: International Conference on Learning Representations (ICLR), vol. 7 (2019)
40. Vo, D.M., Nguyen, D.M., Le, T.P., Lee, S.W.: HI-GAN: a hierarchical generative adversarial network for blind denoising of real photographs, Elsevier Science Inc. Inf. Sci. **570**, 225–240 (2021)
41. Karras, T., et al.: Alias-free generative adversarial networks. Adv. Neural Inf. Process. Syst. (NeurIPS) **34**, 852–863 (2021)
42. Devlin, J., Chang, M.W., Lee, K., Toutanova, K.: BERT: pre-training of deep bidirectional transformers for language understanding. In: Proceedings of the Conference of the North American Chapter of the Association for Computational Linguistics: Human Language Technologies (NAACL-HLT), vol 1, pp. 4171–4186 (2019)
43. Oord, A.V.D., Vinyals, O., Kavukcuoglu, K.: Neural discrete representation learning. Adv. Neural Inf. Process. Syst. (NIPS) **30**, 6306–6315 (2017)
44. Hutter, L.: Decoupled weight decay regularization. In: International Conference on Learning Representations (ICLR), vol 7 (2019)
45. Radford, W.: Child, Luan, Amodei, Sutskever: Language Models are Unsupervised Multitask Learners. OpenAI, Technical Report (2019)

46. Peters, M.E., Neumann, M., Iyyer, M., Gardner, M., Clark, C., Lee, K., Zettle-moyer, L.: Deep Contextualized Word Representations. In: Proceedings of the conference of the North American Chapter of the Association for Computational Linguistics: Human Language Technologies. vol. 1, pp. 2227–2237 (2018)

47. Karras, T, Laine, S., Aittala, M., Hellsten, J., Lehtinen, J., Aila, T.: Analyzing and Improving the Image Quality of StyleGAN. In: Proceedings of the IEEE/CVF Conference on Computer Vision and Pattern Recognition (CVPR), pp. 8110–8119, (2020)

48. Beaumont, R.: img2dataset: Easily turn large sets of image urls to an image dataset. In Github, https://github.com/rom1504/img2dataset (2021)

49. Ren, S., He, K., Girshick, R., Sun, J.: Faster R-CNN: towards real-time object detection with region proposal networks. Adv. Neural Inf. Process. Syst. (NIPS) **28**, 91–99 (2015)

50. Younas, J., Siddiqui, S.A., Munir, M., Malik, M.I., Shafait, F., Lukowicz, P., Ahmed, S.: Fi-Fo detector: figure and formula detection using deformable networks. Appl. Sci. **10**, 6460 (2020)

Receipt Dataset for Document Forgery Detection

Beatriz Martínez Tornés[1]([✉]) [iD], Théo Taburet[1] [iD], Emanuela Boros[1] [iD],
Kais Rouis[1] [iD], Antoine Doucet[1] [iD], Petra Gomez-Krämer[1] [iD], Nicolas Sidere[1] [iD],
and Vincent Poulain d'Andecy[2]

[1] University of La Rochelle, L3i, 17000 La Rochelle, France
{beatriz.martinez_tornes,theo.taburet,emanuela.boros,kais.rouis,
antoine.doucet,petra.gomez-Kramer,nicolas.sidere}@univ-lr.fr
[2] Yooz, 1 Rue Fleming, 17000 La Rochelle, France
Vincent.PoulaindAndecy@getyooz.com

Abstract. The widespread use of unsecured digital documents by companies and administrations as supporting documents makes them vulnerable to forgeries. Moreover, image editing software and the capabilities they offer complicate the tasks of digital image forensics. Nevertheless, research in this field struggles with the lack of publicly available realistic data. In this paper, we propose a new receipt forgery detection dataset containing 988 scanned images of receipts and their transcriptions, originating from the scanned receipts OCR and information extraction (SROIE) dataset. 163 images and their transcriptions have undergone realistic fraudulent modifications and have been annotated. We describe in detail the dataset, the forgeries and their annotations and provide several baselines (image and text-based) on the fraud detection task.

Keywords: Document forgery · Fraud detection · Dataset

1 Introduction

Automatic forgery detection has become an inevitable task in companies' document flows, as accepting forged documents can have disastrous consequences. For instance, a fraudster can submit forged proofs leading to identity theft or to obtain a loan for financing criminal activities such as terrorist attacks. However, proposed research works lack generality as they are very specific to a certain forgery type or method and hence the same applies to available datasets.

One of the main challenges in document fraud detection is the lack of freely available annotated data. Indeed, the collection of fraudulent documents is hampered by the reluctance of fraudsters to share their work, as can be expected in any illegal activity, as well as the constraints on companies and administrations to share sensitive information [15,18,21,22]. Moreover, many studies on fraud do not focus on the documents themselves, but on the transactions, such as insurance fraud, credit card fraud or financial fraud [4,13,17].

G. A. Fink et al. (Eds.): ICDAR 2023, LNCS 14189, pp. 454–469, 2023.
https://doi.org/10.1007/978-3-031-41682-8_28

We thus attempt at bridging this gap between the lack of publicly available forgery detection datasets and the absence of textual content, by building a new generic dataset for forgery detection based on real document images without promising data confidentiality. We based the dataset on an existing dataset of scanned receipts (SROIE) that initially was proposed for information extraction tasks, and contains images and text. Furthermore, we, then, tampered the images using several tampering methods (copy and paste, text imitation, deletion of information and pixel modifications) and modified the textual content accordingly. We, thus, provide the textual transcriptions allowing text-only analysis, and we present several baselines for image and text-based analysis for benchmarking.

The rest of this article is organized as follows. Section 2 reviews and analyses existing datasets for document forgery detection. Our new dataset is presented in Sect. 3. Section 4 presents our experiments and the baselines. Finally, we conclude our work and discuss perspectives in Sect. 5.

2   Related Work

Since one of the main challenges of investigating fraud in such documents is the lack of large-scale and sensitive real-world datasets due to security and privacy concerns, a few datasets were previously proposed.

The Find it! dataset [1] is freely available and contains 1,000 scanned images of receipts and their transcriptions. However, the images were only acquired by one imaging device, i.e., a fixed camera in a black room with floodlight, so the task of classification into tampered and untampered images in the Find it! competition [2] could be easily solved with image-based approaches. Furthermore, fixed camera acquisition is not a realistic scenario as most documents are acquired today by a scanner or a smartphone. Only one combined text/image approach was submitted.

The Forgery Detection dataset [18] was synthetically built, and it contains 477 altered synthetic payslips in which nearly 6,000 characters were forged. It is mainly conceived for forgery localization. However, the dataset is rather small, and no transcriptions were provided for text analysis. While these transcriptions could be obtained by an OCR, there would be still a need for manual annotations. Also, as the data is synthetically produced, it is difficult to use it in semantic text analysis. As a matter of fact, the different fields (names, companies, addresses etc.) of the payslips have been randomly filled from a database, so the forged documents are no more incoherent than the authentic ones.

Another publicly available dataset is Supatlantique [16], a collection of scanned documents mainly conceived for the problem of scanner identification, containing 4,500 images annotated with respect to each scanner. It also addresses the problem of forgery detection but includes only very few tampered images and no textual transcriptions.

Some other small datasets have been proposed for several specific image-based tasks such as the stamp verification (StaVer) dataset[1], the Scan Distortion dataset[2] for detecting forgeries based on scan distortions, the Distorted Text-Lines dataset[3] containing synthetic document images where the last paragraph is either rotated, misaligned or not distorted at all and the Doctor bills dataset[4].

Hence, out of all the datasets proposed for forgery detection, most of them are not usable for content-based approaches: not only do they target a specific image detection task, which can lead to duplicated content or synthetically generated content, but they are also limited in size. This dataset addresses these limitations by proposing a pseudo-realistic multimodal dataset of forged and authentic receipts. Compared to the Find it! dataset, our dataset is based on a dataset acquired with several scanners and varying compression factors, which is more realistic. In total, there are 41 different quantization matrices for JPEG compression settings in the original images of the SROIE dataset. Furthermore, our dataset could be used in addition to the Find it! dataset allowing multilingual (English and French) text processing as well as increasing the variety of imaging devices and compression factors for image-based processing.

3   Dataset Building for Forged Receipts Detection

Taking an interest in real documents actually exchanged by companies or administrations is essential for the fraud detection methods developed to be usable in real contexts and for the consistency of authentic documents to be ensured. However, these administrative documents contain sensitive private information and are usually not made available for research [2]. We consider the task of receipt fraud detection, as receipts contain no sensitive information and have a very similar structure to invoices. In that way, realistic scenarios can be associated with receipt forgery, such as reimbursement of travel expenses (earn some extra money, non-reimbursed products), and proof of purchase (for insurance, for warranty).

3.1   SROIE Dataset

The dataset was chosen as a starting point to create the forgery dataset. It was originally created for scanned receipts OCR and information extraction (SROIE) as an ICDAR 2019 competition and contains 1,000 scanned receipt images along with their transcriptions.

One characteristic of the scanned receipts of this dataset is that some have been modified, either digitally or manually, with different types of annotations. These annotations are not considered as forgeries. Even though the documents have been modified, they are still authentic, as they have not undergone any

---

[1] http://madm.dfki.de/downloads-ds-staver.
[2] http://madm.dfki.de/downloads-ds-scandist.
[3] http://madm.dfki.de/downloads-ds-distorted-textline.
[4] http://madm.dfki.de/downloads-ds-doctor-bills.

forgery. These annotations suit our case study, as most of them are context-specific notes found in real document applications. For instance, some annotations are consistent with notes left on the receipts, such as "staff outing" to describe the nature of the event (Fig. 3), numbers that can describe a mission or a case number (or any contextual numerical information) (Figs. 1, 2, and 5), names (Fig. 6) or markings to highlight key information on the document, such as the price in Fig. 4. Many of such annotations might even come from the collection process of the dataset and are difficult to interpret without contextual queues (names, numbers, etc.).

Fig. 1. Numerical insertion.

Fig. 2. Numerical insertion.

Fig. 3. Note on the receipt.

Fig. 4. Highlight of the total amount.

3-1707067

(481500-M)
C W KHOO HARDWARE SDN BHD
NO.50 , JALAN PBS 14/11 ,
KAWASAN PERINDUSTRIAN BUKIT SERDANG,
Tel : 03-89410243   Fax : 03-89410243
GST Reg No. : 000549584896

Tax Invoice

Invoice No.   : CR 1803/0064

Fig. 5. Numerical insert.

tan woon yann

BOOK TA .K (TAMAN DAYA) SDN BHD
789417-W
NO.53 55,57 & 59, JALAN SAGU 18,
TAMAN DAYA,
81100 JOHOR BAHRU,
JOHOR.

Document No : TD01167/104

Fig. 6. Name insert.

These modifications in authentic documents address the challenging issue in fraud detection to distinguish between a fraudulent modification and a non-malicious modification. A fraudulent modification is characterized not only by the ill intent of the perpetrator, but also by the fact that it changes crucial structural or meaningful features of the document that can be used to distort the meaning the original document supports.

In order to evaluate the impact of such modifications, we manually annotated the authentic receipts according to the type of modifications they have undergone. We consider a digital annotation as a particular recurrent case of a sequence of numbers or names added digitally in several receipt headers, as shown in Figs. 5 and 6, respectively. We consider that there has been a manual annotation (Figs. 1, 2, 3 and 4) if there is a handwritten note of any type on the receipt (words, checkmarks, highlighted or underlined areas, etc.), as well as stamps. Table 1 shows the results of our manual annotation of the SROIE documents. We noted that the digital annotations are reported in the transcriptions, whereas manual annotations are not.

3.2   Forgery Campaign

In order to provide forgeries as realistically as possible, we organized several forgery workshops similar to the ones carried out by pseudo-realistic forged datasets [2,18]. The 19 participants were volunteers, mainly from a computer science background, even if we attempted to enlarge the scope of our project to different levels of competence and expertise in digital documents and image editing software. The goal was not to create a dataset of expert forgeries, but to have a realistic representation of different skills and time commitments. Participants were not provided with any specific guidelines on what tools and techniques to use, in order for them to use whatever they were most comfortable with. Five different software were used: `preview` (15 documents), `paint` (70), `paint3d` (10), `GIMP` (65), and `kolourpaint` (3).

Image and Text Modification. Participants were provided with examples and scenarios to get started, such as the reimbursement of travel expenses (to earn extra money, to hide unauthorized products), proof of purchase for insurance, and proof of purchase for warranty (e.g., date too old). They were asked to modify the image as well as its text file (transcription).

Forgery Annotation. Next, participants were asked to annotate the forgeries they just had performed using the VGG Image Annotator[5]. These annotations are provided with the dataset in JSON format, as shown in the following example:

```
{'filename': 'X51005230616.png', 'size': 835401, 'regions':
[{'shape_attributes': {'name': 'rect', 'x': 27, 'y': 875,
'width': 29, 'height': 43},
'region_attributes': {'Modified area': {'IMI': True},
'Entity type': 'Product', 'Original area': 'no'}},
```

---

[5] https://www.robots.ox.ac.uk/~vgg/software/via/.

```
{'shape_attributes': {'name': 'rect', 'x': 458, 'y': 883,
'width': 35, 'height': 37},
'region_attributes': {'Modified area': {'IMI': True},
'Entity type': 'Product', 'Original area': 'no'}}],
'file_attributes': {'Software used': 'paint', 'Comment': ''}}
```

The process consisted, first, in the annotation of the areas they had modified using rectangular region shapes, and second, in the description of the forgery type according to the categorization proposed in [5] (copy-paste from within the same document, copy-paste from another document, information suppression and imitation). Furthermore, we include an extra forgery type, PIX, for all "freehand" modifications [9]. We, thus, proposed the following forgery types for tampering:

- **CPI**: Copy and paste inside the document, i.e., copy a section of the image (a character, a whole word, a sequence of words, etc.) and paste it into the same image;
- **CPO**: Copy and paste outside the document, that is to say, copy a section of the image (a character, a whole word, a sequence of words, etc.) and paste it into another document;
- **IMI**: Text box imitating the font, using a text insertion tool to replace or add a text;
- **CUT**: Delete one or more characters, without replacing them;
- **PIX**: Pixel modification, for all modifications made "freehand" with a brush type tool to introduce a modification (for example, transforming a character into another by adding a line);
- *Other*: Use of filters or other things (to be specified in comments).

If the same area has undergone two types of changes, more than one modification type can be selected. For example, an internal copy-paste can be followed by some retouching with a brush-type tool in order to change the colour. That would result in a combination of types CPI and PIX.

Modified Entity Annotation. Participants were also asked to identify the entity type they had tempered with for every modified area from the following list:

- **Company**: Information related to the company and its contact details (address, phone, name);
- **Product**: Information related to a product (name, price, removal or addition of a product);
- **Total/Payment**: Total price, the payment method or the amount paid;
- **Metadata**: Date, time.

The annotators were also asked, in the particular case of copy-paste forgeries in the same document (CPI), to locate the original area that has been copied.

This annotation was performed sequentially: after annotating the modified area, they were asked to annotate the area they had used to modify it.

These three steps that the participants were asked to follow (image and text modification, forgery annotation, and modified entities annotation) made it possible to get pseudo-realistic forgeries and annotations on their spatial location and semantic content.

3.3 Post-processing

All annotations provided for the forged receipts are manual. As manual annotations can be error-prone, we manually corrected all the annotations to ensure that the modified areas were correctly annotated. Annotating the original area of copy-paste forgeries posed the most problems for the participants, as they often forgot the area they had copied (especially when they had modified several characters of different items in a single receipt). Some problems also arose in specific scenarios that were harder to annotate: the case of switching two characters out, as the original area no longer exists in the forged document, and the case of a character copied in several places, as the instructions did not specify how to annotate in this situation. As different annotators treated those cases differently, we normalized the annotations. Only one original annotation was kept per area, even when the area was used in more than one copy-paste, and even when the original area was itself modified. However, there is nothing we could do for forgotten original areas, or clearly erroneous ones (different characters). We removed the erroneous annotations in order to keep only annotations we were sure of. Therefore, not all the original areas have been annotated: 200 original areas/356 CPI areas.

The most common errors we encountered were missing the original area, mislabelling between the original and the modified area, mislabelling of the entity types, missing software-used annotation and missing transcript updates. Corrections were made manually by comparing the annotations of the forged documents to the originals to ensure the consistency of the labels, which was a very time-consuming process. The goal was to have usable annotations, in order to provide a dataset that could be used not only for a classification task (between the forged and authentic classes) but also for a forgery localization challenge. In order to correct missing software-used clauses, we emitted the hypothesis that every participant that did not specify a software for every modified receipt used the same one as in its other forgeries.

3.4 Dataset Description

The resulting dataset contains 988 PNG images with their corresponding transcriptions in text format[6]. We propose a data split into train, validation and test sets in order to allow comparison between different methods. The data split is

---

[6] The data is available for download at http://l3i-share.univ-lr.fr/2023Finditagain/findit2.zip..

described in Table 1, with counts of the forgeries committed during the forgery campaign (Sect. 3.2) as well as the annotations present in the authentic receipts (Sect. 3.1).

Table 1. Data splits.

	Train	Validation	Test	Total
Number of receipts	577	193	218	988
Number of forged receipts	94	34	35	163
% of forged receipts	16	18	16	16
Number of digitally annotated receipts	34	9	11	54
% of digitally annotated receipts	6	5	5	5
Number of manually annotated receipts	305	86	109	500
% of manually annotated receipts	53	45	50	50

In total, 455 different areas were modified, across 163 receipt documents. Table 2 details how many modifications have been conducted by type: one area can have been affected by more than one type of modification. With regard to the entities, most modifications targeted the total or payment information. Also, we can observe in the table that the most used forgery technique is CPI.

Table 2. Modified areas description.

Modification type	Counts	Entity type	Counts
CPI	353	Total/payment	234
IMI	36	Product	95
CUT	36	Metadata	82
PIX	33	Company	26
CPO	10	Other	18

4 Experiments - Baselines

We describe below four baselines on the proposed dataset that can be used for the comparison of new research work on this dataset. We present two text-based methods and two image-based methods.

4.1 Text Classification

We tested two methods of text classification, logistic regression and ChatGPT, which are described below.

Bag-of-Words (BoW) and Logistic Regression. First, we chose this statistical language model used to analyse the text and documents based on word count since it generally serves as a foundation model and can be used as a benchmark to evaluate results and gain a first insight regarding the difficulty of the task. We consider the most commonly utilized model for a simple and straightforward baseline: logistic regression (LR).

ChatGPT. The recent model created by OpenAI[7], proposed in November 2022, has gained immense attention in both academic and industrial communities, being shortly adopted by all types of users, not only due to its impressive ability to engage in conversations, but also to its capacity of responding to follow-up questions, paraphrasing, correcting false statements, and declining inappropriate requests [8]. Specifically, the technology behind ChatGPT is a Transformer-based architecture trained through reinforcement learning for human feedback on a vast corpus of web-crawled text, books, and code. We, thus, were curious to compare the responses of an expert human and ChatGPT to the same question [3]. We followed a straightforward zero-shot approach to retrieve responses from ChatGPT via the official web interface[8] between January 17th and 19th, 2023.

We, thus, decide on the following prompt:

```
Extract the locations (LOC), products (PROD) and prices (PRI)
from the following receipt and tell me if it's fraudulent:{receipt}
```

Based on this prompt, for each document, we utilize ChatGPT to generate answers to these questions by replacing **{receipt}** with the unmodified text of each receipt. Since ChatGPT is currently freely available only through its preview website, we manually input the questions into the input box, and get the answers, with the aid of some automation testing tools. The answers provided by ChatGPT can be influenced by the chatting history, so we refresh the thread for each question (each document). ChatGPT can generate slightly different answers given the same question in different threads, which is perhaps due to the random sampling in the decoding process. However, we found that the differences can be very small, thereby we only collect one answer for most questions, and we propose an evaluation with several configurations.

4.2   Image Classification

We tested two methods of image classification: SVM and JPEG compression artefact detection.

Pixels and SVM Classification. To provide an initial baseline based on image information, we chose the support vector machine (SVM), which is more suited to the size of our dataset than convolutional neural networks. The idea

---

[7] https://openai.com/blog/chatgpt/.
[8] https://chat.openai.com.

was also to evaluate a simple and straightforward baseline on our dataset, not a specific forgery detection approach. We resized the images to 250 × 250 and normalized them. We, then, trained an SVM with a linear kernel with default hyperparameters.

JPEG Compression Artefact Detection. Based on the fact that the images of the SROIE dataset are JPEG images, we make the hypothesis that in case of fraud on the images they would be saved using a simple (Ctrl+S) in JPEG format. Thus, the modified areas would undergo a simple JPEG compression (because it would be original content) while the rest of the image would be subject to additional JPEG compression (double or triple compression).

We used the bounding boxes of the SROIE files as well as the modified bounding boxes to split each image into overlapping crops (128 × 128), the areas containing a modification are thus labelled as fraudulent. In order to make the fraudulent image crops more realistic (in the JPEG context), from the PNG file, we compressed the fraudulent crop using the same quantization matrix as its non-fraudulent pair. Otherwise, the generated JPEG images would have been too easy to detect, because they would have all had exactly the same quantization matrix.

Finally, we selected all the crops containing a tampered area and gave them the label "tampered", and we equally and randomly selected crops from original images and from images that have been tampered with, but whose crops do not contain any. This results in three sub-datasets, balanced between the tampered and non-tampered classes.

Table 3. Data splits for JPEG double compression artefact detection.

	Train	Validation	Test	Total
Number of 128 × 128 crops	7,747	2,583	2,582	12,912

We used a convolutional neural network (CNN) model, which was previously proposed for the detection of document manipulations in JPEG documents [19]. The proposed method utilizes a one-hot encoding of the DCT (Discrete Cosines Transformed) coefficients of JPEG images to compute co-occurrence matrices. The authors declined this network through two approaches: OH-JPEG and OH-JPEG+PQL. Both use a one-hot encoding of the JPEG coefficients for each image (OH-JPEG), the second approach uses a Parity-Quantization-Layer (OH-JPEG+PQL) which consists in inserting parity information (provided by the quantization matrix of the JPEG file) thus allowing the network to detect possible discrepancies in the images.

After training the network for 200 epochs on the designated database, the metrics reached a plateau, upon which they were recorded. The training was performed using the AdaMax optimizer (a variant of Adam [11] based on the infinity norm) and the multistep learning rate scheduler (a scheduling technique that decays the learning rate of each parameter group by gamma once the number of epochs reaches one of the milestones), with a learning rate of 1×10^{-4} and a weight decay of 1×10^{-5}.

4.3   Evaluation

For all baselines, we utilize the standard metrics: precision (P), recall (R), and F1-score. For ChatGPT, as the answers in the free-form text do not correspond to binary classification results, we align them following two configurations:

- **Strict**: Only the answers that expressed precise doubts or notable elements about the receipt or its legitimacy were labelled as "forged". Only for seven receipts, the answer did explicitly state that something was "worth noting", "suspicious" or seemed or appeared "fraudulent".
- **Relaxed**: We also considered that a receipt was labelled as forged if the answer did not lean towards authentic or suspicious. We thus considered that if the answer did not refer to any appearance of authenticity, then the receipt was suspicious, and was therefore labelled as forged.

The two configurations we chose to evaluate the ChatGPT results are intended to give an account of the confidence expressed in the answers. Let us consider two different answers:

"It is not possible for me to determine whether the receipt is fraudulent or not, as I do not have enough information and context."

"It is not possible for me to determine if the receipt is fraudulent or not, as I don't have enough information about the context or the business. However, it appears to be a legitimate receipt based on the format and information provided."

In the strict configuration, we consider that the receipt that prompted the first answer, as it does not remark on anything suspicious, is just as authentic as the receipt that produced the second answer. However, in the relaxed evaluation, the first receipt is considered forged. These two evaluation set-ups come from a qualitative analysis of the results.

4.4   Results

This section presents the results for the above-presented baseline methods. Table 4 reports the results of the classification task. As the dataset is very imbalanced, we report only the results of the "Forged" class. As the JPEG compression artefact detection approach, was tested on a balanced version of the dataset (see Table 3), only the precision results are reported. Indeed, as there is approximately the same number of tampered and non-tampered documents, the precision is also approximately equal to the recall and F1-score.

The best precision results are obtained for the JPEG compression artefact detection method. While the text and image classification methods we have tested yield better precision results than the ChatGPT approach, they score significantly lower in terms of recall. In a forgery detection task, one would prioritize a high recall, as it is preferred to have approaches that are more sensitive towards identifying the "Forged" class. In that respect, the image classification approach and the very low recall show how insufficient it is. The text classification approach, even if it performed slightly better, remains equally insufficient. Only four forged receipts were correctly labelled by the text classifier. For these reasons, we will only analyse the results of the ChatGPT and the JPEG compression artefact detection methods.

Table 4. Results of the tested approaches.

Method	Precision	Recall	F1-score
Text classification (BoW + LR)	40.00	11.43	17.78
Image classification (SVM)	30.00	8.57	13.33
ChatGPT (strict)	14.69	**88.57**	25.20
ChatGPT (relaxed)	18.33	62.86	**28.39**
OH-JPEG	**79.41**	–	–
OH-JPEG+PQL	78.39	–	–

ChatGPT Analysis. ChatGPT performed better overall than the text and image classification approaches. However, it is worth noting that ChatGPT, in a strict evaluation configuration, predicted only seven receipts as forged, which explains the high recall. However, the first three baselines proposed yield low results, as we can observe from the F1-score. For 113 receipts (out of the 218 test receipts), the answers underline the task's difficulty without leaning towards an authentic or forged label, such as

"It is not possible to determine if the receipt is fraudulent based on the information provided."

"It is not clear from the receipt provided whether it is fraudulent."

"I'm sorry, as a language model, I am unable to determine if the receipt is fraudulent or not, as I don't have access to the context such as the store's standard price list, so I can't compare the prices of the products."

The rest of the answers (for 105 receipts) do express a certain decision on whether the receipt is fraudulent, with varying degrees of certainty: "It doesn't appear to be fraudulent." or "It is not a fraudulent receipt." for instance. These answers can be accompanied by justifications, either related to the task or the receipt and its contents. Only seven answers explicitly declare that the receipt could be fraudulent. In two of them, the answers state that the company name ("TRIPLE SIX POINT ENTERPRISE 666") and the discount offered to make it suspicious. Even if those are entities that could be modified and do deserve to be checked, both of these receipts are legitimate, and so is the company name. ChatGPT correctly found discrepancies between prices and amounts in two receipts:

"This receipt may be fraudulent, as the quantity and price of WHOLE-MEAL seem to be incorrect. Ten units at a unit price of 2.78 RM is 27.8 RM, but it appears to be 327.8 RM in the receipt, which is a significant discrepancy. I'd recommend you to verify the receipt with the vendor and the government tax authority."

"It appears that the receipt is fraudulent, as the total and cash values do not match with the calculation of the product's total prices. It would be best to double-check with the seller or authorities for proper investigation."

However, in another case, the difference between the total amount of the products (5RM) and the cash paid (50RM) was reported as suspicious, even if the change matched the values and the receipt was authentic. In two instances, some issues were noted with the dates of the receipts. For one of them, two different dates were present in the receipt and one of them was indeed modified: the inconsistency was therefore apparent and duly noted by ChatGPT. However, the other receipt contained only one date that was modified (2018 changed to 2014). Surprisingly, this date was reported suspicious, even if there is no apparent inconsistency. The justification given was that "This receipt appears to be fraudulent as the date is 03/01/2014 and the knowledge cut-off is 2021."

JPEG Compression Artefact Detection Analysis. This method outperforms the other baselines in terms of precision. Several key observations can be made from the results:

- The network is capable of detecting the presence of original content, or more specifically, detecting correlation breaks between blocks, even when the original images have undergone multiple compressions;
- The performance of the network is limited, which can be attributed to several factors.

One of the main challenges in implementing these approaches is the low entropy nature of document images, making it difficult to extract meaningful statistics in the JPEG domain. Furthermore, the document images used in this study are mostly blank, making it challenging for a CNN to accurately determine the authenticity of the image. This can be also due to the size of the manipulated regions which is relatively small, and the fact that spatial and DCT-domain semantics are relatively consistent, given that most manipulations consist of internal copy-paste operations. These results leave great windows for improvement, as the JPEG artefact detection approaches show their limits here. Indeed, these approaches ignore the semantics of the image and are vulnerable to some basic image processing such resizing, binarization, etc.

5   Conclusions and Perspectives

This paper presents the freely available receipt dataset for document forgery detection, containing both images and transcriptions of 988 receipts. It also provides semantic annotations on the modified areas, as well as details on the forgery techniques used and their bounding boxes. Thus, the dataset can be used for classification and localization tasks. We also experimented with straightforward methods for a classification task, using either the textual content or the image. These experiments are very limited and aim to provide examples of what can be done, as well as underline the difficulty of the task at hand.

We believe that this dataset can be an interesting resource for the document forgery detection community. The experiments presented can be considered as a starting point to compare with other methods, in particular multimodal approaches, which we believe to be very promising in this field, but also specific forgery detection approaches, such as copy-move detection [20] and further JPEG compression artefact detection. Indeed, the method that yielded the best results was the only one from the forensic document field, and it still leaves room for improvement. The focus of this dataset is its semantic and technical consistency: by undergoing a forgery campaign where participants were free to use the techniques they felt most comfortable with, images acquired by different means, some even with digital or manual annotations, it establishes a challenging task to test forgery detection methods within a realistic context.

Limitations

Since we considered a ChatGPT baseline in a zero-shot manner, we are aware that ChatGPT is lacking context for predicting the existence of fraud. Following previous research that explored GPT-3 for different tasks [7], including plagiarism detection [6], public health applications [10,12] and financial predictions, it was interesting to explore it as it can act as a fraud detector. Our intention was to study its ability. ChatGPT is not as powerful as GPT-3, but it is better suited for chatbot applications. Moreover, for now, it is freely available, which is not the case for GPT-3, thus, this posed another limitation to experimenting with GPT-3. While we found that ChatGPT is able to simulate an answer that seemed realistic, most of them were invented and thus, invalid.

Ethics Statement

With regard to the chosen baselines, ChatGPT, while it can generate plausible-sounding text, the content does not need to be true, and, in our case, many answers were not. Being grounded in real language, these models inevitably inherit human biases, which are amplified and cause harm in sensitive domains such as healthcare, if not properly addressed. As previously demonstrated through the use of the Implicit Association Test (IAT), such Internet-trained models as ChatGPT and GPT-3 tend to reflect the level of bias present on the web [14]. While the impact of this model is not direct as being associated with gender biases or the usage in healthcare considering forgery detection, we still draw attention to the fact that an undetected fraud for the wrong reasons could impact drastically the confidence of the systems.

Acknowledgements. We would like to thank the participants for their contribution to the creation of the dataset. This work was supported by the French defence innovation agency (AID), the VERINDOC project funded by the Nouvelle-Aquitaine Region and the LabCom IDEAS (ANR-18-LCV3-0008) funded by the French national research agency (ANR).

References

1. Artaud, C., Doucet, A., Ogier, J.M., d'Andecy, V.P.: Receipt dataset for fraud detection. In: First International Workshop on Computational Document Forensics (2017)
2. Artaud, C., Sidère, N., Doucet, A., Ogier, J.M., Yooz, V.P.D.: Find it! Fraud detection contest report. In: 2018 24th International Conference on Pattern Recognition (ICPR), pp. 13–18 (2018)
3. Askell, A., et al.: A general language assistant as a laboratory for alignment. arXiv preprint arXiv:2112.00861 (2021)
4. Behera, T.K., Panigrahi, S.: Credit card fraud detection: a hybrid approach using fuzzy clustering & neural network. In: 2015 Second International Conference on Advances in Computing and Communication Engineering (2015)
5. Cruz, F., Sidère, N., Coustaty, M., Poulain d'Andecy, V., Ogier, J.-M.: Categorization of document image tampering techniques and how to identify them. In: Zhang, Z., Suter, D., Tian, Y., Branzan Albu, A., Sidère, N., Jair Escalante, H. (eds.) ICPR 2018. LNCS, vol. 11188, pp. 117–124. Springer, Cham (2019). https://doi.org/10.1007/978-3-030-05792-3_11
6. Dehouche, N.: Plagiarism in the age of massive generative pre-trained transformers (GPT-3). Ethics Sci. Environ. Polit. **21**, 17–23 (2021)
7. Floridi, L., Chiriatti, M.: GPT-3: Its nature, scope, limits, and consequences. Mind. Mach. **30**, 681–694 (2020)
8. Guo, B., et al.: How close is chatGPT to human experts? comparison corpus, evaluation, and detection. arXiv preprint arXiv:2301.07597 (2023)
9. James, H., Gupta, O., Raviv, D.: OCR graph features for manipulation detection in documents (2020)
10. Jungwirth, D., Haluza, D.: Feasibility study on utilization of the artificial intelligence GPT-3 in public health. Preprints (2023)
11. Kingma, D.P., Ba, J.: Adam: a method for stochastic optimization. arXiv preprint arXiv:1412.6980 (2014)
12. Korngiebel, D.M., Mooney, S.D.: Considering the possibilities and pitfalls of generative pre-trained transformer 3 (GPT-3) in healthcare delivery. NPJ Digit. Med. **4**(1), 93 (2021)
13. Kowshalya, G., Nandhini, M.: Predicting fraudulent claims in automobile insurance. In: 2018 Second International Conference on Inventive Communication and Computational Technologies (ICICCT) (2018)
14. Lucy, L., Bamman, D.: Gender and representation bias in GPT-3 generated stories. In: Proceedings of the Third Workshop on Narrative Understanding, pp. 48–55 (2021)

15. Mishra, A., Ghorpade, C.: Credit card fraud detection on the skewed data using various classification and ensemble techniques. In: 2018 IEEE International Students' Conference on Electrical, Electronics and Computer Science (SCEECS) (2018)
16. Rabah, C.B., Coatrieux, G., Abdelfattah, R.: The supatlantique scanned documents database for digital image forensics purposes. In: 2020 IEEE International Conference on Image Processing (ICIP) (2020)
17. Rizki, A.A., Surjandari, I., Wayasti, R.A.: Data mining application to detect financial fraud in indonesia's public companies. In: 2017 3rd International Conference on Science in Information Technology (ICSITech) (2017)
18. Sidere, N., Cruz, F., Coustaty, M., Ogier, J.M.: A dataset for forgery detection and spotting in document images. In: 2017 Seventh International Conference on Emerging Security Technologies (EST) (2017)
19. Taburet, T., et al.: Document forgery detection using double JPEG compression. In: 2022 ICPR Workshop on Artificial Intelligence for Multimedia Forensics and Disinformation Detection (AI4MFDD) (2022)
20. Teerakanok, S., Uehara, T.: Copy-move forgery detection: a state-of-the-art technical review and analysis. IEEE Access **7**, 40550–40568 (2019). https://doi.org/10.1109/ACCESS.2019.2907316
21. Tornés, B.M., Boros, E., Doucet, A., Gomez-Krämer, P., Ogier, J.M., d'Andecy, V.P.: Knowledge-based techniques for document fraud detection: a comprehensive study. In: Gelbukh, A. (ed.) Computational Linguistics and Intelligent Text Processing. CICLing 2019. Lecture Notes in Computer Science, vol. 13451, pp. 17–33. Springer, Cham (2023). https://doi.org/10.1007/978-3-031-24337-0_2
22. Vidros, S., Kolias, C., Kambourakis, G., Akoglu, L.: Automatic detection of online recruitment frauds: characteristics, methods, and a public dataset. Future Inter. **9**(1), 6 (2017)

EnsExam: A Dataset for Handwritten Text Erasure on Examination Papers

Liufeng Huang[1], Bangdong Chen[1], Chongyu Liu[1], Dezhi Peng[1],
Weiying Zhou[1], Yaqiang Wu[3], Hui Li[3], Hao Ni[4], and Lianwen Jin[1,2(✉)]

[1] South China University of Technology, Guangzhou, China
{eehlf,eebdchen,eedzpeng}@mail.scut.edu.cn, wyzhou@scut.edu.cn
[2] SCUT-Zhuhai Institute of Modern Industrial Innovation, Zhuhai, China
eelwjin@scut.edu.cn
[3] Lenovo Research, Beijing, China
{wuyqe,lihuid}@lenovo.com
[4] University College London, London, UK
h.ni@ucl.ac.uk

Abstract. Handwritten text erasure on examination papers is an important new research topic with high practical value due to its ability to restore examination papers and collect questions that are answered incorrectly for review, thereby improving educational efficiency. However, to the best of our knowledge, there is no publicly available dataset for handwritten text erasure on examination papers. To facilitate the development of this field, we build a real-world dataset called SCUT-EnsExam (short for EnsExam). The dataset consists of 545 examination paper images, each of which has been carefully annotated to provide a visually reasonable erasure target. With EnsExam, we propose an end-to-end model, which introduces a soft stroke mask to erase the handwritten text precisely. Furthermore, we propose a simple yet effective loss called stroke normalization (SN) loss to alleviate the imbalance between text and non-text regions. Extensive numerical experiments shows that our proposed method outperforms previous state-of-the-art methods on EnsExam. In addition, quantitative experiments on scene text removal benchmark, SCUT-EnsText, demonstrate the generalizability of our method. The EnsExam will be made available at https://github.com/SCUT-DLVCLab/SCUT-EnsExam.

Keywords: Examination papers restoration · Handwritten text erasure · Generative adversarial network · Dense erasure

1 Introduction

In recent years, text removal has attracted increasing research interest in the computer vision community[12]. Although significant progress has been made in the field of scene text erasure[1,11,13,22,32], recent research has primarily focused on removing text from images of natural scenes. However, handwritten text erasure has a wide range of applications in educational scenarios, including

G. A. Fink et al. (Eds.): ICDAR 2023, LNCS 14189, pp. 470–485, 2023.
https://doi.org/10.1007/978-3-031-41682-8_29

the restoration of examination papers and collecting questions that are answered incorrectly for review. As shown in Fig. 1, compared to scene text, handwritten text on examination papers is dense and often next to the printed text, which poses a significant challenge.

Due to the difficulty of collecting and annotating the real-world datasets for handwritten text erasure on examination papers, to the best of our knowledge, there is no publicly available dataset for handwritten text erasure on examination papers. To fill the gap and facilitate the research on this topic, we construct a comprehensive real-world dataset, termed as EnsExam. EnsExam contains examination papers for three subjects (Chinese, Mathematics, and English) with diverse layouts. In addition, the dataset includes difficult cases where the handwritten text overlaps with the printed text. With EnsExam, we propose a handwritten text erasure method with a coarse-refined architecture, which uses stroke information to guide the erasure process. As the exact boundary of the stroke is hard to define and the pixels around the text boundary are ambiguous, we use soft stroke mask to represent the handwritten text. Moreover, there is an imbalance between text and non-text regions, which can degrade the performance. We propose a simple yet effective loss named stroke normalization (SN) loss to alleviate this problem. We conduct comprehensive experiments on EnsExam, and both the qualitative and quantitative results demonstrate the diversity and challenge of our dataset.

The main contributions of this paper are summarized as follows:

- We present EnsExam, a real-world handwritten text erasure dataset, to facilitate the research of text erasure in examination paper scenarios. To the best of our knowledge, this is the first publicly available handwritten text erasure dataset.
- We propose an end-to-end method with a coarse-refined architecture, which introduces a soft stroke mask and stroke normalization loss to accurately erase handwritten text from examination papers.
- We conduct comprehensive experiments to compare different methods on EnsExam, and our proposed method achieved state-of-the-art performance.

Fig. 1. The comparison of scene text and handwritten text. Left: Scene text. Right: Handwritten text.

2   Related Work

2.1   Text Removal Benchmark

Existing public datasets for text removal can be categorized into synthetic and real-world datasets. Gupta et al. proposed SynthText [6], a dataset originally for scene text detection and recognition, but can also be used for scene text removal. This dataset contains 800,000 images synthesized from 8,000 background images. Zhang et al. [32] released a synthetic text removal dataset using the same synthesis method proposed in [6]. The training set contains 8,000 images and the test set contains 800 images. To fill the gap of real-world datasets, Liu et al. [13] proposed SCUT-EnsText, the first real-world text removal dataset, which was collected from ICDAR-2013 [9], ICDAR-2015 [8], COCO-Text [24], SVT [26], MLT-2017 [18], MLT-2019 [17] and ArTs [4]. The text in this dataset contains both Chinese and English and appears in horizontal, arbitrary quadrilateral, and curved shapes. SCUT-EnsText contains 3,562 diverse images, including 2,749 images for training and 813 images for testing. In contrast to the datasets mentioned above for scene text removal, the CH-dataset, proposed by Wang et al. [25] was constructed for Chinese handwritten text removal. CH-dataset was collected from contracts, archives, and examination papers and contains 1,423 images for training and 200 images for testing. Unfortunately, CH-dataset is not publicly available. To the best of our knowledge, there is no publicly available dataset for the task of handwritten text erasure on examination papers.

2.2   Text Removal Methods

The existing text removal methods can be categorized into one-stage methods and two-stage methods. One-stage methods erase text using an end-to-end model without external text detector [13,22,28,32]. Zhang et al. [32] proposed EnsNet, which introduced a local-sensitive discriminator to maintain the consistency of text-erased regions. To erase text more precisely, Liu et al. [13] proposed EraseNet with a coarse-to-refine architecture, which introduced a segmentation head to help locate the text. MTRNet++ [22] introduced a mask-refine branch to predict the stroke to guide the erasure process, and it supports partial text removal. Wang et al. [28] proposed PERT, which erases text with several progressive erasing stages. Two-stage methods use a text detector to help indicate the text regions [1,11,12,21,23]. MTRNet [23] used an auxiliary mask to achieve stable training and early convergence. Tang et al. [21] predicted text strokes in word patches and fed both strokes and patches into a background inpainting module to obtain the text-erased results. Bian et al. [1] proposed a cascaded network consisting of two text stroke detection networks and two text removal generative networks. Lee et al. [11] introduced Gated Attention (GA) to focus on both the text stroke and its surroundings. Liu et al. proposed CTRNet [12] which uses the CNN to capture local features and a Transformer-encoder to establish the long-term relationships globally. However, the methods mentioned above are mainly for scene text removal and are not directly applied to examination paper scenarios.

Table 1. Distribution of EnsExam dataset

	Chinese	Math	English	Total
Training	130	189	111	430
Testing	28	69	18	115
Total	158	258	129	545

3 EnsExam Dataset

3.1 Image Collection and Data Statistic

We collected 545 high-resolution examination papers using a scanner. All images were collected from real-world examination papers of various schools with diverse layout distributions. To protect personal privacy, sensitive personal information, such as name, was removed from the examination papers. EnsExam contains examination papers for three subjects, i.e., Chinese, Mathematics, and English. Some samples are shown in Fig. 2. The distribution of EnsExam is listed in Table 1. The dataset is randomly divided into training set and test set of 430 and 115 images, respectively. EnsExam is a challenging dataset, and the characteristics of this dataset are described below:

Layout Diversity: EnsExam is a handwritten text erasure dataset with diverse layouts. As shown in Fig. 3, it contains calculation, graphing, matching, and true or false judgment, etc.

Dense and Massive Erasure Regions: As listed in Table 2, EnsExam contains 22,426 erasure regions in the training set and 5,941 erasure regions in the test set. Compared to the SCUT-EnsText [13], EnsExam has an average of nearly 46 more erasure regions on each image, which significantly increase the difficulty of erasing all handwritten text completely.

High Precision Erasure: As shown in (a), (e), and (g) of Fig. 3, EnsExam contains difficult cases where handwritten text is next to or even overlapping with the test questions. If the printed text around the handwritten text is erased incorrectly, it can lead to incomplete test questions. Therefore, in such difficult cases, high precision erasure is required.

Table 2. Number of erasure regions

Dataset		Scenario	Annotation	Min	Max	Total	Average
SCUT-EnsText [13]	Training	Scene	Word level &	1	117	16460	6.0
	Testing	Text	Text line level	1	38	4864	6.0
EnsExam	Training	Handwritten	Text line level	7	145	22426	52.2
	Testing	Text		8	151	5941	51.7

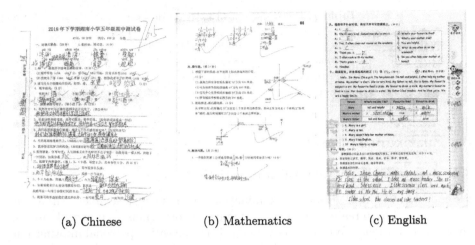

Fig. 2. Samples from EnsExam.

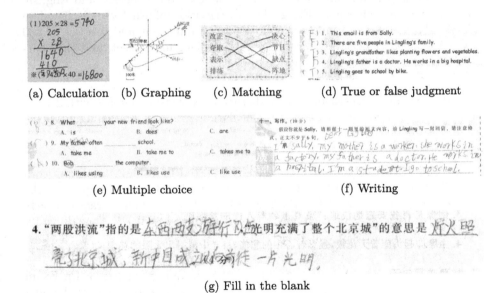

Fig. 3. Test questions in EnsExam, including: calculation, graphing, matching, true or false judgment, multiple choice, writing, fill in the blank, etc.

3.2   Annotation Details

The annotation of EnsExam is shown in Fig. 4, which comprises two components: 1) the ground truth (GT) of text erasure, and 2) the quadrilateral annotation of text detection. To create the GT of text erasure, we manually removed the handwritten text using Adobe Photoshop (PS) and obtained a visually plausible target. Specifically, we used the copy stamp function in PS to replace the pixels of the handwritten text with those of the surrounding background. For cases where handwritten text was located adjacent to printed text, we carefully replaced the handwritten text without altering the printed text. All annotations were double-checked to ensure their quality. For detection annotation, we annotated at the text line level while preserving the structural integrity of the formula. As illustrated in Fig. 4, for vertical math formula that spans multiple lines, we annotated the entire formula as a single bounding box instead of breaking it down into individual boxes. This approach preserves the vertical structure of the formula. In addition, the detection annotation includes two categories of students' answers and teachers' correction marks.

| Input | GT for text erasure | Detection annotaton |

Fig. 4. Annotations of EnsExam. For detection annotations, the green ones are students' answers, while the red ones are teachers' correction marks.

4 Methodology

Our method follows the EraseNet [13] framework with the following two improvements: 1) To erase handwritten text precisely, we propose to use stroke information to guide the erasure process. Different from [1,21,22] which use hard-mask, we further introduce a soft stroke mask to alleviate the ambiguity of the stroke. 2) As text strokes make up only a small portion of the entire image, there is an extreme imbalance between text and non-text regions. To tackle this issue, we propose a stroke normalization (SN) loss.

4.1 Model Architecture

Generator. Our model is built on the generative adversarial network (GAN), and the overall architecture is shown in Fig. 5. The generator consists of a coarse network with segmentation head and a refinement network. CBAM [30] is used in the coarse network to focus on the stroke of handwritten text, in the refinement network, dilated convolution [3] is used to obtain large receptive fields. The output of the proposed method is expressed by Eqs. (1)-(3)

$$M_s, M_b, I_{c4}, I_{c2}, I_{c1} = Net_{coarse}(I_{in}), \tag{1}$$

$$I_{re} = Net_{refine}(I_{in} \oplus M_s \oplus I_{c1}), \tag{2}$$

$$I_{comp} = I_{re}M_b + I_{in}(1 - M_b), \tag{3}$$

where $\oplus$ indicates concatenate. The coarse network $Net_{coarse}(.)$ and refinement network $Net_{refine}(.)$ are U-Net-like FCN with an encoder and decoder. $Net_{coarse}(.)$ encodes the input image $I_{in} \in \mathbb{R}^{H \times W \times 3}$ and obtains feature $F \in \mathbb{R}^{\frac{H}{32} \times \frac{W}{32} \times 512}$, then decodes F to generate text block mask M_b, text stroke mask M_s, and multi-scale coarse output $I_{ci} \in \mathbb{R}^{\frac{H}{i} \times \frac{W}{i} \times 3}$ (i=1,2,4). To erase handwritten text precisely, we concatenate I_{in}, M_s, and I_{c1} together and feed into $Net_{refine}(.)$ to obtain the refined output I_{re}. Finally, we replace the handwritten text regions in I_{in} with I_{re} and generate the composited result I_{comp}.

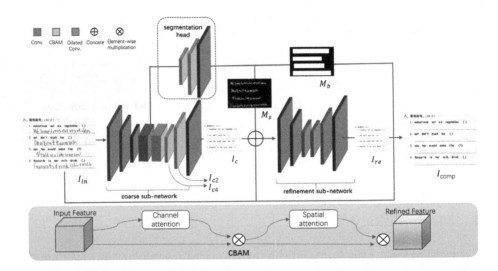

Fig. 5. Overall architecture of the proposed model.

Discriminator. To guarantee the consistency and quality of the results, we adopt the local-global GAN framework proposed in EraseNet [13] to penalize the output globally and locally. For local discrimination, we use the GTs of the block mask to indicate the text regions.

4.2 Soft Stroke Mask and Stroke Normalization Loss

Soft Stroke Mask. To obtain stroke information to guide the erasure process, a stroke mask is generated by directly subtracting the input image from its corresponding GT with the threshold set to 20. However, as shown in (c) of Fig. 6, such a coarse stroke mask contains noise and the pixels around the stroke boundary are ambiguous, which leads to a degradation in performance. Therefore, we introduce a soft stroke mask to alleviate this problem. The process of generating the soft stroke mask is described as follows: (1) We use erosion and dilation algorithm to filter out scattered noise; (2) We shrink the stroke by 1 pixel to obtain the skeleton, and expand it by 5 pixels to obtain the outer boundary; (3) We assign the values of the skeleton and the outer boundary to 1 and 0, respectively, while the values of the middle region are gradually reduced by the SAF [31], as defined in Eq. (4). The process of generating a soft stroke mask is shown in Fig. 6.

$$SAF_{M_s} = C \times \left(\frac{2}{1 + e^{\frac{-\alpha D_{(i,j)}}{L}}} - 1 \right), \tag{4}$$

$$C = \frac{1 + e^{-\alpha}}{1 - e^{-\alpha}}, \tag{5}$$

where C is a normalization constant, $D_{(i,j)}$ is the distance from the pixel (i,j) to the outer boundary, α is a transforming factor, and L is the maximum distance. In our experiments, we set $\alpha = 3$ and $L = 5$.

Fig. 6. Process of generating a soft stroke mask. Image from left to right: original image, GT, coarse stroke mask, denoised mask, soft stroke mask.

Stroke Normalization Loss. Since strokes only occupy a small proportion of the entire image, there is a significant imbalance between text and non-text regions, which can negatively affect the model's performance. To alleviate this problem, we propose a simple yet effective loss function called stroke normalization (SN) loss, defined by Eq. (6).

$$L_{SN} = \frac{\|M_{gt_s} - M_s\|_1}{min(\sum_{x,y} M_{gt_s}, \sum_{x,y} M_s)}, \tag{6}$$

where $\sum_{x,y}(.)$ denotes the sum of all pixels, and $min(.)$ denotes the minimum value. M_s and M_{gt_s} denote the stroke mask and corresponding GT, respectively. From Eq. (6), it can be seen that when the strokes in M_s is missed, $\sum_{x,y} M_s$ is small and therefore the penalty is large.

4.3 Training Objective

Block Loss. We use the Dice loss [15] to guide the block mask.

$$Dice(M_b, M_{gt_b}) = \frac{2 \times \sum_{x,y}(M_b \times M_{gt_b})}{\sum_{x,y}(M_b)^2 + \sum_{x,y}(M_{gt_b})^2}, \tag{7}$$

$$L_{block} = 1 - Dice(M_b, G), \tag{8}$$

where M_b and M_{gt_b} denote the block mask and corresponding GT, respectively.

Local-Aware Reconstruction Loss. Results from different stages are considered, and the handwritten text regions are assigned higher weights.

$$L_{LR} = \sum_n \lambda_n \|(I_{out_n} - I_{gt_n}) * M_{gt_b}\|_1 + \sum_n \beta_n \|(I_{out_n} - I_{gt_n}) * (1 - M_{gt_b})\|_1, \tag{9}$$

where I_{out_n} represents I_{c4}, I_{c2}, I_c, and I_{re}. M_{gt_b} and I_{gt_n} represent GT of block mask and erasure GT. $\{\lambda_n, \beta_n\}$ are set to $\{5, 0.8\}$, $\{6, 0.8\}$, $\{8, 0.8\}$, and $\{10, 2\}$, respectively.

Perceptual Loss. Both refinement output I_{re} and the composited output I_{comp} are considered.

$$L_{per} = \sum_i \sum_{n=1}^{N} \|\phi_n(I_i) - \phi_n(I_{gt})\|_1, \tag{10}$$

where I_i represent I_{re} and I_{comp}. ϕ_n denotes the feature maps of the n-th($n = 1, 2, 3$) pooling layers of VGG-16 [20] pretrained on ImageNet [5].

Style Loss. It is calculated using the Gram matrix $Gr(.)$ of the feature maps in perceptual loss.

$$L_{style} = \sum_i \sum_n \frac{\|Gr(\phi_n(I_i)) - Gr(\phi_n(I_{gt}))\|_1}{H_n W_n C_n}. \tag{11}$$

Adversarial Loss. Following Miyato et al. [16], we use hinge loss as our adversarial loss. The generator's adversarial loss is defined by Eq. (13)

$$L_{adv\_D} = E_{x \sim P_{data(x)}}[ReLU(1 - D(x))] + E_{z \sim P_{z(z)}}[ReLU(1 + D(G(z)))], \tag{12}$$

$$L_{adv\_G} = -E_{z \sim P_{z(z)}}[ReLU(1 + D(G(z)))]. \tag{13}$$

Total Loss. The final objective function of the generator is defined by Eq. (14)

$$L_G = L_{adv\_G} + \lambda_{lr} L_{LR} + \lambda_p L_{per} + \lambda_s L_{style} + \lambda_{sn} L_{SN} + \lambda_b L_{block}, \tag{14}$$

where λ_{lr} to λ_b represent the weights of each loss term. Following the previous work [13], we set λ_{lr}, λ_p, and λ_s to 1.0, 0.05, 120, respectively. For the L_{SN} and L_{block}, due to the crucial role of the stroke information in the erasure process, we assign a higher weight to L_{SN}. Specifically, we empirically set λ_{sn}, and λ_b to 1 and 0.4, respectively.

5   Experiments

5.1   Evaluation Metrics

Following previous work [13,32], we evaluate the performance using both Image-Eval and Detection-Eval. For Image-Eval, the following metrics are included: (1) Peak signal-to-noise ratio (PSNR); (2) Multi-scale Structural Similarity [29] (MSSIM); (3) Mean Square Error (MSE); (4) AGE, which calculates the average of the absolute difference on gray level; (5) pEPs, which denote the percentage of error pixels; and (6) pCEPs, which denotes the percentage of clustered error pixels (four-connected neighbors are all error pixels). Higher values of PSNR and MSSIM and lower values of MSE, AGE, pEPs, and pCEPs indicate better results. For Detection-Eval, OBD [14] is employed as an auxiliary text detector and precision (P), recall (R), and F-score (F) are calculated to further evaluate the performance. Lower P, R, and F indicate that more handwritten text has been removed.

5.2   Implementation Details

We use Adam [10] solver as our optimizer. The learning rate is set to 0.0001, and β is set to (0.5,0.9). The model is trained for 100 epochs with a batchsize of 4. For existing methods, we set the hyperparameters according to the original papers. Given the large size of the images, we crop 4,995 patches into 512 × 512 pixels from the training set of EnsExam during the training phase. Then we apply two inference strategies: 1) Divide the image into 512 × 512 patches for inference, and then combine the inferred results of each patch to reconstruct the original image; 2) Perform inference on the whole image. For the two-stage methods, we use HTC [2] as the text detector in EnsExam.

Table 3. Ablation studies on EnsExam, MSSIM, MSE, pEPs, and pCEPs are represented by %. Net_{refine} represents the refinement network. Concatenating stroke mask into Net_{refine} indicates $I_{re} = Net_{refine}(I_{in} \oplus M_s \oplus I_c)$; otherwise, $I_{re} = Net_{refine}(I_{in} \oplus I_c)$.

Methods	Stroke mask loss			Concatenating	Image-Eval					
	L1	Tversky [19]	SN	into Net_{refine}	PSNR	MSSIM	MSE	AGE	pEPs	pCEPs
Ours	✓				31.76	95.61	0.12	1.89	0.81	0.24
Ours			✓		34.73	96.37	0.06	1.69	0.53	0.13
Ours		✓		✓	35.69	96.57	**0.05**	1.47	**0.47**	**0.11**
Ours			✓	✓	**36.05**	**96.59**	**0.05**	**1.43**	**0.47**	**0.11**
MTRNet++ [22]	✓			✓	32.77	92.64	0.08	2.51	0.74	0.12
MTRNet++ [22]			✓	✓	33.71	92.80	0.07	2.45	**0.70**	**0.11**

5.3　Quantitative and Qualitative Results

Ablation Study. According to the results shown in Table 3, compared with Tversky loss [19], used in MTRNet++ [22], The proposed SN loss can improve the PSNR by 0.94 and 0.36 for MTRNet++ [22] and our method, respectively. In addition, the SN loss can improve the PSNR of our method by 2.97 over the L1 loss. The comparison between the second and fourth rows show that concatenating the stroke mask into the refinement network can improve the PSNR by 1.32.

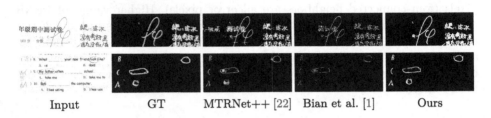

| Input | GT | MTRNet++ [22] | Bian et al. [1] | Ours |

Fig. 7. Comparison of the stroke mask.

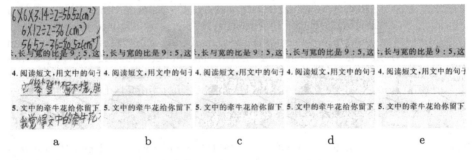

| a | b | c | d | e |

Fig. 8. Composited with block mask and stroke mask: from (a) to (e) is the input images, GTs, results of MTRNet++ [22], our results composited with stroke mask, and our results composited with block mask.

Comparison with the State-of-the-Arts. As shown in Table 4, we conduct experiments to evaluate the performance of our method and the relevant SOTA methods on EnsExam. Except for Pix2Pix [7], performing inference on the whole image obtains higher PSNR than patch because cropping the image into patches intercepts the text, which degrades the performance. Our method obtains the best result on Image-Eval except for MSSIM. Though MTRNet++ [22] obtains the lowest F-Score, it tends to erase printed text incorrectly and obtain a low PSNR.

The outputs of the stroke mask are shown in Fig. 7. Benefits from the soft stroke mask and SN loss, our stroke mask is more precise and smoother, without predicting printed text by mistake. As shown in Fig. 8, different from the result of MTRNet++ [22] which composited with stroke mask, our result composited with block mask and achieve a better visual effect.

The qualitative comparison with previous methods on EnsExam is shown in Fig. 9. The results of Bian et al. [1] and CTRNet [12] suffer from incomplete erasure and mistaken erasure, while our method can erase handwritten text next to the question, as shown in the fourth row.

Some failure cases are shown in Fig. 10. For the difficult cases where handwritten text overlays with printed text, as shown in the first row, our proposed method fails to erase handwritten text correctly. It is a challenge that deserves further investigation.

Table 4. Comparison with previous methods on EnsExam. For the detection-eval, OBD [14] is adopted as an auxiliary detector. MSSIM, MSE, pEPs, and pCEPs are represented by %. Bold indicates SOTA. Underline indicates second best.

Methods	Image-Eval						Detection-Eval		
	PSNR	MSSIM	MSE	AGE	pEPs	pCEPs	R	P	F
Original images	-	-	-	-	-	-	79.4	83.1	81.2
Inference on patch									
Pix2Pix [7]	29.01	89.79	0.16	3.84	2.45	0.44	9.1	37.6	14.6
EnsNet [32]	33.54	95.51	0.07	2.10	0.67	0.13	3.4	25.3	6.0
EraseNet [13]	33.34	93.55	0.07	2.59	0.67	**0.12**	1.6	24.9	**2.9**
MTRNet++ [22]	31.23	92.34	0.10	2.70	1.15	0.34	1.8	24.1	3.4
Bian et al. [1](w HTC [2])	34.10	95.75	0.07	1.90	**0.48**	0.14	<u>1.7</u>	<u>21.7</u>	<u>3.1</u>
CTRNet [12](w PAN [27])	32.89	96.03	0.09	1.72	0.74	0.22	5.1	**18.5**	8.0
CTRNet [12](w HTC [2])	<u>35.34</u>	**96.65**	**0.06**	<u>1.58</u>	0.54	0.15	1.8	26.9	3.3
Ours	**35.44**	<u>96.47</u>	**0.06**	**1.46**	<u>0.49</u>	**0.12**	1.9	27.1	3.5
Inference on whole image									
Pix2Pix [7]	28.99	89.54	0.16	3.89	2.50	0.46	9.5	42.2	15.5
EnsNet [32]	33.87	94.93	0.07	2.25	0.64	0.12	3.1	25.9	5.6
EraseNet [13]	33.84	93.69	0.07	2.55	0.63	**0.11**	<u>1.5</u>	25.1	<u>2.8</u>
MTRNet++ [22]	32.77	92.64	0.08	2.51	0.74	0.12	**0.6**	**12.6**	**1.1**
Bian et al. [1](w HTC [2])	34.27	95.80	0.06	1.89	<u>0.48</u>	0.13	1.6	21.1	3.0
CTRNet [12](w PAN [27])	33.01	96.07	0.09	1.70	0.73	0.22	4.8	<u>17.5</u>	7.6
CTRNet [12](w HTC [2])	<u>35.68</u>	**96.71**	**0.05**	<u>1.54</u>	0.49	0.13	1.6	25.4	3.0
Ours	**36.05**	<u>96.59</u>	**0.05**	**1.43**	**0.47**	**0.11**	1.6	25.8	3.0

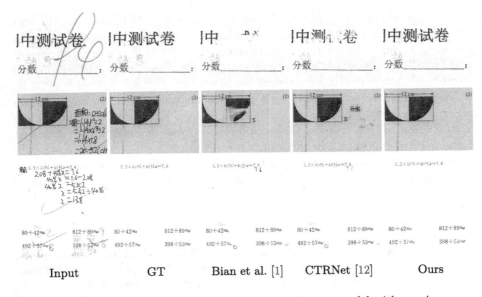

Fig. 9. Qualitative results on EnsExam for comparing our model with previous scene text removal methods. Image from left to right: Input images, GTs, results of Bian et al. [1], results of CTRNet [12], and our results.

Table 5. Comparison with state-of-the-art methods on SCUT-EnsText, One-stage indicates an end-to-end model without auxiliary detector, two-stage indicates using an auxiliary detector, and Syn indicates using synthetic dataset. MSSIM, and MSE are represented by %.

Methods	Type	Training data	PSNR	MSSIM	MSE
Pix2Pix [7]	one-stage	SCUT-EnsText	26.70	88.56	0.37
EnsNet [32]	one-stage	SCUT-EnsText	29.54	92.74	0.24
EraseNet [13]	one-stage	SCUT-EnsText	32.30	95.42	0.15
PERT [28]	one-stage	SCUT-EnsText	33.25	**96.95**	**0.14**
Ours	one-stage	SCUT-EnsText	**33.74**	96.74	**0.14**
Tang et al. [21]	two-stage	SCUT-EnsText+Syn	35.34	96.24	**0.09**
CTRNet [12]	two-stage	SCUT-EnsText	**35.85**	**97.40**	**0.09**

5.4   Extension of the Proposed Method on Scene Text Removal

In this section, we apply our method to scene text removal. As shown in Table 5, although our method is designed for handwritten text erasure on examination papers, it can also be applied to scene text removal and achieve comparable performance to state-of-the-art one-stage methods. In particular, our method achieves the highest PSNR among one-stage methods.

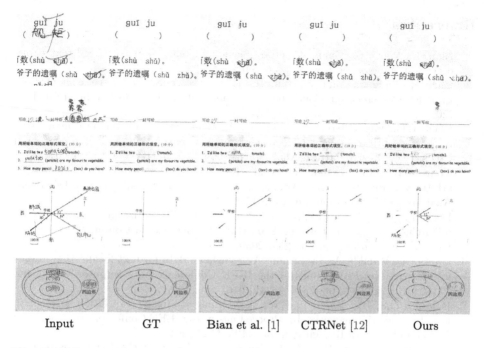

Input	GT	Bian et al. [1]	CTRNet [12]	Ours

Fig. 10. Failure cases. Image from left to right: Input images, GTs, results of Bian et al. [1], results of CTRNet [12], and our results.

6 Conclusion

In this paper, we construct the first public-available real-world handwritten text erasure dataset, EnsExam, consisting of 545 examination paper images. In addition, we propose an end-to-end method that introduces a soft stroke mask and SN loss to erase handwritten text precisely. Numerical results show that training with the proposed SN loss and using the soft stroke mask to guide the erasure process is useful for accurately erasing handwritten text. Moreover, qualitative results show that erasing handwritten text that overlays with printed text is challenging and still not well resolved. Therefore, this remains an open problem that deserves more attention and further investigation.

Acknowledgements. This research is supported in part by NSFC (Grant No.: 61936003), Zhuhai Industry Core and Key Technology Research Project (no. 2220004002350), and Science and Technology Foundation of Guangzhou Huangpu Development District (No. 2020GH17) and GD-NSF (No.2021A1515011870).

References

1. Bian, X., Wang, C., Quan, W., et al.: Scene text removal via cascaded text stroke detection and erasing. Comput. Vis. Media **8**(2), 273–287 (2022)
2. Chen, K., Pang, J., Wang, J., et al.: Hybrid task cascade for instance segmentation. In: IEEE Conference on Computer Vision and Pattern Recognition, pp. 4974–4983 (2019)
3. Chen, L.C., Papandreou, G., Kokkinos, I., et al.: DeepLab: semantic image segmentation with deep convolutional nets, atrous convolution, and fully connected CRFs. IEEE Trans. Pattern Anal. Mach. Intell. **40**(4), 834–848 (2018)
4. Chng, C.K., Liu, Y., Sun, Y., et al.: ICDAR2019 robust reading challenge on arbitrary-shaped text -RRC-ArT. In: 2019 International Conference on Document Analysis and Recognition, pp. 1571–1576 (2019)
5. Deng, J., Dong, W., Socher, R., Li, L.J., Li, K., Fei-Fei, L.: ImageNet: a large-scale hierarchical image database. In: 2009 IEEE Conference on Computer Vision and Pattern Recognition, pp. 248–255 (2009)
6. Gupta, A., Vedaldi, A., Zisserman, A.: Synthetic data for text localisation in natural images. In: Proceedings of the IEEE Conference on Computer Vision and Pattern Recognition, pp. 2315–2324 (2016)
7. Isola, P., Zhu, J.Y., Zhou, T., Efros, A.A.: Image-to-image translation with conditional adversarial networks. In: Proceedings of the IEEE Conference on Computer Vision and Pattern Recognition, pp. 1125–1134 (2017)
8. Karatzas, D., Gomez-Bigorda, L., Nicolaou, A., et al.: ICDAR 2015 competition on robust reading. In: 2015 13th International Conference on Document Analysis and Recognition, pp. 1156–1160 (2015)
9. Karatzas, D., Shafait, F., Uchida, S., et al.: ICDAR 2013 robust reading competition. In: 2013 12th International Conference on Document Analysis and Recognition, pp. 1484–1493 (2013)
10. Kingma, D.P., Ba, J.: Adam: a method for stochastic optimization. In: International Conference on Learning Representations (2015)
11. Lee, H., Choi, C.: The surprisingly straightforward scene text removal method with gated attention and region of interest generation: a comprehensive prominent model analysis. In: Avidan, S., Brostow, G., Cisse, M., Farinella, G.M., Hassner, T. (eds.) Computer Vision – ECCV 2022. ECCV 2022. Lecture Notes in Computer Science, vol. 13676, pp. 457–472. Springer, Cham (2022). https://doi.org/10.1007/978-3-031-19787-1_26
12. Liu, C., et al.: Don't forget me: accurate background recovery for text removal via modeling local-global context. In: Avidan, S., Brostow, G., Cisse, M., Farinella, G.M., Hassner, T. (eds.) Computer Vision – ECCV 2022. ECCV 2022. Lecture Notes in Computer Science, vol. 13688, pp. 409–426. Springer, Cham (2022). https://doi.org/10.1007/978-3-031-19815-1_24
13. Liu, C., Liu, Y., Jin, L., et al.: EraseNet: end-to-end text removal in the wild. IEEE Trans. Image Process. **29**, 8760–8775 (2020)
14. Liu, Y., et al.: Exploring the capacity of an orderless box discretization network for multi-orientation scene text detection. Int. J. Comput. Vision **129**(6), 1972–1992 (2021)
15. Milletari, F., Navab, N., Ahmadi, S.A.: V-Net: fully convolutional neural networks for volumetric medical image segmentation. In: 2016 Fourth International Conference on 3D Vision, pp. 565–571 (2016)

16. Miyato, T., Kataoka, T., Koyama, M., Yoshida, Y.: Spectral normalization for generative adversarial networks. In: International Conference on Learning Representations (2018)
17. Nayef, N., Patel, Y., Busta, M., et al.: ICDAR2019 robust reading challenge on multi-lingual scene text detection and recognition - RRC-MLT-2019. In: 2019 International Conference on Document Analysis and Recognition, pp. 1582–1587 (2019)
18. Nayef, N., Yin, F., Bizid, I., et al.: ICDAR2017 robust reading challenge on multi-lingual scene text detection and script identification - RRC-MLT. In: 2017 14th IAPR International Conference on Document Analysis and Recognition, pp. 1454–1459 (2017)
19. Salehi, S.S.M., Erdogmus, D., Gholipour, A.: Tversky loss function for image segmentation using 3d fully convolutional deep networks. In: Machine Learning in Medical Imaging, pp. 379–387 (2017)
20. Simonyan, K., Zisserman, A.: Very deep convolutional networks for large-scale image recognition. In: International Conference on Learning Representations (2015)
21. Tang, Z., Miyazaki, T., Sugaya, Y., Omachi, S.: Stroke-based scene text erasing using synthetic data for training. IEEE Trans. Image Process. **30**, 9306–9320 (2021)
22. Tursun, O., Denman, S., Zeng, R., et al.: MTRNet++: one-stage mask-based scene text eraser. Comput. Vis. Image Underst. **201**, 103066 (2020)
23. Tursun, O., Zeng, R., Denman, S., et al.: MTRNet: a generic scene text eraser. In: 2019 International Conference on Document Analysis and Recognition, pp. 39–44 (2019)
24. Veit, A., Matera, T., Neumann, L., Matas, J., Belongie, S.: COCO-Text: dataset and benchmark for text detection and recognition in natural images. arXiv preprint arXiv:1601.07140 (2016)
25. Wang, B., Li, J., Jin, X., Yuan, Q.: CHENet: image to image Chinese handwriting eraser. In: Pattern Recognition and Computer Vision, pp. 40–51 (2022)
26. Wang, K., Belongie, S.: Word spotting in the wild. In: Daniilidis, K., Maragos, P., Paragios, N. (eds.) ECCV 2010. LNCS, vol. 6311, pp. 591–604. Springer, Heidelberg (2010). https://doi.org/10.1007/978-3-642-15549-9_43
27. Wang, W., Xie, E., Song, X., et al.: Efficient and accurate arbitrary-shaped text detection with pixel aggregation network. In: Proceedings of the IEEE/CVF International Conference on Computer Vision, pp. 8440–8449 (2019)
28. Wang, Y., Xie, H., Fang, S., et al.: PERT: a progressively region-based network for scene text removal. arXiv preprint arXiv:2106.13029 (2021)
29. Wang, Z., Bovik, A., Sheikh, H., Simoncelli, E.: Image quality assessment: from error visibility to structural similarity. IEEE Transactions on Image Processing, pp. 600–612 (2004)
30. Woo, S., Park, J., Lee, J.-Y., Kweon, I.S.: CBAM: convolutional block attention module. In: Ferrari, V., Hebert, M., Sminchisescu, C., Weiss, Y. (eds.) ECCV 2018. LNCS, vol. 11211, pp. 3–19. Springer, Cham (2018). https://doi.org/10.1007/978-3-030-01234-2_1
31. Zhang, S.X., Zhu, X., Chen, L., et al.: Arbitrary shape text detection via segmentation with probability maps. IEEE Trans. Pattern Anal. Mach. Intell. (2022). https://doi.org/10.1109/TPAMI.2022.3176122
32. Zhang, S., Liu, Y., Jin, L., et al.: EnsNet: ensconce text in the wild. In: Proceedings of the AAAI Conference on Artificial Intelligence, vol. 33, pp. 801–808 (2019)

MIDV-Holo: A Dataset for ID Document Hologram Detection in a Video Stream

L. I. Koliaskina[1] , E. V. Emelianova[1] , D. V. Tropin[1,4] , V. V. Popov[2] ,
K. B. Bulatov[1,4(✉)] , D. P. Nikolaev[1,3] , and V. V. Arlazarov[1,4]

[1] Smart Engines Service LLC, Moscow, Russia
hpbuko@gmail.com
[2] Lomonosov Moscow State University, Moscow, Russia
[3] Institute for Information Transmission Problems (Kharkevich Institute), Moscow, Russia
[4] Federal Research Center "Computer Science and Control" or Russian Academy of Sciences, Moscow, Russia

Abstract. One of the most important tasks related to identity document analysis is to verify its authenticity. A powerful family of security features of ID documents is optically variable devices, such as holograms. The development of computer vision methods for the automatic validation of holograms is complicated due to the lack of public datasets. In this paper, we present MIDV-Holo – a public dataset of synthetically created passports and ID cards, captured on video under various conditions. The dataset features 300 video clips of documents with custom holographic security elements, as well as 400 video clips of presentation attacks of different types. The paper describes the process of dataset creation, available annotation, and a baseline algorithm for the holograms detection method based on an analysis of pixel-wise chromaticity statistics accumulated in a video stream, with an evaluation on the presented dataset. Authors believe that, being the first public dataset in the domain of hologram-based identity document authentication, MIDV-Holo will provide researchers in the field of identity verification with a valuable resource for testing new methods of video-based ID document analysis.

Keywords: identity documents · forgery detection · open dataset · OVD · holograms

1 Introduction

At the current stage of the development of remote services, there are many tasks related to the automatic identification and authentication of individuals. Personal identification is usually performed by presenting a passport or ID card. Automatic document recognition systems help simplify the process of remote provision of services and save time and money resources for companies. Therefore, systems for automatic recognition of identification documents are actively developed [1–3]. Unfortunately, cases of presentation of forged documents are

G. A. Fink et al. (Eds.): ICDAR 2023, LNCS 14189, pp. 486–503, 2023.
https://doi.org/10.1007/978-3-031-41682-8_30

common, therefore, along with the task of automatic recognition of personal data of document owners, an important task is to check the authenticity of the presented document. There are lots of ways to protect identity documents from falsification such as security fibers, watermarks, metameric colors, fluorescent ink, OVD (optically variable device), guilloche, etc. [4]. Many methods for automatic document verification based on different features have been proposed. For example, there are methods for stamp verification [5,6], methods for detection of fluorescent security features [7,8], the method for ink mismatch detection in hyperspectral images [9] and forgery detection method based on font analysis [10].

A powerful method to protect identity documents is OVDs (optically variable devices) – security features that appear in various ways depending on the capture angle and/or lighting conditions [4]. OVD is recommended by the European Union as a protection of the biographical data page of the document against copying [11]. A widely used type of OVD is diffractive optical elements, usually called "holograms", even though most often they are not holograms in a strict physical sense. Their optical effect is based on an interaction between light and diffraction grating with typical periods of 1 micron and height of 100 nm, organized as pixels with a size of 5–50 microns. While illuminated by a non-monochromatic point light source such structures create characteristic dynamic and color effects [12]. An example of a hologram on an identity document is shown in Fig. 1.

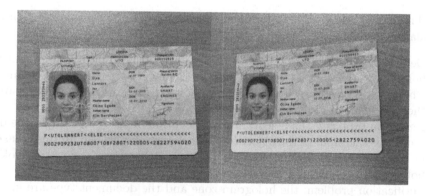

Fig. 1. An example of an artificial passport with a hologram captured from two angles. It can be seen that the holographic pattern changes its color characteristics depending on the camera angle. (Color figure online)

There exist several types of OVDs based on diffractive optical elements employed in the production of identity documents. One of the most prevalent is a security film, covering the identity document and having the general composition, presented in Fig. 2.

The external layer is typically a polyethylene terephthalate base, after which follows a thin layer of a thermoplastic material that holds the diffractive grating pattern (sometimes if the external base layer needs to be removed after production, a thin wax layer is placed between them). To provide reflective properties

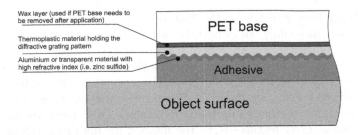

Fig. 2. Structure of a typical film-based diffractive optical element.

to the grating, it is covered with a layer of an opaque reflective material such as aluminum, or with a layer of a transparent material with a high refractive index, such as zinc sulfide. The grating itself is produced by embossing using a matrix composed of a resist material [12]. The original grating pattern matrix is produced using various methods, such as electron-beam lithography [13, 14].

Existing methods for automatic detection and validation of holograms on identity documents and banknotes can be divided into two classes:

1. methods requiring special devices such as a special photometric light dome or portable ring-light illumination module [15, 16];
2. methods designed to work only with the photos or video sequences obtained by the camera of a mobile device [17–19].

There are various ways to formulate the problem of hologram detection and validation, for example:

1. Detection problem: both the document type and the information about the hologram (including its location on the document) are considered unknown. The task, in this case, is to detect the hologram (if it exists) in any part of the document in the video sequence [19].
2. Detection problem: the hologram zone (location of the hologram on the document) is the input of the algorithm. The detection problem, in this case, can be reformulated as the classification problem – the absence or presence of the hologram in the input zone.
3. Verification problem: the hologram zone and the document type are known. The input zone needs to be checked against the reference hologram in this zone for the given document type [15, 17].

The article [15] describes a method for verification the holographic protection based on comparing a sample descriptor of a holographic element with a reference descriptor of a holographic element. Such descriptors are generated using a neural network from images obtained by the camera of a mobile device and a portable ring-light module mountable to a mobile device. In the article [16] authors propose a hologram detection method based on evaluating the behavior of a bidirectional reflectance distribution function (BRDF) on images obtained using a special photometric light dome. In the article [17] the verification of the hologram is carried out by comparing the hologram images with reference images of a given type of hologram prepared in advance under various capture angles. The images of the test hologram under correct angles are obtained by navigating the user with the help of augmented reality. To compare those images such metrics as SSIM (Structural Similarity Index) are used. In the work [18] the standard deviation of RGB values is calculated from a stack of document images in each pixel to construct a hologram map, after that the resulting map is segmented to find peaks that represent potential zones where the hologram is present. A stack of document images is obtained by using tracking the document in a video stream. Authors of the article [19] propose a hologram detection method that doesn't require prior information about the type and location of the hologram using only a camera of smartphone. They propose to process frames separately before considering the whole clip. Firstly, they highlight pixels with high saturation and brightness by analyzing saturation and brightness histograms to get potential holographic zones. Then the connected components in the obtained image and its dilation are analyzed to eliminate "porous" zones which are associated with non-holographic zones. After that the color analysis is performed: the number of hues normalized on the size of the considered zone (limited by the connected component of the dilated image obtained on the first step) is estimated. The authors emphasize that one frame is not enough to extract hologram pixels without false positives and provide the method of video analysis using the per-frame results.

Thus, there are many variants of problem statements for hologram detection task and despite the fact that this area of research is actively developed there is still a lack of clarity in the formulation of the problem and determination of the limits of applicability, which leads to difficulties in the analysis of existing methods and their differences. A variety of methods for automatic hologram detection exists today. However, due to the lack of open datasets containing documents with holograms, it is impossible to compare the quality of these methods. Hence, further development of security hologram detection methods is significantly complicated. The goal of this paper is to introduce a new target dataset with a baseline hologram detection algorithm, and thus facilitate more objective and productive research on this topic.

2   Dataset

In this section, we review the process of creating documents for their usage in an open hologram detection dataset and the dataset preparation process.

Initially, the document blanks of the fictional country "Utopia" were designed. We made 10 types of passport blanks of the size 125×8 mm and 10 types of ID card blanks of the size 85.6×54 mm. The sizes for documents were chosen according to the standard ICAO Doc 9303 [20] as the most common in the world. Document background images taken from the stock images marketplace [21] were used to create unique blanks of "Utopian" documents. The images have been processed to create pale backgrounds in various colors with abstractions and lines.

To complete the document design process we generated text field data (such as name, place of birth, place of issue, etc.), photos of non-existent people, and created signatures roughly corresponding to the generated names. Names and addresses were generated using an online country and nationality data generator [22] and fonts were obtained using an open online source [23]. Artificially generated face images obtained from Generated Photo service [24] were used to provide a unique document holder for each document. Pictures were selected so that they roughly correspond to the age of the holder. Some of the documents contain document holder images not only in color but also in grayscale. We designed 5 documents for each blank (100 in total) in the way described above. The examples of created document templates are shown in Fig. 3.

a) "Utopian" passport

b) "Utopian" ID

Fig. 3. Designed document templates with artificial data

We designed one holographic coating for all passport types and one for all ID card types. The binary masks of the resulting holograms are shown in Fig. 4 and are also present in the dataset. White pixels represent the location of holographic patterns on the document. The designed holographic patterns were produced by the Center "Fine Optical Technologies" [25] using electron beam lithography, using a zinc sulfide layer under the thermoplastic material with diffractive grating in order to obtain a transparent film. Template images were printed to scale on glossy photo paper, laminated, cut out, and covered with the film. All corners of the ID cards and the two bottom corners of the passports were rounded with a 4 mm radius rounder. The resulting documents with holograms we will call "original" documents.

Along with the "original" documents we also prepared samples of several types of threat models in the MIDV-Holo dataset:

1. copy of a document template without hologram ("copy without holo");
2. copy of a document template with hologram pattern drawn in an image editor ("pseudo holo copy");
3. printed photo of "original" document ("photo holo copy").

b) ID card hologram mask

a) Passport hologram mask

Fig. 4. Designed hologram masks for passports and ID cards

Template images prepared in this way were also printed on glossy photo paper, laminated, and cut out. Holographic elements are often located on the photo of the document holder in real documents to prevent photo replacement. Considering this, we also modeled the case of photo replacement – an "original" document with a printed and cut-out photo of a different person attached. The resulting "original" document and 4 types of threat models are shown in Fig. 1 and Fig. 5.

Since the behavior of the hologram (including the saturation of the produced images), highly depends on the lighting condition we captured clips in various lighting conditions, which include:

A. office lighting, avoiding glare on the surface of the document while video shooting;
B. office lighting;
C. lighting of the flashlight of a mobile device, avoiding glare on the surface of the document while video shooting;
D. lighting of the flashlight of a mobile device;
E. outdoor lighting.

We captured clips using only cameras of mobile devices. An iPhone 12 and a Samsung Galaxy S10 were used. 60 clips of documents with holograms and 20×4 of 4 threat models were captured in each lighting condition. Detailed information about the shooting conditions for each clip is presented in the dataset description. Each clip was captured with a duration of approximately 7 s in a resolution of 3840×2160 (Ultra HD) or 1920×1080 (Full HD). Each clip was then split into frames with fps $= 5$ using FFmpeg. The number of produced frames in a clip varies from 33 to 71 with a mean number of 43.

Thus, we got 300 clips of "original" documents and 100 clips of each threat model (700 clips in total). We also added annotation for each frame of each clip in JSON format. Annotation includes manually selected document boundary quadrangle and the unique identifier of a document. The dataset is available for

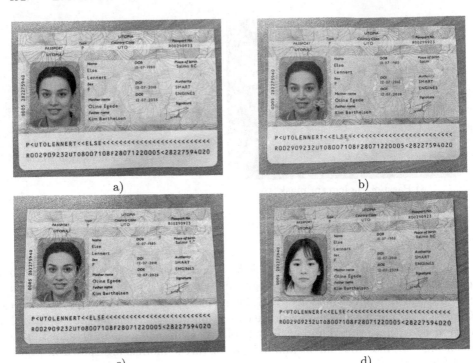

Fig. 5. Examples of 4 types of threat models: a) copy of a document template without hologram ("copy without holo"), b) copy of a document template with hologram pattern drawn in an image editor ("pseudo holo copy"), c) printed photo of "original" document ("photo holo copy"), d) replacement of the photo of document holder ("photo replacement").

free download at [ftp://smartengines.com/midv-holo]. The examples of frames from the dataset for each type of lighting condition are presented in Fig. 6.

3   Benchmark: Hologram Detection Method

In this section, we describe a hologram detection method that will be used as a baseline for further work with the MIDV-Holo dataset.

3.1   Problem Statement

We consider the type of document (and hence its linear sizes) to be known. The input of the algorithm is:

- frame sequence $\{F_i\}$, in which the document is shown;
- document boundary quadrangle for each frame $\{q_i\}$;
- linear sizes of the document page ($t.width$, $t.height$);

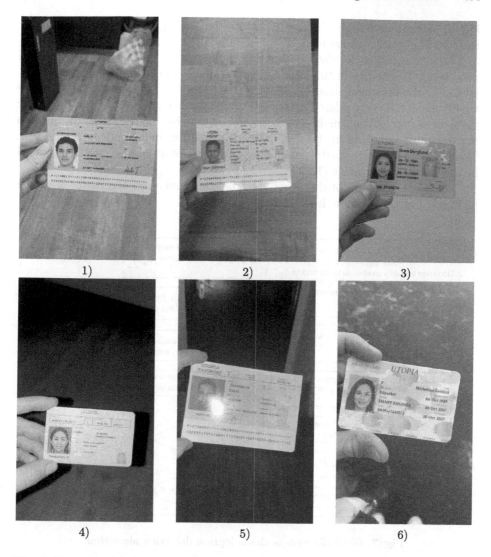

Fig. 6. Examples of clips from MIDV-Holo captured in different lighting conditions (lighting condition codes in brackets): 1) "original" passport (A), 2) "original" passport (B), 3) "original" ID card (B), 4) "original" ID card (C) 5) threat model – passport "pseudo holo copy" (D) 6) threat model – ID card "photo holo copy" (E)

The output of the algorithm is a binary response (hologram is found/hologram is not found).

3.2 Proposed Algorithm

The proposed algorithm requires quadrilateral document borders on each frame as an input. Let us call the localization of the document the detection of quadrilateral document borders in the frame without using information from other frames. There are various methods of document localization, for example, [26–31], but independent localization in each frame can be unstable during the clip processing. Instability in this case can cause small displacements of the quadrilateral of the document from frame to frame. This kind of displacement can be significant for the considered method because further analysis is done pixel by pixel on projectively normalized document images. For this reason, we propose to use document tracking in the video stream instead of independent localization. Tracking can be performed by matching the descriptors of keypoints computed in the consecutive frames [32–34].

The block diagram of the proposed hologram detection algorithm is shown in Fig. 7. Let us consider the key points in detail.

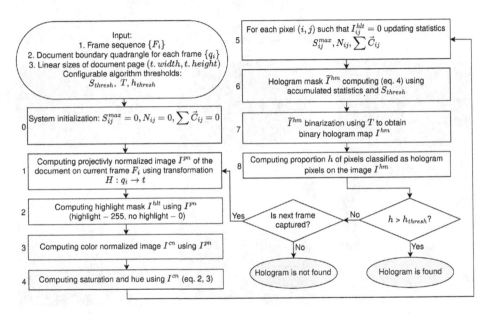

Fig. 7. Block diagram of the hologram detection algorithm

At the beginning a projective transformation of the input frame is performed using the input document quadrangle, to get a projectively normalized document image I^{pn} (block 1). Highlights, that appear on the document image due to lamination, can make a negative impact on the hologram detection process. For this reason, highlights should be detected in the normalized document image. One of the ways to get the highlight mask I^{hlt} (the binary image which represents pixels of highlight) is to convert the color document image to a grayscale image and binarize the obtained image using a fixed global threshold (threshold value 240 was used). As a result, pixels, which are too bright can be found (block 2).

Then we apply color correction to the normalized document image to reduce the impact of an automatic camera reconfiguration on the color distribution during the video shooting. The algorithm based on the Gray-World assumption was used as the color correction [35]. I^{cn} represents the resulting color-normalized image (block 3).

In further calculations, we ignore pixels that represent highlights using highlight mask I^{hlt}. For each pixel (excluding highlight pixels) of the corrected document image I^{cn}, the chromaticity vector C is calculated. The length of C is equal to saturation S, and the angular position around a central point is equal to hue H (block 4):

$$C = \begin{pmatrix} S\cos(H) \\ S\sin(H) \end{pmatrix}, \tag{1}$$

$$S = max(R, G, B) - min(R, G, B), \tag{2}$$

$$H = \begin{cases} 2\pi(\tilde{H}/6 + 1), & \tilde{H} < 0 \\ 2\pi(\tilde{H}/6), & \tilde{H} \geq 0, \end{cases} \quad \tilde{H} = \begin{cases} 0, & S = 0 \\ \frac{G-B}{S}, & max(R, G, B) = R \\ \frac{B-R}{S} + 2, & max(R, G, B) = G \\ \frac{R-G}{S} + 4, & max(R, G, B) = B. \end{cases} \tag{3}$$

Figure 8 shows the chromaticity C of pixels with fixed coordinates on a normalized document image for each frame of a testing clip, showing a hologram on an "Utopian" ID card (40 frames). There are 4 points from different parts of the document selected for demonstration: points A and B were chosen from non-holographic zones, C and D were chosen from holographic zones. By chromaticity diagram we mean vectors with RGB values at their ends which represent C for each frame (circle represents maximum possible saturation). It can be seen that hologram zone pixels produce more saturated colors with greater hue spread.

Then the following statistics are updated for each pixel excluding highlight pixels (block 5):

1. maximum saturation S_{ij}^{max} over a stack of processed frames;
2. sum of chromaticity vectors $\sum C_{ij}$ over a stack of processed frames;
3. processed pixel counter N_{ij}.

After that the hologram map image $\tilde{I}^{hm}$ is generated, with pixel values calculated as follows (block 6):

$$\tilde{I}_{ij}^{hm} = \begin{cases} 255 \cdot \dfrac{\left\| \dfrac{\sum C_{ij}}{N_{ij}} \right\|_2}{S_{max}}, & \text{if } S_{max} > S_{thresh}; \\ 255, & \text{otherwise.} \end{cases} \tag{4}$$

S_{thresh} is a configurable algorithm parameter.

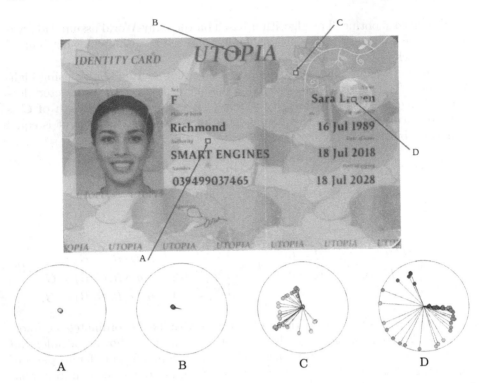

Fig. 8. Chromaticity diagrams for 40 frames of testing clip with "Utopian" ID card for 4 points: A, B – pixels of non hologram zones, C, D – pixels of hologram zones. It can be seen that C and D show more saturated colors with greater hue spread. (Color figure online)

The image $\tilde{I}^{hm}$ is binarized by the global threshold T to obtain the resulting binary hologram map I^{hm} (block 7). Black pixels of I^{hm} are classified by the algorithm as hologram pixels and white pixels are classified as non-hologram ones. The hologram map and its binarization are presented in Fig. 9 for the 5th, 10th, and 15th frames of the clip with an "Utopian" passport showing a hologram.

Then the proportion of black pixels h on the binary hologram map is calculated (block 8). If this number exceeds the minimum threshold h_{thresh}, the response "hologram is found" is returned, otherwise the next frame is captured and the process continues until the hologram is found or the frames run out. If there are no frames left and the hologram is not found, the response "hologram is not found" is returned.

3.3 Experimental Section

We consider the described system to be a fraud detection system, the absence of the hologram can be a sign of a fraud document. Hence, as the task is to detect the absence of a hologram on the document, we will consider that:

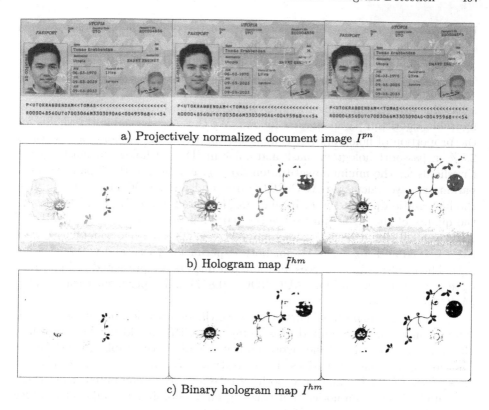

a) Projectively normalized document image I^{pn}

b) Hologram map $\tilde{I}^{hm}$

c) Binary hologram map I^{hm}

Fig. 9. Hologram map computing process: 5th, 10th, 15th frames of the clip from left to right.

- negative (N) is the number of clips with "original" documents;
- positive (P) is the number of clips with an anomaly (i.e. documents without a hologram).

We will evaluate the algorithm by considering it as a binary classifier using such metrics as FPR (False Positive Rate) and *Recall*.

$$FPR = \frac{FP}{FP + TN} \qquad Recall = \frac{TP}{TP + FN} \qquad (5)$$

We consider two different ways of running the algorithm:

- using manually annotated document quadrilaterals which are presented in the MIDV-Holo dataset;
- using document tracking in the video stream (as it was mentioned in the Sect. 3.2), but instead of localization in the first frame we use a manually annotated document quadrilateral as a start for following tracking. To avoid dealing with first frames which do not contain a document we also check if the number of keypoints within the marked quad exceeds the minimum threshold. If it does not, we use the next frame.

Both variations use annotated coordinates of the document boundaries in an effort to make experimental results reproducible.

We used 300 clips of "original" documents and 300 clips of documents without hologram (which include the following threat models: "copy without holo", "pseudo holo copy", "photo holo copy") of the MIDV-Holo dataset to provide a baseline experiment using the proposed algorithm. Firstly, we tuned the configurable algorithm parameters as follows. We chose 3 reasonable thresholds for the proportion of hologram pixels h_{thresh} (the percentage of hologram pixels is 2.8% in passport hologram mask and 4.3% in ID card hologram mask) and 3 thresholds for the minimum saturation S_{thresh}. For each of the 9 pairs of these parameters, we calculated AUC ROC (area under the ROC curve) by varying the binarization parameter T. Results are presented in the Table 1. To measure the quality of the algorithm we use the pair that gives the greatest number of AUC ROC. It can be seen that maximum AUC ROC is reached using parameters $S_{thresh} = 50$ and $h_{thresh} = 1\%$ for both experiments (with and without tracking). The resulting ROC curves for these parameters are presented in Fig. 10 for experiment without tracking (AUC ROC = 0.847) and experiment with tracking (AUC ROC = 0.867).

Using the obtained ROC curves we chose the binarization parameter $T = 80$ (maximum recall was required considering that FPR should not be more than 10%). The final quality of the algorithm is presented in the Table 2, which shows raw numbers of false negatives, false positives, true negatives, true positives (FN, FP, TN, TP), and false positive rate and recall computed using these raw numbers. One can notice that tracking allows obtaining significantly bigger recall with a small increase in false positive rate.

Table 1. Thresholds selection. AUC ROC (value from 0.0 to 1.0) for hologram detection algorithm for each pair of configurable parameters.

	Parameters (S_{thresh}, h_{thresh})								
Input quads	30, 1%	30, 2%	30, 3%	40, 1%	40, 2%	40, 3%	**50, 1%**	50, 2%	50, 3%
no tracking	0.795	0.825	0.832	0.828	0.841	0.832	**0.847**	0.838	0.807
with tracking	0.821	0.851	0.851	0.853	0.856	0.843	**0.867**	0.851	0.819

Table 2. Quality of the hologram detection on MIDV-Holo using the proposed baseline algorithm (parameters: $S_{thresh} = 50$, $h_{thresh} = 1\%$, $T = 80$).

Input quads	FN	FP	TN	TP	FPR	$Recall$
no tracking	119	26	274	181	9%	60%
with tracking	86	31	269	214	10%	71%

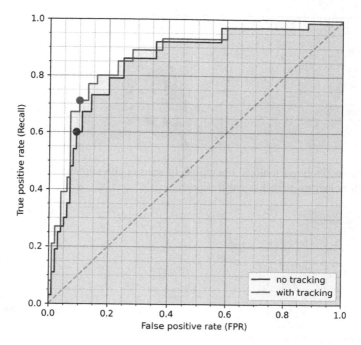

Fig. 10. ROC-curves for hologram detection algorithm, with the variation of the binarization parameter T. The points on the plot correspond to $T = 80$.

3.4 Typical Error Cases

Let us discuss some of the errors of the baseline algorithm. False positives are often caused by low saturation of the hologram under some lighting conditions. An example of this case is shown in Fig. 11. To fix such errors, one can consider color correction algorithms that can increase the saturation of local image elements. False negatives can be caused by saturated highlights on the surface of the document. An example of this case and the resulting hologram map are shown in Fig. 12. To eliminate these errors more accurate highlight detection algorithm can be proposed. Due to small localization errors or tracking inaccuracies black pixels near the edges of the hologram map that are misclassified as hologram pixels may appear. This may lead to additional false negatives. To fix such errors, pixels located near the hologram map edges could be ignored while computing the proportion h of hologram pixels. For example, by ignoring the margins with a thickness of $0.03 \cdot w$, where w is the width of a normalized document image, in the experiment with tracker FPR increases from 10% to 11% while recall increases from 71% to 85%.

Fig. 11. Projectively normalized document images of the clip showing "original" document with hologram with low saturation.

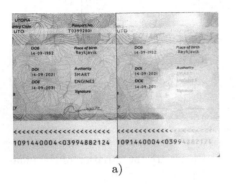
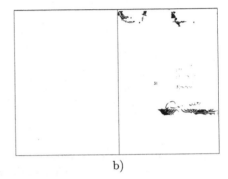

a)   b)

Fig. 12. a) Projectively normalized document fragments of the clip showing a document without a hologram, b) Fragments of binary hologram maps constructed using a baseline algorithm.

4 Conclusion

In this article, the new dataset MIDV-Holo with 700 video clips of artificial identity documents was proposed. The dataset contains 300 clips of documents with holographic security elements and 400 clips with no holographic elements, comprising examples of 4 different types of fraud models, captured in various conditions from simple ones to more challenging ones. We added annotation for each frame that contains information about the type of the document and the document location in the frame. We also presented a baseline method for hologram detection based on the accumulation of pixel-wise statistics representing the change of chromaticity. The method was used to provide baseline hologram presence verification results (71% correctly detected documents without the hologram, 10% "original" documents falsely identified as documents without hologram) on the new dataset. To the best knowledge of the authors, MIDV-Holo is the first large open dataset made specifically for the development of hologram detection methods on identity documents, and we believe that it will be useful for advancing the studies in the field of document forensics and development of secure identification methods. The dataset is available for download at [ftp:// smartengines.com/midv-holo].

References

1. Fang, X., Fu, X., Xu, X.: ID card identification system based on image recognition. In: 2017 12th IEEE Conference on Industrial Electronics and Applications (ICIEA), pp. 1488–1492. IEEE (2017)
2. Attivissimo, F., Giaquinto, N., Scarpetta, M., Spadavecchia, M.: An automatic reader of identity documents. In: 2019 IEEE International Conference on Systems, Man and Cybernetics (SMC), pp. 3525–3530. IEEE (2019)
3. Bulatov, K., Arlazarov, V.V., Chernov, T., Slavin, O., Nikolaev, D.: Smart IDReader: document recognition in video stream. In: 2017 14th IAPR International Conference on Document Analysis and Recognition (ICDAR), vol. 6, pp. 39–44. IEEE (2017)
4. PRADO - Public Register of Authentic identity and travel Documents Online, September 2022. https://www.consilium.europa.eu/prado/en/prado-start-page.html
5. Duy, H.L., Nghia, H.M., Vinh, B.T., Hung, P.D.: An efficient approach to stamp verification. In: Zhang, Y.D., Senjyu, T., So-In, C., Joshi, A. (eds.) Smart Trends in Computing and Communications. LNNS, vol. 396, pp. 781–789. Springer, Singapore (2023). https://doi.org/10.1007/978-981-16-9967-2_74
6. Matalov, D.P., Usilin, S.A., Arlazarov, V.V.: About viola-jones image classifier structure in the problem of stamp detection in document images. In: Thirteenth International Conference on Machine Vision, vol. 11605, pp. 241–248. SPIE (2021)
7. Kunina, I.A., Aliev, M.A., Arlazarov, N.V., Polevoy, D.V.: A method of fluorescent fibers detection on identity documents under ultraviolet light. In: Twelfth International Conference on Machine Vision (ICMV 2019), vol. 11433, pp. 89–96. SPIE (2020)
8. Halder, B., Darbar, R., Garain, U., Mondal, A.C.: Analysis of fluorescent paper pulps for detecting counterfeit Indian paper money. In: Prakash, A., Shyamasundar, R. (eds.) ICISS 2014. LNCS, vol. 8880, pp. 411–424. Springer, Cham (2014). https://doi.org/10.1007/978-3-319-13841-1_23
9. Khan, M.J., Yousaf, A., Abbas, A., Khurshid, K.: Deep learning for automated forgery detection in hyperspectral document images. J. Electron. Imaging 27(5), 053001 (2018)
10. Chernyshova, Y.S., Aliev, M.A., Gushchanskaia, E.S., Sheshkus, A.V.: Optical font recognition in smartphone-captured images and its applicability for ID forgery detection. In: Eleventh International Conference on Machine Vision (ICMV 2018), vol. 11041, pp. 402–409. SPIE (2019)
11. An official website of the European Union, September 2022. https://eur-lex.europa.eu/eli/reg/2004/2252/oj
12. Kaminskaya, T.P., Popov, V.V., Saletskii, A.M.: Characterization of the surface relief of film diffractive optical elements. Comput. Opt. 40(2), 215–224 (2016)
13. Palevičius, A., Janušas, G., Narijauskaitė, B., Palevičius, R.: Microstructure formation on the basis of computer generated hologram. Mechanics 17(3), 334–337 (2011)
14. Girnyk, V.I., Kostyukevych, S.A., Kononov, A.V., Borisov, I.S.: Multilevel computer-generated holograms for reconstructing 3D images in combined optical-digital security devices. In: Optical Security and Counterfeit Deterrence Techniques IV, vol. 4677, pp. 255–266. SPIE (2002)
15. Soukup, D., Huber-Mörk, R.: Mobile hologram verification with deep learning. IPSJ Trans. Comput. Vision Appl. 9(1), 1–6 (2017)

16. Soukup, D., Štolc, S., Huber-Mörk, R.: Analysis of optically variable devices using a photometric light-field approach. In: Media Watermarking, Security, and Forensics 2015, vol. 9409, pp. 250–258. SPIE (2015)
17. Hartl, A.D., Arth, C., Grubert, J., Schmalstieg, D.: Efficient verification of holograms using mobile augmented reality. IEEE Trans. Visual. Comput. Graph. **22**(7), 1843–1851 (2015)
18. Hartl, A., Arth, C., Schmalstieg, D.: AR-based hologram detection on security documents using a mobile phone. In: Bebis, G., et al. (eds.) ISVC 2014. LNCS, vol. 8888, pp. 335–346. Springer, Cham (2014). https://doi.org/10.1007/978-3-319-14364-4_32
19. Kada, O., Kurtz, C., van Kieu, C., Vincent, N.: Hologram detection for identity document authentication. In: El Yacoubi, M., Granger, E., Yuen, P.C., Pal, U., Vincent, N. (eds.) ICPRAI 2022. LNCS, vol. 13363, pp. 246–257. Springer, Cham (2022). https://doi.org/10.1007/978-3-031-09037-0_29
20. Uniting Aviation. A united nations specialized agency, September 2022. https://www.icao.int/publications/pages/publication.aspx?docnum=9303
21. Stock Images, September 2022. https://www.istockphoto.com
22. Data online generator, September 2022. https://www.fakenamegenerator.com
23. Open sources fonts, September 2022. https://www.fontesk.com
24. Generator Photo service, February 2022. https://generated.photos/
25. Center "Fine Optical Technologies" (an official website), September 2022. https://center-tot.ru/
26. Tropin, D.V., Ershov, A.M., Nikolaev, D.P., Arlazarov, V.V.: Advanced Hough-based method for on-device document localization. Comput. Opt. **45**(5), 702–712 (2021)
27. Awal, A.M., Ghanmi, N., Sicre, R., Furon, T.: Complex document classification and localization application on identity document images. In: 2017 14th IAPR International Conference on Document Analysis and Recognition (ICDAR), vol. 1, pp. 426–431. IEEE (2017)
28. Zhukovsky, A., et al.: Segments graph-based approach for document capture in a smartphone video stream. In: 2017 14th IAPR International Conference on Document Analysis and Recognition (ICDAR), vol. 1, pp. 337–342. IEEE (2017)
29. Zhu, A., Zhang, C., Li, Z., Xiong, S.: Coarse-to-fine document localization in natural scene image with regional attention and recursive corner refinement. Int. J. Doc. Anal. Recogn. (IJDAR) **22**(3), 351–360 (2019)
30. Ngoc, M.Ô.V., Fabrizio, J., Géraud, T.: Document detection in videos captured by smartphones using a saliency-based method. In: 2019 International Conference on Document Analysis and Recognition Workshops (ICDARW), vol. 4, pp. 19–24. IEEE (2019)
31. Skoryukina, N., Arlazarov, V., Nikolaev, D.: Fast method of ID documents location and type identification for mobile and server application. In: 2019 International Conference on Document Analysis and Recognition (ICDAR), pp. 850–857. IEEE (2019)
32. Schmid, C., Mohr, R.: Local grayvalue invariants for image retrieval. IEEE Trans. Pattern Anal. Mach. Intell. **19**(5), 530–535 (1997)
33. YAPE implementation, February 2022. https://github.com/jpilet/polyora/blob/master/polyora/yape.cpp

34. Fan, B., Kong, Q., Trzcinski, T., Wang, Z., Pan, C., Fua, P.: Receptive fields selection for binary feature description. IEEE Trans. Image Process. **23**(6), 2583–2595 (2014)

35. Cepeda-Negrete, J., Sanchez-Yanez, R.E.: Gray-world assumption on perceptual color spaces. In: Klette, R., Rivera, M., Satoh, S. (eds.) PSIVT 2013. LNCS, vol. 8333, pp. 493–504. Springer, Heidelberg (2014). https://doi.org/10.1007/978-3-642-53842-1_42

Author Index

Printed in the United States
by Baker & Taylor Publisher Services